THOMAS A. EDISON

A STREAK OF LUCK

THE DA CAPO SERIES IN SCIENCE

WORKS BY ROBERT CONOT

Ministers of Vengeance

Rivers of Blood, Years of Darkness

"Profiles of Disorder" in *Report of the National Advisory Commission on Civil Disorders*

"Riot Profile, the Police Reaction," in *Law Enforcement Science and Technology*, II

American Odyssey

A Streak of Luck

Edison as he appeared in April, 1878, shortly after he invented the phonograph and a few months before he started his search for an incandescent light.

THOMAS A. EDISON

A STREAK OF LUCK

ROBERT CONOT

A DA CAPO PAPERBACK

Library of Congress Cataloging in Publication Data

Conot, Robert E.
Thomas A. Edison: a streak of luck.

(Da Capo series in science) (A Da Capo paperback)
Reprint. Originally published: A streak of luck. 1st ed. New York: Seaview Books, c1979.
Includes bibliographical references and index.
1. Edison, Thomas A. (Thomas Alva), 1847-1931. 2. Electric engineers—United States—Biography. 3. Inventors—United States—Biography. I. Title. II. Series.
TK140.E3C56 1986 621.3′092′4 [B] 86-1982
ISBN 0-306-80261-9 (pbk.)

PHOTO CREDITS

All photographs, with two exceptions, are courtesy of the Edison National Historic Site. The portraits of Sarah Bernhardt and Jay Gould are courtesy of the Library of Congress.

Edison's portrait on the frontispiece first appeared in the *New York Daily Graphic,* April 10, 1878. The sketches of Edison with the Long-waisted Mary Ann and Francis Jehl with the vacuum pump were published in *Scribner's* in February, 1880.

Designed by Tere LoPrete

This Da Capo Press paperback edition of *Thomas A. Edison: A Streak of Luck* is an unabridged republication of the edition, *A Streak of Luck,* published in New York in 1979. It is reprinted by arrangement with the author.

Published by Da Capo Press, Inc.
A Subsidiary of Plenum Publishing Corporation
233 Spring Street, New York, N.Y. 10013

Manufactured in the United States of America

I'll never give up
for I may have a streak of luck before I die.

—Edison, *July 26, 1869*

For Ottilie

Contents

Acknowledgments

I am most grateful to all the staff at the Edison National Historic Site, where I spent many months and conducted the preponderance of my research. Lynn R. Wightman, the manager of the facility, and Elizabeth Albro, the curatorial supervisor, were unfailingly courteous and helpful. I am particularly indebted to Reed Abel, the archivist, and Leah Burt, curatorial associate, who worked closely with me throughout. They were never at a loss in response to a request, and always retained their good nature.

Bernard Finn, curator of the Division of Electricity and Modern Physics, and Elliot Sivowitch, museum specialist in communications, at the Smithsonian Institution assisted me greatly by checking the chapters on the development of the incandescent light, and directing me to the William Hammer Collection.

My gratitude also goes to Douglas A. Bakken, director, Winthrop Sears, Jr., archivist, and David R. Crippen, reference archivist at the Ford Archives; to Joan Gartland, librarian, and Edward Kukla, assistant librarian, at the Tannahill Library of the Edison Institute and the Henry Ford Museum; to John Anderson, archivist, at the Perth County Historical Board, Stratford, Ontario; to Lynne Utley, research aide at the Adrian Public Library, Michigan; and to the staff of the Baker Library at Harvard University.

As for Flick, who during the past five years has typed approximately a million manuscript words—I love you.

Introduction

An author sometimes finds that one book leads to another. In my own case, my work for the National Advisory Commission on Civil Disorders carried me on to *American Odyssey,* a history of the people of Detroit. The interest I developed, during the research on *American Odyssey,* in Thomas Alva Edison, who spent a large part of his youth in and near Detroit, engendered *A Streak of Luck.*

Edison was history's most prolific inventor—1093 patents were issued in his name—and one of the nation's most honored men. At one time he was the best-known American in the world. Yet his life and career seemed peculiarly elusive. Immense gaps existed at critical junctures. The development of his greatest inventions—the phonograph and the incandescent bulb—was shrouded in mist, out of which the finished product emerged as if formed by a flash of lightning. Edison himself dismissed a traumatic turning point in his career—the formation of Edison General Electric and his subsequent loss of control over the corporation—with two sentences: "At these new works our orders were far in excess of our capital to handle the business and both Mr. Insull and I were afraid we might get into trouble for lack of money. . . . When Mr. Henry Villard and his syndicate offered to buy us out, we concluded it was better to be sure than sorry; so we sold out for a large sum."

Edison's stories were droll and sometimes funny. He encouraged the press to inspect and write about his latest inventions. He spoke wittily and in the popular idiom. Superficially he was a reporter's delight. Yet he never allowed anyone to penetrate to the private Edison. His bonhomie and way with words obscured more than enlightened, and acted as a shield to keep the curious at bay.

Edison's life resembled a drama on which the curtain fell whenever a climax approached; and the world was left to wonder what had happened.

I initiated my search for the missing scenes and hidden personality at the Edison laboratory in West Orange, New Jersey, which is maintained as it was during the inventor's life by the National Park Service. When I arrived there, I was taken to a massive underground vault that turned out to be a scientific King Tut's tomb. Packed tightly into over one hundred file drawers and a score of steel cabinets are more than one and a half million letters, notebooks, drawings, contracts, scrapbooks—documents of every nature—spanning nearly seventy years of the inventor's life. Models and machinery are arrayed on shelf after shelf. An inner vault, as formidable as a bank's, holds 3500 laboratory notebooks and other unique documents. The greater part of the material had never been examined by anyone but members of the curatorial staff.

In the winter of 1976 I began to work my way through the files, from the 1860s to the 1930s, sometimes scanning as many as fifteen thousand items a day. Some of the material was uninteresting and unimportant—the most difficult part of the task was occasionally spending hour after hour with little to show, while paraphrasing Edison's words: "Negative results are just as important."

Then, suddenly, sometimes in an entirely unexpected place, significant information appeared. One witness to an incident or an experiment that had taken place in 1870 might show up (within the documents and correspondence) in 1878, another in 1894. Clues and bits of information were scattered over decades. The search became a fascinating combination of detective story and scientific archaeological expedition—it involved, on the one hand, the assembling of witnesses and evidence, on the other the collecting and piecing together of potsherdlike fragments.

Even now, more than forty years after the inventor's death, new documents continue to be discovered. It was only the acquisition by the archives of the Francis Upton papers in the 1960s that makes possible for the first time the tracing of the logical and chronological development of the incandescent light. The papers of John Tomlinson, an important Edison associate, turned up packed away unsorted in an old box during my stay at the laboratory.*

Eventually my pursuit of Edison carried me afield from the archives. At Orange, New Jersey, I was given access to the large volume of family correspondence held by the Charles Edison Fund, and this provides an intimate picture of the inventor's relations with his second wife, Mina Miller, and with his children.

At the Smithsonian Institution, in Washington, D.C., I found the papers of William Hammer, the principal organizer of the Edison Pioneers. The Hammer Collection, the existence of whose documentary portion was virtually unknown, adds many details to the history of electric light and power, and furnishes insights into the inventor's relations with Henry Ford and other friends and associates.

* In the summer of 1978 Rutgers University initiated a twenty-year project to publish Edison's principal papers.

Additional unexplored material existed at the Henry Ford Museum and the Tannahill Library in Dearborn, in the Henry Villard papers at Harvard University's Baker Library, in the National Archives, in the Burton Historical Collection at the Detroit Public Library, at the Adrian, Michigan, Public Library, and in the holdings of the Perth County, Ontario, Historical Board.

It became evident very early in my search that the Edison of popular repute was a person enveloped in a cocoon of myth. The original documents contradicted or revised a major portion of the story of his life, as told in the past. Sometimes the existing accounts had turned events, as they had actually occurred, topsy-turvy.

The Edison that I discovered was a lusty, crusty, hard-driving, opportunistic, and occasionally ruthless Midwesterner, whose Bunyanesque ambition for wealth was repeatedly subverted by his passion for invention. He was complex and contradictory, an ingenious electrician, chemist, and promoter, but a bumbling engineer and businessman. The stories of his inventions emerge out of the laboratory records as sagas of audacity, perspicacity, and luck bearing only a general resemblance to the legendary accounts of the past.

PART I

THE INVENTOR

CHAPTER 1

A Bushel of Wheat

His mother called him Alva. To his friends he was Al. His sister nicknamed him Rinkey. He was frail and prone to respiratory infections. His body was so spindly that when he was twelve years old he exclaimed: "Ma, I'm a bushel of wheat. I weight just sixty pounds."* His neck scarcely seemed able to support his outsized, triangular-shaped head with its broad brow. His mouth, like his mother's, had an odd downturn at the corners. When he was born, on February 11, 1847, the last of seven children, his mother was haunted by death: six-year-old Carlisle had died in 1842, three-year-old Samuel in 1843, and three-year-old Eliza was to be buried before the end of the year.

Nancy Elliot Edison had never been a lighthearted woman, and after the deaths of her three children she wrapped herself in black. The daughter of an American Baptist minister who had gone to preach in Vienna, Ontario, she was a devout member of the Presbyterian Church. Like many of the better-educated teenaged girls, she had taught school as a preparation for marriage—which had come in 1828, when she was seventeen.

The groom was Samuel Edison, Jr., twenty-four years old, six feet two inches tall, handsome and bearded. His grandfather John, a prosperous New Jersey landholder, had been a Tory activist who had fought on the British side in the Revolutionary War, and had narrowly escaped being hanged as a spy by the Colonials. After the war he and his family had emigrated to Digby, Nova Scotia, where they scratched out a living

* A common bushel is equal to eighty pounds, but a bushel of wheat was measured at sixty pounds in the nineteenth century. Some writers "corrected" Edison's remark to read: "I. weigh just eighty pounds."

from the harsh soil. In 1811, John's eldest son, Samuel Ogden, had received a land grant on the Otter River, just inland from the Erie lakeshore in Ontario. When the War of 1812 had broken out, he had been commissioned a captain in the British forces, and had participated in the campaign that led to the capture of Detroit.

Sam Senior had an individualistic flair, and this developed in Sam Junior into iconoclasm and rebellion. Influenced by the writings of Thomas Paine, Sam junior was a freethinker and outspoken antiauthoritarian. During his teens the Erie Canal was being built, and as its construction brought a shipping boom, Sam became a sailor. Then, with indifferent results, he tried carpentering and tailoring. After his first marriage, he turned to innkeeping.

In contrast to the American Midwest, where the conquest and confiscation of Indian lands was fueling a period of enormous expansion and prosperity, in Canada government and wealth were in the hands of an oligarchy, and there was growing dissatisfaction among younger men at the lack of opportunity. In December, 1837, the discontent burst into rebellion. A force under William Mackenzie attacked Toronto, but was driven off. The British militia came looking for Sam junior, who was known to have been one of the participants. Sam fled to Detroit, and after some time there settled in Milan, Ohio.

Milan was a town located on a short canal, which the citizens expected to be the making of their fortunes when it was dug south to the Ohio River. Politics was a continual fount of recreation for the men, and Sam, like almost everyone scheming to make a killing, was a Jacksonian Democrat. He prospered moderately as a lumber miller and carpenter, though his inquisitive mind and exploratory bent kept him from throwing himself wholeheartedly into any one enterprise. He had the most sanguine expectations about any project that came into his head. His opinionated liberalism bordered on radicalism. He tucked scores of stories into his excellent memory, and would spin them out for hours. He proclaimed: "I am a master of smoking, drinking, and gambling. I have smoked and drunk whiskey moderately when I needed it, and have known when to let it alone. I let gambling alone." He was an easygoing, offhandedly affectionate father to his children, but not an easy husband for a woman.

Nancy accepted her tribulations as God's will, and was determined to give her youngest, Thomas Alva, a strict and exemplary upbringing. Until he was two, Alva was cared for much of the time by his twenty-year-old sister, Marion. His earliest memory was of crawling over the floor after a silver dollar given him by Marion's boy friend, whom he never forgave for taking Marion away from him. He cried desperately when she married and left home in 1849, and three quarters of a century later recalled being "held in arms to witness the marriage of my sister to this lover."

Alva's brother, William Pitt, was seventeen at the time, and his other sister, Harriet Ann, sixteeen. Nancy's expectations, not only for her youngest, but for the three dead children, were focused upon Alva—he had, in effect, the upbringing of an only child. As a toddler he was fussed over and a center of attention. But as he grew older he was often forced to amuse himself, and so learned to rely on his imagination and make up his own games. Nancy, protective, strict, and puritan, demanded piety and obedience from him.

In the summer of 1852 the family sailed across Lake Erie to visit the Edison clan in Vienna. Sam Senior was eighty-five (he lived to the age of a hundred and two), a white-haired, tobacco-chewing patriarch. He had fathered eight children by his first marriage and five more by his second, contracted after he was sixty years old. To Alva he seemed forbidding, and the small boy kept a respectful distance from him. Some of Alva's cousins and uncles took the boy fishing and taught him to swim. By the end of the summer he had no fear of the water.

Adventurous and inquisitive, he had more than the normal tendency to get into scrapes. He experimented with starting a fire, and succeeded—his father's barn burned to the ground. Sam rewarded him by giving him a public whipping in the town square. Then Alva and a playmate, the son of Milan's most prominent storekeeper, went to explore the canal basin below the Edison house. Alva decided to go swimming. The Lockwood boy could dare no less.

When it got dark Alva returned home, very subdued, and shortly crawled into bed. Meanwhile, the Lockwoods grew concerned over their son's failure to appear. Neighbors were alerted, lanterns were lit, and a large part of the community turned out to search.

Eventually someone learned that the boy had last been seen with Alva. Alva was awakened. Under hard questioning he admitted they had gone into the canal together. He had seen his friend's head disappear, but had not known what to do. He had waited awhile, then, too frightened to tell anyone, had run home and climbed under the covers.

The canal in which the Lockwood boy had drowned was, in fact, becoming a community disaster. Toledo, to the west, had a canal linking it to Cincinnati, and Cleveland, to the east, a canal connecting to the Ohio at Portsmouth. Milan's short canal still led nowhere, and proved irrelevant when a railroad was built from the lake port of Sandusky, ten miles to the northwest, to Columbus and Hockingport on the Ohio. The town went into a precipitous economic decline. In the spring of 1854, Sam decided to move.

One of his brothers was living near Fort Gratiot, a military post on the outskirts of Port Huron, sixty miles northeast of Detroit. Sam was retained as the fort's lighthouse keeper, general provisioner, and carpenter. Near the eighty-two-foot-high wooden lighthouse, sending its signal out over Lake Huron, was a building containing a steam engine that activated a fog whistle eight seconds out of every minute; and Sam's duties included tending its boiler.*

From his predecessor, Sam bought a substantial two-story, six-bedroom house on the military reservation that surrounded the fort. It was an enormous home for what was now a small family—Harriet Ann married after the Edisons moved to Fort Gratiot, and only twenty-two-year-old Pitt and seven-year-old Alva remained with their parents. Alva's highly active, free-roaming imagination was stimulated by the house and its setting. Looking out, he could see the many white-painted buildings of the fort, the soldiers on parade in the center, and the banners flapping in the wind blowing off the lake. To the north and east were sweeping vistas of the water; to the south and west the yet untouched, dense forest. Nearby was the cemetery where three hundred soldiers who had died in the cholera epidemic of 1832 lay buried. When the wind whirled off the lake and vivified the branches of the trees, Alva could see the ghosts darting in and out among the trunks. In the fierce chill of winter he huddled in bed and shivered at the wail of a wolf. Sometimes lying awake for hours, he tried to shut out the thumps and creaks that wandered through the darkness of the half-empty house.

In the fall of 1854, Alva was enrolled in the school of the Reverend and Mrs. G. B. Engle—the customary age for children to enter school was between seven and eight, and there was as yet no public school in the community. The severe-featured Reverend Engle wore a long frock coat and carried a cane. His children, Willis and Mary, were playmates of Alva. Alva, however, was terrified by the disciplinary atmosphere of the school. The minister, of course, taught by rote, a method from which Alva was inclined to disassociate himself. He alternated between letting his mind travel to distant places and putting his body in perpetual motion in his seat. The Reverend Engle, finding him inattentive and unruly, swished his cane. Alva, afraid and out of place, held up a few weeks, then ran away from the school.

It was midwinter, a time when Alva frequently came down with colds, earaches, bronchitis, and other pulmonary ailments. Nancy, having lost three children, was especially protective of Alva. She decided to keep him home and teach him herself.

Sam and Pitt were often away, so Nancy spent many hours and days alone with Alva. Her relationship with her husband and older son was sometimes stormy—they were irreligious and went on sporadic drinking bouts—and she was determined to bring Alva up to be a better man. She made Alva study the Bible, and every Sunday took him to church, where Alva learned of the eternal hellfire that was the punishment for any of a multitude of sins, some of which he seemed to have a knack for committing.

Nancy rarely forgave a transgression, and had little understanding of

* One of the numerous wild tales that Edison biographers picked up and told about the family was that Sam had built the tower as a tourist attraction.

her son. Even the cookie jar was a bone of contention, for Alva had a craving for sweets, and no matter where or in how inaccessible a place she concealed it, he would track it down and pilfer from it.

She was a hard taskmistress, as intolerant as the Reverend Engle of Alva's flights of imagination. She kept a switch behind the old Seth Thomas clock; and before Alva reached his teens she had worn the bark off it on Alva's legs and buttocks.

Sam, on the other hand, was inclined to let the boy go his own way. Alva trailed after his father when he tended the machine that operated the steam whistle, and watched with fascination as the two-horsepower engine activated an endless screw with 120 cogs. His eyes lit up whenever the lever opened the valve to let the steam shoot through the whistle.* He became interested in mechanics, and when he was nine his parents gave him *Parker's Natural and Experimental Philosophy*, a grammar-school science book. With a quick, absorbent mind, he learned to read well. Books became his favorite entertainment. When he grew older, he delighted in the stories of Sylvanus Cobb, Jr., particularly *The Gun Maker of Moscow*, which inspired him to organize a secret service among his few friends.† Digging a cave with a concealed trapdoor, they fitted it out with a fireplace, table, chairs, games, and a larder filched from their parents' pantries.

After Alva's disastrous experience with the Reverend Engle, he attended another private school, conducted by P. L. Hubbard. This he enjoyed. When he was eleven he journeyed to the two-story public school in Port Huron. But—especially in winter, when his ears periodically became inflamed and filled with fluid—his hearing was uncertain. Sometimes he could hear all right. At other times he had to strain, and was frustrated because the teacher did not talk loud enough. He asked questions when she expected him only to listen, and was not listening when she asked questions.

Two or three children had to share each book, and Alva, a rapid reader, had no patience with his classmates. He tried to manipulate them, and demanded his own way. Since he was puny and not well coordinated, he did badly in games. Odd-looking with his big head, he was unpopular, and was razzed and bullied by his mates.

Teachers thought him a problem child and a mischief maker. Once he and a girl let down a baited fishhook from the second story, and hauled up a squawking, flapping chicken. The whole school was thrown into an uproar, and the black strap that the principal, Alex Crawford, always carried with him was exercised on Alva.

Sam was worried about Alva because "Some folks thought he was a

* Not coincidentally, perhaps, many of his early inventions featured cogwheel devices.

† Twenty years later, when Edison rocketed to prominence, one of his more recent acquaintances wrote an article for the *Chicago Tribune* in which the boy was transformed into a child prodigy who read Newton, Gibbon, Hume's *History of England, The History of the Reformation*, and the like.

little addled." Canadian-born Jim Symington, who was now Sam's best friend—though, ironically, he had been a member of the militia that had chased him across Canada—believed Alva's mind was too active for his body, and advised Sam not to let him read so many books.

In fact, Alva's imagination and liking for books were counterbalanced by an abhorrence of physical labor that would have made a sloth seem like a long-distance runner. Sam had a ten-acre truck garden, and Alva was sent round with a horse and wagon to peddle the produce. He was also expected to help hoe and weed, tasks that he disliked intensely. Once he undertook to plant six acres in turnips. This he and another boy accomplished in two and a half hours by the expedient of placing the seeds no less than fourteen feet apart.

It was evident that Alva was no more inclined toward agriculture than Sam or Pitt. Pitt, as restless and visionary as his father, had trouble settling down to a job, and went off on hunting and trapping expeditions. As Alva grew up, Pitt took a brotherly interest in him, and Alva became closer to him than to any other member of the family.

For his father Alva had respect and affection. Sam was not a man to spend a great deal of time with a boy, but Alva listened attentively whenever Sam discussed the fortune he intended to make. Michigan was in the throes of a boom fueled by land speculation and vast resources of lumber, iron, and copper. Railroads were branching out in all directions. More and more ships were plying the Great Lakes. Newspapers told of millionaires sprouting like trees out of the land. It seemed all that one needed was a little capital and luck. And Sam, a man of rough-hewn literacy but great imagination, devoted hours to scheming with his cronies.

But while he was planning how to make and spend his fortune, the family was chronically in debt. Sam never paid the bill for Alva at the Reverend Engle's school, nor the seventeen dollars he owed to Engle's sister-in-law, who taught Alva piano and patiently worked with him on "The Detroit Schottisch."

Nancy, sorely pressed, tried to counteract Sam's influence by emphasizing to Alva that work, not dreams, made money. From observation, Alva knew his mother was right. Instinctively, however, he wanted to believe Sam. He felt that, in any case, he never would be able to live up to his mother's standards, and was afraid to confide in her. When he failed in school or in some other endeavor, he learned that by small prevarications or sheer bluff he could sometimes avoid a confrontation. He was uncomfortable with his actions; but they were easier to live with than her tongue-lashings and whippings.

During the winter of 1859 Alva suffered another bronchial infection. Once again his hearing was adversely affected by the inflammation of his mastoid bones. His repeated absences aggravated his problems in school. "Teachers told us to keep him in the streets for he would never make a

A sketch of the Reverend Engle's schoolhouse was sent to Edison by one of his classmates, William Brewster, for Christmas, 1891. The Reverend Engle, wearing a frock coat and holding a cane, stands in the doorway, while a cow looks on from the right.

scholar," Sam related; and when Port Huron was connected to Detroit by railroad, he got Alva a job hawking newspapers and sundries on the train.

It was quite normal in those days for boys to go to work at the age of twelve; only the larger cities had a high school, and enrollment beyond the sixth grade consisted mostly of the offspring of the well-to-do. For Alva the job came like a deliverance—afterward he looked back upon the period he spent on the railroad as the happiest of his life. The train, consisting of three small wooden coaches and a combination smoker-baggage car, left Port Huron at seven-fifteen each weekday morning and arrived in Detroit, after numerous stops in between, four hours later. Alva had to get up at six o'clock—the first morning Nancy took him to the station and anxiously entrusted him to the safekeeping of the conductor, Alex Stephenson.

The job itself was not too demanding. In addition to selling his wares, he sometimes had to tend the oil lamp that hung overhead, make sure the water tank with its large brass cup was filled, and throw a log into the small stove. In the intimate confines of the cars—the seats consisted of wooden benches running down each side—he had an unequaled opportunity to meet and mingle with people.

The train did not return to Port Huron until eight in the evening, and he did not get to bed until eleven o'clock. He had, however, the chance to take a nap during the daily five-hour layover in Detroit. Stretching out on one of the wooden benches in the cars, he acquired a lifelong habit of daytime slumber.

Every night when he returned home from the station he had to pass by the cemetery. Once, when he was driving a horse and wagon, he was so terrified he slackened the reins, shut his eyes, and let the horse proceed on its own—even if the ghosts saw him, it was better if he did not see the ghosts. Then, in the winter of 1860, he picked up his father's copy of Thomas Paine's *The Age of Reason.* A good deal of the language was above his head, but for all of his life he would remember "the flash of enlightenment." Paine was a man of letters and of liberty; a freethinker. Alva focused on "The world is my country; to do good my religion." The death-and-hellfire, sin-and-damnation sermons that Alva had grown up with and hated were swept away; afterlife and ghosts were simultaneously dissolved. Paine was an inventor—he had conceived the hollow candle, the central-draft burner, the iron bridge. He seemed to Alva a greater man than George Washington. The thirteen-year-old boy could see no better course than to follow in Paine's path.

Knowledge to Al was like water to a cactus; he absorbed and stored whatever came his way.* Detroit's population was nearing fifty thousand, and it was one of America's great industrial centers. The waterfront was a gallery of sailing vessels and steamers. Half a dozen nationalities bumped shoulders. Drunkards, prostitutes, and beggars shared the streets with ladies of fashion and men of substance. Imposing churches stood next to horrible dives. The city was a university of practical knowledge, and for Al Edison it was a much more effective medium of education than the classroom had ever been.

Chemicals and pharmaceuticals were among Detroit's principal products, and Al, who had the instincts of a pack rat, started assembling a collection. Half of the baggage car was partitioned off as a smoker, but since it was only eight feet long, had no window, and only a table with two benches for furnishings, no one ever used it. Al took it over by default. First he kept his newspapers and goods there; then, gradually, he filled it with his other acquisitions. Soon he had a shelf full of chemicals. Experimenting with them on the table helped occupy the long hours in the city. A lonely boy existing in an adult world, he relied on himself for his entertainment.

He applied himself with a passion to whatever caught his attention. But his attention was easily diverted. He paid little heed to details. On the shelf in the baggage car he had a bottle of phosphorus. When the water covering the chemical evaporated, Al forgot to replenish it. The phosphorus caught fire, the bottle fell off the shelf, and the flames

* When he was fifteen he joined the Young Men's Society of Detroit and decided he would read his way through the association's library shelf by shelf. After scanning about ten rather dry volumes, however, he changed his mind and gave up on the encyclopedic enterprise.

spread over the floor of the baggage car. The baggage master arrived in the nick of time to prevent a disaster. He was unable, however, to douse the blaze with water, and when he tried to pick up the phosphorus, it stuck to his fingers and burned him severely. At last extinguishing the flames, he boxed Al's ears and heaved the whole collection of chemicals out the door.

What Al managed to salvage he added to the growing stash he kept in the basement of the house. His mother did not approve—the chemist had a habit of conducting his experiments on the main floor and leaving a big mess behind. Sunday was his only day at home, and Nancy wanted him to observe it strictly as a day of rest and prayer. But now that he was contributing to the family income and was a boy of independent means, he rebelled at going to church, and responded less and less when she, stout, prematurely aged, dressed in black, came out to the stoop of the house and severely called: "Alva!" She thought he should be more careful of both his money and his life—repeatedly she told him and his friends, Jim Clancy and Michael Oates, that they would blow their heads off. More than once he tried to fulfill her prophecy. He built water mills, experimental devices, and cannons. Hollowing out a large hole in a log near the icehouse, he packed it with gunpowder, drove a plug in, and told Oates to put his cap over the plug. He then touched a match to the short fuse. The instantaneous blast hurled the three boys to the ground, sailed the cap to the top of the light tower, and caved in the side of the icehouse.

Deprived of his chemicals in the baggage car, Al cast about for other means to occupy his time. As the war continued, the price of scrap metal rose rapidly. Al, thinking he might go into the junk business, initiated the venture by stripping the zinc off the bottom of his mother's stove. When this enterprise was immediately discouraged—a harbinger of future times when other people would have difficulty appreciating his vision—he bought a batch of obsolete type and a small printing press. At the newspaper offices in Detroit where he picked up the papers he sold on the train, he watched the printers set the type and listened to the articulate and profane speech of the newsmen. Seeing no reason why he could not produce and sell his own newspaper, he set up his press in the baggage car. In the winter of 1862 he began publishing the *Grand Trunk Herald*.

The sheet sold for three cents and carried news of occurrences along the line between Port Huron and Detroit. Setting the type by hand was a tedious process, and Al put out only a few issues. Instead, he promoted a partnership with William Wright, a printer's devil at the *Port Huron Commercial*. Al was the editor and Will the printer of the new publication, *Paul Pry*. Every Saturday evening after the newspaper office closed,

Will sneaked Al into the plant and they took over the facilities. Around midnight they stopped, and Al would go out and bring back bologna, crackers, and cider.

Paul Pry was a peppery sheet, quite different from the *Grand Trunk Herald,* and it acquired the reputation of being "hot stuff." "Young F., who has just returned from the army, had better keep out of J. W.'s saloon," advised an item. A doctor, mistakenly thinking the line referred to his son, strode fuming into the office and threatened to toss the editor into the St. Clair River if he could be found—but of course he could not.

Al was a hustler. Under normal circumstances, the trainboy's job was not lucrative, and he was always on the lookout for ways to supplement his income. When, on April 6, 1862, the Union and Confederate forces engaged in a great battle at Shiloh, on the Mississippi, he had to push his way through the throng at the door of the *Detroit Free Press* office—each such battle generated enormous interest, as well as great anxiety among families who had men at the front. Al decided he could sell 1000 newspapers instead of the usual 100. Since he had money enough for only 300, he asked Wilbur Storey, the editor, to let him have the others on consignment.

Storey was a rabid, Negro-baiting, antiwar, pro-Southern Democrat in a state where the conflict was distinctly unpopular with small farmers and the working class.* The initial reports of the battle were unfavorable to the Union forces. Sixty thousand men were said to have been slain, and Storey was all for spreading the word as widely as possible. Al got his 1000 papers. He bribed the Grand Trunk and Western Railroad telegrapher in Detroit to wire a bulletin about the battle to the stations along the line. Crowds awaited the arrival of the train at each town. The papers were snapped out of Al's hand. At the first station, Utica, he sold 35 papers instead of the usual 2. At Mount Clemens he raised the price per copy to a dime. Those he had left at Port Huron went for a quarter each.

On another occasion, Eber Brock Ward, a Detroit shipping magnate and millionaire industrialist, appeared at the station. The captain of one

* The border with Canada had to be closed to prevent draftees from fleeing, and the *Port Huron Commercial* had the following message delivered to its subscribers on January 1, 1863:

> O what a land is this where madness rules
> Whose scamps are officers, whose statesmen fools
> Whose armies wither in the grasp of knaves,
> Whose rogues are honored, and whose poor are slaves,
> A very byword for the wide world's scorn,
> Pitied, derided, wretched, and forlorn.

of his large ships had died unexpectedly, and Ward wanted a message delivered immediately to a former captain who had retired and lived about fourteen miles from Ridgeway,* a railroad station twenty miles from Port Huron. Ward offered Al fifteen dollars. But Edison, even at the age of fifteen a shrewd bargainer, said he was afraid to go alone and would have to engage another boy as a companion. He negotiated Ward up to twenty-five dollars.

It was nine o'clock on a dark, rainy night when Al and his hired boy trudged off along the trail into the forest. Before going far they convinced each other that the woods were full of bears. Every stump took on the form of a beast. The boy suggested that they climb a tree and wait till dawn. But Al had heard enough about bears to believe that any tree they climbed might already be occupied. The message had to be delivered in time for the captain to catch the morning train, so they continued on. After a while one lantern went out. Then, about two miles before they reached their destination, the other died. Leaning against a tree, Al and his companion burst into tears. If he ever got out of the escapade alive, Al thought, he would be as accomplished a woodsman as existed. After a time the rain slackened, and there was just enough light for them to make their way. Eventually, at the first glimmer of dawn, they entered the captain's yard. The experience diminished Al's slight love of nature even further. "In my whole life, I never spent such a night of horror as this," he recalled.

Three years of daily travels made Al a familiar figure in stations and telegraph offices along the line. Men laughed over his enterprises and misadventures. Like most boys, he was fascinated with locomotives and fantasized himself at the throttle. He dispensed gifts of apples and cigars to engineers and firemen so that they would allow him to climb into the cab. He volunteered to polish the brass. He ran errands. He grunted and sweated as he lugged logs from the tender and pushed them into the fire, while the fireman looked on in amusement.

Operations of the railroad were relaxed. The train stopped at any crossing to pick up and discharge passengers. Some of the crew carried rifles, and when a flock of wild turkeys was sighted, the locomotive belched to a stop for a turkey shoot. One day the engineer was weary from partying the night before. Going back to the caboose to take a nap, he left the fireman to run the engine and Al to stoke the fire. About halfway to Grand Trunk Junction, a railroad crossing just outside Detroit, the fireman started to nod off too.

Al professed he would be glad to take over and let the fireman rest. He knew the danger of allowing the water level to get too low, and so

* The name of Ridgeway was later changed to Richmond.

kept pumping water into the boiler. Soon, black mud was belching out of the stack, covering not only every part of the engine but the youthful engineer as well. Unable to fathom the cause of this phenomenon (it stemmed from the excess of water blowing the soot out of the stack), Al was too proud, too stubborn, and too satisfied to awaken the fireman. "As the great train thundered on under my hand (at the breathtaking speed of 12 miles per hour) I was supremely happy," he recalled.

He had noted that at one point on the route the fireman always went out to the cowcatcher, opened the oil cup on the steam chest, and poured oil in. But when he undertook the task, the steam roared out. Hot oil splattered over his hands and face, blistering him and nearly knocking him off the engine—he had failed to observe that the engineer always shut off the steam before opening the cup. The engine had started on its journey with its woodwork beautifully painted and its brass bands glistening. It arrived at the Grand Trunk Junction covered with black mud. Its battered operator looked like a chimneysweep who had tumbled off the roof. Word of the odd sight swept over the yards, and workers gathered with great hilarity and caustic comments.

So ended Al's dream of being an engineer. Uncoordinated and absentminded, he was never in his life able to master the knack of running machines of locomotion. His attempts at chemistry and publishing had had their repercussions. That left telegraphy. The rapid growth of commerce, railroads, and newspapers was bringing about an enormous expansion of telegraph lines at the same time that the army was employing hundreds of telegraphers at premium pay for its own communications network. In railroad telegraph offices from Detroit to Port Huron, Al badgered operators to let him practice. When, in the winter of 1863, he was seriously ill again and his mother insisted on keeping him home, he made a set of instruments and drafted a schoolmarm who was boarding at the house to help him study Morse code.

In February he turned sixteen. It was time he learned a trade. At Mount Clemens, midway between Port Huron and Detroit, the telegrapher was James Mac Kenzie, a tall, heavyset, red-bearded twenty-five-year-old Scotsman who had become a friend of the Edison family while telegrapher at Point Edward, across the river from Port Huron. He had supplied Al with the telegraph news for the *Grand Trunk Herald*'s "Hot off the Wires" column. Mac Kenzie had a lot of time on his hands. He was teaching one youth telegraphy in return for guitar lessons, and offered to take Al on as an apprentice, too.* Al found another boy to take over the portion of the route from Mount Clemens

* After Edison became prominent, Mac Kenzie related he had taken Al on as an apprentice in gratitude for plucking his three-year-old son out of the way of an approaching boxcar. This incident, however, had apparently occurred many months earlier, in the summer of 1862, and Al's fellow apprentice with Mac Kenzie knew nothing about it.

to Detroit, and daily dropped off the train at the halfway station to study telegraphy.

Al's initial lessons in Morse were both auditory and visual, since originally the telegraph, as invented by Samuel Morse, recorded the messages with pencil on a continuous strip of paper. The telegrapher had read the letters off the strip to a second man, a recorder, who wrote them down in roman characters. As telegraphers had become more experienced, however, they had discovered they could recognize the letters by the clicks of the instrument, and had dispensed with the helper. Although this was a practice that Morse had tried to stamp out (he feared it would endanger the validity of his patent), it had, because of its economics and the increase in the speed of transcription, soon gained acceptance.

Many telegraph offices still contained the old transcribing instruments, and they were of great aid in learning. To enable his apprentices to practice, Mac Kenzie ran a line from the station to the tank house, a hundred feet away. There Al worked the telegraph key, and finally mastered MISSISSIPPI—his first long word. An incentive to his efforts was the cooking of the young Mrs. Mac Kenzie, whose apple pie he craved.

In order to practice at home, Al converted the cave in his backyard into a telegraph office, and strung stovepipe wire through the trees to Jim Clancy's yard, a mile away. Organizing Clancy and Wright into a telegraph corps, he taught them what he himself was learning at Mount Clemens. Some nights on arriving home he took the newspaper to Clancy and had him transmit the stories from it over the line.

By summer he had progressed enough to obtain a part-time job in Port Huron's telegraph office, located in a corner of Thomas Walker's jewelry store. He was a miserable transmitter—he never did learn to send well—and an uncertain receiver. But he had the old Morse transcribing instrument to help him out, and what he couldn't take off by ear he could read from the strip. There wasn't much traffic on the line, and the pay was low—twenty-five dollars a month, less than one fourth of what a journeyman telegrapher commanded. With plenty of time to pursue his diversions, he returned to his chemical experiments. The electricity for the telegraph was supplied by primitive batteries—they consisted of jars of acid in which metal electrodes were immersed. Among the tasks of the telegrapher was to mix and replenish the acid periodically. Nothing could have suited Al better. Often he invited his friends in to observe as he added this and subtracted that, stirred in a new compound or concocted a novel substance. Thus it came about that one day the citizens of Port Huron observed three youths staggering out of the building amid a confetti of splintered boards, broken glass, telegraph keys, and shredded paper.

Badly singed, Al had blown himself out of a job. The citizens of the town looked upon him as a misfit. The womenfolk, especially, considered him a troublemaker, and didn't want their sons, much less their daughters, associating with him.

Accepting an invitation from a quartermaster lieutenant who had once been stationed at Fort Gratiot, Al went off on a visit to Fort Lyons, near Washington, D.C., in the hope of becoming a military telegrapher. When his youth and inexpertise dashed that aspiration, he returned to Port Huron. In the fall of 1863, Mac Kenzie helped him get a job as a telegrapher at the Stratford, Ontario, station of the Grand Trunk Railroad. Sixteen and a half years old, he was on his own.

CHAPTER 2

A Drifter and a Dreamer

Stratford was a rough-hewn town in which eight churches, numerous saloons, and three weekly newspapers competed for the attention of the 3500 citizens. It was the crossing point of the Grand Trunk Railroad routes from Toronto to Sarnia, and from Buffalo to Goderich Harbor on Lake Huron.

Al was made night operator and station agent on the latter line. It was neither a demanding nor a glamorous job. His combination ticket office and telegraph table consisted of a large dry-goods box. During his entire shift only one train passed in each direction on the line, and messages were so sparse it was difficult to retain an interest in the proceedings. Since he had a habit of nodding off, he devised a clockwork mechanism to send . . . . . . (the number 6) to signal his presence every half-hour.

All trains stopped at Stratford for twenty-five to forty-five minutes to take on coal and water. Because of the conjunction of the two routes, the station was considered hazardous, and every train had to come to a complete halt before passing over the crossing. The engineer was prohibited from proceeding until a semaphore that usually indicated "Danger" was lowered to "Caution."

Setting the semaphore was one of Al's duties. On the night of Friday, December 11, 1863, the 11:15 freight had just steamed in from Goderich, and the switch had been thrown to shunt it to the side track. A few minutes later another freight approached on the Toronto-Sarnia line from the east. Too late Al realized he had forgotten to change the semaphore, and dashed out with a lantern. The engineer, seeing the signal at "Caution," chugged ahead at about walking speed. When the locomotive hit the switch, it flipped off the tracks, and was followed by the tender and one boxcar. No one was hurt, and the damage was minimal.

Al and the supervising agent at Stratford were summoned to Toronto by the Grand Trunk's general manager, W. J. Spicer. When Spicer saw Al and heard the explanation, he was impressed by both Al's youthfulness and his evident unreliability, and threatened to have him and the supervisor shipped off to jail together.

Before matters could progress further, Spicer was interrupted by the arrival of some important English visitors. He ordered the two miscreants to wait in an outer office. Al, not one to stick around when he scented disaster, availed himself of the opportunity to slip away. Abandoning the twenty-eight dollars in pay owed him, he hopped a train for Sarnia and made his way across the frozen river to safety in Port Huron.

The incident ended the possibility of further association with the Grand Trunk Railroad or its affiliates in Canada and the United States. Al scarcely had need to worry about future employment, however. The Civil War was at its peak, and the shortage of manpower, and especially telegraph operators, was critical. Jobs were available everywhere, for telegraph technology was still so crude that messages could not be sent more than 200 miles without being taken off the wire and retransmitted; to send a telegram from Boston to St. Louis required a half-dozen telegraphers.

Telegraph companies were, in fact, invariably local or regional operations interconnecting, at most, a few sizable cities. Only Western Union operated a national network; and that had not had its inception until seven years before, when railroad magnate Ezra Cornell had combined the New York, Albany and Buffalo Telegraph Company with other lines stretching to Wisconsin, Minnesota, Ohio, and Illinois. Western Union's link to San Francisco, subsidized by the Federal government, was only two years old.

One of the Western Union lines ran along the Lake Shore and Michigan Southern Railroad, which hired Al as a night operator at Adrian, thirty-five miles northwest of Toledo. There Al met Ezra Gilliland, a fellow telegrapher a year older than himself. Ezra was the son of Robert Gilliland, who was a cousin of Ezra Cornell and an executive with the Michigan Southern Telegraph Company. With Robert Gilliland's help, Ezra and Al obtained a few pieces of equipment and set up a small workshop. Al's friendship with Ezra was to be of great importance in his life and continue for a quarter century before ending finally in a tempestuous blowup.

Al's stay at Adrian proved short. One day he became embroiled with the railroad's superintendent of telegraphy over the sending of a dispatch, and found himself out of his third job in less than a year. Going to Toledo, he was hired by the Pittsburgh, Fort Wayne, and Chicago Railroad. The railroad soon transferred him to Fort Wayne. There, after two months, he heard from Gilliland, who had moved on to Cincinnati, about an opening in the Indianapolis telegraph office.

He was hired as a plug (or novice) operator, paid seventy-five dollars a month. Experienced operators received at least 50 percent more, and looked down on and made fun of the plugs. Edison was assigned the day shift on a "way wire," linking small towns. On the way wire he could get by on a speed of ten to fifteen words per minute. Of course, the ambition of every plug was to improve to the expert rate of twenty-five to thirty—and occasionally even forty—words per minute. Night after night Edison returned to the office and sat next to the man taking the press dispatches for distribution to the city's newspapers. But he was unable to distinguish the signals, much less copy down the messages.

Lying about the office were some of the old Morse embossing registers which had been in use before operators had learned to receive copy by sound. Taking two of them, Edison hooked them together and attached a clockwork mechanism so that the embossed strip recorded by one could be fed through the second at reduced speed. Activating a lever, the device reproduced the clicks, and enabled Al to take off the copy by sound at a speed he could handle. When he had attained proficiency at one speed, he stepped up the clockwork to the next. It was a simple but clever device. It involved no great invention; yet it became the fountainhead of the most productive inventive career in the history of the world.

Every night Al and a fellow plug, Ed Parmalee, practiced together, one taking copy while the other kept the instruments in synchronization. Soon the regular operator, watching the two youths duplicate the copy he was transcribing, wondered what he was there for. Letting them take over, he went to a theater or saloon every night, and only returned at 1:00 A.M. for the final two hours of the shift. This convenient arrangement lasted until the newspapers filed a dual complaint. They were receiving the pre–1:00 A.M. copy later and later; but, since Edison took pains in his writing, it was infinitely more readable than the post–1:00 A.M. copy, which reached them in the usual telegrapher's scrawl. They wanted the early copy at the speed of the late copy, and the late copy as neat as the early copy. This, of course, was impossible. Investigating and discovering what was going on, Sam Wallick, the office manager, put a halt to the operation.

Edison, however, had been set off. In the largely isolated years of his childhood and youth, he had learned to explore, and to do things for himself. He had come by his knowledge pragmatically. He had had little guidance, and had never been inculcated with any great respect for dogma or prevailing wisdom. He was stubborn, confident in his own ability; he was a skeptic, unwilling to believe that just because something hadn't been done it couldn't be done; he was ingenious, quick to perceive an opening or uncharted path; he was reckless and undisciplined, willing to expend himself and others in pursuit of his goals.

Since messages between cities more than 200 miles apart had to be taken off at intermediate stations and retransmitted, why, Edison asked himself, couldn't an old Morse register be linked with a transmitter so as to form an automatic repeater? The embossed strip taken off the register could be fed into the transmitter, whose key would be rigged so as to be activated by the dots and dashes. After working on the machinery for several weeks, Edison enlisted Ed Parmalee to help him carry it to Union Station. It failed. Edison was so disgusted that he threw the whole lot down in front of the Palmer House and strode off. Parmalee, less volatile, gathered up the remnants. After calming down, Edison returned to work. Eventually he put the instruments on the line, and was able to repeat a message automatically from Pittsburgh to St. Louis.

The manager nevertheless chided Edison for not paying stricter attention to his job and wasting time on useless experiments. The first-class operators derided his inventive pretensions. After four or five months, he was distinctly unpopular. From Ezra Gilliland he heard of an opening in Cincinnati. With scarcely more than fifty cents in his pocket—for wartime inflation had shrunk the purchasing power of money to half of what it had been, and Edison was a careless spender—he arrived in the Ohio River city in February, 1865. He wore a frayed blue flannel shirt and unpressed trousers. Though he talked sparingly, his jaws were in constant motion—like his father and grandfather, he chewed tobacco incessantly.

Assigned to the Portsmouth, Ohio, way wire, he moved in with Gilliland and Nat Hyams, the comedian at the Woods Theater. He was so fond of reading turgid novels that he was nicknamed "Victor Hugo." He now acquired a taste for tragedy, especially *Othello,* and he and the "City Jerkwater" operator, Milt Adams, three years older, regularly went to the theater. Stagestruck, Edison decided he would become a tragedian, but was restrained by two handicaps—his voice was high-pitched and shrill, and he was terrified of facing an audience.*

With Gilliland, Edison continued his habit of going to the office at night to practice. Gilliland would transmit parts of plays, and Edison would copy. They sold the copies to actors, thereby turning the practice to slight profit. Occasionally they visited the German district, "Over the Rhine," where they drank beer and listened to band music at Loewen Garden. When the weather warmed, they went swimming in the Ohio. Generally, however, Edison's simultaneous brashness and introversion,

* On April 8, 1878, Jot Spencer, a telegrapher friend, wrote Edison: "I cannot avoid a tinge of regret when I think of the loss suffered by dramatic art when you turned your back on tragedy."

his uncouth manners, and his fondness for playing practical jokes led most operators to shun him.

The end of the war brought an upsurge in the union movement. Three delegates from Cleveland, a center of the telegraphers' union, arrived to form a chapter in Cincinnati. The eight first-class operators of the night shift went off with them to a meeting that, Edison declared, "resulted in tne imbibing of a large quantity of brewery serum, although most of the men was immune to that anesthetic." Only two of them showed up for work, and Edison perceived his opportunity. Stationing himself at an instrument, he began taking press off the wire. The copy was supplied to four newspapers, so he had to imprint the letters with an agate stylus through five layers of oil tissue, which worked much like carbon paper. "The individual words would not bear close inspection," he recalled. "When I missed understanding a word there was no time to think of what it was so I made an illegible one to fill in, trusting to the printers to sense it. I knew they could read anything because Mr. Bloss, an editor on *The Inquirer,* made such bad copy that one of the editorials in manuscript was posted up on the board with an offer of $20 to any man who could read 20 consecutive words. Nobody ever did it."

Edison worked until the wire closed, at 3:00 A.M., then was too wrought up to go to sleep, but waited for the day manager, Stevens, with great anticipation. There was considerable hubbub around the office, but Stevens, an austere man of whom Edison was afraid, paid no note. Not until 3:00 P.M. did Stevens go over to the hook and examine the copy. Turning to Edison, he declared: "Young man, I want you to work the Louisville wire nights."

So Edison earned his first-class rating and a near doubling of his salary, to $125 a month. The additional money enabled him to expand his experiments, which now took a new direction. The great increase of telegraphic activity was placing an enormous burden on the wires, and a number of experimenters were trying to devise a *duplex*—an apparatus by means of which two messages could be sent over one wire simultaneously. Edison obtained equipment and made several sketches for such a duplex. But, having alienated the other operators by his union-busting opportunism, his situation was so unpleasant that in the fall of 1865 he departed for Memphis, where, an acquaintance informed him, work was available in the military telegraph office.

Memphis, like other Southern cities, had suffered a complete breakdown of its social order. If the eighteen-year-old Edison arrived with any remnants of innocence, which is doubtful, Memphis stripped them away. Gamblers, gunmen, whores, carpetbaggers, riverboat men, Union soldiers, and footless Confederate veterans filled the streets. Over twenty keno rooms were in operation. One was in a former Baptist church, where the man with the wheel dispensed numbers in place of psalms from the pulpit. Every night Edison visited a gorgeously furnished faro bank which served free suppers. He was able to obtain only part-time work with the military telegraph, and his pockets were emptier than ever. He slept on the floor of a commandeered building, in and out of which drunk telegraphers wandered all night.

The war having ended, the military telegraph force was rapidly being reduced, so it would have been wise to be prudent. Edison, however, installed his repeater in an effort to link Louisville with New Orleans, and kept the office in an uproar with his duplex experiments. Colonel Coleman, the officer in charge, was almost in tears: "Any damn fool ought to know that a wire can't be worked both ways at the same time," he cried, and ordered Edison to direct himself one way: out of the office!

Edison was in such straits that he joined a railroad gang laying track across the river in Arkansas. It was the kind of physical, aching labor he hated. The laborers were mostly Irish and German immigrants, and he felt so humiliated working alongside them that he never spoke of the experience afterward.

Once he had accumulated enough money to last a few days, he headed east. Dressed in a linen duster and old army shoes, he rode the rods and walked along the railroad. He bunked and ate at telegraph offices along the way, and finally reached Louisville in a snowstorm during the late fall of 1865.

Louisville was one of the cities on the American Telegraph Company network, which now stretched from Boston to New Orleans, and Edison was known there from his stint on the Cincinnati-Louisville wire. The high turnover of operators had the office in turmoil—one night it was left completely unmanned because one operator had fallen off his horse and broken his leg, another had been stabbed in a keno room, and a third had gone to see a hanging. After a few days, Edison was hired.

Though he was paid $110 a month, he lived, like most of the other telegraphers, in "Poverty Flat," where he had a room over a lager beer saloon. The office itself was on the second floor of a dilapidated building infested with rats. The place was never cleaned. The filth on the floor was complemented by the dried plugs of tobacco stuck to the ceiling, onto which they had been hurled by the men. The instruments were arrayed on a dozen tables about the room, and connected to a two-foot-square switchboard by a spiderweb of wires. Adjacent was the battery room. One hundred cells of nitric acid, which supplied the power, occupied a stand in the center. Spilled acid had almost eaten through the stand, as well as the floor. The remainder of the room had gradually filled up with piles of old books and messages, which furnished unlimited forage for the rats. The place had a peculiar attraction for lightning, which was drawn in by the wires and exploded like cannon shots from the switchboard.

All in all, it was a fairly typical telegraph office of the period. Sixteen telegraphers and a dozen clerks and messengers composed the staff. They were a hard-drinking, tobacco-chewing, foul-mouthed, iconoclastic, peripatetic, literate, and intelligent lot. Edison fit the mold, except in one respect—he could not stomach hard liquor. He tried drinking corn whiskey, but one teaspoon of the stuff put him to sleep. Even small quantities of alcohol produced such a negative reaction that, aside from taking an occasional glass of beer or wine, he became an abstainer.

Because Edison's poor coordination made him a jerky, spasmodic, "York and Erie" sender, he was used primarily to take press. This suited him perfectly, for he was still attracted by journalism. After getting off work at 3:00 A.M. he would pick up all the newspapers in the city, and read them before the noon meal the following day. He haunted the office of the Associated Press manager, a Harvard graduate named Tyler, and listened for hours to his discussions with George D. Prentice, the *Louisville Journal* editor, who was one of the developers of humor in newspapers. Storytelling was in great vogue, and everyone tried to top everyone else. Telegraphers occupied idle time by transmitting jokes. Edison systematically collected items from the newspapers and the wire and compiled them in a scrapbook. It was typical of his methodical though nonselective approach to any subject that interested him.

One day the New York and Boston operators issued a challenge through the *Telegrapher* to determine the fastest sender in the United States, the winner to receive a gold key. The Louisville office wanted to enter George Everett of Nashville. Edison, who was continuing his experimentation with the repeater, perforated MISSISSIPPI in Morse through a strip of brown paper. He then inserted the repeater in the circuit, and drew the strip through again and again to activate the key. Repeatedly and with enormous rapidity, —— .. ..——. .——. .. flashed over the wire, creating a veritable *brrrrrrrr* at the other end.

"Who in the hell is that?" the New York operator asked.

"That is our man Everett," Edison replied.

With the growth of traffic, the managers of the telegraph lines expressed increasing interest in the development of a duplex. Such a device, if possible to perfect, promised great economies. Edison responded to whatever was topical like a reporter to a murder. To the rear of the telegraph office was a two-story brick building housing a mattress factory and an instrument maker's shop. Edison strung a wire across the yard to the shop, and began rearranging all the telegraph sounders, keys, and relays in the office in order to test his ideas.

His capacity for generating chaos was once more leading to friction with a manager when, in the spring of 1866, he saw an advertisement for telegraphers by the Brazilian government. The telegraph was in its formative stage in South America, and salaries offered were high. Edison set about to learn Spanish—which, of course, would have done him little good—and after a few weeks departed for New Orleans with two fellow operators. There he was stranded for some time. When he could obtain only spot work at the telegraph office, he was again reduced to sleeping on floors. Finally, in July, he obtained passage on a steamer that was to carry Confederate veterans and their families to a new life in South America.* Before they could sail, however, violent antiblack riots erupted, and the military government commandeered the vessel to bring troops to the city.

Edison gave up, and by hook and crook made his way back to Port Huron. Arriving broke, he discovered his father and mother had scarcely a penny either. His friend Jim Clancey, who had just married, invited him to dinner. Afterward, Al explained his plight. Clancey, the possessor of thirteen dollars, lent Edison ten dollars so he could return to Louisville.

Edison, who had learned from his father to be cavalier about financial obligations, failed to pay the sum back, although he was rehired by the Western Union office. His intensive reading of the newspapers and popular magazines gave him an excellent grasp of current affairs, and enabled him on occasion to make up his own dispatches. "I knew every member of Congress and what committees they were on and all about the topical doings," he declared. "I was in a much better position than most operators to call on my imagination to supply missing words and sentences, especially on stormy nights. I had to supply in some cases one fifth of the whole matter but I seldom got caught. Except once there had been some kind of convention in Virginia in which John Minor Botts was a leading figure. There was no doubt but that the vote the next day would go a certain way. A very bad storm came up about 10 o'clock at night. My wire worked very bad. There was a cessation of all signals. Then I made out the words 'Minor Botts.' I filled in a paragraph about the convention and how the vote went as I was sure it would, but the next day I learned that instead of there being a vote the convention had adjourned for one day."

Edison's transgression in concocting his own news was viewed less seriously than his assault on the office's elegant new quarters. The instruments had been fastened down, and the operators received strict instructions they were not to monkey with any of them. In the case of Edison, that was like putting the catnip off limits to the cat. At one time he had all the wires scrambled and every instrument in the room connected together. One night he went into the battery room to extract some sulfuric acid for an experiment. Unfortunately, he lost control of the

* According to a story told by his father twenty-five years later, Edison may have actually left New Orleans on an earlier ship. But because of seasickness and the outbreak of yellow fever, he got off at Vera Cruz, and returned to Louisiana via Cuba. Edison himself never spoke of this abortive sally, but did mention that his two companions died in a Mexican port.

heavy carboy—slipping through his hands, it smashed to the floor. The acid spread like a searing flood, and made Swiss cheese of the ceiling of the bank below. Dripping onto the expensive furniture and carpet, it destroyed those, too. This catastrophe, added to another Edison disaster in attempting to concoct daulin (dynamite), strained the manager's good nature beyond endurance. Once more Edison was fired.

For three years Edison had wandered from place to place. In 1863 he had gone from Port Huron to Stratford. In 1864 he had moved on to Adrian, Toledo, and Indianapolis. In 1865 to Cincinnati, Memphis, and Louisville. In 1866 to New Orleans, then back to Louisville. Now, toward the beginning of 1867, he went for the first time to Boston.

He had, during his travels, become associated with an older telegrapher, Sam Ropes. Together they were attempting to devise a telegraphic printer. The basic concept was Edison's, but Ropes had put up the money. After the money ran out, Ropes made his way to Boston. There he interested two Cambridge businessmen in providing additional funds. On January 1, 1867, Edison signed a contract for developing the invention. Upon its successful completion he was to be paid only a small sum—$250. But for the first time his imagination and mechanical bent promised to be rewarded.*

Ezra Gilliland had remained in Cincinnati, where—as was the fashion—he financed a workshop from his operator's wages. Edison returned to Cincinnati and, despite his misadventures, managed to get himself rehired by the Western Union office there. With Gilliland's assistance, he rented a room on the top floor of an office building, bought an old cot and oil stove, and moved in a foot lathe and some tools. He was now as dedicated to fashioning a printer as he was bored by receiving and transmitting. Interested only in the mechanical and electrical aspects of telegraphy, he looked on the instruments primarily as experimental devices. Frequently he feigned illness so that he could spend his days and nights reading in the Mechanics' Library. He cultivated the acquaintance of Sommers, the superintendent of telegraphy for the Cincinnati and Indianapolis Railroad; and Sommers, a witty man who was fond of experimenting himself, let him scavenge among the worn instruments and tools.

The latest, most prized possession Edison had acquired was a secondhand Ruhmkorff induction coil, which transformed the current from a battery into high voltage. With Sommers he went to the railroad roundhouse, connected the coil to a long wash trough, and then climbed up to watch the fun through a hole in the roof. When the first man put his hands in the water, his arms shot up and were nearly wrenched

* Edison was not granted the patent for the printer until the summer of 1869, and it did not make an overwhelming impact on the telegraph industry.

from their sockets. The next man entering was similarly shocked. The crowd grew. The appearance of each new victim was eagerly awaited. Everyone speculated about the cause of the phenomenon. Electricity was still an unknown force to all but a few initiates, and none of the men had any inkling or understanding of what was going on.

Edison's irresponsibility and lack of sobriety soon wore out his welcome in Cincinnati. In the late fall of 1867 he was back in Port Huron, once more without a job and without money. His mother, now fifty-six years old, was drifting into senility and had periods of mental derangement. She had lost another child—their thirty-year-old daughter, Harriet Ann, had died four years before. Only thirty-seven-year-old Pitt remained at Port Huron. Not only did he appear destined, like his father, to pursue unprofitable business ventures, but, to the chagrin of Sam and Nancy, he had married an Irish Catholic girl.

Most shattering of all for Nancy was the failure of Alva, for whom she once had had great expectations, to settle down, hold a job, and behave responsibly. Despite his sporadic school attendance, he had attained a knowledge and an education far beyond her own and Sam's. He had learned a skill that paid good money. Yet he spent every cent he earned, and borrowed more; he conjured, like Sam, visions of fortune; he played practical jokes that frightened her (that Christmas he knocked down every relative and friend he was able to lure into contact with his Ruhmkorff coil). He seemed a drifter and a dreamer who lived from hand to mouth. She had pretty much given up hope he would ever amount to anything.

CHAPTER 3

The Walking Churchyard

Al's future, in truth, did not seem promising. He was a clever dabbler at invention, but there were scores of telegraphers who were passionate tinkerers. Though he had a rough-hewn likableness that attracted people upon initial contact, he had no hesitation about exploiting his friends. All too often those who had been his most ardent partisans became his most bitter detractors.

The swath he had cut through the Midwest made further employment in that area doubtful. Sending out inquiries to contacts in the East, he wrangled a pass on the Grand Trunk and Western Railroad. After four days' journey through a blizzard, he arrived in subzero weather in Montreal. He found a telegrapher named Stanton, with whom he had worked in Cincinnati. Stanton took him to an unheated, slum-district boardinghouse. The water in the basin was frozen solid, the beds offered only a thin blanket, and the staple at the table was watery gruel—the best that a dollar and a half a week would buy. It was January, 1868, and he had reached a nadir.

Then his friend Milt Adams, who had moved from Cincinnati to Boston, wired that there was an opening in the Western Union office, if Edison could come immediately. Edison materialized as if transported on a beam of light. Carrying a small satchel, he appeared at the office dressed in his usual blue flannel shirt, a rough jacket, a pair of jeans several inches too short, and a broad-brimmed hat with a torn rim. Nervously he pulled at his upper lip, where a hazy moustache was sometimes mistaken for a streak of dirt. His Midwestern twang was accentuated by some odd pronunciations, such as "deef" for deaf. When the Boston telegraphers, who dressed like gentlemen, beheld Edison, they considered having him cast in bronze as the figure of a tramp operator.

George F. Milliken, the manager, told him to report back for a tryout on the night shift.

Operators were probably the most mobile and well-informed group in the world. They continually talked with one another over the wires. There were few secrets among them. Edison was known as a loner. He drifted off, sporadically, into daydreams that resembled trances. He had a reputation for going to bed with books, while other operators went to bed with seasoned whiskey and young women, or vice versa. His addiction to experimentation and offbeat behavior had earned him a nickname, "the Looney," that was well known throughout the fraternity. The men in the office laid plans to salt him.

On returning, he was directed to take a special dispatch for the *Boston Herald*. The operator in New York was Hutchinson. One of the fastest on the force, he had been told about the reception for the new man. Edison's instrument began clicking at twenty-five words per minute. Since he was not exceptionally speedy, he had developed several tricks enabling him to keep up. The faster the copy came, the smaller he wrote, and he could make letters so minute a magnifying glass was required to decipher them. He concentrated on the content and "read" the message as it arrived, so that he could drop one or two sentences behind, yet still continue. In a tight spot he could, of course, always fall back on his imagination. The clicks crept up to thirty, then thirty-five words per minute. The room was very quiet, and all eyes were on Edison. Gradually the New York operator began abbreviating the words and slurring the signals. About halfway through the report Edison had to "break."

"Say, young man," he clicked on the key, "change off and send with your other foot."

Everyone broke up. He had passed the test. Yet he was not long in reaffirming his reputation as an unsettling spirit. He wired the water bucket and fixed the Morse keys so that they stuck. Lacking the judgment to know when a joke had lost its zest, he so enraged another operator that the man hurled a heavy glass insulator at his head. Edison ducked barely in time to save him from the hangman.

Yet whereas Edison had been merely an oddball in the offices of the Midwest, in Boston he was one of numerous young operators bending their imaginations to invention. New England was the cradle of the American Industrial Revolution, and the city was the crucible of scientific and especially electrical research. Charles Williams, a purveyor of electrical devices, encouraged young inventors. Many outfitted small rooms as laboratories in the same building as his shop, and the place was the fulcrum of electrical experimentation in America. Displayed in the window was a working platinum incandescent lamp, designed by Moses Farmer, the nation's foremost electrical engineer.

Edison went to the shop on Court Street to buy equipment, listen, and exchange ideas.* During the years in the Midwest he had acquired skill in working with telegraph apparatus, and he had a knack for analyzing and going to the heart of a problem. In Boston he was able to draw upon the experience of others, and he was not loath to adapt whatever he perceived as valuable to his own designs.

Within three months of arriving in the city, Edison constructed an improved model of his duplex. Milliken, who had a long-standing interest in the technical improvement of telegraphy, introduced him to Frank L. Pope and James Ashley, who had been operators under Milliken on the American Telegraph Company before its absorption by Western Union in 1866. Pope, though only seven years older than Edison, was one of the half-dozen most respected experts in telegraphy. His book, *The Modern Practice of the Electric Telegraph,* which was to pass through fifteen editions in twenty-three years, was scheduled for publication in a few months. Edison demonstrated his duplex apparatus for sending two messages simultaneously over the same wire to Pope, and Pope was impressed. Together with Ashley, Pope had been instrumental in founding the *Telegrapher,* the industry's independent journal, and Ashley was the paper's editor. In April, 1868, the *Telegrapher* ran a long article on Edison's duplex.

Others had preceded Edison. Farmer had successfully tested a duplex between New York and Philadelphia in 1856, and Joseph Stearns used one wire to send two messages at the same time between New York and Boston late in 1868. The article, nevertheless, focused considerable attention on Edison. And Edison, in conjunction with several other telegraphers, established a small workshop.

They were freewheelers with more than moderate suicidal impulses. Adams boiled ether over an open flame. Edison, reading *Scientific American,* came across the formula for making nitroglycerin. Unable to resist, he soon had a jar prepared. To test the solution, he let a couple of drops fall to the floor. They went off like the opening rounds at Bunker Hill. The magazine warned that if the mixture was brown it was bad and unstable. Before Edison's eyes the liquid was turning browner and browner—which was very bad indeed. Hastily but gently Edison capped the container, and lowered it into a sewer in the heart of the city. He preserved, nevertheless, a small vial that he took to the office, where he gleefully announced that the contents were enough to transport everyone to a better climate. The operators came near panicking. After an earlier experiment, when two liquids of different colors had blown up, Edison had walked into the office unhurt but two-toned.

Living frugally, he was devoting most of his $120 monthly salary to books, chemicals, and electrical equipment. He bought the three-volume

* A half-dozen years later, Alexander Graham Bell was to have his rooms in the same neighborhood.

set of Michael Faraday's works, and set about reading them in the hall bedroom* he paid $12.50 a month for at 9–10 Wilson Lane, a residence occupied also by MIT students. Browsing among the bookstalls on Corn Hill, he concentrated on whatever subject was of latest interest, so that he could keep one step ahead of everybody else. Though he was little older than his fellow boarders, his articulateness and wide experience enabled him to dominate the conversations that lasted long into the night. At the office he often bested Pat Burns, a Harvard Law School graduate and fine orator, in debates over the Bible—a book with which Edison was well versed, but to which he was ill reconciled.

Opportunities were opening all about him as the result of the impetus given to telegraphy by the Civil War. The industry was branching out in new directions; lines were being installed for police- and fire-alarm systems, and for private communication by businesses, which previously had had to rely on messengers. Edison was soon involved in all the different branches—the most important of them the stock-ticker system, to which he was introduced by Pope.

In New York, Pope was the superintendent of the Laws Reporting Telegraph, whose proprietor, Samuel S. Laws, was the inventor of the gold ticker. The forty-five-year-old Laws, a brilliant, mercurial, eccentric man, had been president of Westminster College in Missouri until ridden out at the the start of the Civil War because of his Southern sympathies. After a brief sojourn in Paris, he had returned to New York to handle the financial affairs of wealthy relatives. In 1863, on the organization of the Gold Exchange—which laid the foundation for Wall Street—Laws had been chosen presiding officer. Since the price of gold determined the prices of all other commodities, hundreds of messenger boys, constantly running in to obtain the latest quotations, turned the room into a bedlam. To keep the boys out of the room, Laws devised a simple, clocklike instrument, electrically controlled. Placed in the window, it registered the minute-by-minute quotations. In 1866 Laws had improved the system so the line could carry additional instruments. These were distributed to the offices of brokers and commodity dealers. Pope produced a further elaboration with an arrangement for putting the quotations in roman characters on flowing paper, and the stock ticker was born.

Edison, in partnership with a number of other telegraphers, notably D. C. "Bob" Roberts and Frank Hanaford, decided to establish a gold-ticker system in Boston. Going to New York, Edison returned with a transmitter and several tickers that he obtained from Pope, and set up an office in the Boston gold exchange. When he and his partners made their first sale, to the banking house of Kidder & Peabody, they were jubilant. That night, after finishing work at Western Union, they went around the corner to George Young's Coffee House, famed for its "buttered Santa

* Hall bedrooms were cubicles whose only window opened onto the corridor. In poorer tenements they often contained no windows at all.

Cruz rum." Roberts expressed the opinion they were on their way "to make Rome howl!"

Gradually they were able to obtain other contracts, and spent their days clambering over rooftops to install private telegraph lines for such concerns as the D. N. Skillings Oil Company and the Continental Sugar Refining Company. Municipalities did not regulate the stringing of wires, property owners did not exercise control over their roofs, and each entrepreneur laid lines wherever his business carried him; so that Boston—and other cities—little by little took on the appearance of being inhabited by giant spiders.

While Al was earnest, dedicated, and determined to make his fortune, several of the others looked upon the enterprise and the small workshop primarily as a lark. "I squandered and you experimented," one of them (to whom Al lent a hundred dollars, never paid back) summed up.

The headmistress of a finishing school who thought her charges should know something of the new art of telegraphy was referred to Edison, and he agreed to give a demonstration. But when the day arrived, he had forgotten—an appointment with Edison was always likely to turn into a disappointment. Milt Adams, whom Edison was employing to help string wires, had to haul him down from a rooftop, and they appeared before the ladies panting and disheveled. Edison, who could talk so well with two or three people, found that "when there are more they radiate some unknown form of influence which paralyzes my vocal cords." He put a wire across the room and clicked and clacked with Adams, but said little to the ladies. The girls nevertheless thought him droll. From then on whenever the young ladies were walked two by two down the street on their constitutionals and encountered Edison, they would greet him demurely—greetings much enhancing his reputation with his fellow telegraphers.

Edison routinely slept till mid-morning, experimented and tended to business during the day, then returned to Western Union in the evening. Stubborn and willful as always, he continued to experience difficulties with his employer. When printers complained his handwriting was too minute to read, he enlarged it until one line filled an entire page. When "Jesus Christ" appeared an infinite number of times in a long-winded report of a Methodist Church synod, he substituted the abbreviation "J.C.," and pretended not to understand what the subsequent flap was about—after all, no one objected to "B.C." When one of the frequently magnificent discussions broke out in the Boston office and he was unable to follow it because of his impaired hearing, he grounded his instrument under the table, then summoned the chief operator. While that man, cursing, looked for the trouble amid the debris on the floor, Edison idled up close to the speakers and listened to their talk.

His habit, like that of other experimenters, was to try out his devices on the lines during the post-midnight hours, when traffic was sparse. But then William Orton, the company's recently named president, banned "the flood of capricious inventions." At the end of January, 1869, Edison departed—as he had left or been fired from every other job he had ever held.

Supremely confident, he decided he would make his living as a deviser and purveyor of telegraphic equipment. His imagination was boundless. He believed he could solve any problem. In looks and conversation he seemed older than he was. He joined a whimsical charm to the sales ability of a snake-oil pitchman. He had one overriding goal—he wanted to make money. That was the driving force behind his experimentation then and throughout his life. A contemporary described him as "a singular specimen of man having the highest natural inventive facility coupled with the playfulness and artlessness of a child."

Since both private telegraphy and the stock ticker had limited utility until a printer containing all the letters of the alphabet could be devised, Edison applied himself to the creation of such a machine. He attempted to win the contract to install a fire-alarm system in Cambridge, but was rebuffed because the town's elders lacked confidence in his ability. Financed by Bob Roberts, who supplied one hundred dollars, he assembled an electrochemical vote-recording machine.

His most ardent endeavor, since it promised to be the most lucrative, continued to be the fashioning of a duplex. He met a Boston businessman, E. Baker Welch, who—bit by bit—contributed a total of eight hundred dollars in return for half interest in the invention. By early May, 1869, Al was ready to test the device. Pope agreed to help him, and secured permission to try out the duplex on the Atlantic & Pacific Telegraph Company line from New York to Chicago. Edison dropped one set of instruments off with Frank Pope, then took the train to Rochester, where he connected the other set. As happened frequently in telegraphy, the mechanism performed quite differently on the long, exposed wires than in the protected confines of the shop. After several days of fruitlessly trying to get the instruments working, Edison headed back to Boston.

Welch refused to advance further funds. Within a few weeks, Edison spent his last $160 on experimenting and buying equipment for the erratically performing stock ticker and telegraph lines. He badly needed more printers, but had no funds to buy them. When he had but a dollar or two left, he caught a steamer for New York in the hope of obtaining assistance from Pope and his acquaintances. To tide him over, Pope bedded him down in the battery room of the company, advanced him money for the apple dumplings and pie that formed the major part of his diet, and employed him as a piecework mechanic.

During the next several months, while Hanaford attempted to run the Boston operation, to which Edison theoretically was committed to return, Edison was, in reality, sinking his roots deeper and deeper into

New York. When he had arrived in Boston, a year and a half before, he had been an unsophisticated aspirant of considerable but unproven talent. As he moved on to New York, he had learned enough to be an accomplished artisan.

In Boston, one thing after another was going wrong with the lines—the batteries were giving out, the printers were breaking down, there were bugs of various dimensions. In mid-July, Hanaford, who was in despair, pleaded with Edison to return. On the twenty-sixth, Edison wrote back, blaming Charles Williams, whose shop had made the instruments, for shoddy workmanship. "It would appear from what has happened already that the gray-eyed spectre of destiny has been our guardian angel for no matter what I may do I reap nothing but trouble and the blues," he lamented.

"Please do all you can," he continued, "till I get the Printers ready, and put them on the line, you ought to know me well enough to know that I am neither a dead beat or a selfish person and that I always do as I agree without some damnable God damnable ill luck prevents it. However, I'll never give up for I may have a streak of luck before I die."

A few days later the prediction seemed fulfilled. A spring broke on the Laws transmitting instrument, fell into the gears, and jammed the works. As the transmitter halted with a crash, the instruments in the brokers' offices ran wild. Within minutes the gold room turned into a scene of shouting men and boys. Edison, in the absence of Pope, examined the transmitter and ascertained the difficulty. Laws rushed in, his arms waving, and yelled: "Fix it! Fix it!"

Edison responded that he would. In a couple of hours he had the transmitter repaired. Operations could not be resumed, however, because the receiving instruments had to be reset by hand. Edison told Laws he could construct a device to halt all instruments in unison and prevent them from running wild. (This became one of Edison's first successful inventions.) The excitable Laws took Edison to his book-crammed office. After subjecting Edison to intense questioning on a variety of subjects, while occasionally lapsing into Latin, Greek, and Hebrew, Laws promised him a supervisory position with the company at $225 a month if he perfected his "unison stop."

Unfortunately for Edison, Laws's competition, the Gold & Stock Telegraph Company, had an instrument superior to the Laws-Pope ticker. Despite Edison's good work, the breakdown of the Laws equipment resulted in an additional shift of customers. On August 27, Laws, who was studying law at Columbia and had just been admitted to the bar, sold out to the Gold & Stock Company.

The consolidation, Edison told Hanaford on September 17, "upset all my calculations." He had applied at Western Union, but no telegrapher positions were available. He was working on a job for the Gold & Stock Company and expected to be paid soon, when he would send Hanaford money to buy a new battery for the Skillings line. "A barber told me yesterday the roots of my hair were all coming out gray and that I would be gray in ten months, by the trouble which my Double Transmitter give me and Skillings and other things."

Edison's lamentations, perhaps influenced by his ringside seat at the frenetic spectacle occasioned by Jay Gould's attempt to corner the gold market, were slightly overdramatic. Pope had entree to most of the business world concentrated on Wall Street and lower Broadway. With Ashley, he formed the Financial and Commercial Telegraph Company and went into competition with the Gold & Stock Company. Pope took Edison in as a partner, and they announced themselves in the *Telegrapher* as "Electrical Engineers, and General Telegraphic Agency." The Pope-and-Edison printer—primarily the work of Edison—became the mainstay of another new concern, the American Printing Telegraph Company, organized to erect private lines for business houses, as Edison had done in Boston. It was the first such venture in New York.

Pope boarded Edison at his parents' house in Elizabeth, New Jersey, across the river from New York, and arranged for him to have work space at the shop of an old electrical manufacturer in Jersey City.* Edison was to be the principal mechanic; Pope would be the businessman; and Ashley would assure plenty of publicity.

Although it was a shoestring venture, it was an important step ahead for Edison. His Boston enterprises were fizzling out. His electrographic vote recorder, for which he had been issued his first patent on June 1, turned out like the first duplex to be a failure, and Bob Roberts was unable to elicit any interest from the Massachusetts legislature.

Yet Edison was well advanced on the development of several printing telegraph devices. Through the frigid winter months he commuted between Elizabeth and Jersey City. Often he left home at seven o'clock in the morning and did not return until after midnight.

The end of January found Edison morose but hopeful. "Keep your courage up and it will come out all right," he wrote Hanaford. "Think I can get you a red hot situation here where you can make some stamps. My hair is damned near white. Man told me yesterday I was a walking churchyard."

* Edison's sojourn in Jersey City had its odd and even nauseating moments, as when the manufacturer, Dr. Bradley, attempted to devise an apparatus for the quick, electrical aging of whiskey, but produced instead only a cloud of hydrogen sulfide gas that routed everybody from the premises.

CHAPTER 4

The Proliferating Inventor

The "red hot situation" which boosted Edison's morale was a deal that Pope was working on with Marshall Lefferts, one of the most successful of the pioneering telegraph executives. Lefferts had been president of the American Telegraph Company, where Pope had risen to be his assistant, and when the firm had been bought out by Western Union, Lefferts had become the president of the Gold & Stock Company.

Pope, Ashley, and Edison, in their endeavor to compete with the Gold & Stock Company in the distribution of financial news, were scarcely a threat; and, lacking capital, they would probably soon fail. But they had applied for several valuable patents, the most important of which were Edison's. The negotiations between Pope and Lefferts culminated on April 18 in the signing of a mind-boggling, eleven-page agreement. Pope, Ashley, and Edison sold their patents and merged their stock-quotation firm into the Gold & Stock Company. They were paid $5000 down, and were to receive $5000 on validation of the patents, plus $5000 more in three monthly installments starting in September. Edison's portion of the initial payment was to be $1200, which he told Lefferts he did not think was fair. Furthermore, he was worried about his share being withheld from him. Laughing, Lefferts had a check for $1500 made out to him.

Edison had no experience with a sum like that. Self-consciously, he took the check to a bank on Broadway. His diminished hearing tended to deteriorate even more under stress, and he had a premonition of disaster. When the teller took the check, glanced at it, then shoved it back at Edison with some words he could not comprehend, Edison felt the eyes of everyone in the bank on him. He fled.

After some time, he gathered the courage to return to Lefferts's office. There, he complained he had been unable to obtain his money. Lefferts, examining the check, was sympathetically amused. He explained to Edison that, first, he had to endorse it.

Edison used the money to pay off his six-month, $150 boarding bill to the Popes, and the $20.50 Frank Pope had advanced him for pocket money. The remainder he deposited in the Mechanics National Bank in Newark. He didn't bother to send Hanaford the sum he owed him, or pay the rest of his outstanding bill to Charles Williams. The way to stretch money, Edison believed, was never to pay today what could be put off until tomorrow.

To manufacture the devices for the Gold & Stock Company, Edison needed his own workshop. Lefferts, a bluff, hearty man who took a paternal liking to the young inventor, arranged for Edison to go into partnership with William Unger, a Newark machine-shop operator distantly related to Lefferts. The firm of Edison and Unger was established at 15 Railroad Avenue on the Passaic River.

Edison, however, was soon unhappy again. He was contributing most of the ideas and doing most of the work. But Pope and Ashley kept such a tight rein on the money that he felt he did not have adequate funds for experiments. Writing Hanaford in Boston, he asked what prospects the future seemed to hold, and what had happened to one of their "antiquated fossil parietal" backers?

"Does he still pursue the ignatus fatus of Private Telegraphy, that bright beacon light upon the headlands of eternity where the bright prospects of two sanguine young men were totally wrecked and obliterated, alas that the bright vision of untold wealth should vanish like the fabric of a dream, that two ambitious mortals gifted by the genius of enterprise, and strength, should drink the bitter dregs of premature failure in their grand dreams for the advancement of science."

In such a state of mind, Edison during the summer of 1870 had a chance meeting in Lefferts's office with Daniel H. Craig, a man fully his equal in fulminating hyperbole. Gray-eyed, with wispy gray hair and a pointed beard punctuated by an asymmetrical mouth, Craig was an old associate of Lefferts and had played a leading role in the development not only of telegraphy but of modern newspaper techniques.* In

* As a news agent Craig had, in the 1840s, met European vessels off Nova Scotia and employed carrier pigeons to fly the latest dispatches to Boston and New York, thus getting a beat on the opposition. When an Englishman, Alexander Bain, invented an automatic system for telegraphy, Craig introduced it into the United States. (The Morse manual system of telegraphy was limited in practice to the transmitting of about twenty-five words per minute. Theoretically, a wire could handle forty or fifty times that number of words. With Bain's system a strip of paper was prepared beforehand and fed into an automatic transmitter. At the receiving station the characters were transcribed chemically.) In 1849 Craig induced Lefferts to install the Bain machinery in his newly formed New York and New England Telegraph Company; but, given the primitive state of electrical technology, the Bain apparatus failed in commercial application in the United States.

1851 he had organized the New York newspapers into the Associated Press, which two years later had expanded into a national news service. This had brought him into conflict with telegraph companies, which had tried to take over the organization. Finally, in 1866, he had been ousted as manager during a bitter dispute with Western Union.

Western Union, which had come into the dominion of New York Central Railroad baron Cornelius Vanderbilt, was now viewed by its smaller rivals as an octopus that threatened to bankrupt or absorb all of them. In fifteen years the corporation had become the nation's most profitable business. It handled 90 percent of the telegrams sent, and averaged $2000 per stockholder in profits yearly.

Craig, plotting revenge against Western Union, obtained control of the National Telegraph Company, a largely paper organization planning to operate between Boston, New York, and Washington. To equalize the competition, he intended to automate his line. Though the Bain system had proven impractical in the United States, Craig based his hopes on a new automatic system devised by a Passaic, New Jersey, inventor, George Little, who was financed by Lefferts.

When Edison met Craig, Craig had just bought an interest in the Little system. He was, however, greatly dissatisfied with Little's machine for perforating the paper strip fed into the automatic transmitting device. The machine would perforate only seven words per minute, a ridiculous rate when copy could be sent at a hundred words per minute or more. Edison told Craig he could make a machine four times as fast as Little's. Craig jumped at the offer. On August 3, 1870, he signed a contract with Edison for development of a perforator costing not more than $50. Edison was to receive $1300 in cash and $3700 in the stock of the National Telegraph Company.

Edison applied to Pope and Ashley for the additional funds required to devise the perforator. But they, not involved in the deal, turned him down. Edison complained to Craig about the parsimony of his partners. On August 12 Craig sent Edison a "confidential" message:

"At the end of a few days I hope to be in funds beyond my immediate necessities, and if you should then be in want I will most cheerfully aid you. It evinces a mean disposition and great want of common sense for any one who has control of your very superior talents to keep you in trouble about a few paltry dollars. When you and I get into close business relations I will assure you of better treatment."

One of the problems in automatic telegraphy was the necessity of employing one copier for every 500 words per hour received. In the hopes of speeding up the writing of the messages, Craig had paid $5000 for the rights to a type writing-machine invented by Christopher Sholes, a Wisconsin newspaper editor. It was a primitive device with a piano keyboard consisting of eleven white and ten black keys. Craig showed the wooden model to Moses Farmer, who had come to Manhattan to test Craig's new line from Boston to New York, and asked Farmer if he would perfect it. Farmer told Craig it was quite practicable to have a letter writing (or printing) machine with which at least 300 words per hour could be written in plain roman characters—he had worked the thing out years ago but had never had time to finish the machine. He had no time now. However, he was willing to contribute "all [my] ideas and help any competent mechanic to perfect the thing."*

Craig thought he could sell 10,000 such machines, and proposed to Edison that he should have "The exclusive manufacture of the machines. You, Farmer and myself sharing equally in all expenses and all profits. I really wish you could oblige me in this matter." He wanted Edison to come to New York Saturday morning and meet Farmer at the Astor House, the city's most elegant hotel. To Farmer he explained that Edison was "A genius only second to yourself, who will give us the benefit of his brains and machine shop, to work up your ideas for the writing machine."

Edison, however, was not disposed to be the tool for Farmer's ideas. He was already doing the work for Pope's and Ashley's "ideas," while his own mind was agitated by inventions he had no time for. He declined Craig's offer, but then made Craig a counteroffer.

Because of static, induction, resistance in the wire, and lack of power, an automatic system had no hope of working over a line longer than 200 miles. A message from Boston to Washington would have to be received in New York and transcribed onto another tape for relay, a tedious and uneconomic process. Edison proposed to invent an automatic relay. On August 17 Craig agreed to provide funds for the development of a "Repeater for Automatic or Fast Telegraphy" in return for half interest in the invention.

Craig had, in fact, very little money. A schemer all his life, he was financing the building of the National Telegraph Company line by such devices as staging a raffle among telegraphers for his Hudson River property, then manipulating the drawing so that no one held the winning ticket. He was urging the New York newspapers to provide him with funds so that they could form a new press association and escape the domination of Western Union, but was getting little response.

Edison's association with Craig marked the onset of a period of immensely complex business relationships for the inventor. Any new idea or proposition for an invention was like candy to him; he gobbled up

* Though the fifty-year-old Farmer was the nation's leading expert on electricity, he lacked the ambition and aggressiveness to transform his ideas into commercial instruments, and had found a comfortable niche as electrician of the United States Naval Torpedo Station at Newport, Rhode Island. He was birdlike, cheerful, well liked, a lover of music, who, a contemporary explained, "has been unfortunate in being almost always in advance of the necessities of the times. He thus became the feeder of others, while his own platter remained empty."

whatever came within reach and gave scarcely a thought to how he would digest it all.

There was now no question of his returning to Boston, so he simply abandoned his venture in that city. He was still in partnership with Pope and Ashley on a half-dozen devices he was developing for Marshall Lefferts and the Gold & Stock Company; he was in partnership with William Unger as a manufacturer; and he had just agreed to provide Daniel Craig with several items for the radically different branch of automatic telegraphy.

Most of the inventions Edison was working on involved an intricate linkage of electricity and mechanics—the current activated the clocklike machinery of gears, cogwheels, and levers. The automatic relay he proposed to fashion was not in this class, but harked back to his days as a plug telegrapher—it would be a more complex version of the relay he had tinkered with five years before. Notably, none of the inventions, although they were for telegraphy, required the exercise of auditory perception.

Since Edison could not proceed without money, Craig, in late August, enlisted George Harrington, another foe of Western Union, in his scheme.

Harrington, a man of means, had been assistant secretary of the treasury under President Lincoln and had just returned from four years as American minister to Switzerland. He was linked to a group of Philadelphia financiers who were bitter rivals of Wall Street; and since Western Union was controlled by New York bankers and railroad men, the Philadelphians opposed the company almost as a matter of course.

Harrington's principal associates were William J. Palmer, Josiah Reiff, and Uriah Painter. Palmer, though still in his early thirties, had been a cavalry general in the Civil War, and was now construction manager of the Kansas-Pacific Railroad, which had just reached Denver in its quest to become a rival of the Union Pacific (completed the year before). Reiff, thirty-three years old, was a dashing figure who had been breveted colonel for bravery during the war. He had inherited considerable wealth, and was the railroad's treasurer. Painter had built the New York, Philadelphia, and Norfolk Railroad, and was an intimate of many Republican leaders and members of President Grant's cabinet. An erudite man of diverse interests,* he was, under guise of capital correspondent for the *Philadelphia Inquirer*, a Washington lobbyist and wheeler-dealer.

Harrington had studied the operation of automatic telegraphy in Europe, and he and his associates (who soon were to play major roles in Edison's life) were interested in acquiring the Little system as an adjunct to the new Pacific railroad. Harrington persuaded Lefferts to

* He subsequently organized the first telephone company in the capital, and was the proprietor of the Lafayette Square Opera House.

sell controlling interest in the automatic system, and incorporated the Automatic Telegraph Company. (A few months later he also purchased the National Telegraph Company from Craig for $522,000—$105,000 in cash, the remainder in stock.)

To manufacture the equipment for the Automatic Telegraph Company, Harrington turned to Edison, whom he had heard about from Craig. In September, Harrington put up $6000 to form the American Telegraph Works with Edison. Edison, contributing his time and talent in lieu of $3000, had a one-third interest and was to receive a salary of $1350 a year.

Craig had promised Edison "to keep all my arrangements with you private—strictly so," yet, he supposed, "it will be impossible to prevent a few people around us from suspecting we are putting our heads together." Toward the end of September, Lefferts heard from Unger that Edison was spending less and less time at the shop but was in the process of equipping an entirely new establishment.

Lefferts summoned the proliferating inventor and demanded an explanation.

Edison protested that his inventive powers were being inhibited by his partnership with Pope and Ashley. He was doing all the work for them at the expense of his own interests. On October 3 he had delivered five new devices to his patent solicitor. He wanted to preserve friendly relations with the Gold & Stock Company. In fact, he was almost ready to demonstrate a small, private-line printer that was by far the best yet devised, and would give the Gold & Stock Company two months to experiment with it. He completely restored Lefferts's confidence. Lefferts drafted a memorandum, and on October 19 Edison signed it:

"I have no intention of doing the Gold & S. Tel. company injury if I can help it—they will however recognize my right as also my necessities to invent new and improve existing telegraph machines, but in doing so I am willing they should be benefited by such inventions as I may devise."

Simultaneously, Lefferts signed a new contract with Edison, excluding Pope and Ashley. Edison was to furnish his Universal Printers and accessories to the Gold & Stock Company. A hard though not always a sharp bargainer, Edison asked for $40,000, but settled for $30,000, to be paid in stock of the company if the printer proved all Edison claimed. Edison made sure that it would not be watered stock by including a clause that the payment was to be made on the basis that the company's stock did not exceed $1.25 million.

Within six months Edison had become party to five contracts, two with the Gold & Stock Company, two with Craig, and one with Harrington. He had committed himself to Herculean labor. He was an expert in the fields of standard telegraphy, the stock-ticker system, and private telegraphy; and he was now on the threshold of initiation into the technology of automatic telegraphy.

Edison and Unger's shop at 15 Railroad Avenue was a small place for which they paid $15 a month. For the American Telegraph Works, however, Edison had grand designs. He rented a huge building running from 93 to 109 Railroad Avenue for $1000 a year. He bought the finest machines and the best tools—some of them by the score. He spent more than $30,000 the first week, and sent the bills to Harrington, noting on them "to be accounted for on settlement" or "to be accounted for hereafter." All his life Edison had had to scratch for money; never had he had enough to buy even a small portion of the things he wanted for experimentation. Suddenly he had discovered his genie—rub the lamp and Harrington paid: a total of $1368 on October 31, $450 on November 3, $230 on the eleventh, $3489 on the twenty-sixth, $725 on December 1, $3926 on the thirty-first. By the first week in December Harrington was wondering whether Edison was under the misconception he was still assistant secretary of the treasury. Edison had spent three times as much as Harrington had contemplated. Harrington, who divided his time between New York and Washington, rushed to Newark to find out what was going on.

Edison was devoting sixteen to eighteen hours a day to his two establishments. Often he didn't bother to go home to Elizabeth, but slept on a table or curled up around a machine. He concentrated on inventing and the building of models—he did not like the routine tasks of supervising production or running a business. So it happened that when Harrington arrived one morning he discovered the shop operating at a leisurely pace under the foreman. The boss was absent—he was having a good sleep.

The men were in consternation as Harrington, asking questions, minutely examined everything in the place. He wanted a demonstration of the models. But the workers were all thumbs. Nothing seemed to operate right. Harrington grew more and more nervous and irritated. Finally, threatening to close down the shop, he departed.

When Edison returned to discover what had happened, he exploded into a rage primed by fear—he could not lose what he had just attained. He wrote Harrington he would be able to show him the copying printer and the transmitting and receiving apparatus within a week, and the perforating machine within two weeks. "You quite misunderstood my men here, the machine that they attempted to show you work had been taken apart by me in experimenting, and put together again but with not all the parts in, that is the reason it did not work." It was a story that was to be resurrected periodically at times of other crises.

To Craig, Edison complained that Harrington "by his fidgety manner got them so excited that they didn't know which end they stood on. I have eight men working night and day and it is an utter impossibility to hurry it more than I am, I and the draughtsman and men were up nearly all last night = Galieo discovered the principle of accurate Hologogy in the swinging lamp of Pisa = It wouldn't be a very sage remark to say—why damn it that lamp aint a clock. The Manufacture of Mechanism is a slow operation being legitimate, and all legitimate Businesses they are slow but sure, and the *slower the more sure.*"

It was Edison's way of declaiming in the dark, but in his heart he knew he had better produce working machines; and it was to just such a challenge that he responded superbly. Subsisting mainly on strong coffee, cigars, chewing tobacco, and pie, he seldom left the shop in December. The regular workweek was sixty hours (ten hours a day, six days a week). But the week starting December 5 one man worked sixteen hours on Monday and Tuesday, eighteen on Wednesday, sixteen on Thursday and Friday, and ten on Saturday—a total of ninety-two hours, for which he was paid $27.60. (The standard pay was 30 cents an hour.) Almost all of the men worked between seventy and eighty hours. The tremendous drive continued over Christmas. When New Year's arrived, Edison had the models in shape. The men went out, got drunk, and collapsed. The first week in 1871 the shop was practically empty.

"Hurrah for us!" Craig, informed of Edison's success, exulted on January 12. "Your confident face always inspires us with new *vim.*" A few weeks later he was even more emphatic: "Indeed, if you should tell me you could *make babies by machines,* I shouldn't doubt it."

The joyous air at the New York office of the Automatic Company caused stirrings of unease at Western Union headquarters, a short distance away on Broadway. The giant company had started negotiations to absorb the Gold & Stock Company, its only competitor in the distribution of commercial and financial news; and from Tracy R. Edson, the Gold & Stock vice-president, Western Union president William Orton heard astonishing stories of the Newark inventor. Edison's stock printer was far superior to any devised before. His perforator for the automatic system was almost ready; and he had made rough sketches for an automatic roman-character printer that would operate at 200 to 300 words per minute at the receiving station, thus eliminating that bottleneck.

Western Union had succeeded by not only squeezing but preempting its rivals. Talented inventors were too valuable, and dangerous, to be allowed to roam outside the fold. So it was that Tracy Edson dropped a discreet note to Edison, and sent him on to meet the president of Western Union. Orton told Edison the telegraph company was interested in his duplex—would he like to see if he could perfect the device? It was a low-key approach, a lure. The twenty-four-year-old Edison, an itinerant telegrapher three years before, had arrived in the corporate offices of Wall Street and Broadway.

CHAPTER 5

A Tool of Wall Street

Older men like Lefferts and Orton were fascinated by Edison. On first contact he seemed reserved, almost withdrawn. But when discussion turned to invention and the techniques of telegraphy, he became animated, his eyes sparkled, his imagination took flight. There were tens of thousands of hustlers with Bunyanesque ambition, thousands of fine artisans and mechanics, and scores of talented inventors—but only in Edison were all the characteristics combined. For most men imagination was a vehicle of escape. For Edison it was the catalyst for work. New York was full of men hungry for recognition and money—what set Edison apart was that, with all his bounding exaggeration, he conveyed the feeling he would succeed, that no matter what the obstacles he would pound away until he had demolished them.

Among the employees Edison hired at the American Telegraph Works was a black-bearded, handsome Englishman named Charles Batchelor, who was to have a major impact on Edison's career. Two years older than Edison, Batchelor was the son of a Manchester traveling salesman who for years had had a solid, unremarkable career. Then, on one trip, he had lost, squandered, or gambled away the receipts of 200 pounds (1000 dollars). Afterward, he had never again been able to obtain decent employment, and had frequently left his wife and seven children to their own devices—which included hunting over the heaths on Sundays to catch sparrows for sparrow pie.

Charles, still a boy, had been placed in a cotton mill. In 1865 he had been dispatched to America to introduce a new cotton machine. Subsequently, after returning to England, he had been sent to install machinery in the Clark thread factory in Newark. Deciding to remain in America, he had come to work for Edison.

Batchelor was neither the most experienced nor the highest paid among the men, but he and Edison were cut from the same pattern. Like Edison, Batchelor had no family, he was lonely, he too was a creature of the night. He knew little of telegraphy, but more of mechanics than Edison. Well read and articulate, he shared with Edison a love for Shakespeare. While Edison was handicapped by poorer-than-average hand coordination and was a primitive sketcher, Batchelor was a meticulous, trained draftsman with dexterous fingers. The grand strategy of Edison was transformed by Batchelor into the minutely detailed plan of operations.

The third member of the inventive trio appeared now in full-bearded John Kruesi, born in Switzerland. Twenty-nine years old, Kruesi had been orphaned at an early age and raised in an asylum. After serving an apprenticeship with a locksmith, he had become a journeyman machinist and for three years had traveled through Holland, Belgium, and France. In 1870 he had taken a ship for New York and joined the Singer Sewing Machine Company. Kruesi became Batchelor's intimate friend, and it was he who took the Batchelor drawings and transformed them into models.

Though Edison, Batchelor, and Kruesi formed an efficient inventive and production team, Edison had difficulty translating their efforts into a successful machine shop. He lacked the training, aptitude, and interest necessary for operating a business. He did not know how to estimate and balance income and outlay. He did not, in truth, want to know, for he was a compulsive spender. He was not profligate in his personal habits, though he did not live spartanly, either. But when it came to books, or chemicals, or machinery, or anything to do with experimentation, he indulged without regard to cost. When he received a large amount of money he went on a spree—it was his way of rewarding himself. To pay accumulated bills provided no gratification, so they were left to yellow till payment was forced by threat of court action. Edison's concept of balancing the books was to develop whatever income was needed to cover expenditures.

From the very beginning at Edison and Unger, the pattern was established. A $428 bill was paid $240 in cash, then $50 more, then nothing —until the patient, pleading creditor finally sent the sheriff. A $335 bill was paid $150 in cash, $145 later, and $40 much, much later. A $140 bill was paid in driblets of $10, $50, and $20. Edison even deferred the wages of his workers, paying them the $15 to $18 they earned for their regular sixty-hour week, then talking them into letting him carry the overtime on his books. Sometimes the sums mounted into the hundreds of dollars.

At the American Telegraph Works balancing the books was much easier—Edison just kept rubbing the lamp. Harrington had expected to invest $6000, but by the end of March Edison had spent well over $30,000. Craig pleaded: "If you don't let me see your Perforator and

Printer working soon, I think I shall hang myself—for things are getting too hot."

Soothingly, Edison responded: "Expenses may be a little heavy for the present but results will be tremendous."

That was too much for Harrington. He stormed into Newark the first week in April. In view of the heavy expenditures, he demanded that Edison sign a new, personal contract with him.

To Edison what mattered was the cash, not the contract. On April 4 he gave Harrington power of attorney for five years, and assigned to Harrington "two thirds interest of all my inventions (relating to the Little System), including therein all my inventions of mechanical or copying printers, and all the patents that may be issued applicable to automatic telegraphy mechanical printers." It was a contract lawyers were later to debate for years in court.

Though Harrington momentarily felt better, the problem of accounting remained. Harrington sent one man to audit the books in May, and another two months later to take charge of them.

"About time!" Edison ejaculated. "That department is in a very slack condition! *You cannot expect* a man to invent and work night and day and then be worried to a point of exasperation about how to obtain money to pay bills. If I keep on in this way six months longer I shall be completely broken down in health and mind."

Having gotten that off, Edison's mind turned to a new system of telegraphy. A week later he made the first notation for a "Dot and Dash and Automatic Printing Translating System, Invented for myself exclusively and not for any small brained capitalist."

Edison's confidence was expanding. His devices were proving successful, but the automatic system of George Little, for which they were supposed to be adjuncts, was giving nothing but trouble. As summer turned into fall, Edison planned replacements for all of Little's machinery. For Craig and Harrington this posed a dilemma. They had in Edison an explosive inventor, more talented than any other in the telegraphic field. But they also had a large investment in Little. They were in imminent danger of losing the railroads if they could not make the system operate. Craig put the matter to Edison as diplomatically as he could:

"Our great wants a year ago were a rapid perforator and a good copying printer. I take credit to myself for having discovered in you the genius we required. [But] your program is not calculated to advance the immediate introduction of Automatic Telegraphy. There is the most *urgent necessity* for immediate operations by the company."

At Craig's insistence, the system was given a full-scale test in mid-October. It was a disaster. The Little machinery actually decreased "the natural speed of the wire"—that is, the number of impulses that could be transmitted. Harrington, however, could not tear himself from Little.

He reproached Edison bitterly over the amount of money spent, the elaborate equipment, and the inefficiency of the American Telegraph Works. On October 27 Harrington appointed a supervisor to take charge of the shop.

Edison walked out.

He was now in want of neither money nor a place to work. Edison and Unger was doing very well. In May Western Union had acquired controlling interest in the Gold & Stock Company, but simultaneously turned over its commercial and financial department to the affiliated concern, so that the Gold & Stock Company now had a monopoly on Wall Street news. The Gold & Stock Company expanded its order for Edison's Universal Printers to 1200, worth $78,000—the largest single order in the telegraphic field to date. The instruments were to become the basis for the modern financial network.

Edison's companion of the night, Charles Batchelor, married in May. Consequently, Edison often found himself alone during his creative vigils. And though he generally was steeped in his own thoughts, he did not like solitude. His experience as a telegrapher had accustomed him to the feel of people about him. He worked best when he had someone on whom he could try out his ideas.

With Batchelor less available, Edison went more often to the theaters and the music halls, for which he retained lifelong enthusiasm. He increased the number of his trips to New York, where he roamed the streets, conducted business, and checked up on the operation of his printers at the Gold & Stock Company office.

To work the transmitters for its expanding network, the Gold & Stock Company conducted a training program for young girls. One of the workers in Edison's shop was John Ott, a cousin of Unger. Ott had an acquaintance, Charles Stilwell, whose sixteen-year-old sister, Mary, was looking for a job. With Ott's and Edison's assistance, Mary was taken on as a trainee by the Wall Street concern.

For Edison, women were as strange as inventing was natural. Certainly he was no innocent—he had had his boyhood sweetheart and his youthful acquaintance with low life. Having knocked about in a man's world since his early teens, he had picked up sex education in the rough; he had heard all the lewd stories; the jokes and tales that he collected in his pocket notebook were anatomically explicit. Yet his lack of close association with women increased his natural reserve; he did not know how to approach girls or what to say to them.

Thus, though his attention at the Gold & Stock Company was more and more drawn to the cascading blond hair and slim figure of the earnest, concentrating Mary, it was essentially a silent admiration. Yet his interest was so obvious that the girls in the office had a hard time trying to suppress their giggles. Mary herself was so embarrassed she was unable to coordinate her fingers with the keys at all.

Since her home was only ten blocks from Edison's shop in Newark, he had the opportunity to travel on the ferry with her. She was, as well as being flustered, flattered by his interest. His moustache gave him a rakish appearance. Courteous and generous, he could, when his tongue was loosened, also be funny. He was acknowledged to be an extraordinary man of immense talent, an inventor, the proprietor of his own machine shop. Most assuredly, the gossips said, he was a man of wealth.

She, on the other hand, came from a crowded, middle-class home—she had not only a host of brothers and sisters, but stepbrothers and stepsisters as well. Though her mother was well educated and she herself was intelligent, her father was a sawyer—a semiskilled occupation halfway between woodsman and carpenter—and the family was always pinched for money.

Her background, in fact, was not unlike Edison's. At sixteen, she was of prime marriageable age—many girls married at thirteen or fourteen. She liked musical comedy. As their relationship warmed and he grew bolder, he took her to Taylor and Sylvester's Music Hall. He bought expensive presents for her. The latter part of November they began looking at stately houses in the $10,000-to-$20,000 range. Eventually, however, Edison decided he could not afford to buy a house, and rented one instead.

On Christmas Day, 1871, they were married.

The next day Edison was back at work. Soon he was once more staying at the shop until midnight and later. He was not only a night person but a night creator—his mind worked best in the shadows. Mary, who had during her courtship gotten to know an attendant, entertaining, theatergoing Al, found that as a husband he displayed quite different characteristics. She tried to please him, but he was not observant and took little note. She wanted to be taken out, but he now had few hours to spare. She liked to dance, but he was of the opinion "If a chimpanzee had a sense of rhythm he would dance himself to death."

She tried to lure him home with stratagems. They called each other "Popsy Wopsy," after a song they'd heard a popular music-hall comedienne, Ada Alexander, sing:

"I am an 'elpless female, an unprotected female, my husband's been and gone. I'm left alone to sing my Popsy Wopsy's vanished from my sight and we might have been so happy so we might. But I mean to go and join him, go on the sly and join him, just drop upon him unawares and stop at once his fun. But if they will neglect us, to love they can't expect us, for what is sauce for goose you know is sauce for gander too."

It was bantering, bittersweet. Yet six weeks after the marriage there was already an undertone of serious discord. One night Edison was transcribing the result of an experiment in his lab book when he exploded:

"Mrs. Mary Edison my wife Dearly Beloved cannot invent worth a Damn!!"

Accustomed to a houseful of people, she was so fearful of staying by herself at night that he agreed to let her sister Alice move in. He tried to compensate her for the lack of his companionship by giving her plenty of money. He established accounts at stores for her. But she was even less capable of budgeting than he. After a youth of plain living, she could now buy almost anything she wanted. And she bought candy and oysters, brandy and cherries, fancy clothes and feathered hats. The accounts grew in size; the tradesmen had to plead for their money. Mary turned out to be the domestic counterpart of Edison in business.

Three weeks after Al's marriage, Sam came to inspect his new daughter-in-law. He was sixty-seven years old, with snowy hair and beard, but still as spry as a mountain goat. Nancy, her spirit broken and her mind deteriorated by a lifetime of worrying about money, sickly children, and a blithe-spirited husband, had died the previous April. Sam had hardly buried her than he let his eyes stray after pretty girls, who made good chasers for his whiskey. To the citizens of Port Huron, he seemed to be cavorting like a twenty-year-old.

Sam called his son's wife "dear little Mary," and simultaneously overwhelmed and irritated her. Steeped in nineteenth-century Midwestern prejudice, he was pleased she was not a Catholic. "There is one other that I do not own," he told her, referring to Pitt's wife. "She is Irish. The Pope is her father."

To Al, Sam brought a request from Pitt. Six months before, Pitt had become involved in the organization of a horsecar line, the Port Huron and Gratiot Street Railroad Company. Pitt, who had all of his father's faults but little of his charm, had the innate business savvy of a hustler selling shares in a dry oil well and convincing himself to buy the stock. The street railroad was already in trouble up to its horses' tails. Pitt needed a substantial sum to avoid bankruptcy.

Edison went to see Josiah Reiff, who now was secretary of the Automatic Telegraph Company as well as treasurer of the Kansas-Pacific Railroad. Reiff was astonished by Edison's ability, he liked him personally, and was to be a lifelong friend. He resembled Edison in his eternal optimism and somewhat frivolous approach to the dispensing of money. Edison left his office with $3100; and Sam conveyed the sum back to Port Huron.

As winter turned to spring, Edison's own financial improvidence was bringing him to a parting of ways with his partner, William Unger. Since

Edison and Unger paid wages neither more nor less liberal than those of other employers—a boy started at seven and a half cents an hour, a master mechanic earned thirty cents, a foreman fifty cents—they should have been showing a substantial profit on the Gold & Stock contract for 1200 tickers. But Edison drained the money no matter how swiftly it flowed in. He always had several men—usually the most skilled—working on experimental devices. His own hours were so erratic that he was nearly useless as a manager—some days he showed up at ten or eleven in the morning, on others he went home when everybody else came in. He disliked giving orders, and did not have the stomach for the face-to-face disciplining of men.

Toward the end of June, when the last machines were being completed, Unger decided that his bright but disorganized partner was of but slight profit to him. He told Edison he wanted to dissolve the partnership.

Edison was upset. His quarrel with Harrington had deprived him of the superbly equipped shop of the American Telegraph Works. Edison and Unger, since its start in the small shop on Railroad Avenue, had expanded twice, and was now located in a large, four-story building that dwarfed its neighbors on Ward Street. Yet Unger wanted to liquidate the business, sell off the machines, and divide the proceeds—thus leaving Edison without a place to work. Their combined equity was $22,878, but of that Unger owed Edison $3649.

Edison decided he would take over and run the shop by himself. He paid Unger $2500 in cash, assumed $6045 in debts, and gave William's brother Hermann, a Newark jeweler, a $5000 promissory note. The rest of the money needed to conduct the business he borrowed from the Republic Trust Company at 18 percent interest on a two-year note. It turned out to be one of the more important transactions in American business and legal history.

CHAPTER 6

The Night of Suspense

Loaded with debt, Edison worked ferociously through the summer of 1872. He formed a new partnership with Joseph T. Murray, the thirty-year-old son of a prominent Newark police judge and scholar of Indian history. Murray had been a sailor, an adventurer in Africa, an abolitionist, and a soldier in the Civil War—not particularly appropriate training for a machine-shop operator. He regarded Edison with awe and was devoted to him. But he was neither a good craftsman nor a businessman.

Edison turned daytime operations of the shop over to Murray, then, after the men went home and darkness settled, threw open the windows. Ideas danced through his mind like shadows from the breeze-stirred gaslights. Often he verbalized his thoughts as they took shape, so that for Batchelor (who also had reverted to his old habits) they became three-dimensional projections. Batchelor contributed his own ideas. After an interchange, Edison's rough sketch was transformed into a careful drawing by Batchelor. Kruesi made the model. Edison checked it, tinkered with it, and determined if it looked and worked as he had visualized.

So, part by part, with frequent alterations and backtrackings, Edison built his machines. During the first six months of the year he received thirteen patents, nearly all of them for printing telegraphs, and during the last half of the year he obtained twenty-one more. Working over a hundred hours a week with Batchelor and Kruesi, he was putting together an Edison System for Automatic and Chemical Telegraphy far superior to Bain's or Little's. He was constructing the implements for a domestic telegraph system linking houses to a central office so that firemen, police, a doctor, or a messenger could be summoned. He took the concept of the Sholes device, and developed his own type writing-machine. It had a letter keyboard, an electric-powered revolving type wheel, and an electromagnet furnishing the impulses for the impressions on the paper. It was a primitive prototype of the IBM Selectric, which would not appear until nearly a century later. But at a time when power had to be obtained from batteries, it was impractical for general use; and the first commercial typewriter was produced by the Remington Arms Company working from the Sholes model.

As impressive as these inventions were, they cost money and did not bring immediate returns. By fall, Edison had borrowed $4000 from the Gold & Stock Company, and was defaulting on the notes he had given Hermann Unger and other creditors. By October 9 he had accumulated nearly $9000 in unpaid bills. Mournfully going over them, Edison noted: "Soon as possible"; "must pay"; "quick"; and, repeatedly, "very sad case." Looking at a bill for $1970.75, he instructed Murray, "You can fix this."

Some relief came in the form of a new contract with the Gold & Stock Company. But Edison's real hope lay in the sickly state of the American Telegraph Works and the Automatic Telegraph Company. The manufacturing concern was being run no more efficiently by Harrington's man than it had been by Edison. Little was making no progress on his automatic system. Reiff was fed up, and convinced the only way to put the system in operation was to get Edison back working on it. On November 5 he called Edison to his office on Broadway and made a personal deal with him, promising to pay him a salary of $2000 a year.

By early January, 1873, Harrington, too, was disenchanted with Little. Little even accused the general manager of the Automatic Telegraph line, Edward Hibberd Johnson, of being the cause of failure because he had insulated the line too well—the Little system needed "leakage" to work. Johnson laughed at him. A twenty-six-year-old member of the Philadelphia group, Johnson had started as a telegrapher with Western Union in 1863, and in 1866 had become General Palmer's secretary and telegraphy expert. When Palmer and Reiff had bought into the automatic system, Johnson had been sent from Colorado to take charge.

With Edison's return, Johnson extended the Automatic Telegraph Company's wire by obtaining the use of lines from Pittsburgh to New York, and from Washington to Charleston, South Carolina. He and Edison began testing at night, Johnson at one end of the wire and Edison at the other. Yet as Edison brought the speed of transmission up, still another part of Little's system failed.

Alexander Bain, trying to circumvent Morse's patent, had discovered that when an electric current is passed through moistened paper the water is decomposed, and oxygen released. The oxygen instantaneously and minutely rusts a metal stylus. If a paper strip is impregnated with certain chemicals, the chemicals react with the rust to discolor the paper and form a mark. This reaction takes place at such infinite speed that it is possible to record a message chemically much faster than mechanically. In order not to infringe on Bain's patent, Little had had to devise a new chemical combination. But . . .

"Little's solution is a failure. The worst we have ever thought of and got a mark from is better," Batchelor exclaimed disgustedly. "I can make as much mark with my fingernail."

Neither Batchelor nor Edison had much knowledge of chemistry. Edison, however, was a firm believer that starting out with a clear head was not a handicap but an asset. He bought what chemistry books were available, and enrolled in a class at Cooper Union in New York.* He and Batchelor set to work concocting solutions that would react with a metal stylus. At times they seemed more like cooks than chemists:

"Add teaspoonful of nitrate ammonia. To this add what is held on small knifeblade of aurichloride of sodium."

They tried all the known, and some unknown, iron solutions, but the highest speed they could attain was 150 words per minute, and the marks faded too quickly for practical recording. Experimenting with hundreds of different solutions over the period of a month, they found ferric cyanide of potassium superior to anything in use.

"Bully experiment!" Batchelor cried on February 9 as they recorded perfectly on chemical paper at 250 words per minute, a speed that required the stylus to rust and be derusted 63 times a second.

"Now is the winter of our discontent made glorious summer by the son of York," Batchelor transcribed, quoting from Shakespeare's *Richard III*. It was a phrase that would run over and over again through the experiments of Edison and Batchelor.

Edison's endeavors with the automatic system caused his mind to turn back to the duplex, on which he had done little work since talking with Western Union president Orton two years before. Why, Edison asked himself, could not the automatic system be combined with a duplex, so that two messages could be sent on one wire simultaneously with perforated transmitting paper and chemical receiving paper? Early in February, Edison informed Orton that he had a duplex ready to demonstrate, and would like to test it on the Western Union lines. On February 6 Orton replied that Edison could set up his duplex in the Western Union office any time, and that the company would be glad to consider other propositions also.

Three or four days later, Edison was in Orton's office. He brought with him drawings of seventeen possible duplex combinations, and a

* Cooper Union had been established by Peter Cooper, whose career as an inventor and manufacturer, predating Edison's by half a century, had many similarities. The school provided free instruction in the arts and sciences.

basketful of actual devices. Edison explained the technical details, and said he could make a bushel of duplexes—they were of no particular account.

"Very well," Orton replied, "I'll take all you can make. A dozen or a bushel."

But when Edison tried to sell Orton on the virtues of an automatic duplex, Orton demurred. He had no confidence in automatic telegraphy. What Western Union really would like Edison to do was work on and improve the Stearns duplex, which the company had bought.

"What about a *diplex*?" Edison asked. So far, experimenters had confined themselves to *contraplex* duplexes, which sent two messages simultaneously over the same wire in opposite directions. The complexities of a diplex—which would transmit two messages at the same time in the same direction—seemed insuperable. Would a diplex be more valuable than a contraplex, Edison wanted to know.

Orton answered he supposed it would. Like rush-hour traffic, he said, everybody seemed to want to go in the same direction at the same time.

Orton not only granted Edison permission to test over the Western Union wires, but authorized him to draw on the Western Union shop for parts. Since Edison posed a potential threat, Orton was happy to have the opportunity to keep an eye on him. Harrington and Reiff, conversely, were not concerned over Edison's dalliance with their archrival, for Harrington's contract with Edison seemed to have him sewed up tight—whatever he produced at Western Union would be owned by themselves.

When Edison discovered he was unable to enter the tightly guarded Western Union building at night, he requested and received permission "to feel the pulse of my patients." He was less concerned about the pulse of his wife, Mary, momentarily expecting their first child. Between testing at Western Union and at the Automatic offices, taking trips to Washington and Pittsburgh, and continuing work at the shop, Edison had so devised it that he was never home at night, and seldom in the daytime. The details of the appearance of a baby girl on February 18 were left to Alice, Mary's mother, and the doctor.

By April 4 Edison had experimented twenty-two nights at Western Union and tried twenty-three different duplex combinations. "Nine were failures, four partial successes, and ten were all right," he wrote. "Please inform Mr. Orton I have accomplished all I agreed to with one exception [an attempt to transmit 2400 miles without a relay]. I have full records of all experiments to the minutest detail, with dates. The duplex shall be a patent intricacy, and the intricacy owned by the Western Union. Please ask Mr. Orton what I shall do next."

What Edison, to his surprise, was to do next was sail for England. Even while working at Western Union he had, in a brief span, taken the automatic system, improved it immeasurably, and for the first time made it work over a distance as long as 200 miles. On that basis, Reiff and Harrington were negotiating not only with American railroads and independent telegraph lines, but with British interests. The third week in April, Reiff suddenly told Edison he had booked passage for him on a steamer.

Taking with him a small satchel of personal belongings and three large boxes of instruments, Edison sailed on April 23. The Cunard vessel was aptly nicknamed "The Jumping Java," and by the time it reached England, some two weeks later, Edison and his assistant, Jack Wright, looked as if they had been on a roller-coaster. They were met by Colonel George Gouraud, a distinguished Civil War veteran who was General Palmer's financial agent in London.

In England, Bain's automatic system was part of the established telegraph network operated by the Post Office. Since the Bain machinery would not work over distances of more than forty or fifty miles, the postal authorities were searching for a system that could be installed on the lines to the Midlands, Scotland, Ireland, and France. Gouraud had induced the Post Office to give Edison's system a trial, and had interested several potential investors.

Edison moved into a small hotel in Covent Garden and set up his equipment at the Post Office on Telegraph Street. Wright went to Liverpool. The first tests were frustrating. The standard English Sand batteries were weak, and in the damp air the corroded wires spilled electricity into the night like a leaky hose. Gouraud was frantic. Edison told him the only cure for the trouble was a stronger battery. So Gouraud laid out a hundred guineas to buy the one-hundred-cell British Royal Institution battery, one of the most powerful in the world. The transmission instantly perked up. At a hundred words per minute and more the stylus jotted down the signals on Edison's blue chemical strip. British telegraphy experts were impressed.

The next step was to test the system's performance over an underwater cable. Edison was given a cable coiled in a tank to experiment on. When he sent a dot supposed to be 1/32 inch, he got a mark "27 feet long" —give or take 26 feet 6 inches. Neither he nor anyone else yet understood much about induction.*

Edison's usual diet of pie and coffee was being undermined by British cooking. He was eating nothing but flounder and roast beef. "My imagination was getting into a coma," he complained. "I found a French pastry shop in High Holborn Street and filled up. My imagination got all right."

Keeping his new British acquaintances guffawing with such tales of an innocent Yankee's adventures in London, he continued his experiments.

* Induction is a complex phenomenon that occurs when electricity passing through a wire generates a magnetic field that, in turn, produces a secondary current in either the same wire or another, unconnected, nearby.

Sir James Anderson, head of the Indo-European Telegraph Company, began discussing with Gouraud the installation of the Edison automatic system on the line between Falmouth, Marseilles, and Egypt. There was talk of employing the system on a new Atlantic cable. Edison, his imagination overstuffed with French pastry, wrote Murray he expected to sell the system for £100,000 ($500,000). An elated Johnson wrote Edison:

"Heres my hand, old Boy on the Extraordinary work you have accomplished among em Britishers. No other man living could have brought success from out of such laborynthine *complexity*."

Edison, however, had been gone from America for nearly six weeks. Murray's reports were so contradictory Edison found it impossible to divine the true state of affairs. In one sentence Murray would say, "Don't feel uneasy on account of shop I shall keep it square till you return." In the next: "Believe me I have had a hard time since you left." Nothing, it seemed, worked without Edison. "Automatic is dead here only when you are present to give it life business is dull money is worse than ever. I will meet all notes without selling machinery but I shall be left poor as it takes all profits."

Before he left, Edison had filed patent applications for the devices he had been testing at Western Union. But his application on the duplex was rejected because one claim was old, two had been anticipated by other inventors, and for the rest the descriptions were inadequate. Murray reported that Edison's four-month-old baby was "fat as she can be," and that he was taking care of all the family finances. But he was indulging Mary outrageously. He had given her $200, and she had not paid one bill out of the sum.

Edison had an acute attack of anxiety. Not only did he feel things slipping away in America, but, despite his expressions of optimism, he really did not understand what was happening on the underwater cable. He needed to think matters through, and experiment in a less pressurized atmosphere. To a startled Gouraud he announced he was leaving for home. The chagrined protestations of the Britishers could not dissuade him. He promised them that as soon as he had settled affairs at home he would return. In the meantime, his assistant, Jack Wright, could continue the experiments.

While Edison was gone, Johnson had been thinking about some of the problems that they had experienced with transmission. He wrote Edison a long, technical letter offering new insights. Edison, cogitating on Johnson's hypotheses, decided that the diplex could be made to work if a *quantity* of current was used to control one set of instruments and a *change of polarity* (direction) the other.

Mary, who had scarcely seen her husband since the first of the year, saw him little more after his return. By late September, 1873, Edison had a diplex model in operation.

The Automatic Company's negotiations with the Baltimore & Ohio and the Pennsylvania Railroad, and with telegraph lines running south to New Orleans and west to Chicago and St. Louis, were nearing culmination. If all the contracts could be pulled in, Western Union would, for the first time, have a national rival.

Uriah Painter, Automatic's Washington man, came to New York. On the evening of October 1 he showed up at Edison's shop with a quartet of visitors, including the head of the Canadian telegraph system and, most astonishingly, General Thomas T. Eckert, the general superintendent of the Eastern division for Western Union.

Painter had known Eckert since the Civil War, when the general had been in charge of the military telegraph headquarters in Washington. He was forty-eight years old, efficient, punctilious, and military, with the powerful build and moustachioed face of a German brewmaster. When he had joined Western Union, in 1866, he had expected soon to assume the company presidency. The following year, however, Cornelius Vanderbilt had installed his own man, Orton, in the post. Eckert was chafing. If the Automatic Telegraph Company could induce Eckert to desert Western Union and manage the new concern, success seemed assured.

To demonstrate the diplex, Edison placed sets of instruments at opposite ends of the room and sent messages back and forth. Leading the group upstairs, he showed them his nearly completed automatic receiver, which took the messages off in roman letters, so eliminating the need for transcription.

It was a good exhibition. Everything was developing favorably. Even Edison's precipitate departure from England was redounding to the company's advantage, for the Englishmen, feeling fortune tingling their fingertips, longed for Edison's return. Reiff and Harrington promised to send him back once a contract was signed. The London company agreed to make an initial payment of $50,000.

Of the total, Edison received one third, almost equal to a skilled workman's wages for twenty years. The sum's value was enhanced further by the advent of the Panic of 1873. Millions of people were reduced to beggary. Edison, however, spent money with no thought of tomorrow. He paid off some of the accumulated bills and notes; he splurged on chemicals and instruments; he was generous with Mary and her family, and with his father and his brother, all of whom he helped support. Pitt he more than supported, for Pitt's business acumen would have drained the Mother Lode. In December Edison sent him another $1000 to keep the Port Huron horse railroad trotting along.

At Western Union, Orton felt Edison's hot sparks on his neck. Orton had been collector of internal revenue for New York before entering the telegraphic field in 1865. His forte was financial management, but he had little money of his own. His job depended on his staying in the good graces of the Vanderbilts. So far, by driving employees' wages down, and the feeble opposition companies up their poles, he had satisfied the commodore's insatiable demand for profits. Now, however, he was imperiled from two directions. In addition to the Automatic Company, there was Postmaster General John Creswell, who favored nationalization of telegraph lines on the European model and berated Western Union for "high rates, enforced with a strong hand. When new associations have been formed for the purpose of reducing rates, the Western Union has at once entered the lists to destroy its rivals. It has resorted to artifices and triumphed by making gold its weapon."

Responding to Creswell, Orton said Western Union had halved telegraph rates in six years. To demonstrate Western Union's efficiency, and show up the Automatic, he arranged to have President Grant's message of December 6, 1873, sent simultaneously over the company's eight wires from Washington to New York. Sixteen operators transmitted the 11,300-word message in seventy minutes. The Automatic would need seventy-five people to equal the Western Union feat, Orton said sarcastically.

It was a challenge to which Harrington and Reiff felt compelled to respond. On January 27, 1874, with Edison hosting a group of Western Union observers in the Automatic's New York office, the company sent the same message in sixty-nine minutes. They used two telegraphers, ten perforators, and thirteen copyists, and required only one wire.

The triumph, nevertheless, was hollow. The panic had halted the Automatic's negotiations with the railroads, and on January 1 Orton had snipped the company's Southern connection by buying the Southern and Atlantic Company for $2 million, even though the line had never made a profit.

Still, while Edison was at large Orton could not feel easy. Edison's experiments at Western Union the past spring had not produced anything of particular value. Since then, however, he had made great progress with the diplex, and during the past few weeks the diplex had multiplied into a mind-boggling concept—the quadruplex. The quadruplex was to be an arrangement of diplex and contraplex enabling one wire to carry two messages simultaneously in one direction and two in the other. If it worked, it would mark a great breakthrough not only in the technology but also in the economics of telegraphy: one wire would be able to perform the work for which four previously had been required.*

Elatedly and quixotically, Edison informed President Orton: "I have struck a new vein in duplex telegraphy. Two messages can be sent in the

* The Stearns duplex, or contraplex, was, as previously indicated, limited in utility because it could be used only to send two messages simultaneously in *opposite* directions.

same direction. In opposite directions. My shop is so full of nonpaying work I should like to saddle this on Western Union, where they are used to it."

The truly remarkable powers of Edison's mind, which multiplied devices from a single idea like a dividing amoeba and then compartmentalized the creations and endeavors, were now being tested to the limit. Still in his middle twenties, he was a one-man conglomerate, spanning the entire field of telegraphic development.

In addition to his work in the multiplexing of wires, he was the leading inventor of printing telegraph devices, used in the stock-ticker network and private telegraphy. On the night of February 3, 1874, he successfully tested a one-wire roman-letter telegraph, and Batchelor crowed: "Built a machine on the principles set forth by Edison for this extraordinary feat!" A month later he perfected the instruments for a police, fire-alarm, and messenger system to the point that Tracy Edson of the Gold & Stock Company agreed to raise $20,000 for formation of a Domestic Telegraph Company, of which Edison was to be president.

The firm of Edison and Murray, of which Edison was the proprietor, made electrical equipment of every kind: galvanometers, resistance coils, condensers, submarine keys, electrometers, ink recorders, batteries, polarized relays, Morse registers and sounders, and an Edison *Inductorum*—a Ruhmkorff induction coil that Edison guaranteed would delight (the practical joker), shock (the recipient of the charge), and cure rheumatism, gout, sciatica, and nervous diseases.

Edison's principal endeavors were (at least in the minds of Harrington and Reiff) still directed toward the advancement of automatic telegraphy. Yet though Edison had carried the technology forward further than anyone else, his mechanisms were still far from dependable. In England, the Post Office had made the London-Dublin line available for tests, but the system was not working. The British financiers demanded Edison be dispatched immediately, in keeping with his, Harrington's, and Gouraud's promises. Until he came, the Britishers would not invest in the American lines, as Reiff was urging them to do.

But what was now uppermost in Edison's mind was the $10,000 note on his machinery and shop, along with the accumulated interest, due in July. And, having accomplished the remarkable feat of disposing of a lump sum of $17,000 in a few months, he had no money to pay it. He depended on the quadruplex, the Domestic Telegraph Company, or some other new invention for a further influx of funds.

Reiff reassuringly promised him $10,000 out of the first remittance from England, and exhorted him to take ship for the British Isles: "My well has never yet quite run dry, and the interest of Edison is the interest of Reiff. Take your wife and baby with you if you desire."

But Edison balked; and once Edison made up his mind a mule was, in comparison, complaisant. Gouraud messaged that Edison's presence was "immediately and unconditionally necessary. It is no longer a question of should he come, or is he more important in America, he must either come, or we must give up all hope of further progress."

"Now, my dear boy," Reiff coaxed Edison, promising to pledge his railroad stocks and bonds as security. "Your property shall be protected and the mortgage cancelled. The name of Edison in Europe and America will be greater than Morse."

Edison was capable of superhuman effort over long periods. But when things weren't going right and the pressure kept building, he had a habit of collapsing. Falling seriously ill, he settled the issue of his going. "Have had the most interesting features of 4000 nightmares in the daytime," he recounted afterward. "Cause—root beer and duplex."

Recuperation always brought new will and fresh ideas. Before his illness he had been working at Western Union but getting little cooperation—Orton had gone on a two-month European trip, and Eckert, whose loyalties were not where his job was, had offered subtle resistance. Edison was no Candide—he knew what the problem was. As matters stood, no one at Western Union had anything to gain from his success. Meanwhile, the day of reckoning was approaching. On May 19 Edison wrote George Prescott, the Western Union electrician:

"Mr. Orton's sudden departure took the bottom out of my boat, and I can do nothing without his or your cooperation.

"I make this proposition—

"That you give me facilities and personal help to test, and then take the patents out in our joint names, and then present them to the company for purchase on their merits alone; profits, if any, to be divided equally."

During the nineteenth century the term "electrician" denoted the manager of technical affairs. It was a position of great power. No sooner had Edison made the proposition than the doors were flung open and he was embraced like a convert. He was given a key to the ultra-secret marble-floored experimental room, which, he discovered with some chagrin, "was a very hard floor to sleep on." He nevertheless had a habit of crawling under desks to study and read. There he sometimes dropped off into a sound slumber, not even awakening when some startled telegrapher, stretching his leg unawares, knocked against his body on the floor.

Harrington and Reiff might have shown concern; but with the depression following the Panic of 1873 tattering their finances and the possibility of British investment lost, they were themselves trying to sell the Automatic Company to Western Union. After some preliminary sparring, Orton met Reiff at an attorney's office on the sixteenth of June. It was a hot day. Stripping off his coat, Orton said:

"Let's get down to business."

Cautiously, Reiff felt Orton out about Western Union's inclinations. Orton wanted an accounting of who controlled Edison's patents. Reiff responded by detailing the history of the relations between Little, the National Telegraph Company, the Automatic, Craig, Harrington, and himself.

Orton waved him off. The only things of value were Edison's patents, he said. Edison was "a very ingenious man, but very erratic," and he wanted to get Edison firmly into the Western Union fold and under the control of Prescott. "What will induce you and Harrington to transfer all of Edison's patents? They are all I care anything about. Give me a price," he requested.

"I'm prepared to discuss anything but that," Reiff responded. Negotiations were in progress with certain parties for formation of a new telegraph company, he explained, and he had an obligation to them.

"Then what are we here for?" Orton asked.

"I don't know," Reiff said unhappily.

After a few more remarks, the meeting broke up. If Edison perfected the quadruplex for Western Union, Orton would have less concern about competition from the Automatic Company than ever before—and Edison was making excellent progress. Orton assigned eight of the best telegraphers to Edison for the tests. As they assembled in the experimental room, they talked among themselves about Edison's purported expertise as a telegrapher. Skeptical remarks passed back and forth. When Edison entered, the conversation died like a tap shut off but still leaking a few drops. He could divine the drift. "Well, boys," he said, "why don't we get warmed up."

He sat down at one instrument, while seven operators took places at the others. Edison turned on the automatic transmitter, starting it at a speed of twenty-five words per minute. After a few minutes he turned it up to thirty; then to thirty-five. At forty all the operators except Edison were sweating—he still copied steadily and coolly. At forty-five most dropped out. After a few minutes at fifty the last threw down his pen.

Edison kept moving along. Everyone gathered to watch. He was 200 words behind, but never hesitated, finishing perfectly several minutes after the tape had run out. It was the most amazing performance the operators had ever seen, and after it Edison had their complete respect.

Not till twenty years later did he reveal he had run the strip many times before, and knew every word on it.

By the end of June Edison was ready to demonstrate the quaduplex to Orton. The Western Union president, however, was away on another trip, and not due back until the seventh of July. The note was due on the first. If Edison failed to meet it, the bank would foreclose and sell the shop and machinery at auction. Edison went to Reiff. But despite

Reiff's earlier protestations that Edison need not worry and the note would be taken care of, Reiff was in financial straits himself. He held considerable railroad securities, but they had been battered down to a price at which Reiff would have to sell them at great loss. The best Reiff could do was give Edison a $3500 note to cover the interest.

Chafing, miserable, Edison rumpled his hair and poured out his troubles to Prescott. Edison's dyspeptic stomach was giving him even more trouble than usual. His jaws worked steadily on a wad of foul-smelling tobacco. He seemed in such a state that Prescott thought something had better be done, and telegraphed Orton. Orton arranged for the Western Union–controlled Gold & Stock Company to give Edison a $3000 advance. But that still left Edison $7000 short. Paying a $5-a-day fee to the sheriff to keep him from foreclosing, Edison haunted the Western Union building.

As soon as Orton was back, Edison laid siege to his office. Privately, Orton was a pleasant, good-hearted man who, like Edison, was a habitual storyteller. But, hounded by Commodore Vanderbilt, in business he was like a spider, luring the victim, getting a thin thread attached, then slowly reeling him in and sucking him dry. Edison told Orton he desperately needed $10,000. Orton asked Edison what security he had.

Edison offered to sell Western Union his one-third interest in the automatic system.

"I won't give $10,000 for all those traps!" Orton snorted. Knowing full well the difficulties Reiff and Harrington were in, he advised Edison to borrow the money from them.

Edison had mentioned to Reiff that he was ready to sell the quadruplex rights to Western Union. Reiff had retorted that he must under no circumstance do that, that it would be giving Western Union a knife to place at the Automatic's throat. To Edison, however, it seemed a matter of survival. He told Orton the system was ready to be exhibited, and he was prepared to deal.

Orton asked him if he had put the contract with Prescott in writing. Edison replied he hadn't. Orton said that would have to be attended to.

The exhibition was scheduled for July 9. Edison had the system working well to Albany and back in good weather, but when a storm came up the bad side got a little shaky. Edison told the eight operators that if the weather turned against them they should draw freely on their imaginations. Orton, William Vanderbilt (the commodore's son), and several other company officers and directors came in. Everything worked beautifully. Then a thunderstorm broke. Edison's heart began keeping pace with the key clicks. But the system pulled through.

That afternoon Edison and Prescott signed the agreement giving each half interest in the invention. When the deed was accomplished, Edison walked the short distance down Broadway to Reiff's office to tell him.

Reiff was furious. He told Edison he had done exactly what he had been warned not to do. Edison, defending himself, made light of the matter. The automatic system was far superior to the quadruplex, and the quadruplex could not hurt the automatic. Besides, he continued: "I have put in there that nothing can be sold by the Western Union or anybody else without I agree, and I will not agree to have anything sold without you and Mr. Harrington agreeing to it."

The clause Edison referred to actually prevented either partner from disposing of his interest without the consent of the other, and thus handcuffed Edison and Prescott to each other. Reiff told Edison that he had made a great mistake and the thing must be stopped.

That evening Orton arranged for Edison and Prescott to meet with a reporter from the *New York Times*. Orton told the reporter that during the past three years Western Union had had to erect 60,000 miles of wire, but the quadruplex would reduce the mileage needed the following year to 2000. "The discovery may be called the solution of all difficulties in the future of telegraphic science," he declared.

After writing the story, the reporter read it back. The quadruplex was the third great step in telegraphy, he noted, the first being the invention of Morse, the second the development of the duplex. "The invention is the result of the joint labors of Messrs. George B. Prescott and Thomas A. Edison." Edison, hearing Prescott's name placed first, felt a knot twisting in his stomach.

Harrington, who had been out of the city, returned that evening, and immediately became the recipient of rumors. At midnight he scrawled a desperate note to Edison:

"Having learned of what is going on, *beg* of you to see *me* before you sign any more papers, take any more money or go to any other place. Come to 80 Broadway. I am in hopes that I can relieve you. At this moment adverse action shall cause a loss of $100,000."

In the dark morning hours Harrington dispatched the letter to Newark with his son Chase. Edison had moved twice, in keeping with the vagaries of his finances, since his marriage. It was 6:00 A.M. before Chase was able to track him down in an apartment above a drugstore and, awakening him, hand him the letter.

Edison's response to Harrington was that the whole matter could be easily settled if he were supplied with the money to pay off the mortgage. Scrounging frantically, Harrington and Reiff induced one of their Philadelphia associates, William Seyfert, a banker involved in the financing of the Kansas-Pacific Railroad and the Automatic Company, to lend Edison $6600 on two personal notes.

The crisis behind him, Edison happily plunged back into his multifaceted existence. A large part of the time he spent in New York, either at Western Union or at his own Domestic Telegraph Company, which began operations at the end of July. He loved the gaslit, shadowed streets, the coffee and cakes served in some grimy, smoky joint, the showpeople, newsmen, and bummers who glided about the darkness. He stepped gingerly past the awful dives centered about Crosby and Houston Streets, where drunkards and prostitutes swirled in and out of the House of Lords, the Bunch of Grapes, and the Dew Drop Inn. He darted out of the way as dogs and pigs fought over the garbage. He stopped to watch the performances of bears, which, led around on chains by their trainers, were more common on the streets than in circuses or the Central Park Zoo. He frequented Oliver's, the café on Printing House Square, where he was able to rub elbows with Horace Greeley of the *Tribune* and other newspapermen. Sometimes he telegraphed Batchelor in Newark to come and meet him in Manhattan, adding parenthetically: "Inform wife will not be home tonight."

He was a drop-in husband and a pop-up proprietor at his electrical shop. Customers complained that their orders were not attended to, that they would order one thing and be sent another, that they were quoted one price and charged more. Johnson in Philadelphia could not contain his irritation and asked Reiff to investigate what was going on, because "No attention whatever is paid to our dispatches of late."

On August 19 Edison took Prescott to his patent solicitor, Lemuel Serrell, and asked Serrell to file an application for the quadruplex listing himself and Prescott as co-inventors. After asking Edison a few questions, Serrell told him the patent office frowned on naming someone who had not contributed to the invention as a co-inventor. Discovery of the deception would invalidate the patent.

"I guess it will be better to enter the application in Edison's name only," Prescott agreed.

Edison's cavorting in the Western Union web panicked the Automatic Company's investors. Orton, who had already been approached by Harrington and Reiff, now received an offer to sell the company from Craig. (Harrington had maneuvered Craig to one side, but Craig had two contracts with Edison, and under another still held an option to buy one-third interest in the Automatic patents.) While Orton contemplated that the Automatic Company seemed to have more owners than telegraphers, Reiff was forced to cast about elsewhere to unload the Automatic and its growing debts. By September 11 he had found someone.

"Smiles will wreath your lips and enthusiasm fill your brain," he wrote mysteriously and exultantly to Edison. "Our night of suspense is over. *We will shake the foundation of things.* The *money* and satisfaction is from another direction. Batchelor the faithful and Murray the persistent will be duly cared for. The old wheelhorse TAE and I will pull together."

The opinion of the *Telegrapher*, however, was that "The automatic invention fades into insignificance alongside of the great 'quadruplex invention,' which is bound to at once revolutionize telegraphy, and carry the inventors, the Western Union Telegraph Company, and all concerned on to glory and fortune."

On October 19 Orton ordered the Western Union shop to prepare several sets of quadruplex instruments. During the course of the next few weeks they were installed first on the New York–Boston line; next on the longer New York–Buffalo line; finally, on the New York–Chicago line. They worked beautifully. At the annual stockholders' meeting in November, Orton proudly reported that Western Union's net profits during the last eight years had been $23 million. "The past year," he said, "has produced an invention that will solve satisfactorily the most difficult problem which has ever been presented: how to provide for the rapidly increasing volume of business without an annual expenditure for the erection of additional lines and wires that would prevent the payment of reasonable dividends."

Edison, of course, was listening. So far he had not received a dollar from Western Union. As usual he was short of cash. Unpaid bills were piled up in the shop. His brother, Pitt, was importuning him that an unparalleled opportunity existed to expand horsecar operations to Sarnia, Ontario, across the river from Port Huron. It would take only $5000. He needed the money by December 9.

It was actually not until December 10 that Edison was able to see Orton. Edison said he was ready to make the deal for the quadruplex. Until the contract was signed, "I would like to have 10 . . . 9 . . . 8 . . . 7 . . . 6 . . . 5 . . . 4 . . . 3 . . . 2,000 dollars, as you may choose to advance."

Orton said he could not negotiate that way. However, he would place an order with Edison for twenty sets of quadruplex instruments, and give him $5000 on account. Edison should come back in a few days with a concrete proposal, and they would talk.

After meeting with Prescott, Edison returned to Orton's office. On December 16 he handed the Western Union president a memorandum. The quadruplex would, in effect, create 50,000 additional miles of wire for Western Union immediately, and countless numbers in the future. "For protecting the company in the monopoly, we will take 1/20th of the cost of maintenance of 50,000 miles of wire for 17 years, one third down and the balance in yearly payments."

Orton asked Edison to reduce that to figures.

Edison said he supposed it would be about $450,000. (Actually, it would have been $368,000.)

Orton smiled. Edison had a lean and hungry look, and immense talent. He drove himself ruthlessly. Orton launched into the story of a steamboat captain who was requested to take on a steward like Edison. "The captain had only one question. Did he expect to own the boat at the end of one season, or would he be satisfied to wait two?"

Orton suggested Edison temper his demand, and come back. The next day Edison handed Orton a memorandum containing two proposals: 1) $25,000 down and $25,000 in six months, and an annual royalty of $166 for each quadruplex circuit; or 2) $25,000 down and an annual royalty of $233 per circuit. Orton read the memo, nodded, said Edison was doing well, and told him to keep on going lower and lower.

Orton actually liked Edison. And, in light of the quadruplex's value, Edison's propositions were not excessive. But Orton was confident Edison had nowhere else to go, and the lower Edison went the better Orton himself would look before Western Union's executive committee.

After playing with Edison for a few minutes, Orton made a counter-proposal: Western Union would pay $20,000, and $10,000 a year for ten years in lieu of royalty. Since half the sum would go to Prescott, Edison would net $60,000. Edison, who in the early years of his career had a habit of asking too much and taking too little, inquired why the $10,000 a year should not run for seventeen years, the life of the patent. Orton replied he was going to Chicago, but the difference could be resolved upon his return.

What Orton did not take into account was an old link between Thomas T. Eckert and Jay Gould, one of the sharpest speculators in the history of finance. During the Civil War Eckert, by relaying news of Union victories received on the military telegraph, had helped Gould in his early speculations. By 1865 Gould had, over a span of thirteen years, parlayed $50,000 into several million. Winning control of the Erie Railroad, he had bilked Cornelius Vanderbilt out of $8 million. By corrupting President Grant's brother-in-law and beguiling the secretary of the treasury, he had cornered the gold market, generated a frenzy unsurpassed in Wall Street history, and pocketed $11 million. Once again, in the crash of 1873, Gould had anticipated the market. For the past year he had been buying Union Pacific shares for a fraction of their former price, and now had working control of the railroad.

Since operation of the railroads and the telegraph went hand in hand, Gould had acquired ownership of the Atlantic & Pacific Telegraph Company. By adding the lines of the Automatic Telegraph Company, Gould would be in control of a network stretching from coast to coast; and if this network were to utilize Edison's quadruplex, as well as his automatic telegraphy equipment, it would be a fair match for Western Union. Gould was the mysterious angel to whom Reiff had referred in September. He was the archrival of the Vanderbilts, and always ready to spike the commodore.

On the twentieth of December, the slight, dark, and balding Gould crossed the icy Hudson River and was taken to Edison's shop by Eckert. Edison, distinctly unhappy at his treatment by Orton, was impressed by anyone who could generate money rather than dissipate it. While Edison worked the quad for Gould, Eckert worked on Edison, and assured him he would never get another penny from Orton.

On December 24 Prescott left for a ten-day vacation; and, with Orton also absent, the plotters struck. Between the twenty-seventh and the thirtieth, Reiff negotiated the sale of the Automatic Telegraph Company and Edison's patents to Gould. Reiff assured Gould that Edison's contract with Prescott was invalid, because Harrington owned Edison's right to all automatic or "fast" telegraphy inventions.

Edison had the same respect for contracts as for witches' incantations, which he thought they resembled, and he had signed so many he no longer had any idea what he owned. Eckert was breathing over him, pressuring him to forget the agreement with Prescott. On the evening of January 4 Eckert took Edison to Gould's mansion on Fifth Avenue. Gould was an art collector and a bibliophile. Edison's gaze fell on painting after painting, on paneled halls and walls of morocco-bound books. His feet sank into thick Persian rugs. In his ordinary suit—for once neatly pressed—he seemed as drab as a mudhen next to the elegant Gould.

A lengthy discussion ensued about who had title to the quadruplex, and what Edison wanted for the invention. Gould offered Edison the position of electrician of the Atlantic & Pacific Telegraph Company—a major plum!—and $100,000: 3000 shares of A & P stock, valued at $75,000, and $25,000 in cash.

"That's too small," Edison said, referring to the cash.

"I will give you $30,000," Gould suggested.

"All right," Edison agreed.

"Do you want it now?"

"Part of it."

"Come downstairs."

They descended to Gould's office, where Giovanni Morosini, a former Italian sailor who was Gould's bodyguard and secretary, sat behind a desk. Edison gave Gould a power of attorney for the quadruplex, and received $10,000 from Morosini. Three nights later, after the entire contract was drawn up, he returned. He was paid another $10,670, and given $16,500 in Union Pacific bonds, valued at about $10,000 on the current market.

From the cash, Edison paid the long-accumulated bills of the butcher, the grocer, the milkman, the newspaper boy, and the doctor. He made presents of $200 to his wife, $15 to her sister Alice, and $50 to Murray. He paid back two loans totaling $2200. He lent $225 to Johnson, and smaller amounts to other men, some of whose names escaped him. He bought books, and woodcuts of Samuel Morse's drawings. He spent $200 on a quick trip to Port Huron. While there, he gave his father $170, and his brother, Pitt, $150. On January 11 he bought $7900 worth of stock in the Sarnia horsecar line, and presented Pitt an additional $1600 for various expenses connected with the road. He did not, however, concern himself with the notes he had given to Seyfert six months before in return for the $6600 to save his shop. Reiff indicated they would be taken care of in the overall settlement. But they were not. A decade later one would return, in the possession of a woman scorned, to haunt Edison.

The same day that Edison bought controlling interest in the Sarnia horsecars, Eckert resigned from Western Union. Rumors swirled like leaking gas around the Broadway headquarters, and then exploded. When Orton, who was still in Chicago, heard what had happened, he got sick to his stomach and made a frantic dash for the train to New York. On the fourteenth Eckert was named president of the Atlantic & Pacific Telegraph Company, and five days later the Western Union executive committee, for the purpose of establishing their good faith and laying the groundwork for a suit, authorized acceptance of the second of Edison's proposals made to Orton a month before. One of the Western Union officers went down the street to notify Edison.

Edison, however, was in no mood for further overtures. Orton hadn't treated him right, he stormed. Never again would he cross Western Union's portals. Western Union could not use the quadruplex, and if it did there would be trouble. He was going for the Western Union Company "red hot." He thanked heaven he was now "involved with business men. Men that sleep with their boots on!"

CHAPTER 7

The Tangled Web

In fact, it was a deal over which no one was to rest easy. Within a few months the tangled web became entirely unraveled. A half-dozen lawsuits were filed, with each party in the dispute suing one or more others. Not until thirty-eight years later, when most of the participants were dead, would the final decision be rendered.

On January 23, 1875, the Atlantic & Pacific Telegraph Company, controlled by Gould, initiated use of the automatic system. Gould was to reimburse the stockholders in the Automatic Telegraph Company with 40,000 shares of A & P stock, including the 3000 for Edison.* The delivery was held up pending determination of ownership of the quadruplex rights by the patent office, and the assignment of Harrington's contract with Edison to Gould.

The quadruplex, it became apparent, was the key to the automatic system; without it, the automatic was of marginal value. The automatic could achieve great economy over short lines with heavy traffic. But its speed decreased geometrically with distance,† and its economy according to the proportion of the time the wire was utilized. Twelve thousand words, for example, could be transmitted manually in an hour and over the automatic in less than fifteen minutes, yet the automatic had no advantage if the wire sat idle the other forty-five minutes. Operators were required to perforate the strips with Edison's typewriterlike device,

* Harrington was to get 20,000 shares, Reiff 7000, the Philadelphia financial firm of Seyfert, McManus and Company 4700, and General Palmer 540. Three hundred and twenty shares were to go to William Seyfert in liquidation of his July loan to Edison.

† Theoretically, it could transmit 1000 words per minute over 200 miles, and 100 words per minute over 1000 miles. In practice, the speed was 80 to 90 words per minute.

and the maximum speed an expert perforator could achieve was sixty words per minute. At the receiving station a copier had to transcribe the signals—Edison had not yet perfected his Automatic Roman Character Printer to the point of practical operation. One telegrapher still had to be present at either end, so that there could be manual communication in case trouble developed with the equipment—and trouble developed frequently. The automatic system, therefore, was of advantage only in special situations. Combined with the quadruplex, it could offer enormous flexibility. Without the quadruplex, it was impressive technically but limited practically. And so long as Western Union could utilize the quad, A & P had little chance of developing the volume of traffic required to make the automatic system economic.

On March 20, the commissioner of patents ruled that the agreement between Edison and Prescott on the quadruplex was valid. Edison's 1871 contract with Harrington did not cover the quadruplex, so neither Harrington nor Edison could transfer the rights to Gould.

While the Atlantic & Pacific Company appealed the decision to the secretary of the interior, Gould put pressure on Harrington, who was not well, to sign over his controlling interest in Edison's patents. On April 16 Harrington gave in, and Gould paid him $106,000—the approximate sum Harrington had invested in the automatic system in 1871. Harrington departed to recuperate in Europe.

A few days passed before Harrington's associates realized what had happened—Harrington had sold them out. Gould had the automatic and the patents, but they did not have the A & P stock. Legally, of course, Gould was committed to deliver the shares. But, in practice, he lifelong pursued the policy that possession was superior to any number of judgments.

Gould justified his action by pointing out Reiff had assured him that Edison's contract with Prescott was invalid. Furthermore, with respect to the Kansas-Pacific Railroad, Reiff had pulled a classic flimflam on Gould. On a Reiff-arranged inspection trip of the railroad, Gould had seen train after train of full boxcars, sidings stacked with goods awaiting transportation, and platforms filled with people. Impressed, he had invested heavily in the financially strained railroad. The drastic decline in business that followed had gradually enlightened him that all he had seen on the tour had been orchestrated by Reiff. Not one to accept such deception by writing another check, Gould now used his advantageous position to press Reiff for a renegotiation of the Automatic Telegraph Company purchase.

Reiff responded by demanding, on May 3, return of the deeds and assignments. Two months later Gould, instead, transferred the documents to the Atlantic & Pacific Telegraph Company, and asserted the matter was out of his hands. In August, Reiff and the other investors in the Automatic Company filed suit.

Edison's situation was curious. He was a party to the suit, yet he continued as the Atlantic & Pacific Company's electrician. Despite holding this key office, he absented himself almost entirely from the company's affairs. The personality differences and antagonisms between him and Eckert were too great to be bridged by common enterprise—if anything, they increased with Edison's elevation to power. Edison believed he should be able to run technical affairs as he pleased, and Eckert shuddered at the thought of Edison turned loose among the instruments and wires. Edison periodically visited Gould's mansion, where Gould tried to interest him in the complexities of railroad organization and Wall Street. Edison reacted by unrolling some of his droll but often dull stories, to which Gould responded with a fishy stare. Gould told Edison he was ready to hand him his 3000 shares of A & P stock any time, if Edison would get to work. Edison and Reiff, however, had promised to stick by each other no matter what, and while Edison generally was as sentimental in business as a hungry rattlesnake in a rabbit hutch, he never wavered in his friendship for Reiff. Edison refused to accept the shares until Reiff received his, and in the meantime did nothing to complete development of the roman-character printer, the automatic repeater, or the other devices that would make the system competitive. On July 26 he made a final pitch for authority to Gould, informing him he had the instruments ready for initiation of direct service between New York and Chicago.

"I have done all I could to reduce this delay but I have no authority to make the changes necessary to bring it about without General Eckert makes some sweeping reforms. That boat will be a long time in reaching port." He proposed several innovations; among them a "*Letter Telegram Service,* operated at night only, with rates so low that it will attract urgent letters."* He offered to lease the wire between Chicago and New York himself from 2:00 to 6:30 A.M. daily for $4000 a year. He suggested all employees be put on a piecework basis to reduce labor costs. He wanted "to cut down high rentals and put money in advertising, drumming, etc. Aged men [Eckert was fifty] and telegraphy are incompatible," he concluded.

Edison's work with the automatic telegraph had taken him ever deeper into the field of chemistry; and chemical experimentation was to be a major part of his endeavors for the remainder of his life.

During the spring and summer of 1874, while experimenting with a stylus and chemically impregnated paper for automatic reception, he and Batchelor had observed a fascinating phenomenon: when the paper was wrapped around a cylinder, and both the cylinder and the stylus

* This, of course, was an idea that Western Union later implemented with its Night Letter service.

were connected to a battery, friction increased or decreased according to the strength of the current. The stylus dragged against the chemical paper when there was no current, but slid over it when electricity was applied.

Edison received his patent on the electromotograph, as he named the device, in January, 1875, and devoted most of the spring to attempting to discover uses for the new machine. He thought it would serve as a substitute for a magnet in telegraphic transmission, and concocted one mixture after another with which to impregnate the paper. His search led him into the virgin area of semiconducting materials, and he eventually focused on tellurium, an element related to selenium (now used for electric eyes).* For months he hunted the rare chemical and tried to order barrels of it, only to discover it was almost unobtainable even at thirty dollars an ounce.

Undismayed, Edison and Batchelor tried coating the electromotograph with every element and known compound from aluminum to zirconium, and samples of most of the world's flora—fleabane, Solomon's seal, chickweed, wild cherries, dandelion, heal all, rattlesnake, plantain, catnip, ad infinitum.

"After trying some 15262842981 different solutions of Brazilwood we've come to the conclusion it is not worth a damn," Batchelor noted in mid-March. On May 10 they even sacrificed a portion of their breakfast for the electromotograph. A mixture of coffee, eggs, sugar, and milk resulted in a "Phenomenon. Decreased friction on oxygen."

Aside from Edison and Batchelor, only three workers remained in the laboratory. They were Charles Wurth, the longtime shop foreman, John Ott, and Jim Adams. The most important of these was Adams, a thirty-year-old Scotsman who had once been a sailor on a slaver, and had wandered into the shop about a year before seeking Edison's "Inductorum" to sell to seamen. When this had proved an unprofitable venture, Edison had hired him as a night watchman. As a night watchman he had watched mostly Edison and Batchelor, and soon had taken an interest, then a hand in the experiments. He was a hard-swearing, hard-drinking, tubercular man of considerable talent, and fit comfortably into the milieu.

Edison was paying Batchelor $60, and the others $30 a week. He was making some instruments for the Atlantic & Pacific Company, but his break with Western Union had sharply cut his manufacturing business. He owned about $20,000 in Union Pacific bonds and horsecar stock, but by late spring he was out of cash again. He and Murray talked matters over and decided to separate the laboratory from the machine shop. Murray took over operation of the latter, though Edison remained a silent partner. The machinery was divided, with Murray

* Semiconductors revolutionized the electronics industry after World War II, when the transistor was developed. The most commonly used material is silicon.

to pay Edison $18,000 over a period of nine years for the portion going to him.

Murray was to prove as incompetent as Edison at running the business. But Edison, at last freed from the humdrum burden of the shop and elated he could now devote his full time to experimentation, celebrated by compiling a twenty-item "WANTED" list of new products and processes for industry. Among them were a method of making malleable iron out of cast iron; a method of making sawdust soluble to form a cheap substitute for ebony, hard rubber, or celluloid; a kerosene or other oil lamp which burns without a chimney and gives a bright light; a cheap process for the extraction of low-grade ores, decomposed earth like CARB Ag. or H_2S Ag ores; a sexduplex telegraph; and a cheap process of printing.

Several of the items, like celluloid,* were a passing fancy; others, like the mining of low-grade ores, were to occupy the inventor for decades. Together with Batchelor and Adams, he spent the first nine nights of June in the laboratory working on the electromotograph, the kerosene lamp, celluloid, and the sexduplex. On the morning of the tenth, the trio, heeding the plaints of their wives, solemnly "resolved to work day times and stay at home nights." It was a resolution that would have been as easily kept had it been made by three owls. Every night was the boys' night out; they reveled in their companionship and the atmosphere of the gaslit lab; they sent down to the raucous, sawdust-floored saloon, and drank beer from a broken-nosed white pitcher. They told racy stories, drew rough cartoons, and composed doggerel. Adams sketched a "flying machine recommended for Debtors, Defaulters, clergymen & others." Batchelor suggested a "Photograph of Satan's Privy," "Lucifer's fire alarm Telegraph," and "Hell's Automatic Half Acre," to go along with "A Lefthanded Horseturd." Edison lamented:

"Who woke me from my little cot
And put me on the cold, cold pot
Whether I wanted to or not
My Mother."

They were worldly, eager for knowledge, and interested in a wide variety of subjects. Batchelor's mind was the most searching: "Explain Newton's law of gravitation." "What is the theory of combustion?" "Either light exerts an attraction power and lifts organic matter against gravity as for instance the rise of a potato vine in a cellar to a hole in the roof of same, or else a potato vine is endowed with intelligent energy." "Those nations whose *people live upon a mixed diet are the most*

* The first celluloid products came on the market in 1874 as a replacement for ivory in articles like piano keys and billiard balls. The Celluloid Manufacturing Company, financed by Marshall Lefferts, was close to Edison's shop.

Who took me from my little cot
And put me in the cold cold pot
whether I wanted to or not
My Mother

I tell you what, an operator
Is a most mysterious Crathur
With taste + elegance they dress
But how they do it I Cant guess

High diddle diddle the Cat & the fiddle

If a man who 'turnips' cries
Cries not when his father dies
Tis a proof that he would rather
Have a turnip than his father

If a man who does not well
To tell another go to hell
His nose gets mashed his eye bunged up
He's rightly served the Cursed pup

How nice it is in panic times
To have a pocket full of dimes
I used to know how twas myself
But now alas I'm out of pelf

When you owe your Livery stable
And to pay you are not able
Obliged to walk off on your ear
I tell you it feels awful queer

Doggerel and sketches enliven this notebook from Edison's Newark workshop.

civilized. A mixed diet gives a diversity of ideas, ideas coming from the oxidation of food. Instance the greater learning in cities where the diet is more mixed than in the country; and the superiority of the French, English and German speaking people, & per contra the Chinese who live mostly on rice who will continue to think in the same channels as long as they are confined to this one cereal."

They had the prevailing popular prejudices, they could take facts and deduce the wrong conclusions, or reach the right conclusions for the wrong reasons, and they lacked the knowledge and discipline of better-educated men. But they were distinguished by their drive to explore new territory, to discover new products and devise new machines, to escape the poverty of their childhoods. They reinforced one another's egos, and took exaggerated and almost childish glee in their successes.

"This is a specimen of the new copying press just invented by the renowned trio, Edison, Batchelor, & Adams," Edison wrote Reiff early in June.

The copying press was another offshoot of the automatic telegraph. Edison and Batchelor had noted that sometimes when the stylus punctured the paper the chemical solution left a mark beneath; and it was but one more step to hit upon the principle of the stencil, and conceive a waxlike mixture that would make the paper impermeable except where it was punctured. To prepare the stencils, they invented an ingenious electric perforating pen with a needlelike tip and a cord connected to a battery.

The pen became the first electric appliance ever to go into production. Edison, needing capital for manufacturing and agents to market the pen and autographic press (or mimeographic, as it ultimately became known), enlisted the aid of Ezra Gilliland, the friend of his youthful days, with whom he had lately renewed acquaintance.*

Early in the fall of 1875, Ezra's father, Robert, agreed to finance production in return for a 30 percent interest in the invention. Edison sublet part of the Newark facilities to Gilliland for a manufacturing plant. Batchelor and Edward H. Johnson joined the new concern, and set up a sales office in New York.

Edison himself, quite sure in his heart that Western Union would win the quadruplex case, busied himself concocting a variety of schemes for getting back the property and patents the Atlantic & Pacific Company had locked up in its safe. A half-dozen years of intimate association with Wall Street had provided him with an education in high finance,

* Gilliland, operating a small electric and telegraphic machine shop in Cincinnati, had visited Edison in late February to talk to him about Harold C. Nicholson, who had filed a rival claim to the quadruplex. Gilliland, who still possessed Edison's 1865 drawings and notes on the duplex, said he had provided Nicholson with the information and assistance that had enabled Nicholson to file his claim. After a dispute of many years, the courts ruled for Nicholson on one count, but upheld Edison on four other counts.

and he had absorbed a good deal of the conniving qualities of the financiers as well. The several alternative plans he drew up were as complicated as any issuing from a corporate attorney's office, and worthy of Jay Gould himself.* In general, they all envisioned eventually selling control of the Automatic Company to Western Union. Since Edison's experience had imbued him with the notion that ethics was alien to business, he was concerned only with providing for the interests of himself, Reiff, and one or two associates. "And let the other fellows whistle!" he suggested.

Still, drawing up plans and scheming put no money in his pocket. On the third of August, Edison approached Prescott and bewailed the fact that the Atlantic & Pacific Telegraph Company was using the automatic equipment and Western Union the quadruplex, but he himself was profiting from neither. He proposed that the dispute be settled by swapping his remaining rights in the automatic telegraph to Gould for whatever interest Gould had in the quadruplex, so that he and Prescott could accept the Western Union royalty offer.

Orton, who in Western Union's brief to the patent commissioner had charged, "Edison is insolvent and of no pecuniary responsibility," was delighted. On Wall Street, personal feelings were not allowed to interfere with business, and money could forge stronger links than steel. Orton knew of Edison's experimentation with the sextuplex, but technical interest in the multiplexing of wires had shifted in a new direction. On March 17, Elisha Gray—a thirty-eight-year-old Chicago electrician who had been a partner in the firm that developed into Western Electric—had read a paper, "Transmission of Musical Tones," before the American Electrical Society.† Two years before, Gray had taken up investigation of the work of Charles G. Page, an American physicist who in 1837 had discovered that the rapid magnetization and demagnetization of an iron bar by electric currents produces musical sounds, and of Phillip Reis, a German experimenter, who in 1861 had succeeded in transmitting musical sounds over an electrified wire.‡ On July 27, 1875,

* For example: "2nd Plan A & P to pay Automatic interest 4400 shares Franklin and the Quad patents 1) I taking the Quad at 40,000 in part payment and turning it over to W. U. on the old offer. This will make them all safe from lawsuits in that direction. They can readily pay 68½ for the 4400 shares Franklin [or] have A & P issue 2 million more stock to pay Automatic. Contract to sell the whole to W. U. in a block at 18. This with amount held by Union Pacific (if they hold any?) will control."

† Gray had been a twenty-two-year-old grade-school dropout when a professor at Oberlin had encouraged him to return to school. After three years at a secondary school and two at Oberlin, Gray had turned his attention to telegraphy, and in 1867 had been issued his first patent.

‡ Reis's apparatus was exhibited at Cooper Union in 1868. Three years earlier Joshua Coppersmith had been arrested in New York for fraud and extortion when he tried to raise money for an invention. "He calls the instrument a telephone," a Boston newspaper declared. "Well-informed people know it is impossible to transmit the human voice over wires."

Gray was issued two patents for an electro-harmonic telegraph, embodying the principle that any number of signals can be sent over one wire simultaneously by splitting up the transmission into different musical tones. Orton immediately put the Western Union hook on Gray. Gray's apparatus was tested on the company's wires between Boston and New York, September 11, 1875, and the next month put on exhibit in the Western Union building in New York.

Early in August, Orton hinted to Edison that the electro-harmonic telegraph might soon make the quadruplex obsolete. But, to show that there were no hard feelings, Orton sent Edison a copy of French author J. Baille's book, *Wonders of Electricity,* describing the Reis experiments. Within ten days Edison made sketches of an acoustic telegraph. By mid-September his incipient reconciliation with Western Union had advanced to the point where the company paid the patent fee for an improvement on the quadruplex. The company also evinced interest in the autographic telegraph, a crossbreeding of the autographic printing press with the telegraph and the electromotograph. The device Edison proposed would have the ability to send handwriting, maps, and drawings. Sensors reacting to the friction of the ink would control transmission of the impulses.

Edison's mind was like a bloom ripe for pollination. Each invention germinated ideas for another. He was trying to devise a pocket lamp, a postage-stamp defacer, a cigar lighter, and a better battery for the electric pen. Yet he paid no heed to Tracy Edson, who, from the offices of the Domestic Telegraph Company (of which Edison was still the president), pleaded that he had promised to perfect a repeater and alarm gong for the fire alarm. Without these the company could not attract enough customers to make a profit.

From London, Gouraud continued to pressure Reiff to dispatch Edison back to England. "Cable early positive date of your and Edison's sailing, golden finish if Edison succeeds," he wired.

Yet Edison, whose experience on the "Jumping Java" had soured him on ocean voyages for life, ignored the summons. He had, in truth, already consigned the automatic system and Domestic Telegraph Company to the past.* He had little patience with improving an invention once he had given birth—if it had handicaps, that was too bad. His mind was always focused on the future.

What he could not ignore was that, despite his prodigious creative feats, the tills in which he had his fingers were all nearly empty. Except for a few hundred dollars for experiments from Western Union, his principal source of income was from equipment prepared for the Atlantic & Pacific Company. But this produced only $3245 between mid-July, 1875, and February, 1876. Equipment manufacturers were

* Edison eventually sold his interest in the Domestic Telegraph Company to Tracy Edson and Thomas Eckert for $16,500.

Thomas Alva Edison (center), shown at about age four, was a member of a prolific and long-lived family. Clockwise, from upper left, are his grandfather, Sam Edison, Sr.; his father, Sam Edison, Jr.; his mother, Nancy Elliott Edison; and his brother, William Pitt Edison, fifteen years older than Alva.

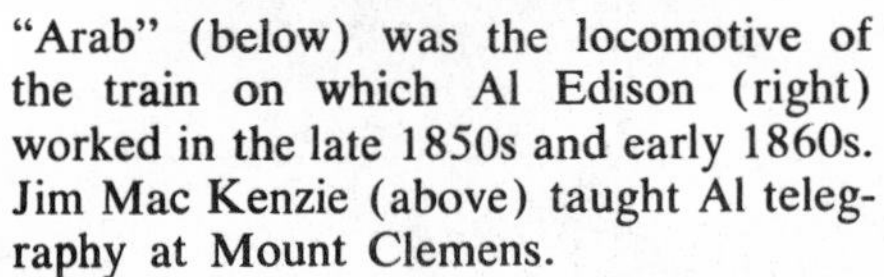

"Arab" (below) was the locomotive of the train on which Al Edison (right) worked in the late 1850s and early 1860s. Jim Mac Kenzie (above) taught Al telegraphy at Mount Clemens.

Clockwise (from upper left): Noted electrical engineer Frank Pope introduced Edison to Marshall Lefferts, president of the Gold & Stock Company, who retained Edison to manufacture telegraphic machinery. Through Lefferts, Edison met William Orton, president of Western Union, and Thomas T. Eckert, manager of Western Union's Eastern division. Eckert, a secret ally of Jay Gould, persuaded Edison to sell his patents to the rival financier.

Thomas Alva Edison, twenty-four, and Mary Stilwell, sixteen (facing page, top), are shown at the time of their marriage, in 1871. Four years later they moved into a house in Menlo Park (facing page, bottom), which was photographed in 1880 with members of the Stilwell family and the Edison children gathered in front. The pictures of Mary (right), and her children, Dot, Will, and Tom (below), were taken about 1882.

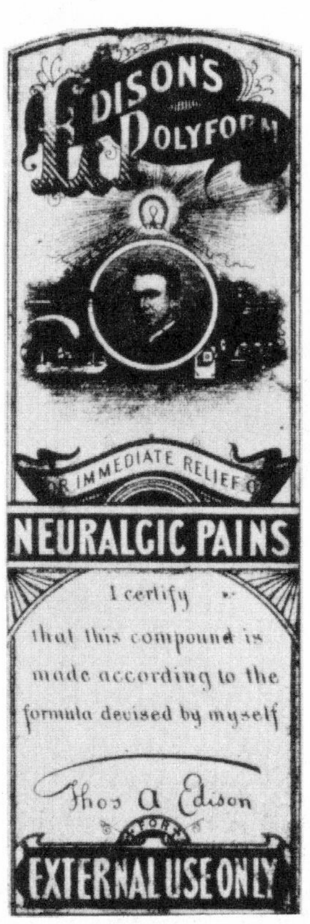

At Menlo Park, Edison concocted the Polyform patent medicine, and invented the phonograph (above right), which he exhibited in Washington, D.C. Edison is seated, while Charles Batchelor (right), his co-inventor, and Uriah Painter (left) stand behind the table. It was while Edison (below, farthest right) was on the expedition to observe the solar eclipse that he began thinking about inventing an incandescent lamp.

Above (left to right) are Francis Upton, Edison's associate in the invention of the incandescent light; John Kruesi, the head of the machine shop; and Edward H. Johnson, involved in numerous Edison enterprises. Menlo Park in winter (below) is shown with the laboratory in the center, the library and office in the right foreground, and the machine shop in the rear.

The photograph above shows the upstairs of the laboratory in 1880, with incandescent lights placed atop the gas pipes. Edison, wearing a glassblower's hat, is seated in the center. Below is Edison's electric railroad of 1880, and at left Henry Villard, Northern Pacific Railroad president, who advanced Edison money to build a commercial model of an electric locomotive.

Sarah Bernhardt (above left) came to see Edison at Menlo Park. The Paris Opera (above right) was lighted with Edison's incandescent lamps. The Edison General Electric Works (below) was established at Schenectady.

Edison and his second wife, Mina, were photographed shortly after their marriage in 1886. Facing page: Edison purchased the estate, Glenmont (bottom), for $235,000. Mina is pictured (top left) with her first child, Madeleine, in 1888. Edison sits with Charles and Madeleine (top right) in 1892.

Facing page: At West Orange, New Jersey, Edison built a huge new laboratory (bottom), where he is pictured (top) with his new phonograph in 1888. Seated to the left and right of Edison are Fred Ott and Colonel George Gouraud. Standing, left, are W. K. L. Dickson and Charles Batchelor. The man in the light suit on the right is John Ott. Edison engaged in a huge magnetic ore-mining project in the New Jersey highlands (below), and experimented with X-rays (right).

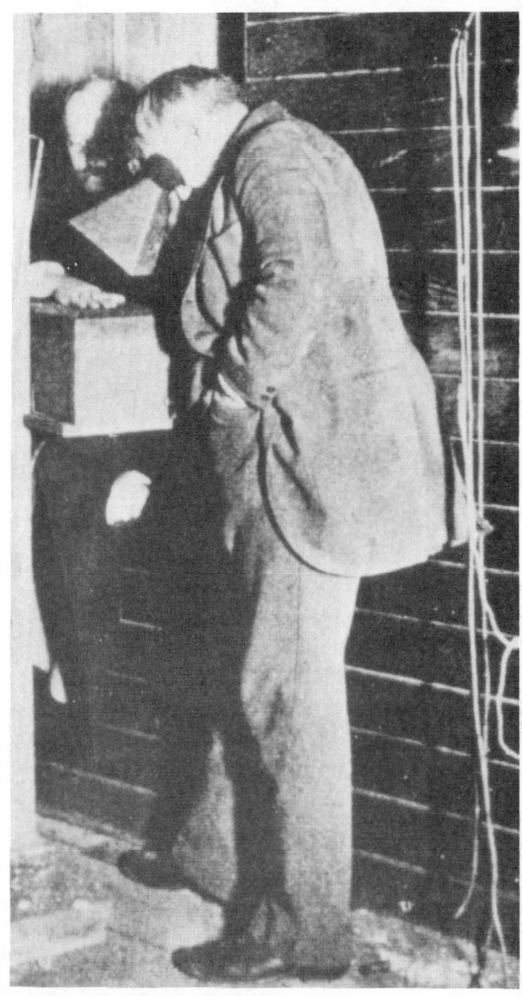

Edison had a bitter quarrel with his old friend Ezra Gilliland (left) in 1888. Edison sent his father, Sam, and Sam's sidekick, Jim Symington (below), to Florida yearly. Facing page: He developed the kinetoscope, and the Black Maria (top) became the world's first motion-picture studio. In it (bottom) world champion Jim Corbett (left) fought Tom Courtney (right) in 1894.

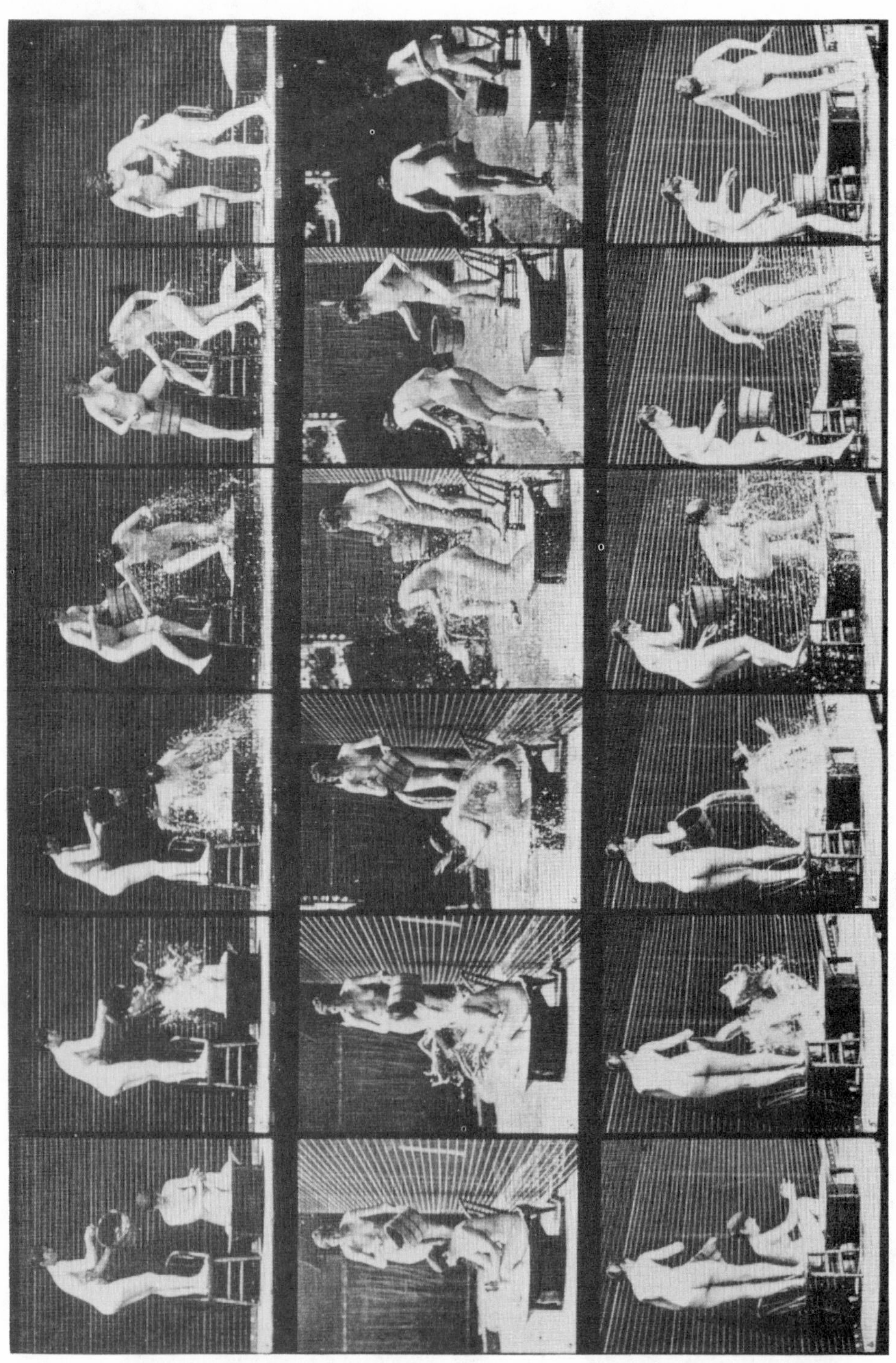

"Animal Locomotion" was the title Eadweard Muybridge gave to his plates. From them Edison got the idea for developing the kinetoscope.

once more dunning him. Mary was expecting their second child in January, and spending money liberally. Doctor bills were allowed to accumulate not for months, but for years. Solvency was a day-by-day and sometimes curious proposition. "I lent Gilliland $2 last night and borrowed $5 leaving me indebted to him for $3," Edison noted to Batchelor.

On November 11, Edison appealed to A. B. Chandler, Eckert's deputy at the Atlantic & Pacific Company: "A.B.C. have you any idea? Of course you have. Do you think? Of course you think. Can you pay one of the smaller of my bills tomorrow. I think under the benign influence of the comely greenbacks this beautiful world of ours would enhance in beauty. New Wonders in the never ending evolution of, revolution & cycloidical transformation of things inorganic into things organic would become conspicuously appealing to my optical nerves."

Edison filed a caveat* on his acoustic telegraph, but the patent office promptly placed it in interference with Gray. He took an apparatus with three reeds transmitting different pitches to the Western Union office. But even to Edison, "The results were eminently unsatisfactory." He complained he had no money to spend on experiments, and could not afford to continue to go in the red. That was music almost as good as electro-harmonics to Orton's ears. Western Union, having used the quadruplex for a year without an agreement, was in a shaky legal position. It was an opportunity, Orton perceived, to extricate the company from liability.

On December 14, Edison signed an agreement drawn up by Grosvenor P. Lowrey, Western Union's counsel, in which the parties released each other from all suits and renounced all claims to damages. Western Union was to pay Edison $200 a week, plus actual costs of experimentation, to perfect a system of acoustic telegraphy. Edison gave Lowrey power of attorney, and Orton virtually all other powers. Edison was to report weekly to the Western Union president, who would judge whether progress warranted payment. If Orton was dissatisfied, he had the right to abrogate the agreement. On acceptance of the system, Edison was to receive $6000, but any additional sum was to be left to "the fairness and judgment of the said Orton."

Once more Edison had found a hand to pull him back from the brink of insolvency—though he had, in fact, assets that on liquidation might have netted him as much as $30,000 or $40,000. What Edison always sorely lacked was *cash*. Not only his career but his family life was being frayed by the repeated financial crises in Newark. Mary could not

* A caveat, a preliminary description of an invention sent to the patent office to establish precedence, was held secret for a year. If by the end of that year the inventor had not filed for a patent on the principles delineated, the caveat lapsed. The patent office has since discontinued the practice of allowing caveats to be filed.

understand his knack for making tens of thousands of dollars vanish. She was unhappy that he was more fascinated with shunts, traps, and polarized magnets than with her, and berated him for diverting large sums to his father and brother in Port Huron. Had he managed his money prudently, she pointed out, he could already be a wealthy man. Yet they did not even possess their own home.

Edison, in turn, was intolerant of Mary's improvidence. Though he gave her at least $2000 a year for household needs, she frittered it away on frills and luxuries, as well as on loans and gifts to members of her family. The necessities of life she charged, and often left the bills unpaid.

Edison decided that, for him and Mary, Newark had become a luxury. They would move out of the high-rent city and cut down on their overhead. From William Carman, the young accountant and purchasing agent of the Electric Pen Company, whose family owned considerable land in the environs of Menlo Park, twelve miles south of Newark, Edison learned of the availability of a substantial home near the railroad tracks that led directly to Jersey City. The community, settled in the eighteenth century by Cornish copper miners, was now a farming village. But the farmers were suffering badly from the depression, and land was cheap.

On December 29, Mary—who twelve days later was to give birth to a son, who was named Thomas Alva Edison, Jr.—signed a contract and a mortgage to buy the house. From the Carman family Edison purchased a parcel of several acres, including a gentle hill with a commanding view of the rolling countryside. To supervise construction of a laboratory on the hill, he summoned his father from Port Huron.

CHAPTER 8

A Remarkable Kaleidoscopic Brain

Sam, at the age of seventy-one, was as agile and full of life as he had ever been. When a train failed to stop at Menlo Park, he jumped off as it puffed along, and when a Hudson River ferry pulled off without him, he leaped across six feet of water to its deck. His wit had not left him, and when his son asked for the legal papers he had sent to Port Huron for safekeeping, Sam wrote: "This Packedg contains all mast entr Discription of Dockaments know to the civilised world." Shortly after the death of Nancy he had started sleeping with the Port Huron vamp, a very pretty but not-too-bright teenager named Mary Sharlow. In December, 1873, the nineteen-year-old Mary had borne him a daughter, Marietta. The affair was the scandal of the town and of the Edison family. On April 13, 1874, a cousin wrote Al: "He is conducting himself fearfully he told me he was going to marry that thing on Sunday and I am fearful he will do it as he is perfectly insane." The marriage did not come off, but the affair continued. In May of 1878, when Mary Sharlow was pregnant again, Jim Symington, at the time deputy sheriff and town constable, suggested to Al that he finance a trip to Europe for Sam. Symington thought that "By managing the matter shrewdly he might be detached from his present position permanently." Symington offered "To get the girl married to some widower or some low rascal who could easily be tempted to marry her for 300 dollars. I have accomplices in Canada shrewd enough fellows who would manage this." Sam, however, would have none of it. Maude was born in July, 1878, and Mabel followed in October, 1882, when Sam was seventy-eight. The girls adopted the name Edison. But Al steadfastly refused to acknowledge their relationship.

By March, 1876, Sam had a rough-hewn 100-foot-long two-story wooden building constructed on the hill 300 yards from the railroad station. Al carted his machinery from Newark to Menlo Park at a cost of some $2700. During the next five years Edison, separated from the distractions of the city, would turn this remote but accessible spot surrounded by blackberry brambles and a few apple trees into the wellhead of the most concentrated outpouring of invention in history.

In addition to his proclivity for financial embarrassment, Edison had a knack for involving himself in scientific and technical controversy. He and Batchelor were continually coming across what seemed to them new things, or as they labeled them, "Phenomena!" They were impulsive and often "unscientific." Edison had a tendency to conduct an experiment, form a conclusion, announce it, and only then try to check it out. He lacked the carefully cultured diplomacy and sophistication of the academician. He irritated scientists by making discoveries in their fields, then publishing them with off-the-cuff pronouncements.

On November 22, 1875, Edison, Batchelor, and Adams had been working with a vibrating electromagnet for the acoustic telegraph. Sparks were emanating from the core of the magnet, a manifestation they had noted before in working with both the magnet and the electric pen. They had always attributed the occurrence to induction, but what they saw now were such extraordinary sparks that they thought something else must be afoot. By connecting one end of the wire to the magnetic rod and the other end to the gas pipe, they were able to draw sparks not only by touching the wire to the pipe anywhere in the room, but also by placing it in contact with an otherwise unconnected metal box. Edison was ecstatic: "This is simply wonderful, a *true unknown force*."

His fifteen-year-old nephew, Charley (Pitt's son), who was on a visit, hung from a gas pipe and drew brilliant sparks with a knife blade, even though he had no contact with the ground. By charging a gas main, Edison was able to obtain sparks from the fixtures in his house, several blocks away. The electrical energy even traveled on a telegraph wire from Newark to New Brunswick, on to New York, and back to the laboratory. Edison and Batchelor attached foot-square pieces of tinfoil to the ends of the wires, and watched sparks leap across—an action beyond static electricity of even the highest potential. Nonconductors like air, glass, and rubber failed to impede the force. Edison built a small, wooden dark box with separated points of carbon, and through a peephole watched electricity arc between the points.

Edison thought that since energy can take various forms, and it was possible to change electricity into magnetism, magnetism might be transformed into something else. Since that "something" seemed to be akin to the manifestations of the aurora borealis, Edison called it an "etheric force." Because the force appeared unaffected by contact with the earth, Edison foresaw the possibility of doing away with poles and insulators in telegraphic transmission.

The scientific community almost unanimously rejected Edison's deductions. Edwin J. Houston and Elihu Thomson of Philadelphia, whose rivalry with Edison was to span two decades, spoke for the scientists when they said: "All the manifestations classed as 'etheric' are due solely to inverse currents of induced electricity." Only a few, like Dr. George M. Beard, thought Edison had made a genuine discovery. Beard theorized "that this is a radiant force, somewhere between light and heat on the one hand and magnetism and electricity on the other." Not until the experiments of Heinrich Hertz a decade later would it become evident that observations of the mysterious "etheric force" marked a step toward the discovery of electromagnetic waves.

All through the winter of 1875–76 the etheric-force experiments were ancillary to and distractions from Edison's work for Western Union on the acoustic telegraph. On a large table in the main room of the Newark lab Edison had an array of tuning forks, telegraph keys, and resonators. Edison himself used a small, partitioned room where he could obtain as much quiet as possible for his tortured ears. Reiff and Edward H. Johnson were frequent visitors. Edison used Reiff to send dots and dashes (mostly dots, since Reiff wasn't very good at dashes), and excoriated Johnson for his inability to distinguish sounds he should hear from noise he shouldn't. Through December and January Robert Spice, a Brooklyn high-school professor, was at the lab nightly to furnish expert advice on acoustics.

To test the instruments on the wires, Edison applied once more for a pass to the Western Union operating room, and received the gruff but cordial reply from Orton: "Bring in all your traps." On February 19, Edison advised Orton: "I have at last got the animal tamed. A Seance will be held Monday or Tuesday whereat the professor of 'Acousticity' will exhibit the latest theft from the German telegraph books."

Gray, however, was obviously well in advance of Edison. By the first of the year he was able to transmit eight messages on one wire over two hundred miles. In mid-January he went to Washington to file a number of patents. On February 14, he submitted a caveat that was to turn the entire harmonic telegraph competition in an entirely new direction. "I claim," Gray stated, "the art of transmitting vocal sounds or conversations telegraphically, through an electric circuit."

A few hours before Gray, Alexander Graham Bell had filed an application on a similar claim. Bell, three weeks younger than Edison, had emigrated to Canada from Scotland with his family, then moved to Boston. His father was a prominent teacher of the deaf, and Bell had followed in his profession. One of his pupils was the young daughter of Gardiner Hubbard, the heir to one of America's oldest and largest fortunes. Hubbard was a bitter opponent of Western Union. An ally of Harrington and Postmaster General Creswell, he had devoted his energies for years to lobbying for establishment of a U.S. Postal Telegraph Service. When Bell, in 1874, fell in love with his daughter, Hubbard encouraged him to turn his knowledge of acoustics to the invention of a harmonic telegraph; in fact, Hubbard virtually made successful development of a system a condition for approving the marriage.

Bell had little knowledge of mathematics or electricity. He first hired one of Moses Farmer's assistants as a helper, then retained twenty-year-old Thomas Watson, a journeyman in Charles Williams's shop, on a part-time basis. In February of 1875, he took out his first patents, and the next month tested a crude but workable apparatus on the Western Union wires in New York. Orton downplayed the achievement and, sitting on the office couch with his legs up on a chair, used much the same tactics on Bell as he had on Edison. Gray's apparatus was far more advanced, Orton told Bell, but that did not matter, for whoever had Western Union's backing would emerge the winner. Western Union would purchase Bell's invention, if perfected, but Bell should not try to make a deal with the Atlantic & Pacific Telegraph Company—that would only lead to litigation.

For the next nine months Bell worked sporadically on the telegraph. He was chronically short of money, he had periods of lassitude, and he had repeatedly to be prodded by Hubbard. During the summer he turned his attention more and more from the harmonic telegraph to other Reis experiments dealing with the transmission of the human voice. He studied the phonautograph, an instrument built in 1857 by Leon Scott, who hypothesized sounds would register as identifiable shapes on paper. On one or two occasions Bell and Watson, employing a membrane diaphragm immersed in water, thought they could hear each other's voices over the wire between two rooms. Bell, however, had done little more than retrace Reis's work when, in mid-January, 1876, he had one of the great insights in scientific history. He postulated that the key to transmission lay in the use of a continuous, undulatory current, modified by sound, rather than a "make-and-break" current that was activated (turned on and off) by the voice.

The chief examiner of the patent office's mathematical-philosophical instrument division, to which Gray's and Bell's applications were directed, was Zenas Fisk Wilber. A graduate of Kenyon, he was a member of a prominent Ohio family, and the cousin and ward of Rutherford B. Hayes, who was to be elected President in the fall of 1876. American patent laws were so structured as to invite perjury, bribery, misdating, and other trickery. And Wilber, who was an alcoholic, chronically in need of money, was easily corruptible.

(Wilber had been acquainted with Edison since 1871, and was a frequent visitor to Newark. Harrington, Painter, and the anti–Western Union group also knew him well, and had prevailed upon him to execute a crude forgery in an attempt to invalidate Edison's agreement with Prescott on the quadruplex. In the copy of the 1871 Harrington-Edison contract filed with the patent office, Wilber inserted "and" in the phrase "automatic fast telegraphy." This changed it to read "automatic and fast telegraphy," making it appear that the contract embraced not only automatic but multiple telegraph apparatus.)

Wilber was deeply in debt to Major Bailey of the firm of Pollok and Bailey, employed by Hubbard as Bell's patent solicitors. When Bell's and Gray's applications reached him, Wilber placed them in interference with each other. Major Bailey thereupon took Bell to see Wilber and convinced Wilber, on cursory evidence, that Bell's papers had been filed a short time before Gray's. Wilber dissolved the interference and gave precedence to Bell. Bell had submitted no models. His experiments with the electro-harmonic telegraph were elementary compared to Gray's and Edison's. His description of the apparatus he intended to use was crude—it had not, in fact, any practical value for transmitting speech. Gray had not as yet constructed any of the devices he intended to utilize either, but Wilber let Bell examine all of Gray's papers and explained in detail Gray's proposed method of transmitting and receiving. Wilber then allowed Bell to file an amendment to his application.

Bell remained in Washington until March 6, 1876. When he returned to Boston, he and Watson constructed a membrane diaphragm from which a platinum needle dipped into a dish of water. The dish was connected by wire to a reed receiver, similar to those used in harmonic telegraph experiments, in another room. On the afternoon of March 10, Bell shouted: "Mr. Watson, come here, I want to see you." And Watson, his ear pressed to the receiver, heard and came.

The news was soon out. Edison received it through Wilber and Painter, a close acquaintance of Hubbard, as he was in the process of moving to the Menlo Park laboratory. He was astonished, and at first refused to believe it, though he discovered later that one of the devices he had tested in December was capable of receiving speech.* Basing his work on Bell's and Gray's, Edison gave Adams the task of developing a water telephone, and by summer was able to transmit a few articulated sounds.

But Bell, despite the crudeness of his apparatus, had clearly outdistanced his rivals. Though the telephone ("far speaker") produced only a whisper even under the best laboratory conditions, it became the sensation of the Philadelphia Centennial Exposition.

* Moses Farmer, with tears in his eyes, characteristically told Watson: "That thing has flaunted itself in my face a dozen times during the last ten years and every time I was too blind to see it."

A few months later, on October 6, Bell and Watson held the first really distinct conversation on a short line between two rooms of the Exeter Place house. Three days afterward, using a private telegraph line, they were able to understand each other over a distance of two miles. By November they could communicate between Boston and Cambridge.

Hubbard tried to sell Bell's telephone rights to Orton for $100,000, and Painter attempted to peddle them to Gould. Neither Wall Streeter was interested. If the device had any value, Orton was confident he would have to pay far less to Edison and Gray to produce a similar apparatus that would circumvent Bell's patent. Besides, Gray—who was skeptical the telephone had any commercial applicability—seemed to have at least as good a claim to the basic patent as Bell. One thing Orton and Gould agreed on: inventors were flighty and undependable creatures, and investing in them was only slightly more risky than sinking money into racehorses.

Gould continued to be stuck with a balky Edison who refused to do any more work on the sputtering automatic system, no matter how he was coaxed.

"I have always felt very kindly toward you and assisted you financially when but for it you would have been ruined. At least so you told me," Gould appealed to him. "How have you requited me for this kindness? I leave your own conscience to answer."

But Edison was adamant that the choice was between him and Eckert: "Mr. Eckert does not want me to be connected with the A & P. It was not necessary that a man should be knocked down and have the fact forced down his throat with a crow-bar. For instance, keep an impatient man like my self waiting 3 weeks to decide about removing a partition, doubting my honesty in a transaction of four dollars, granting no money to conduct experiments, & then say I am not doing anything for the company. I have been made to suffer for personal spite of Mr. Eckert against others. [Eckert did not like Reiff.] I had no recourse but to take hold of some experimental work for the W. U. I must live, Mr. Gould, and if your agents have not perception enough to deal with an inventor in a different way than with a business man, then you must blame your agents, not me."

Painter tried to reason with Edison, but with no better success. Reiff had debts of $162,000 and Painter thought, "If you or Reiff or both can get a reasonable sum of money out of it [the automatic] by your expending a little time and labor, it would be better to do it than abandon it." Painter pointed out that Edison stood to gain more from two competing telegraph companies than from "one Shylock in the field who has publicly and privately heralded you as a thief and pretender. Nothing can be got out of A & P for Automatic except you go and set it up on its legs and make it walk."

Reiff, in truth, had been reduced from the generous capitalist to the debtor sliding into bankruptcy. He pleaded with Edison to come to a written agreement with him. But Edison had put the whole automatic system into a locked closet of his mind. He didn't want to do any more work on it, either in England or in America.

Through the summer of 1876 Edison and Adams, with a handful of assistants, worked on the electromotograph, the acoustic telegraph, the autographic telegraph, the speaking telegraph, and the electric pen and mimeograph.

The technical success of the electric pen acted like an elixir on Edison, and he perceived a ready market for a whole host of electrically operated devices. He proposed designing electric shears, a revolving display case, a dental plugger (drill), an artificial flying bird, a chimneyless kerosene lamp for which a fan would provide the draft, and, most ambitiously, an electric sewing machine. He thought, after observing the actions of tuning forks in his acoustic telegraph experiments, that more power could be obtained from an electromagnet by using it to place a tuning fork in vibration; and with such a tuning fork motor he succeeded, on October 22, in driving a sewing machine at eighty-two stitches per minute through six layers of cloth.

Aside from these devices, he proposed to duplex the Atlantic cable, and was developing, or hoped to develop, a twenty-cent pocket spectroscope, an artificial, perfumed rose for buttonholes, and "Edison's Perpetual Segar," a refillable cigar that was to drive pipes out of existence. To market these diverse inventions, Edison on November 28 incorporated the American Novelty Company. Edward H. Johnson, contributing one or two minor inventions of his own, became the general manager and factotum of the new concern.

But while Edison envisioned an electrical world, the Electric Pen Company was struggling. The basic problem was that electricity derived from the crude batteries was expensive, messy, and undependable. The first typewriters were beginning to appear in offices, and posing competition for the pen. Edison himself devised a stencil typewriter for use with the press, and gave it to Jim Mac Kenzie, his Mount Clemens telegraph instructor, to work on. (Mac Kenzie, after becoming manager of the domestic telegraph system in Detroit, had taken the same position in Washington, D.C., and was now a frequent visitor to Menlo Park.)

For all of 1876 the pen company's income was only $3880. The Gillilands, following Edison's example of abandoning the high-rent area of Newark, moved the factory to a building along the railroad tracks in Menlo Park. Yet Ezra, managing the New York office, borrowed so consistently from Edison that he ran up a debt of $3700.

When George Bliss, general agent for Western Electric, offered to purchase pen and mimeograph rights for five years, Edison and Robert Gilliland gladly accepted. Bliss, paying $1500 down and guaranteeing royalties of $3000 a year, transferred production to Western Electric in Chicago. But he had no more success than the Gillilands in generating business, and another decade was to pass before the mimeograph came into its own.*

The most important aspect of the transaction was that it freed Batchelor, who for the past year had been the Electric Pen Company's general manager, to return to work in the lab.

During the months of Batchelor's absence, Edison had buzzed about like a bee distracted by every colorful idea that blossomed from his imagination; but he had made little progress in developing any of them. Batchelor was more down-to-earth than Edison, he was more orderly, he was the transformer of conceptions into drawings, he stuck to a task until it was done and did not continually hop from one thing to another. Yet Batchelor did not have Edison's imagination or his instinct for grasping the crux of a problem. He lacked Edison's ability to express himself, he was uncomfortable in the presence of the cosmopolitan New Yorkers, he did not have Edison's knack for dealing with financiers. Thus, Edison and Batchelor were ideal, complementary partners. Edison made an oral agreement giving Batchelor 10 percent of the gross received for all inventions, and one tenth of the stock in all companies formed. Since Edison reckoned profits at 20 percent of gross, he agreed, in effect, to split 50–50 with Batchelor. What's more, Batchelor was sure of his cut whether there was a profit or not. Recognition was as important to Edison as money. So, during the two decades of their association, while Edison's name became famous throughout the world, the self-effacing Batchelor quietly grew rich.

The move of Batchelor's family to Menlo Park was welcomed by Mary, for the members of the small Edison colony were largely isolated from their farmer neighbors. Edison frequently lounged about the combined general store–post office–telegraph office–railroad station that was the community gathering place, and sometimes dropped in at Davis's Lighthouse, the local tavern, where he swapped stories with the proprietor, who spun out tall and tangled yarns in a Scotch brogue. If a story appealed to him, Edison's quizzical expression would change to a broad smile, he would reach behind his neck with his left hand to scratch his right ear, and his laugh could be so long and hearty that tears coursed down his cheeks.

There was, however, little visiting between farm women and wives

* Edison sold British rights for $15,000, but within two or three years European demand for the pen and press all but evaporated.

of Edison men. The Edison women were citified, and by their airs emphasized the social and cultural gap. Mary, especially, behaved as aristocratically as if she were the spouse of a British squire. She ordered many of her groceries from Park and Tilford in New York. She bought her clothes at Lord and Taylor's. Dressed in the latest New York fashions, she drove about the rutted roads in a snappy carriage bought for her by Al. She had three Negro servants to care for the five-bedroom house, which she furnished elegantly. Lace curtains were on the windows. Velvet, Belgian, and Oriental rugs covered the floors. The parlor was filled with a piano, overstuffed furniture, vases and mirrors, and Victorian geegaws. The family physician was Dr. Leslie Ward, Newark's most prominent M.D. and a founder, vice-president, and first medical director of the Prudential Life Insurance Company of America. Ward, summoned by telegram, made frequent trips to Menlo Park, for not only the children but Mary were often sick. Edison, too, had periodic bouts with respiratory infections and his unruly stomach. Ward did not come cheaply, but he was indulgent about collecting the bills. This suited Edison fine, for he did not care how much something cost so long as he was able to postpone the payments.

Although Edison could walk from his house to the lab in less than three minutes, Mary saw him little more in Menlo Park than she had seen him in Newark. Sometimes he would not appear for days. Beneath the stairwell in the laboratory was a small storage space. Into this Edison would crawl any time of night or day, pull the door shut, and like a dog curl up and go to sleep on a pile of old newspapers.

He had a theory that taking off one's clothes changed the body chemistry and induced insomnia—even on days when he trotted home in the early-morning hours he frequently slid into bed without removing one battered item of his wear. Mary became so disgusted she exiled him to the guest room. Though he did not cultivate a beard, he was often unshaven—when he did shave, he squeezed the straight razor in his fist like a lemon, and onlookers held their breath lest he decapitate himself. He had the Victorian aversion to water, and throughout his life took at most one bath a week. When he went to the city he dressed in silk coat and top hat, as befitted a man of his position, but in the lab his rough-hewn, chemicals-stained appearance was often accentuated by a pungent odor of things organic and inorganic.

He was truly at ease only in the world of the inanimate, with companions who shared his interests. A ceaseless prober into the unknown of chemistry and physics, he never failed to express boyish wonder at what he found. "Dip your hand in ether then in boiling water or molten lead—it feels cold," he recorded.

The winter of 1876–77 was harsh—in December a storm howled about the hilltop lab with such force that it nearly toppled the building, and Edison was able to save it only by having twenty old Western

Union telegraph poles propped against the sides. The next day, when Edison went into the lab in five-degree-below-zero temperature, he found everything frozen. Bottles had cracked. Chemicals were congealed into startling crystals and brilliant multicolored designs. Forgetting the cold, Edison took out his microscope and went from bottle to bottle, examining the substances in a state in which he had not seen them before.

As a self-taught chemist, he was repeatedly surprised by some of the reactions he generated. Working in the photographic darkroom, he diluted a batch of sulfur chloride by pouring in a beakerful of water. Almost instantly the mixture exploded and splattered over his face and into his eyes. Nearly blinded, he dashed to the hydrant, leaned over backward, and let the water run over his eyes for several minutes. For three days he was afraid he might lose his sight, and for two weeks he did very little work.

Branching out in the application of his knowledge, he regularly prescribed and mixed chemicals to cure the ailments of his workers. With the state of medicine as it was then, they were probably just as well off to have an inventor for a doctor. For neuralgia he concocted "Polyform," a mixture of morphine, chloroform, ether, and chloral hydrate (knockout drops) suspended in alcohol and scented with spices. This he found—not surprisingly—effected instant relief when applied to a person's face.

One cold and stormy night Edison was graced with the appearance of a human guinea pig. A shivering tatterdemalion, attracted by the light shining through the darkness, knocked and asked if he might have a bite to eat and sit by the stove. His head was of abnormal size, his face intelligent, his body emaciated. He told Edison that he was suffering, and wondered if Edison had any morphine. Edison replied he had about everything in chemistry that could be bought, and got out the morphine sulfate. As Edison watched, he poured out enough to kill two men. When Edison suggested the amount might be excessive because the laboratory was not a "hotel for suicides," the man bared his legs and arms. They were covered with sores and pitted with scars from hypodermic syringes. Edison let him go ahead, and shortly he seemed like another person. He commenced to tell stories, and everyone sat around listening till morning. He was a man of great intelligence and education. He said he was a Jew—though Edison, who shared the popular conception that Jews had hooked noses, glinting eyes, and carried a pound of flesh in a knapsack, could descry "no distinctive feature to verify this assertion."

"He continued to stay around," Edison related, "until he finished every combination of morphine with an acid that I had, probably ten ounces all told. Then he asked if he could have strychnine. I had an ounce of the sulfate. He took enough to kill a horse and asserted it had as good effect as morphine. When this was gone the only thing I had left was a chunk of crude opium, perhaps two or three pounds. He chewed this up and disappeared. I was greatly disappointed because I would have laid in another stock of morphine to keep him at the laboratory." A few days later he was discovered dying in a barn, and Edison contributed a dollar to telegraph the county poor commissioner to come and pick him up.

At the same time that Edison was chronically short of cash, he had more than $11,000 tied up in the Sarnia and Port Huron horsecar lines. Pitt, who had borrowed money on stock he did not own, was importuning him to come to Port Huron with $4000 more. If Al did not show up with the money or the stock, the directors of the Sarnia line threatened to sue. Edison, however, had had enough of the affair, and turned the matter over to a Wall Street attorney who was representing him in the quadruplex case—the one-room lawyer in Sarnia suddenly discovered, to his stunned chagrin, that he was embroiled with one of the country's most prestigious law firms.

Yet his surprise was mild compared to that of the Fort Gratiot commander, who revoked Pitt's license to operate on the reservation and transferred the franchise to a new, competing line. Edison appealed to Painter in Washington, and Painter buttonholed the secretary of war. A stricken major, who never would have believed a horsecar franchise could become of cabinet concern, was confronted with a demand for a full report from the War Department.

By the spring of 1877, Edison was in the process of extricating himself from the horsecar business. He was succeeding also, after trying for nearly a year, in maneuvering Western Union into a renegotiation of its financial support. Though he was supposed to receive $200 a week for experimentation on the acoustic telegraph, he collected the sum only sporadically, and complained to Orton: "At the end of each week if the experimenting has not been satisfactory to myself I find it impossible to screw up my courage to the point of asking for the $200, & consequently have to suffer. Whereas when it has been satisfacting I do not always find you in & then I have to suffer."

In actuality, there were many weeks when Edison let the acoustic telegraph lie idle—what he wanted was a contract that would pay him an unqualified retainer or salary. In the late spring of 1876 he had approached his old mentor, Marshall Lefferts, at the Gold & Stock Company, and Lefferts had replied genially: "Grand Laboratory & general investigating Department of Spiritual and Material Matters. Electro Motive force I × U + G ÷ A & P = W. U. Consolidated. Am prepared to talk about your salary for the next five years." But the proposed consolidation between the Western Union and the Atlantic &

Pacific Telegraph companies had not come off, and on July 3 Lefferts had collapsed and died.

Orton was not disposed to be as indulgent as Lefferts, but in the winter of 1877 Edison's bargaining position improved dramatically. The Atlantic & Pacific Telegraph Company received an infusion of $600,000 from a group of railroad men, cut its rates, and "declared war" on Western Union. Legally, Edison was still the A & P's electrician, and Orton was anxious to keep him from active participation in the enemy camp. Furthermore, the quadruplex trial was scheduled to get under way in a few weeks, and Orton had learned how dangerous and impulsive a dissatisfied Edison could be.

On March 22, Western Union agreed to pay Edison one hundred dollars a week unconditionally for five years.* Edison, in turn, gave Western Union an option "to all inventions and improvements capable to be used on land lines of telegraphs or upon cables." This, of course, included both telegraph and telephone equipment.

Less than a month later, on April 19, the "omnibus trial," in which all the various parties suing were pitted against one another, opened in New York. A dozen lawyers, a fair cross section of the highest-priced legal talent in America, gathered in the courtroom. Edison's counsel was General Benjamin Butler, a Boston lawyer who had earned the nickname "Bloody Ben" for his hard-line, no-nonsense command of the occupation of New Orleans in 1862. Butler, with his large paunch, droopy eyelids, and huge bald head, was clearly the most imposing figure in the case. When he threw back his head and shoulders and expanded his chest, the unlit cigar constantly in his mouth took on the nature of a pointed cannon, and everyone's attention focused on it as if they expected momentarily to see smoke and flame belch forth. With his histrionic performance Butler overshadowed even the huge, tall E. N. Dickerson, who, with Grosvenor P. Lowrey, represented Western Union. Dickerson's broad shoulders and corrugated face were remnants of his former profession, blacksmithing, and he blasted his words out as if they were sledgehammer strokes.

Edison took the stand on April 26. He was called as a witness for the Atlantic & Pacific Telegraph Company. But every day after court the electrician of the A & P met furtively with Orton and other Western Union executives so they could coordinate their testimony. Reiff wanted Edison to testify for him, but Western Union executives warned him not to, "for fear that some old and ugly affidavit of yours will come out."

Making excuses to Reiff and Johnson, Edison told them his posture was one of "neutrality." This caused Johnson to protest: "Your desire

* Edison insisted that Reiff also receive an income from Western Union, and wrote a memorandum: "JCR to be retained by the company for outside work of various kinds at a salary of $2500." Orton, however, didn't want Reiff's name associated with Western Union, and modified the agreement so as not to include Reiff directly. In fact, the payments to Reiff were never made.

for neutrality is commendable within certain limits [but] your absenting yourself at all times except when called for by some Western Union party has already placed you in their ranks."

Reiff, who was being made the fall guy by the Western Union lawyers, lamented to Edison: "I know I am a squeezed orange, or an equally valuable relic. They [Western Union] gained almost all they wanted when they induced you to sign your contract for the future without insisting on a consideration that all our matters with the quad should be fixed." Although Western Union was using forty-eight quadruplex circuits, which according to the Edison proposition that Orton had "accepted" would have netted Edison $11,000 a year, Edison was receiving nothing.

On the stand, Edison listened to the questions with his eyes half closed, then answered them slowly and composedly, displaying an impressive knowledge of technical matters and science. When it came to perception of time, however, he was lost. Experiments were timeless: they went on for days, weeks, and months. Though on the notebooks dates were carefully jotted down as evidence for patent applications, in Edison's mind events blended and telescoped into each other, and he misplaced even fairly recent happenings by as much as three months. "I have got no memory at all for dates," he confessed.

Ashley, in the *Telegrapher,* delighted in calling Edison "the professor of duplicity and quadruplicity." Dickerson, in his brief for Western Union, depreciated his "remarkable kaleidoscopic brain. He turns that head of his and these things come out as in a kaleidoscope. Many of them will work and many of them will not. It is a turning of the kaleidoscope, producing new permutations, but very seldom anything practically valuable developed in any of them."

Butler, in contrast, depicted him as an easy mark manipulated by Gould, "whispering in his ear, the Mephistopheles of this drama, as his master in the Garden of Eden whispered in the ear of Eve, and holding out to him $30,000 in money if he would do the deed. If Jay Gould had not a heart of stone, when that young man was struggling in the toils of Temptation he would have withdrawn that offer and let him go in peace. [It] is one of the most fearful exhibitions of degradation of the day, and the immorality that comes out of that nest in Wall Street, which is the source of most of the crimes we see committed in the community today."

Though Edison was one of the best-known men in the telegraph industry, his name had been anything but a household word. But the drama of some of the nation's foremost men and its most powerful corporation grappling over rights to the quadruplex generated considerable curiosity about the thirty-year-old inventor.* Yet it was not so

* Within a few months the heavy costs of competition and litigation, with the likelihood that the issue would not be settled for years, led the two archenemies to conclude that armistice and accommodation were to be preferred to battling

much the trial as events set in motion concurrently which were to rocket him to worldwide prominence before another year had passed.

---

to insolvency. In meetings between August 15 and 20 a price-fixing and revenue-sharing agreement was worked out: Western Union was to get 87½ cents and the Atlantic & Pacific 12½ cents of each dollar taken in by the two companies. The A & P dropped use of the automatic equipment as unprofitable. On September 11 Edison, without ever receiving the 3000 shares of stock due him, finally resigned as the company's electrician. The following spring the court ruled in favor of Western Union on the quadruplex patent, but by then it was largely a dead issue.

CHAPTER 9

The Speaking Telegraph

For nearly two years Edison had been drifting in an inventive backwater. His last significant creation (a joint effort with Batchelor) had been the electric pen. During the year and three months of Batchelor's absence he applied for only eleven patents, the lowest number during a span of a quarter century.

But with Batchelor's return from New York, the situation changed dramatically. Edison now entered a period that was to culminate in an explosion of creativity unparallelled in inventive annals.

While experimenting on an underwater cable for the automatic telegraph after returning from England in 1873, Edison had found that the electrical resistance and conductivity of plumbago (as carbon was then known) varied according to the pressure it was under. This, as scientists subsequently recognized, was a major theoretical discovery. Edison and Adams employed the principle during the spring and summer of 1876 to construct a "pressure relay," in which carbon supplanted the magnets normally used.

In early February, 1877, Edison told Batchelor that he thought the range of the telephone could be greatly increased by the incorporation of a pressure relay. Bell's instrument, because of its weak articulation and the fact that it could be used only over very short distances, was still imperfect and commercially impractical.

None of the first devices constructed by Edison, Batchelor, and Adams was much good. By March 18, however, they obtained some slight improvement by having the diaphragm strike against two disks of plumbago. The strength of the current, they observed, varied in precise proportion to the pressure generated by the sound waves—the discovery was to bring about the replacement of the electromagnet by carbon in the transmitter. A week later—two days after signing his new agreement with Western Union—Edison took one instrument to New York to demonstrate to Orton. Orton didn't think much of what he heard (which was very little), but encouraged Edison to continue. On April 27, Edison, confident as always, applied for a patent on his "Telespecan," a name he devised to distinguish his apparatus from Bell's telephone.

It might more appropriately have been called the "Telespecan't," for while it could transmit tone and intensity, 90 percent of the quality and articulation were lost. The trio was having much greater success with a new receiver, which they constructed on the electromotograph principle. A brass arm with spring was attached to an old-style dulcimer, which served as a sounding board. The spring pressed the arm firmly against a sheet of bibulous paper soaked with sodium sulfate and wrapped around a brass drum. The variations in the electric current passing through the arm controlled the friction of the paper against the rotating drum, causing the arm to pull upon or be pulled back by the sounding board, so reproducing accurately the notes sung into the transmitter. The device would also transmit the sounds of some musical instruments, notably the cornet. The receiver was so loud that the singing and music could be heard clearly throughout a large hall, even when the point of origination was a hundred or two hundred miles away.

Edison gave the first exhibition of the musical telephone at the Newark Opera House on April 21, and it immediately captured the fancy of the public. The first full-scale concert, piped in over telephone lines to the strange-looking gadget on the stage, followed on May 23. Edison franchised the musical telephone to Edward H. Johnson, who was a natural promoter, and Johnson took it on tour. On July 19 he drew more than 5000 people to a concert at the Permanent Exhibition Hall in Philadelphia. Other concerts were scheduled for resorts: Atlantic City, Cape May, Long Branch, Saratoga. Johnson tailored the programs to the audiences, which numbered as many as 8000. Along the shore he mixed rousers like "Yankee Doodle" and "Columbia, the Gem of the Ocean" with the romantic melodies of "The Last Rose of Summer" and "Annie Laurie." At staider Saratoga, romance was joined to sacred music, like "We Will Meet Him at the River." In the public's mind Edison's invention far surpassed Bell's telephone, a fact emphasized whenever Johnson wanted to communicate with the musicians at the point of origination, and had to click the message out in Morse code.

Orton, however, considered the musical telephone irrelevant and a distraction. Though Edison assured him, "You need have no alarm about Bell's monopoly," Western Union needed an apparatus for transmission of messages—it was not musically inclined. In late May Orton irritably complained: "Our desks are littered up by Edison's traps. He hasn't done a thing in some weeks."

Under pressure from the Western Union president, Edison focused his efforts once again not only on the *speaking* telegraph, but on the autographic telegraph and the harmonic printing telegraph.* After an all-night session with Edison on May 27, Batchelor sketched a telephone transmitter for Kruesi and noted: "Edison wants this bad."

The regular routine continued to be for Edison, Batchelor, and Adams to work from dusk till dawn, then have Kruesi make the devices during the day for testing the next night. The speaker they designed consisted of a bent tube with a mouthpiece at one end and an adjustable diaphragm at the other. The diaphragm vibrated against one or more disks of plumbago, so altering the resistance of the current. Night after night they mixed plumbago with scores of different substances—plaster of paris, furniture varnish, flour, Bermuda arrowroot, glue, tragacanth—to determine what combination would give the best result. Sometimes they lost track of the compound. On June 13, Batchelor chronicled: "Unknown Plumbago mixture simply elegant—It now becomes necessary to find out what this combination is. We believe it is Plumbago with a large proportion of Isinglass [mica] or Gelatin."

By June 25 the three were confident, and Batchelor noted: "Speaking Telegraph is a Success, You Bet." But when one of them read a newspaper article over the wire, the recipient could hardly understand a word. In early July, Edison, slightly affected by the heat and tormented by the huge mosquitoes that caromed about the lab, had one of the eccentric notions that periodically emanated from his head. "I propose to use the gas pipe system in factories and houses to convey the human voice, even to great distances in different parts of a city," he wrote. A bell or a "singing gas jet" would summon a party to a flexible tube over whose end a diaphragm was fastened, and the person would listen and speak into this, meanwhile trying not to breathe too hard.

Though this oddity never reached fruition, a half-dozen other versions of the "Speaking Telegraph" were submitted to the patent office during the summer and early fall. All worked on the principle of the variable resistance of carbon, including the "Inertia Telephone," in which the entire housing, not just the membrane, vibrated and impinged upon the carbon.

Since a brass diaphragm produced a metallic ringing in the ears of the listener, Edison tried substituting membranes made from various woods, rubber, metallic oil, ivory, bamboo, dried lemon peel, tobacco leaf, fish bladder, and a host of other substances.

But while Edison was probing for alternatives and improvements to Bell's apparatus, the Bell Telephone Company was formed. Not only Orton but executives from other telegraph companies prodded Edison

* The latter was a device intended to print messages received via multiplex acoustic telegraphy in the same fashion that the never-completed roman-letter machine would have printed signals received on the automatic.

to provide them with a countering device. William H. Preece, the head of the British Postal Telegraph system, spent several weeks in the United States and visited the lab a number of times. On July 30 Edison applied for a British patent on one of the earlier versions of the rapidly transmuting transmitter, and shipped Preece a pair of samples. Another pair went to Orton. Then Edison discovered that a sponge filled with fine plumbago produced much better articulation than the hard disks, and a few days later Batchelor made a further improvement by taking tiny pieces of silk, soaking them in plumbago, then rolling them up and pressing them together.

"Delay cannot be helped," Edison expostulated on August 29. "Bell had a very easy job compared to what I have had. At the Centennial May 10 he exhibited his speaker and there ain't much improvement or change been made; over a year; whereas I had to create new things, besides I am so deaf I am debarred from hearing all the finer articulations and have to depend on the judgment of others. I had scarcely got the principle working before there is pressure in New York to introduce it immediately. I have two or three pair but found they were unhandy after they were made; that delayed. I have finished a new pair and they have been working two days with no change or adjustment, and everyone concurs that the articulation is perfect and they do not see what improvement is required. I have my man making a model for the patent office which is essential I should get in, and I and Batchelor must go to New York to show it there so you see I have my hands full even working 22 hours per day."

The next day, Thursday, Batchelor and Edison took the instruments to New York and hooked them onto a telegraph line between Broadway and South Street. They were able to speak with Orton, Eckert, and others, though Bell's instruments on the same line could not be understood. Friday, Edison wrote a memo to Orton and promised he would send another pair of telephones over Saturday. "That will make the 3 pair. If you are in no haste for any more those 3 pair can remain on test, while I go ahead with the *'embosser.'* Which shall it be = more Telephone or the embosser!"

CHAPTER 10

The Sound Writer

In the innocent-sounding "embosser" Edison was gestating one of the most significant inventions in the history of man. Its conception was directly related to Edison's multifaceted creative approach, for it was an invention that developed not from a single strand of thought, or a deliberate design. Rather, it was a unique manifestation of serendipity, an outgrowth of all Edison's telegraphic, telephonic, electromotographic, and mimeographic endeavors; an invention that was the offspring of entirely different inventions; an inimitable mutation.

Edison was concerned with four problems: one, the most pressing, a speaker for the telephone; two, a copying machine based on the electromotographic principle; three, the technology and the devices for the autographic telegraph, to be used for transmitting facsimiles of drawings and of handwriting; and four, how to employ the telephone in Western Union operations.

Common to several of the endeavors was a cylinder, or roller, such as used on the electromotograph. For the telephone speaker Edison thought he might be able to adapt the principle of the electromotograph. He ordered Kruesi to make a smooth, platinized wheel which he intended to turn at great speed in front of a diaphragm with a spring and platinized point. "Hence the louder you talk the stronger it presses and you get articulation," he hypothesized.

For the copying press, Edison constructed an elongated cylinder. Wrapped around one end was a master of tinfoil or wax paper, inscribed with the material to be copied.* A sensor was supposed to pick

* Alternately, Edison tried using plain paper containing writing consisting of raised letters produced by arsenic ink.

up the impressions and transmit them to a synchronized pen writing on a sheet of paper wrapped around the cylinder's other end, thus duplicating the original.

For the autographic telegraph Edison intended to use the same basic apparatus. The electromotograph would convert the mechanical movements into electrical impulses. From these, the original would be re-created at the far end of the line.

The question of how to employ the telephone in Western Union's operations might, from the retrospect of a hundred years, seem absurd. But the telephone, in its inchoate and semideveloped stage, was viewed as a device suited primarily for use on private telegraph lines. Western Union had never been in the private telegraph business, and the concepts of switchboards and of telephones in every office—not to speak of in every home!—was far off in the as-yet-inconceived future. So what was Western Union, which was in the business of sending and delivering messages, not the human voice, to do with the telephone?* Even if the company were able to transmit telephonically over the long lines between cities, it would need a means of recording at the receiving end, since it would be uneconomic to tie up the wires by speaking at a speed slow enough for a copier to follow. That a mechanical device for recording voice messages could be constructed seemed inconceivable. Perhaps the solution would be to train a corps of phonographers, or "sound writers" (*phono:* sound, *graphy:* writing), as practitioners of the art of shorthand were then known.

To Edison, of course, the word "inconceivable" was alien—with his sometime selective deafness, he made certain that he never heard it. A European experimenter, Leon Scott, had demonstrated twenty years before that each sound produced a distinctive shape. Possibly, Edison thought, a way could be found to register and print those shapes, so that afterward they could be transcribed into roman letters.

He mentioned his idea to Johnson and to General Butler, with whom he had occasional meetings in the Fifth Avenue Hotel, the hobnobbing place of upper-crust Republicans. But as June blended into July, he was far too occupied with his principal projects to give phonography more than an occasional thought. Testing various devices for the autographic telegraph, he tried tinfoil and wax paper to activate the sensor on the transmitter. On July 5 he noted: "For autographic Tel write your message in parafined paper & transmit by a spring contact."

During these weeks, Edison, Batchelor, and Adams were wandering near the threshold of one of the great principles of physics, and it was a question of whether they would stumble through. They were experimenting with diaphragms of various materials and designs, some of

* The problem was discussed by E. H. Johnson in "The Telephone: Its Origin and Development," in the *Telephone Handbook*, published by Russell Brothers, New York, 1877.

which were coupled with an electromotograph cylinder. They were using a similar cylinder, over which stiffened wax paper or copper foil flowed, in attempting to transmit messages via the embossing autographic telegraph.

One day in mid-July, the two lines of experiment merged. Edison decided to try transmitting on the telephone speaker by placing a sheet of paraffined paper around the electromotograph cylinder. This failed to produce any positive effect. But when the experimenters continued to turn the cylinder and the point of the diaphragm passed over certain portions of the paper, the earlier sound waves seemed to have left some slight impressions which now vibrated the diaphragm and generated a humming or even a musical sound.

For a few days Edison kept thinking about the phenomenon and going back to the experiments. If writing could be converted into electrical signals and re-created by embossing at a distant point, why not the human voice? The problem of how Western Union would employ the telephone would be solved. On July 17 Edison jotted down:

"Speaking telephone reproduced slow or fast by a copyist & written down. Sheet after received is sent to copyist who'll put it in machine similar to that shown or other and copied at rate of 25 words per minute whereas it was sent at rate of 100 per minute."

It would be a device performing a parallel, if infinitely more complex, function as the Morse embossing register he had rigged up to repeat messages at a slow speed when he had been learning telegraphy in Indianapolis thirteen years before.

The next day Edison and Batchelor were once more working on the telephone speaker. But the sounds that had bounced back off the sheet of paper haunted Edison. He mulled over what might happen if he did not attempt to transmit, but simply ran a paraffined strip of paper beneath a stylus-tipped speaker.

Edison halted the telephone tests, and bellowed toward a diaphragm as the strip was run through. The faint line impressed upon the paper was disappointing and manifested scarcely any differentiation. A swirling breeze might have left as distinct an imprint in the dust.

Without much hope, Edison pulled the strip back through. Startlingly, the paper band, despite the faintness of the indentations, generated a murmuring sound. It was a murmur no more intelligible than the buzzing of an insect. Yet the inventors had heard similar, vague, semihuman emanations issuing from telephones. Excitedly Edison shouted, "Halloo," into the diaphragm, and ran the strip through again.

"Batchelor and I listened breathlessly," he related. "We heard a distinct sound, which a strong imagination might have translated into the original Halloo!"

Interrupting his account of the speaker test, Edison wrote: "Just tried experiment with a diaphragm having an embossing point & held against parafined paper moving rapidly the spkg vibrations are indented nicely & theres no doubt that I shall be able to store up & reproduce automatically at any future time the human voice perfectly."*

Within a few hours of the discovery, Edison received a long letter in which Johnson parenthetically discussed the concept of printing sounds: "Now as to the latest Idea of mechanically speaking the Letters of the Alphabet, Prof B. [George F. Barker of the University of Pennsylvania] is delighted & Says it looks as if you might reach by a Short Cut the End Sought by Scientists for ages, viz—the ascertainment 1st of What Constitutes a Vocal Sound of a Letter & 2—How to mechanically reproduce it."

Barker was to be Edison's staunchest champion in academic circles. He already pronounced Edison's telephone the best of any developed, and held Edison's discovery of the change in the electrical resistance of plumbago under pressure to be of unparalleled importance. He intended shortly to bring a party of scientists to visit the lab.

"The Speaker [for the telephone] must follow in the wake of all this *Sure*," Johnson closed.

In truth, Edison, working sleeplessly to produce the telephones for Orton and for Preece, could for the moment give only peripheral attention to the new discovery. During the second week in August, he experimented with a number of different modes of recording: paraffined paper; paper with a ridge folded in the center, so that the point of the diaphragm indented the ridge; metallic foil wrapped around a cylinder. On August 12 he made various sketches, including one of a cylinder with an attached speaker for Kruesi.† Five days later, still thinking in terms of the recorder as an adjunct to the telephone, he referred to it as a "Repeater for talking Telegraph."

After delivering the telephones to Orton at the end of the month, Edison gave increasing attention to the sound writer, or "phonograph," as he now called the proposed device. He considered several new means for making the impressions or indentations on the band of paper. The paper might be coated with plumbago, a rough salt, or arsenic acid. The strip could be turned vertically, and the edge incised. Perhaps thread from a spool could be run down the center of a strip and impressed upon it in a wavering, graphlike line synchronized to the vibrations of the membrane.

On September 7, Edison wrote out the press release he intended to issue as soon as he had coaxed the first intelligible word out of one device or another:

* On July 20, 1877, Edison filed a patent application, Number 141, in which he directed himself to utilization of the discovery in telephone transmission.

† Several researchers associated this sketch with the phonograph cylinder that was not machined until three and a half months later, and so concluded erroneously it must have been misdated.

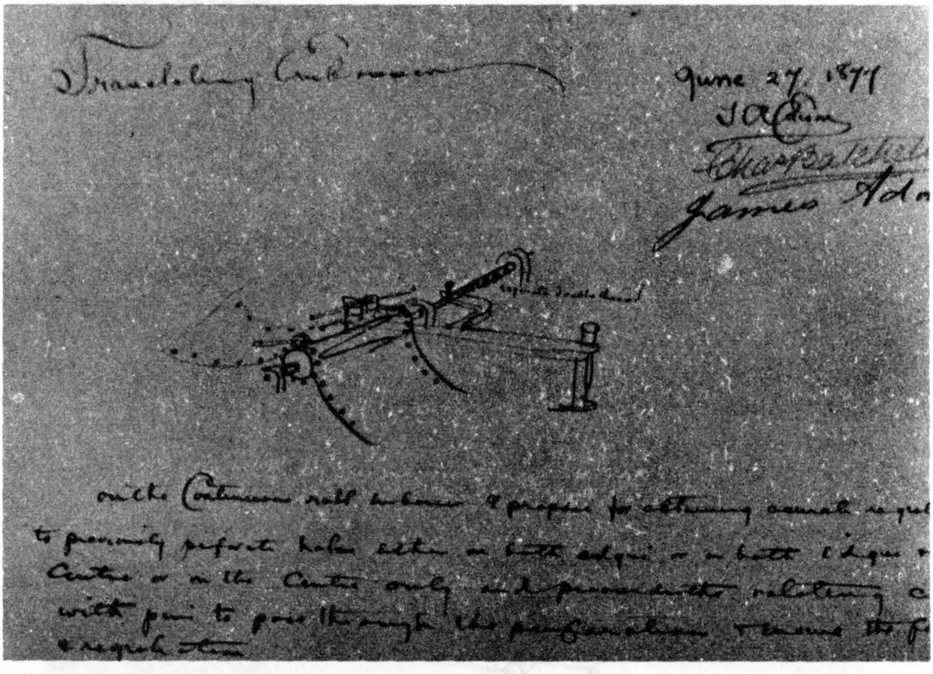

The invention of the phonograph can be traced in the laboratory notebooks during the last half of 1877. Edison was working on an autographic copying press (top) and an autographic, or embossing, telegraph (center), as well as trying to perfect a speaker for the telephone. On July 18 (overleaf) he attached a stylus from an embossing telegraph to a telephone speaker, and shouted into it while running a band of paraffined paper rapidly beneath. By August 12 he had named the device the phonograph. During late summer and early fall (September 7) he was trying various ways to emboss paper strips or employ the roller from the embossing telegraph and copying press (September 21). On December 3, the day before the first phonograph was machined, he was still considering three alternate means of recording: cylinder, disk, and paper band.

Spkg Telegraph

July 18 1877

Chas Batchelor

James Adams

Just tried experiment with a diaphragm having an embossing point & held against paraffin paper moving rapidly the Spkg vibrations are indented nicely & theres no doubt that I shall be able to store up & reproduce automatically at any future time the human voice perfectly

189

Phonograph

September 7. 1877

T A Edison

Chas Batchelor

James Adams

J Kruesi

Record

Plumbago on a spring on diaphragm makes light & heavy marks which have more or less lubrication

Receiver worked by difference friction.

Receiver worked by difference of friction in the ink

Embossing.

Edge embossed

Phonograph

Aug 12 1877

T A Edison

any power to rotate

Roll paper

Phonography Sept 21 1877

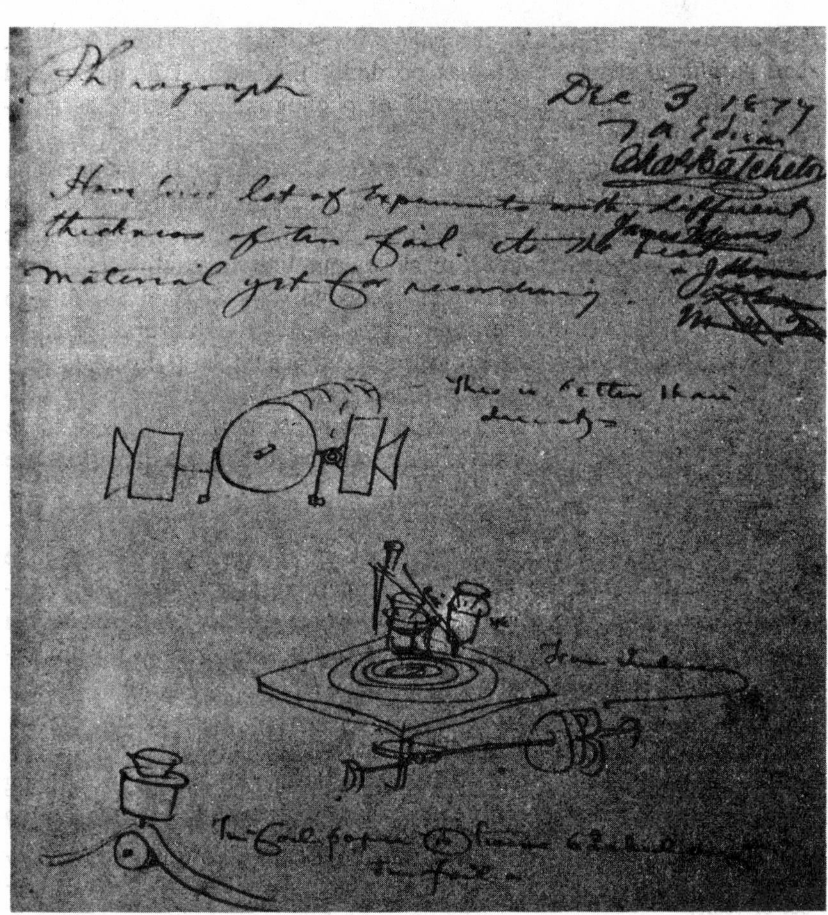
Phonograph Dec 3 1877
T A Edison
Chas Batchelor

Have tried lot of experiments with different thickness of tin foil. It's the best material yet for recording.

Edison Phonograph

An apparatus for recording automatically the human voice and reproducing the same at any future period.

Mr. Edison the Electrician has not only succeeded in producing a perfect articulating telephone, far superior and much more ingenious than the telephone of Bell but has gone into a new and entirely unexplored field of acoustics which is nothing less than an attempt to record automatically the speech of a very rapid speaker upon paper from which he reproduces the same speech immediately or years afterward.

It would seem that so wonderful result as this would require elaborate machinery on the contrary the apparatus although crude as yet is wonderfully simple. I will endeavor to convey the principle by the use of an illustration which although not exaxctly the apparatus used by Mr. Edison will enable the reader to grasp the idea at once.

Edison filled the rest of the page with sketches of ridges, speakers, and paraffined strips—there was no doubt that he had hit upon, and also understood, the basic principle of phonograph recording as it remains to this day.

By the latter part of the month Edison decided that the phonograph might be used for other purposes than recording messages from the telephone. On the twenty-sixth, Stephen Field and Cornelius Herz came to the lab from San Francisco to close contracts for continental European rights to the quadruplex and telephone.* (They paid $18,500 to Edison and Prescott for the quad rights to Austria-Hungary, Belgium, France, and Spain.) In the proposed telephone contract Edison specified: "The portion of the patents which pertain to apparatus for recording & reproducing the Human Voice & their Sound Locally is not to be sold with the telephonic apparatus . . . and when perfected shall be the subject of another contract."

Edison did not tell General Butler about recording, rather than printing, sounds until mid-October, and when the lawyer heard the news he reacted with astonishment: "Tell me something more about your wonderful invention in recording the human voice. I need not say that you had better keep it perfectly secret. It is so remarkable I do not understand it at all."

Edison, of course, had not yet recorded the human voice, or anything else. He only believed that he *could* do it. He had, first of all, still to

* Stephen Field, the nephew of Cyrus Field (layer of the Atlantic cable), knew Edison well. He and Edison had a business and inventive relationship for a number of years.

devise a respectable speaker for the telephone. The plumbago-impregnated silk-fluff speakers were better than anything Bell or Gray had produced, and Murray was making one hundred of the instruments for Western Union. But the fluff kept dropping out and rearranging itself, so that the speaker was full of bugs and needed continual adjustment. Tests of the instruments on the New York lines gave poor results. In the latter part of September Edison abandoned the fluff and tried impregnating the silk cloth with a syrup of plumbago and dextrine. But he found it difficult to concoct any solution that would cause plumbago to stick to cloth.

He and Batchelor were still experimenting with various pastes of plumbago on November 9, when Edison's attention was attracted by the glass of a kerosene lamp. In August, Edison had had a gas-making machine—given him free by a Detroit manufacturer who wanted the publicity—installed in the lab. Gas pipes and jets replaced the kerosene lamps that hitherto had provided light. The lamps were being knocked about and broken. As Edison examined the cracked glass chimney, he noted the unusual blackness of the carbon deposit. Scraping some of the lampblack off, he rubbed it between his fingers, examined it beneath a microscope, and found the particles to be especially fine. Giving the glass to Batchelor, he told him to see what he could do.

Batchelor mixed 1 gram of the lampblack with 250 milligrams of rubber dissolved in carbon bisulfide, and pressed the mixture into a cake. When the cake was reshaped into a button, and incorporated into the speaker, the improvement in articulation was miraculous. It was the breakthrough they had been seeking. On top of it came the discovery that the telephone worked best when placed in a polarized circuit in which the line was worked inductively, so that a low-voltage speaker current could be used and then amplified as it was picked up by the main line—all the horsing around Edison and Adams had done with induction coils was, in the end, to pay dividends. They now had a practical transmitter.* On December 1 the American Speaking Telegraph Company, combining the patents of Edison, Gray, and A. E. Dolbear,† was incorporated by Western Union investors. A contract was signed with the Gold & Stock Company for the manufacture of telephones.

Yet as significant as the month of November was to the development of the telephone, it was even more important to the birth of the phonograph. The *Scientific American* issue of November 3 reported, in an article titled "Graphic Phonetics," the work of two French scientists, Professor Étienne Marey and Dr. Rosapelly, in graphically recording the vibrations made by different sounds. On reading the story Edison

* Essentially the same carbon button is still in use in the telephone transmitter. The scientifically curious reader will find he can easily unscrew the covering of the speaker and drop the loose carbon disk into his hand.

† Dolbear, a Tufts University professor, suggested replacing the electromagnet used in the early Bell instruments with a permanent magnet.

enlisted Johnson as the agent for the press release he had written two months before. Johnson disclosed to the *Scientific American* "the still more marvelous results achieved by Mr. Thomas A. Edison, the renowned electrician. Mr. Edison conceived the highly bold and original idea of recording the human voice upon a strip of paper, from which at any subsequent time it might be automatically redelivered. As yet the apparatus is crude, but . . . he has already applied the principle to a speaking telephone, and will undoubtedly be able to transmit a speech, made upon the floor of the Senate, from Washington to New York, record the same in New York automatically, and by means of speaking telephones redeliver it in the editorial ear of every newspaper in New York."

Edison's audacity bordered on folly. When Johnson wrote the letter Edison had yet to reproduce any identifiable sound, much less speech. He had tried various substances, including wax and chalk, for indenting, but concluded "tin foil over a groove is the easiest of all." For the recording apparatus he was considering either a foot-long grooved cylinder, the revolving horizontal plate designed for the acoustic telegraph, or a band of paper.

When Edison's imagination latched on to a concept, ideas tumbled through his head and he was unable to sleep. He emptied his mind as if it were a barrel of apples:

"I propose to apply the phonograph principle to make Dolls speak sing cry & make various sounds also apply it to all kinds of Toys such as Dogs animals fowls reptiles human figures to cause them to make various sounds to Steam Toy Engines exhausts & whistles = to reproduce from sheets music both orchestral instrumental & vocal the idea being to use a plate machine with perfect registration & stamp the music out in a press from a die or punch previously prepared by cutting in steel or from an Electrotype or cast from the original or tin foil = A family may have one machine & 1000 sheets of the music thus giving endless amusement I also propose to make toy music boxes & toy talking boxes playing several tunes also to clocks and watches for calling out the time of day or waking a person for advertisements rotated continuously by clockwork," and so on and on.

By the last week in November, Edison had settled on tinfoil as the recording material, and was experimenting with various thicknesses of foil for all three modes of recording under consideration. On the night of Monday, December 3, Edison, Batchelor, and Adams drew sketches of the different phonographic apparatuses—the band, the disk, and the cylinder. The next morning Edison handed Kruesi a drawing of a cylinder about half the length but twice the diameter originally conceived, with a speaker mounted on one side and a reproducer on the other, and told him to machine it.

That evening Kruesi brought the model to Edison. Everyone in the lab gathered round as Edison fitted a piece of tinfoil to the cylinder.

Kruesi ventured the opinion that the device would not work. He was backed by Adams, who stirred from the nap he was taking on a table, and by William Carman, who had moved to the lab as Edison's accountant and secretary. Edison bet Kruesi two dollars and Adams and Carman a box of cigars that they were wrong. Turning the handle of the cylinder, he put his lips almost directly against the speaker and shouted into it. Disengaging the speaker, he moved the reproducer against the cylinder and turned the handle again. He heard nothing.

"I guess you've won the cigars," he said glumly.

Kruesi's mouth, however, hung open in astonishment; and, after a moment of shock, the others erupted in elation. Faintly but distinctly, in a metallic, nasal tone, the machine had repeated Edison's words. Through the joyous and boisterous night everyone tried shouting into the phonograph. The next two days and nights, while Batchelor returned to his labors on the telephone, Edison prepared a patent application and a presentation for the *Scientific American.*

Friday, Edison and Batchelor went to New York to test the latest telephones, and to deposit one of the two existing phonographs on the desk of the editor of the *Scientific American.* (The other went to the patent office.) Edison, the editor reported to his readers, "turned a crank, and the machine inquired as to our health, asked how we liked the phonograph, informed us that *it* was very well, and bid us a cordial good night." The dozen persons in the office flocked to the desk. Their amazement could not have been greater had Edison unveiled Aladdin's lamp. Word spread through the building, and people packed themselves into the office until the editor feared the floor would collapse. Over and over they demanded that Edison play the phonograph for them.

The principle involved was beyond the comprehension of almost everyone, as indicated by *Puck,* the American humor magazine. "You do not know what the Phonograph is? Well, Puck will tell you. A quadruplex, double-driving, osculatory cog-wheel, gyrating in a fluted pedestal by the positive and negative current from a cautery voltaic battery strikes the atmospheric tympanic diaphragm. The rheotone depending on the vibratory armature of the secondary coil produces dynamic Faradization. Ahem!"

If the musical telephone had intrigued the imagination of the public, the phonograph captured it. Through the remainder of December, Edison experimented with different versions, and Batchelor designed a "Parlor Phonograph." The difficulties encountered in putting the tinfoil on the cylinder caused Edison to give considerable attention to the disk variation.* At the end of the month Kruesi took a machine in to Orton, and on January 2 it went on exhibit at the Western Union building.

The Reverend John Heyl Vincent, who was the editor of all Methodist

* On December 29, Edison wrote an acquaintance, Frank Foell, that because of the problems "I had to adopt a revolving plate." He then added, with routine overenthusiasm: "The reproduction of the voice is now absolutely perfect."

Sunday-school publications and lived in Plainfield, only a few miles from Menlo Park, came to the lab to see the machine. He spoke into it a long string of abstruse Biblical names that no one but an Old Testament scholar would be familiar with. When the reproducer regurgitated the names, he was satisfied the phonograph was indeed what it was represented to be. In conjunction with Lewis Miller, an Akron, Ohio, inventor and manufacturer, Vincent four years before had founded the summer Chautauqua Institute for the training of Sunday-school teachers. This coming summer the institute was initiating a Literary and Scientific Circle, at which prominent men from all walks of life would speak. Vincent invited Edison; and Edison committed himself to appear the latter part of August.*

Edison's patent application showed a phonograph operated by a clockwork motor. On January 7 he licensed two Brooklyn men to incorporate phonographs in clocks and watches for the purpose of calling out the time, waking people, and advertising messages. He and Batchelor made several trips to the Ansonia Brass and Clock factory in Connecticut, where they worked through the nights trying to adapt phonographs to clocks.

They discovered they were able to get better articulation from copper than from tinfoil, but the weakness of the reproduction was a major problem. Edison devoted much of February to finding a means for amplifying the sound. On March 4 he filed a patent application on an "aerophone," which used "air, gas, steam, or other fluids under pressure" to increase the volumes of sound in recording. In the near future, when the "Goddess of Liberty" was to be erected on Bedloe's Island, he proposed to give her a voice by putting a phonograph in her mouth, and her talking and whistling would be heard all over New York harbor.

Edison, Batchelor, Kruesi, and Adams all had small children, so it was natural that nursery rhymes became a vehicle for the endless testing: "Mary had a little lamb, Its fleece was white as snow, And everywhere that Mary went, The lamb was sure to go." Since, however, the experimenters had more spice than sugar in their personalities, they made up their own variations:

"Mary has a new sheath gown, It is too tight by half. Who cares a damn for Mary's lamb, When they can see her calf!"

Through the winter, Edison and Batchelor worked on a host of devices for all three versions of the phonograph. They designed a plumbago-coated cylinder to replace tinfoil for recording; they experimented with several different methods for duplicating records; they sketched an amplifier for taking down court testimony; they proposed to market a toy phonograph that would help children learn the alphabet;

* Like a number of other men, Vincent thought that the designation "sound writer" was inapt, and suggested Edison rename the phonograph the "tautophone," or "sound repeater."

they devised a two-sided disk, on which both sides could be played simultaneously. Johnson added phonograph exhibits to his telephone concerts, and promoted the marvel throughout the East.

Through the offices of Professor Barker, Edison was invited to demonstrate the phonograph at the National Academy of Sciences meeting in Washington on April 18. The session was sandwiched between readings of papers on "The Effective Force of Molecular Action" and "A Report of Progress on the Subject of Oxygen in the Sun," but the hall was jam-packed for Edison's appearance. The crush became so great the doors had to be taken off their hinges. The thirty-one-year-old Edison, looking, with his shaggy brown-gray hair standing out at all angles, like a mechanic straight from the workshop, sat shy and withdrawn, a rubber band twisting around his fingers.

An awful hush preceded the demonstration. "The Speaking Phonograph has the honor of presenting itself to the Academy of Sciences," a metallic voice uttered from the instrument. Following that introduction, Batchelor shouted, sang, whistled, and crowed like a rooster into the diaphragm. When the machine repeated the sounds, two or three girls in the audience fainted.

"It sounds more like the devil every time," one onlooker remarked, and his words were reinforced by the staring eyes of two seemingly disembodied heads floating over the top of a large photograph in one corner of the room—they belonged to two men who had climbed up to a window.

That evening Edison showed the phonograph to a throng of newspapermen that Painter assembled at the Washington office of the *Philadelphia Inquirer*. Later, Edison joined academy members at the U.S. Observatory, where they looked at the stars through the big telescope. Wilber, who two weeks before had asked Edison to lend him $250, called the White House, where President Hayes liked to answer the telephone himself. (There were only a few score telephones in the capital.) Wilber told Edison the President would like to hear the phonograph, so shortly before midnight the inventor and his party were driven to the mansion. Mrs. Hayes was roused from bed and came downstairs, and Edison stayed at the White House until 3:30 A.M., entertaining and being feted.

The next day he was taken to the Capitol, where, with the same astonishment, congressmen assembled to observe the phonograph. Edison gave a private demonstration to two senators, Beck and Blackburn, in a committee room. Beck recited one of Robert Burns's poems into the machine. When Edison adjusted the reproducer and turned the handle, the phonograph twanged Beck's words back at him. The senator was convinced Edison must be a ventriloquist. He insisted Edison leave the room while he operated the machine without him.

(A few weeks earlier, French scientist Sainte Claire de Ville had assured a gathering of the French Academy that the phonograph was nothing but the trick device of a clever ventriloquist. A Yale professor called a *New York Sun* writer "a common penny a liner in the incipient stages of delirium tremens. The idea of a talking machine is ridiculous.")

Reporters discovered Edison had patented 158 inventions. His models occupied an entire case at the patent office. He was hailed as the nation's greatest inventor. Gardiner Hubbard, who was now residing in Washington, invited Edison to his house. Painter and several other men were present, and an agreement was reached to organize the Edison Speaking Phonograph Company. There was a consensus the invention needed a great deal of work. (In March, the London Stereoscopic and Photographic Company had bought British rights for $7500, but specified that they were doing so in the confidence that Edison would soon improve the phonograph and send one or two disk machines.) Hubbard and his associates were to raise $50,000 capital, of which Edison was to get $10,000 immediately to enable him to perfect the apparatus. He would receive 20 percent royalty on all machines sold. Johnson was to be general manager of the concern.*

The men were all members of the anti–Western Union circle, and were elated at snatching the phonograph rights away from the telegraph company. Painter, Hubbard, and Johnson had been trying for months to get Edison to "bust his contract" with Western Union. Edison himself was steaming—Orton had turned down his offer to sell the telephone transmitter to the company for $102,000 ($6000 a year for seventeen years). The Western Union president was not about to pay for the transmitter what he had refused to pay Bell for the basic invention. Edison's only recourse, according to his contract, was to ask for arbitration.

Though the transmitter was superior to the Bell and Gray devices, and had been tested successfully on Western Union lines between Philadelphia and Menlo Park, it nevertheless was undependable—it worked beautifully on some occasions, but was drowned out by static on others. In England, Preece was having such difficulty with the instruments that Edison sent Adams over with a new set and told him to stay there until he had them working.

Then, toward the end of March, Edison and Batchelor discovered that the bug was in the piece of hard rubber used to pass the vibrations of

* Two of the principal investors were New Yorkers Hilborne Roosevelt and Charles A. Cheever, who held the Bell Telephone franchise for Manhattan. Edison had become acquainted with them while using their lines for testing his transmitter and electromotograph receiver. Roosevelt, a cousin and mirror image of Theodore Roosevelt, had patented an advanced electric organ, and had become the largest organ manufacturer in the United States; in January he had sent Edison one of the instruments to use in his phonograph experiments. Cheever, born to wealthy parents, had been dropped by his nurse when he was a baby. His legs had been crushed and had never developed, so that he had to be carried about by an attendant.

the diaphragm on to the carbon button. The rubber expanded and contracted with the slightest change in temperature; impinging on the carbon button, it generated extraneous variations in pressure. Eliminating the rubber piece, Edison found to his surprise that the diaphragm also was unnecessary. The pressure of the sound waves themselves was enough to compress the carbon particles and alter their resistance to the electric current. The simplified transmitter worked elegantly, and on April 2 Orton in New York was able to converse as easily with Henry Bentley, president of the telegraph company in Philadelphia, as if the two men had been in the same room. During the next few months the carbon button replaced Bell's magneto transmitter on telephones everywhere—including the instruments of the Bell system, which pirated the Edison invention.

On April 22, a day or two after Edison returned to Menlo Park from Washington, Orton suffered a stroke and died. Until a new president was chosen, oversight of Western Union operations was assumed by Hamilton McK. Twombly, a patrician Bostonian who had married one of Commodore Vanderbilt's nine daughters. Where Orton had had to claw to keep his tenuous hold on the rocky ledge of Wall Street, Twombly had a comfortable niche. Orton had treated Edison with avuncular condescension, and Edison had allowed himself to be buffaloed by him. Twombly, to the contrary, was impressed by Edison's achievements and the enormous newspaper stature he had gained in a few months. He was much less concerned whether Western Union paid a few thousand dollars more or less in dividends yearly. He wanted to settle the question of the transmitter, and made his desire clear to Edison. When Edison realized he had Twombly wrapped up, he treated the millionaire like anyone else, failed to show up for appointments, and continually put him off. Finally, on May 31, the contract was signed on Edison's terms, raising the total monthly payment to him from Western Union to over $900.

Although Edison habitually stood people up and failed to keep promises, he did not deliberately intend to slight them. His intentions were usually good. But he had trouble turning anyone down, and kept committing himself to far more than he could accomplish. Human relationships meant little to him, clocks and schedules were nuisances, and he never compromised his egocentricity to accommodate himself to anyone. His overriding desire was to be left alone to do what he wanted. And though he did not object to intruders, and in fact made them welcome, he did not feel they had reason to expect anything from him.

These were characteristics of his personality his acquaintances had difficulty understanding and reconciling themselves to. In the fall of 1877 Professor Barker, who believed his glowing tributes to Edison had earned him a favored position, had asked to borrow some equipment to use at a lecture. Edison had told him he was welcome to it; but it had seemed to Edison a matter of no great importance, and, with his mind on a dozen different experiments, he had forgotten about it. Barker had been bitter: "I am more disappointed than I can tell you at your failure to send me even the least thing which I can use to illustrate your telegraph. I reason that what the past has not permitted you to do the future will not either. I would not have gone back on you this way for the world."

But since Barker, like most other people, was knocking on Edison's door, not Edison on his, the professor swallowed his disappointment and accommodated himself to Edison's eccentricities.

Reiff complained, "When you desire anything done, you expect everyone to run and jump and nothing scarcely can satisfy your impatience, but when any part is due from you and it is unpleasant, you delay."

Reiff had put his finger on Edison's character perfectly. Edison had little tolerance for frustration, and if events and men seemed to conspire to thwart him he could explode into cyclonic cursing and tantrums. He was open-minded about everything until he came to a conclusion, and then his dogmatism bordered on the fanatic. Once he decided on a goal, he pursued it with dogged single-mindedness and stubborn belief in his own vision.

Perseverentia Omnia Vincit was his and Batchelor's motto; or, as Edison later philosophized: "Genius is 1 per cent inspiration and 99 per cent perspiration."

Batchelor shared a house with Kruesi, and was a friend of Adams. He was devoted to his wife and two daughters, and took long sleigh rides—sometimes as far as Newark—with them. In summer he went berry picking. Often he hunted with Adams. The Sunday after taking the phonograph to the *Scientific American* office, Batchelor mundanely shot two rabbits.

For Edison, however, the laboratory was work, play, and companionship combined. His only regular diversion was the theater. On Saturday nights he would take Mary to New York. If she was away from Menlo Park, he might invite one of the laboratory workers. In the city he often met Reiff or another of his friends. They would dine at Delmonico's, and Edison frequently spent the night at the Astor House.

At Menlo Park his outings were confined to the small pond, a stone's throw away, where occasionally he fished for "Montezuma trout," as he euphemistically called the catfish. Sometimes he fed the small black bear, detoured from the streets of New York and chained to an apple tree by the side of the laboratory.*

His principal recreation continued to be reading two, three, and more newspapers a day. If for some reason the papers were delayed, he took

* Edison's small zoo also contained a fox and a raccoon. The bear was shot to death when it escaped and started rampaging about the laboratory.

on the appearance of a boy who has lost his dog, and seemed scarcely able to function.

When, in the spring of 1878, he himself became a prime subject for the newspaper reports, his reading expanded. He was a delightful performer when he did not have to face an audience of more than four or five persons, and his unpretentiousness, Midwestern folksiness, and exaggerated, almost Bunyanesque imagination produced lively and sometimes sensational stories. He was headlined as "The Napoleon of Science" and "The Wizard of Menlo Park." And he loved the attention.

On April 1 the *New York Daily Graphic* bannered: "Edison Invents a Machine that will Feed the Human Race—manufacturing Biscuits, Meat, Vegetables, and Wine out of Air, Water, and Common Earth." It was, of course, an April Fool's story, but other newspapers around the country picked it up and ran it straight. Nothing seemed impossible for a man who could make a machine that talked. Newspapermen flocked to Menlo Park in the hope of being the first to catch the latest marvel to emerge from the magician's workshop. The latter part of May seven Boston reporters came—one from each of the city's newspapers. Edison showed them the new aurophone and telescopophone, which he had devised to concentrate sound for the phonograph. The telescopophone was a trumpet five and a half feet long, with which, Edison said, he had heard a cow chewing grass two thousand feet away. Batchelor and Painter went to a hill two miles distant, and their voices were heard as they shouted, "Now is the winter of our discontent. Flour fourteen dollars a barrel, and I haven't got a cent." Edison planned a trip to the White Mountains, to test the device between peaks ten miles apart, and predicted it would be a boon to the deaf. Combining the aurophone with the phonograph, he "bugged" the lab above his office, and was able to both record and broadcast. "What do you think of the aurophone?" a startled visitor would be addressed etherically, then subjected to a bloodcurdling laugh.

A reporter declared: "Hereafter there can be no actual certainty of privacy in any conversation unless held in a desert. A maiden's sigh can be given the magnitude of an earthquake."

Scores of people visited the lab every week. The mayor and councilmen of Newark came en masse to urge Edison to return to their city. Everyone tugged at him. Hilborne Roosevelt wanted him to go on a vacation trip through New England, and promised free railroad passes. Reiff, who held interests in a number of mining properties, had been after him for a year to see if he could "do anything with the use of electricity in hydraulic gold mining where much of the gold dust is carried off by the current." Edison agreed he would take a trip to the Upper Great Lakes during July to inspect one of the Canadian mines.

As the newspapers carried his fame around the country, thousands of letters poured in. He was inundated with requests from deaf people for his miraculous ear trumpet, which he had talked about as if it were ready to be marketed, and not, what it was in reality, an unwieldy laboratory curiosity. Everyone assumed his inventions were bringing him wagon-loads of money—in requests for assistance the mail deposited the miseries of the world at his doorstep.* Edison had not the heart or the stomach—much less the time—for such correspondence. He hired Stockton L. Griffin, a telegrapher who had been with him in Cincinnati and was now with Western Union in New York, as his full-time secretary. It became the job of Griffin, and of others subsequently occupying the position, to sift the correspondence and pass on to Edison only those letters they thought he wanted to see. Even then, Edison often let the mail pile up for days, and sometimes weeks.

The end of May brought news from England that David Hughes, a noted British electrician and telegraph expert, claimed to have a new invention: the microphone.† Hughes, who was a close acquaintance of Preece, had learned from him about the difficulties with the Edison telephone transmitter, and had started experimenting upon the instrument. Two months after Edison found that the diaphragm was unnecessary and that the carbon button itself would accurately vary the electric current according to the sound waves, Hughes, without knowing of Edison's finding, unveiled the same discovery. Hughes realized more closely than Edison that the device also acted as a sound amplifier and could be employed for other purposes than the telephone—hence the name "microphone." But he did not fully comprehend the importance of the carbon particles, and thought the same effect could be obtained through the use of other substances. He furthermore appropriated the entire concept to himself, giving Edison no credit whatever.

Edison and Batchelor were outraged. In matters of money Edison was an infighter who, from his observations of Wall Street, had concluded that there was no such thing as morals in finance. But he was touched to the quick by any intimation that someone was trying to filch credit he believed belonged to himself. His short fuse gave little opportunity for deliberation, and the explosion followed almost immediately. He accused Hughes of "piracy," and Preece of "abuse of confidence" and "perfidy."

"My instrument is patented all over Europe, so they can't do anything with their alleged invention but deprive me of the credit of inventing it. I can only fight it out in the newspapers," he told a reporter.

* One request came from the wife of A. D. Coburn, who had worked two years for Edison in Newark as an inspector and tool keeper, but had fractured his elbow. The fracture had not healed, and the arm was infected and would have to be amputated. He and his family were penniless in a Brooklyn tenement. Edison sent them five dollars.

† Hughes had been a music professor in Kentucky a quarter century before, when he invented a telegraphic printer for use with the Bain system. Craig bought the machine for a large sum, but it proved unworkable.

Having acquired a keen appreciation of the press, he subscribed to a clipping bureau, and while his mail often went unread, the clippings were devoured every morning. On June 8 he sent copies of a *New York Tribune* article supporting him to Sir William Thomson, Count Du Moncel, Ludwig Helmholtz, Sir Humphrey Davy, Thomas Huxley, John Tyndall, William Siemens, William Crookes, Stewart Balfour—fifty-one of the world's leading scientists in all.

Sir William Thomson, who ranked with Faraday and Davy, had been a friend of Edison ever since he had been introduced to Edison's automatic telegraph at the Philadelphia Exposition in 1876. Thomson declared, "Last September a method of magnifying sound in an electric telephone was described as having been invented by Edison which was identical in principle and in some details with that brought forward by Hughes." Ultimately, Thomson concluded that there was no question of Edison's priority. But Edison's public blast took him and everyone else in the British scientific community aback. Gentlemen, members of the club, simply did not behave this way; they settled their disputes quietly among themselves, not before the world. Edison, by including Preece, the Post Office's electrician, in the charges against Hughes, had assaulted the establishment itself.

"By his violent attack in the public journals on Mr. Preece and Mr. Hughes," Sir William wrote *Nature,* the scientific journal, "he has rendered it for the time impossible for either them or others to give any consideration whatever to his claims. Mr. Edison will see that he has been hurried into an injustice. He will therefore not rest till he retracts his accusations."

Edison, of course, would do no such thing, and the controversy split the scientific community. Coming on top of Edison's failure to honor his commitment to the British purchasers of his automatic telegraph system, the effect was to sour his relations with English scientists and electricians for years.

For months Edison had been under enormous pressure, and the news of Hughes's claim was enough to make him ill. The state of his health was now a matter of public interest, and one newspaper reader's diagnosis seemed as good as any doctor's: "You have got your system in a state of vibration so that you now find it quite impossible to calm and quiet down." Once the counterattack against Hughes was launched, however, Edison felt better, and by mid-June he was once more trying to cope with the ceaseless demands being made on him.

When he had discovered that the cause of the static in the telephone transmitter was due to the expansion and contraction of the rubber piece, the thought had occurred to him that this might be a means to measure changes in temperature—Professor Samuel Langley, astronomer at the Allegheny Observatory, had told him in October that he was in need of a highly sensitive heat-measuring device. At the meeting in

Washington, Edison informed the scientists he had designed a "micro-tasimeter," which would register changes in temperature as small as 1/50,000 degree. With it, he expected to measure the heat emitted by the stars.

When academy members expressed great interest, he set about refining the instrument. A three-inch horn focused heat on a hard rubber rod. As the rod expanded, it pressed down on a carbon button, altering the resistance to an electrical current registering on a galvanometerlike scale.

On July 23 a total solar eclipse was to occur along the Rockies, and a number of astronomers were planning expeditions. Professors Charles F. Young and Rufus Brackett of Princeton visited the lab and pleaded with Edison to provide them with a tasimeter that could be used to measure changes in heat from the sun's corona. Edison finally completed an instrument for them July 1, the day before their party left for Colorado.

Professor Barker was scheduled to depart with Henry Draper of New York University on another expedition two weeks later. Barker was chagrined that the Princeton party had a tasimeter while his did not. Edison was acquainted with Draper, and had a desultory interest in astronomy. Ever since his discovery of etheric force, he had tried to find some relationship between it and the aurora borealis. On the second-floor porch of the lab he had placed a high-powered telescope. (The instrument, however, was focused as much on the buttresses and cables of the East River Bridge, twenty-five miles away, as on the stars.) Barker persuaded Edison to provide a tasimeter for the Draper expedition also. To make sure the commitment was kept, he invited Edison himself to come along.

Edison had been retained to research means for reducing the noise and smoke pollution generated by the Manhattan elevated railroad. He was supposed to go to Lake Superior to look at Reiff's mine. Mary was five months pregnant and not feeling well. In his office stacks of bills so old the paper was crumbling were piled up. But he was tired of problems, and enthused by the prospect of a trip west.

CHAPTER 11

The Napoleon of Invention

On Friday, July 12, Edison sat for a bust commissioned by a phrenological journal. His head, supported by his nephew, Charley, disappeared beneath the plaster cast. Communicating with a Morse sounder, he told Charley that if he let go, "It will break my damn neck," and went on to crack jokes and make quaint suggestions. When the time came to remove the mold, the back half stuck to Edison's hair and refused to come unglued. Edison called to Kruesi to make a thin, pliable steel ribbon. The sharp band was run beneath the cast, and, with a good deal of hair pulling and some serious nicks, Edison was finally separated from the mold. Tufts of his hair lay scattered about as if uprooted by a storm, and his hands and face were streaked with grease and blood.

The next evening, his hair still matted, Edison sat in the special car of the Draper expedition and watched the reporters who had come to see him off disappear in a swirl of smoke on New York's Pennsylvania Railroad Station platform. His honorary title of "Professor" had been augmented by an honorary degree of Doctor of Philosophy from Union College in Schenectady. Puffing on a cigar, he was filled with anticipation. He and the tasimeter were turning the expedition from a humdrum scientific outing into a newspaper attraction. The *New York Herald* even sent a correspondent, Edwin Fox, along. Although a lawyer by occupation, Fox was a regular contributor to the paper, and had known Edison since telegraphy days.

It was more than a week before the dozen members of the expedition, which included two women, straggled hot and begrimed off the train in Rawlins, Wyoming. Rawlins, a place of a score of unpainted clapboard buildings, served as an engine-switching point for the Union Pacific, and the emphasis of pronunciation was on the "raw."

The tenderfoot Easterners and the railhands and cowboys who boistered through the streets at night stared amazedly at each other. Actually, Edison, who thought "manliness" a cardinal virtue, looked more the gunfighter than most of the Westerners—with graying hair, lean, craggy features, slightly ridged nose, and piercing eyes he might have produced a hush had he strode up to the bar. Instead, he went to rubberneck a horse thief and a train robber incarcerated in the jail.

It was a case of the famous ogling the infamous. One evening after he and Fox, who roomed together, had gone to bed, a thunderous knock on the door shook them upright. In strode "Texas Jack," the top of his head brushing the door frame, his eyes bloodshot, and his hands on his gunbelt. Which one, he wanted to know, was Edison? When Edison manfully identified himself in a quavering voice, Texas Jack said it was a pleasure: he himself was the boss pistol shot of the West, and he wanted to meet the great inventor of the phonograph. Whereupon he pulled out his six-shooter and, firing through the window, caused the weather vane across the street to clang into a dizzy spin.

After that, the solar eclipse was an anticlimax. On July 25 the astronomers and Edison set up their instruments in the yard of the railroad superintendent's house. Aiming the tasimeter at the sun, Edison prepared to measure the heat from the corona.

The moon crept over the sun like a shutter blotting out a lens. As the untimely dusk descended, chickens went to roost, dogs barked, and tiny whirlwinds stirred up dust. When all that remained of the sun were fiery gases billowing out from behind the disk of the moon, Edison connected the battery. The needle of the tasimeter jumped, then gyrated back and forth across the scale. The instrument was far too sensitive for its crude design, and once it was turned on could not be controlled. It would register heat—any heat, even the warmth of a person's body a few feet away—but one could not tell *how much*. It generated great scientific interest but little commercial, even though a steamship line wanted to place a tasimeter on the bow of a vessel to determine if it would signal the presence of icebergs. Edison could see no money in it, and didn't even bother to patent the device.

With the scientific portion of the expedition over, Edison and Fox bought Winchester rifles and joined a hunting and fishing party led by Henry Draper. Escorted by a major with thirty troopers, they moved deep into Indian territory. Over the campfires in the high plateau country, where the stars seemed to melt into the earth, conversations about nature and science lasted long into the night. The sensation of the summer were the reports from Paris of Paul Jablochkoff's "candles," a new form of arc light.

The employment of electricity for lighting had fascinated and baffled experts since the very dawn of the electrical age. In 1808 Humphrey Davy had brought two pieces of charcoal connected to a battery near each other and produced a brilliant arc four inches long. In 1822 Professor Silliman and Dr. Hare had repeated the experiment in New Haven, and twelve years later they had generated the brightest light yet seen. By 1844 arc lights were used in a production of the Paris Opéra. In 1858 an arc lamp was installed in the South Foreland Lighthouse in England. During the Franco-Prussian War both sides employed the lights. The carbons were, of course, consumed in the burning of the light, and gave off acrid, noxious fumes. The electricity was obtained from batteries, and was enormously expensive.

By the latter 1860s the gradual development of the electro-magneto, or generator, had reached the point where the machine was usable for lighting. In 1873 a Russian, M. Lodyguine, created a sensation by erecting in the St. Petersburg dockyards 200 lights whose carbons were immersed in an inert gas so as to retard their consumption. Lodyguine was followed by his countryman Jablochkoff, who received the inspiration for his "candle" while visiting the Philadelphia Exposition of 1876. Jablochkoff's improvements were twofold: his light contained a cluster of carbon rods, and when one was consumed another was automatically moved into place; and his lamps were independent of each other, so that if one went off the entire series did not blink out.

The press reports of the Jablochkoff light were ecstatic. The *Telegraphic Journal* noted that the globe of opal glass "gave a peculiar silver luster to the light, which resembled that of the moon," and was in sharp contrast to the dull yellow gaslights. The journal praised Jablochkoff for striking out "in an entirely new direction—an example to those who persistently follow in the footsteps of others instead of taking a distinct line of their own."

In the United States Charles F. Brush was on the verge of duplicating Jablochkoff's success with carbons regulated on a different principle. The lamps were installed in Wanamaker's store in Philadelphia, and there was little doubt that the arc light had reached the stage of commercial practicality.

But arc lamps, because of their enormous power and the gases they generated, were limited in use to open spaces and large halls. Attempts to produce a small light had been going on since 1838, but the problems seemed insurmountable. In 1845 an American, J. W. Starr, had heated a rod of plumbago (carbon) in a vacuum, but he died in the midst of his experiments. A British druggist, Joseph Swan, reading the patent, began twelve years of experiments that culminated in 1860, when he brought a quarter-inch-wide strip of paper to incandescence in a vacuum. But his vacuum was imperfect, the carbon crumbled, and he gave up and turned to photography. Another Englishman, W. E. Staite, tried carbon, then switched to iridium, and by January, 1850, produced a lamp that *Mechanics Magazine* predicted soon would replace gas and oil. It was introduced into railroad tunnels. But since it gave off a blinding light and after a few hours burned itself out, it quickly fell into disuse.

Most other experimenters, beginning with De Moleyns in 1841, tried platinum, which had the highest melting point of the metals then known. Among the most prominent of these electricians was Moses Farmer, who in 1859 had lighted his home with the platinum lamp that Edison later saw at Charles Williams's shop in Boston. Farmer, after a characteristic period of dormancy, had roused himself by demonstrating, in 1875, that electric light could be *subdivided* into small units like gas, and thus perhaps have utility for home illumination. Until then all lights had been placed in series so that the full current passed through each lamp, and if one was turned off all went out. Farmer was the first to arrange lights in multiple arc or parallel. In parallel, each lamp operated independently of every other.

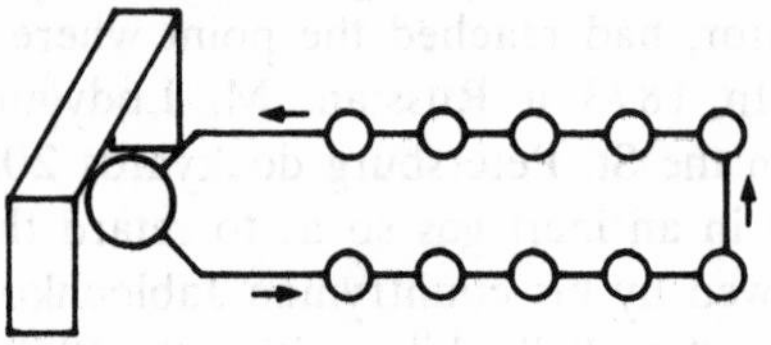

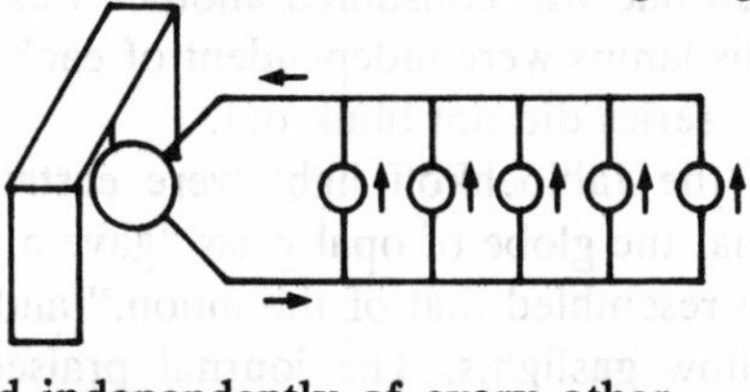

The remarkable party of scientific hunters tracking through the wilds of Wyoming and Utah were all familiar, in greater or lesser degree, with the history of the electric light. Henry Draper, forty-one years old, was the foremost chemist-astronomer in the world—his 1877 discovery of oxygen in the sun ranked him first among pure scientists in the United States. Draper's father, John, had pioneered the study of the chemistry of light. In 1847 he had constructed one of the first workable platinum lamps. He had discovered, like other experimenters, that the platinum overheated and the lamp burned out. Henry, nevertheless, could relate vividly what his father had written:

"An ingenious artist would have very little difficulty in making a self-acting regulator, in which the filament should be maintained."

There had never been a mechanical artist more ingenious than Edison. It seemed to his companions that the tasimeter embodied the essence of the principles needed to construct a regulator that would shut a lamp off and on automatically, so as to keep it from overheating. Could not a rod of rubber (or some other material) that expanded as its temperature rose be employed to press upon a switch and interrupt the flow of the current? When, subsequently, the rod cooled and allowed the switch to close, the flow of electricity would be resumed. What could be simpler?

As a hunter, Edison was miscast—he fell off his horse, and hit nothing more than a tin can. While Draper bagged thirteen tons of buffalo, bear, antelope, and bighorn sheep, Edison spent most of his time fishing.

In Omaha, Edison had obtained a note from the Union Pacific's superintendent of telegraphy: "Please permit him and members of his party to ride on the locomotive or where else they may desire." Edison's desire, when he resumed his journey west, was to ride atop the cow-catcher. Borrowing a pillow, he propped himself up, and from the Continental Divide on traveled in spendid isolation. Ahead of the smoke from the engine, he had an unobstructed view of the vast prairies and magnificent mountains, and could engage uninterruptedly in one of his favorite activities—thinking.

It seemed almost a trick of fortune that he, with his impaired hearing, should have been led ever deeper into the realm of sound. His telegraphic experiments with the repeater, the printers, the duplex, the quadruplex, and the automatic system had placed little demand on his hearing—far less, in fact, than his duties as an operator. But the shift in direction stemming from the new technology of acoustic telegraphy had altered the situation radically. The last two years had produced the musical electromotograph receiver, the telephone, and the phonograph—inventions that Batchelor was physically far better equipped to experiment with than he.

Edison, consequently, had often become tense and frustrated. The idea of developing an incandescent light promised to release him from his auditory travail. A few months before, he had referred an inquiry to Moses Farmer with the notation: "I am fearfully ignorant of the electric light business." But the more he thought about fashioning a lamp, the more enthusiastic he became.

By the time he reached Virginia City, Edison was contemplating all kinds of new uses for electricity. Going down 1800 feet into a silver mine, he proposed employment of the current to probe for veins of metals, thus eliminating the need for drilling and tunneling. From San Francisco he went to Yosemite and the gold country, and verbalized ideas for extracting gold from tailings and from the black sand that could be found along Pacific beaches.

On August 13 he was back in Ogden, Utah, and communicated with Griffin over the telegraph—Edison at one instrument in the Union Pacific office, and Griffin at a sounder on the line that ran directly into the lab. Griffin told Edison that Mary was extremely nervous and upset. The trains rumbling by kept her awake at night. In the daytime she imagined the children were playing on the tracks. She had fainted and seemed to be losing strength. Griffin reminded Edison that he had promised the Reverend John Heyl Vincent he would speak at the Chautauqua Institute's Literary and Scientific Circle on August 20.

But that only gave Edison, who swore he would rather have seven teeth pulled than appear before an audience, added reason to explore the mines of Utah and western Colorado. On August 19 he was in Laramie, Wyoming. In response to Griffin's exhortations to return home because of Mary's health, he said he was on the way, but would be unable to attend Chautauqua—a fact that was by then self-evident.

Instead, he accompanied Professor Barker to the meeting of the American Association for the Advancement of Science in St. Louis. Enthusiastically received, he displayed the tasimeter, put in his claim to invention of the microphone, and unveiled the "sonorous voltameter" Batchelor had worked up during his absence. (The instrument measured electricity by decomposing water into hydrogen and oxygen bubbles, the sound of which was magnified.) He was elated by the news that the display of his inventions in Paris had earned him the grand prize of the exposition.

Edison reached home on August 26, and he found Mary in a condition of nervous collapse. In her loneliness, frustration, and anxiety she was eating as much as a pound of chocolates a day. She was bearing a huge child, and had taken on the look of a double-chinned tub.

Edison telegraphed Dr. Leslie Ward, who, continuing his shuttle to Menlo Park, examined Mary and reassured Edison that she was in no immediate danger.

Hordes of well-wishers and newspapermen came to hail Edison's return. Looking like a cowboy with the Winchester at his feet, he spent hours telling stories like Mark Twain. He complained that on his visit to Yosemite the springless stage had jolted his bones out: "If they had only fastened a good stout plank on the seat of a fellow's trousers and employed an able-bodied mule to kick him uphill and over the canyons, it would have been a big improvement." Plinking away with the rifle at the glass insulators on telegraph poles, he nearly bagged a farmer. He tried to be solicitous to Mary, and went riding in the buggy with her. He played with sprightly five-year-old Marion and quiet, shy two-year-old Tom Junior—nicknamed Dot and Dash. Like most fathers in the nineteenth century, he took no active part in their upbringing. That was considered women's work, and Mary had her sister Alice, and occasionally her mother, as well as three servants, to help her out. Edison liked his children, but it was difficult for him to hear them, and he lacked the temperament to spend time on them.

After a day or two of undirected activity he was chafing to get back on the job. All summer Batchelor had been working more than a hundred hours a week on telephones, phonographs, quadruplexes, resistance coils, reversing sounders, voltmeters, and an ink for the blind. (The arsenic acid ink, intended to produce raised letters, was a product of experiments to make a solution for the autographic telegraph.)

Hamilton McK. Twombly came out to check on Edison's progress with Western Union's telegraph and telephone apparatus. He was surprised when Edison, instead, spread before him a sketch of an "electric draft lamp" combining a platinum spiral with an automatic cutoff regulated by the expansion of heated air. The device, Edison assured him, meant the death knell for gas lighting—all that was required was the money to perfect it.

Hardly had Twombly left when a telegram arrived from Professor Barker announcing he had arranged for Edison to visit the factory of William Wallace in Ansonia, Connecticut, on Sunday, September 8.

Ansonia at that time was the leading copper and brass manufacturing center on the continent. Wallace only four years before had built the first dynamo made in America. He had since constructed what he called a "Telemachon"—a generator coupled to a water-powered turbine from which electricity was conducted a quarter mile to the huge arc lights in the factory. As Barker, Edison, and Wallace stood gazing at the frothing stream, Wallace illustrated the power of the water by saying he had invented a device that would throw a jet of water with such force as to tear the skin off a man's hand. Edison, who had a morbid streak, responded: "If a person cut a man's throat with such a stream of water, I don't believe a jury could be found that would convict him of murder."

The remark constituted a good part of his conversation for the day. Wrapped up in thought, he took little part in the intercourse of the others. Occasionally Batchelor relayed a joke to him. Edison would laugh, respond with a "mosquito whopper" or similar story, then return to his meditation. Wallace had a lavish dinner prepared for his guests, after which Edison diplomatically told his host: "I don't think you're on the right track."

Edison felt sure in his own mind he could design an electric lighting system better than Wallace's. And since Wallace's was the most advanced in America, Edison returned to Menlo Park in a blaze of euphoria. On such occasions, Edison was too wound up to sleep. He headed directly to his retreat, the second floor of the laboratory building. He did not want to know what time it was—during his whole life he had an aversion to watches. He was as surprised when the rising sun shadow-patterned the spiderweb of wires as when sunset tinted the 2000 bottles of chemicals arranged on shelves along the walls. In the middle of the floor was a table holding a large induction coil which would generate a spark more than a foot long; a phonograph with a double mouthpiece for duets; a carbon relay, the progenitor of all existing carbon telephones; and the "electro-harmonic engine," intended to operate sewing machines and consisting of a tuning fork vibrating thirty-five times per second between two small electromagnets. Chemicals fumed and lamps smoked fiercely beneath a wooden hood which only partially snared the odors and funneled them away. With a peculiar grace despite his chronic stoop and shuffling gait, Edison wove in and out of the equipment. His head bobbing, he slid beneath coiled, spiraled, and kinked wires growing out of jars and hanging from invisible supports. Through the night he sat at a table in the extreme rear, where a single gas jet played over his features.

Puffing on cheap cigars—he smoked twenty a day—he clouded the already chemicals-impregnated air. Periodically he stopped for coffee, then took a bite from the huge golden cake of tobacco sent him by the Lorillard company, a cake serrated by the teeth marks of Griffin and other privileged chewers with whom Edison shared it. His jaws chomping, Edison mottled the floor with spittle and labored away.

Five days after returning from Ansonia, Edison wired Wallace: "Hurry up the machine. I have struck a big bonanza." The same day he wrote out his first caveat on electric lighting. This described a spiral of metal heated to incandescence and regulated by a magnet. As heat increased, the magnet broke the circuit, cut off the electricity, and kept the spiral from melting. When the temperature dropped, the circuit closed again.

While waiting for Wallace to send his "light intensity machine" (a small generator), Edison worked out variations on the theme. On September 25 he filed his second caveat, "To Subdivide the Electric Light." "Figure 5," Edison wrote, "shows a stick of carbon acting by its expansion to regulate its temperature—the carbon being placed with the other apparatus in a vacuum tube to prevent oxidation." In the third caveat, a week later, he described an airtight glass tube. As the air inside was heated and expanded it was forced downward, acting on a mechanism in a chamber beneath the lamp to interrupt the circuit. Caveat 4, at the end of the fourth week, suggested a jacket containing a solution of alum and water to control the heat. The next week brought Caveat 5.

So far, the devices existed mostly in Edison's head. But while he was uncertain *what* would work, he had sublime confidence that if one idea did not, another would.

"I came back home and made continuous experiments two nights in succession," he related to the press, "and discovered the necessary secret, so simple that a bootblack might understand it. It suddenly came to me, the same as the secret of the speaking phonograph. I made my first machine. It was a success. The subdivision of the light is all right. I am already positive it will be cheaper than gas, but have not determined how much cheaper." Had the statement come from anyone else, it would have been dismissed as the declaration of an inventor talking through his imagination. But never before had anyone made such an impact within the span of a few months as Edison—the mystique of a wizard had developed about him. He himself was so convinced that mind could conquer matter that he hung up a pendulum to see if he could think it into movement. Many of his acquaintances would not have bet against him. Wall Streeters who perceived the value of their securities threatened by the proposed Edison invention were disposed to hedge their investments. The Vanderbilts were heavy backers of not only Western Union but gaslight companies. On September 25 Twombly took the train to Menlo Park again. He was the family's business manager, and his object was to discuss with Edison the formation of a company to develop and exploit the subdivided incandescent light.

The next day Grosvenor P. Lowrey came out. Lowrey was a bluestocking New Yorker and the attorney for Western Union, but over the course of the past year he had also become Edison's lawyer and advisor. He was capable, shrewd, and a hard bargainer on the one hand; but witty, congenial, and sympathetic on the other. Completely honest, he gave Edison for the first time the capability to meet the financiers on equal footing. His relationship to Edison was almost avuncular, and he became one of the best friends Edison ever had.

Edison authorized Lowrey to negotiate for him. Five days later Lowrey wired that he had "good news." There was general agreement on the terms. A stock company capitalized at $300,000 would be formed. Edison was to sell half his rights for $150,000. He would receive one third on signing the contract, one third for experimental expenses, and one third at the end of the first year if the investors decided to continue. One third of the stock would go to Western Union people, one third to gas-company investors, and one third to other insiders. Lowrey met with Twombly on October 1 and 2 at the Vanderbilt house. He told Twombly: "I understand that all serious difficulties have been overcome. He has discovered the means of giving an electric light suitable for every means at a vastly reduced cost."

Edison, who was beginning to have some second thoughts, was a little shaken that Lowrey should be negotiating with such galloping confidence. He had remarked in an aside that perhaps Lowrey was asking too much, and report of the remark circulated about the gathering at the Vanderbilts. Lowrey was requesting an annual royalty of twenty-five cents per light for Edison, to which William Vanderbilt replied that such a royalty would net Edison a fortune equal to John Jacob Astor's every year. When the discussion ended without definite agreement, Lowrey figuratively shook his finger at Edison and chided: "Don't you tell people any more that I demand too much for you."

Ten days later Lowrey had the first $30,000 in hand for Edison. Such an injection of fiscal adrenaline made it difficult for Edison to remember he had moved to Menlo Park for economic reasons, although William Kirk, the landlord of the Edison-Murray machine shop in Newark, was still pounding on the laboratory door demanding payment of the notes on which Edison had defaulted. The Newark grocer sent his clerk to collect Edison's bill, months overdue, and Edison refused to see him. The New York tailor wondered if Edison would not like to pay for the clothes he had bought for his trip west. A machine-shop operator dunned him for $70 owed since the first of March. The unpaid veterinarian inquired how the cow he had treated in December, 1877, was getting along. Edison, Mary, and some members of the laboratory habitually paid for train tickets, telegrams, and other small items with chits—scraps of torn paper that carried Edison's IOU and became like secondary currency in Menlo Park.

Rather than pay off his old debts, Edison considered the time propitious to contract new ones. The laboratory, though having served well for two and a half years, seemed to Edison to lack style. He needed a place where he could greet his wealthy visitors from New York. He ordered the construction of a two-story building containing an office and an elegant library, replete with plush leather furniture. He felt, furthermore, that the small machine shop on the first floor of the laboratory building was no longer adequate, and so designed a second, brick building to shelter a powerhouse and a greatly expanded shop.

Considering, simultaneously, the equipment he would need for his new venture, he wired Wallace to inquire the price of a duplicate of the machine from which Wallace obtained his big arc. Wallace responded that the normal price was $1000, but he would let Edison have it for $750.

"Send it," Edison ordered.

On October 5 Edison promised Lowrey he would have five lights ready to show in two days. But he was struggling to keep just one platinum burner going for a few minutes. Every three or four days another inquiry came from New York, and Edison turned it aside. The doors of the laboratory were barred to all visitors—except any reporter who happened to show up.

"Of all the things we have discovered, this is about the simplest," Edison told a *New York Herald* man. "We have got it pretty well advanced now, but there are some improvements I have in my mind."

To a writer from the *New York Sun* Edison exclaimed: "I have it now, and singularly enough I have obtained it through an entirely different process than that from which scientific men have ever sought to secure it. They have all been working in the same groove, and when it is known how I have accomplished my object, everybody will wonder why they have never thought of it, it is so simple. When ten lights have been produced by a single electric machine, it has been thought to be a great triumph of scientific skill, with the process I have just discovered I can produce a thousand—ten thousand—from one machine. Indeed, the number may be said to be infinite. Illumination by carburetted hydrogen gas will be discarded. With 15 or 20 of these dynamo-electric machines recently perfected by Mr. Wallace I can light the entire lower part of New York city, using a 500 horsepower engine."

The British *Telegraphic Journal* noted: "The Napoleon of invention delights in startling surprises, and his fertile and daring imagination runs at once through the whole gamut of possibilities as soon as the key note of a new idea is struck. Future developments are to him as if they already existed; but for all that the great mechanical genius of Mr. Edison is so well attested, that any report of this kind is to be taken seriously."

But while Edison was putting up an optimistic front, he felt things closing in on him. He repeatedly broke appointments with Lowrey and Twombly in New York by working all night and then going to bed. Nervous and frustrated, he intensified his habit of plucking at his left eyebrow. Oblivious of the weather, he sloshed through the rain and mud while Dot stretched futilely and tried to hold an umbrella over his head. By Friday morning, October 19, when he was supposed to attend a board meeting, the nerves on his face were raw, and he took to bed with an agonizing case of neuralgia and a bad cold.

In another room Mary was in the final stage of an excruciating pregnancy. Grossly overweight, she was carrying a twelve-pound baby. On October 24 she thought she was going into labor, and Edison telegraphed Dr. Ward to come from Newark. The pains were false. But Ward did not consider her condition favorable. The next day Edison sent another telegram: "Don't go away. May want you today." This was followed at 5:30 P.M. by "Please come at once."

The house was freezing, because the heating system had gone on the blink, and Edison huddled miserably in bed and refused to see anyone. The members of the New York syndicate, concerned about their investment, could get no specific information about what ailed him. The newspapers picked up the reports. Dr. Ward was joined by another physician, and the newsmen, noting the coming and going of the doctors, began hour-by-hour reports on Edison's illness. While Mary thrashed and screamed in pain, and the doctors were not sure she would pull through, the papers carried stories of Edison's supposed critical illness. When Mary was safely delivered of a boy, named Will, on October 26, the press overlooked the event. But the *New York Daily Graphic* warned:

"If he [Edison] turns night into day and refuses to go to bed till sunrise for the frivolous reason that he 'has got an idea,' he will have to pay for his imprudence in neuralgia, brain disease, fever and early death. Edison's engine runs at too high a pressure, and apparently without any safety valve."

While Edison was incommunicado in bed, William Sawyer, a capable though unstable and alcoholic electrician formerly with Western Union, and Albon Man, a prominent New York lawyer, paid a visit to Lowrey. Sawyer told Lowrey he had made a platinum light with a magnetic regulator more than a year ago. He now had developed a far better lamp, however. It consisted of a small piece of carbon immersed in nitrogen, which kept the carbon from being consumed. The lamp was ready for development, and Edison was far behind.

Lowrey, who had been trying for a week to see Edison, thought that the news called for a conference. Edison replied he was still too sick to see anyone, but it was obvious there was a swindle afoot. He consented to send Griffin in to talk to Lowrey. Lowrey told Griffin he and other board members believed immediate steps should be taken to buy the Sawyer-Man patents, or to bring Sawyer into the organization.

Griffin dared not relay to Edison all Lowrey had said. But Edison heard enough to be jolted from his indecision and inactivity. Cursing and spraying tobacco juice, he exclaimed it was the old story—the lack of confidence! Of course everyone who had been experimenting with the electric light was going to set up their claims as soon as they realized his system was perfect! But he had no fear—the line he was developing was entirely original and out of the rut. Any kind of combination or consolidation was unthinkable!

When Lowrey was apprised of Edison's reaction, he hastened to assure the inventor: "My confidence in you as an infallible, certain man of science is absolutely complete."

Nevertheless, an investigation of existing patents seemed called for. Though Edison did not equivocate about his skepticism of scientific history—"When I start in to experiment with anything, I do not read the books; I don't want to know what has been done"—somebody had to find out what had already been accomplished.

The search for an expert researcher turned up twenty-six-year-old Francis Upton, a member of an old and well-to-do New England family. He was a graduate of Bowdoin, had obtained an M.A. in science from Princeton, and then had spent a year with Ludwig Helmholtz, the world's foremost physicist, in Germany. His parents were friends of the Hubbards' and the Butlers', and Upton had just been given a desk in Benjamin Butler's New York office, where he hoped to learn about and make a fortune in the telephone business. Upton was assigned the task of digging out all the patents pertaining to lighting and, in general, summarizing the history of incandescent development.

The directors now demanded to see what Edison had; and with further procrastination impossible, Edison psyched himself up for the visit on November 12. He showed them one new lamp consisting of two rods, one of carbon and the other of platinum—it was similar to Jablochkoff's arc light in that the carbon was replaced as consumed by another rod fed from a magazine. A second new design carried a spiral within a spiral of platinum, titanium, or other nonfusible metal. Edison joked, he told stories, he put on one of the best performances of his life. The New Yorkers laughed so hard they had tears running down their cheeks. When they left they weren't quite sure what they had seen, but they agreed Edison was a capital fellow who deserved their support. Three days later, with the early enthusiasm tempered somewhat, the Edison Electric Light Company was incorporated and the contract with Edison signed. Of the $300,000 in stock, Edison would hold $250,000, and of the remaining $50,000 no more than the $30,000 already subscribed would be paid in immediately. Edison would receive a royalty of 5 cents per light annually, but this could be commuted by the company to a yearly sum of $30,000. The twelve-member board contained, in addition to Edison and Lowrey, some of Edison's old associates from the Gold & Stock Company, representatives from Western Union, surrogates for the Vanderbilts (who, because of their gas-company interests, did not want to be associated openly with the enterprise), and two members of the banking firm of J. P. Morgan: Eggisto P. Fabbri, an Italian-American financier, and J. Hood Wright, related by marriage to Josiah Reiff.

Morgan himself was interested in acquiring British and continental European rights. Since Edison continued to claim to be too busy to go into the city, Morgan caught the train out from New York, and his partner, Anthony Drexel, came from Philadelphia. Early in December, in a clapboard building rising from a muddy field, the elegant, reserved Drexel and the balding, moustachioed, bulbous-nosed Morgan negotiated an agreement for the lighting of Europe.

In the meantime, Baron S. M. Rothschild of Vienna wrote his New York representative, August Belmont, "It would greatly interest me to learn whether really there is something serious and practical in the new idea of Mr. Edison, whose last inventions, the microphone, phonograph, etc., however interesting, have finally proved to be only trifles." Lowrey brought Belmont to Menlo Park the first Saturday in December—a whistle-stop on the Pennsylvania Railroad had turned into a center of international finance.

Feeling better, Edison became expansive again. The light was "all completed now so far as the principle is concerned," he said. He proposed to illuminate Menlo Park with 2000 lights placed on telephone poles—the cost, he estimated, would be *only* about $100,000 to $125,000. Popping onto the newspaper pages like a vaudeville announcer going onstage to usher in each new act, Edison set off a scientific brouhaha of unprecedented proportions. The half-dozen or so experimenters in the world working on the problem of electric lighting were frenzied wondering whether Edison had discovered something they had overlooked all along. Every new pronouncement of his spurred them on.

Scientists, meanwhile, considered Edison's statements absurd. In England, William Siemens, who had worked on electric lighting for a decade, declared: "Such startling announcements as these should be depreciated as being unworthy of science and mischievous to its true progress." He doubted that the electric light would ever be successfully subdivided, at least in the nineteenth century.

Edison's contention that "The same wire that brings the light will also bring power and heat—with the power you can run an elevator, a sewing machine, or any other mechanical contrivance, and by means of the heat you may cook your food"—was called "sheer nonsense" by Professor Silvanus Thompson in London. Edison, he said, was exhibiting "the most airy ignorance of the fundamental principles both of electricity and dynamics."

"The scientific men," Edison countered, "base their assertions on a well-known rule of electric lighting which is that intensity of the light increases to the square of the current. On this they figure an enormous loss in the subdivision, but fortunately"—he smiled enigmatically—"there is another law which is not known to these scientific gentlemen which law if certain conditions are brought about compensates for the loss. These conditions are exceedingly difficult to obtain."

Edison, in reality, was bluffing more than ever before in his life. The enigmatic "law" which he intended to apply consisted merely of spiraling the wire in the "burner" so that it would present less radiating surface and instead concentrate the heat. It was true that the science and technology of electricity were still in preadolescence—the electron was not to be discovered until the late 1890s. Even so, Edison had only the most general concept of what he was talking about, and seemed to be flinging himself into the face of some of the basic principles of electricity.

Three elements are present in the generation of electricity: 1) voltage, the pressure produced by mechanical or chemical means that causes the current to flow; 2) resistance, the friction or opposition that the current must overcome; and 3) amperage, the rate of flow (or current strength) established in consequence of the pressure's overcoming the resistance. The energy output is the product of the pressure and the current strength. One volt multiplied by 1 amp equals 1 watt.*

A person may visualize, analogously, the behavior of water in a pipe. A pump initiates the flow; the diameter of the pipe establishes the resistance; and the volume of flow is a factor of the pressure divided by the resistance. If there are two pipes, one with double the capacity of the other, both will deliver the same amount of water if a pump puts twice the pressure on the liquid in the smaller as in the larger. Applying the principle to electricity:

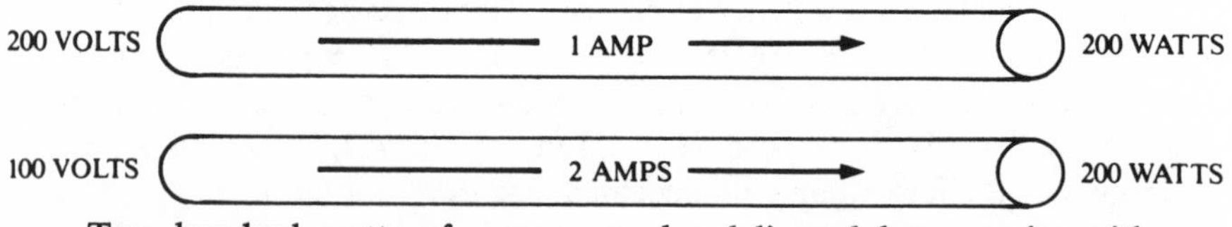

200 VOLTS —— 1 AMP ——→ 200 WATTS

100 VOLTS —— 2 AMPS ——→ 200 WATTS

Two hundred watts of power can be delivered by pumping either 1 amp under pressure of 200 volts or 2 amps under pressure of 100 volts. But in order to economize on wire, which is often the most costly

* Of these terms, none was in general use in 1878. Amperes were known as "webers," and at Menlo Park voltage was commonly expressed in terms of "cells of a bluestone battery." The watt was not devised as a unit of measure for another ten years. (One horsepower is equal to 746 watts.)

element in a system, it is clearly better to couple high voltage with low amperage.

Arc-light systems, arranged in series so that the current passed through all the lamps in turn, employed this principle. A typical ten-light installation used a current of 10 amps under 550 volts pressure. Each lamp had a resistance of 5.5 ohms and operated on 55 volts. The total pressure and resistance in the circuit, 550 volts and 55 ohms, was the sum of the individual voltages and resistances of the lamps.

A German physicist, Georg Ohm, had composed the formula for electrical calculations in 1826:

$$\text{voltage} = \text{current (amps)} \times \text{resistance (ohms)}$$

In the instance of the arc lights:

$$550 \text{ volts} = 10 \text{ amps} \times 55 \text{ ohms}$$

A diagram of the circuit would look as follows:

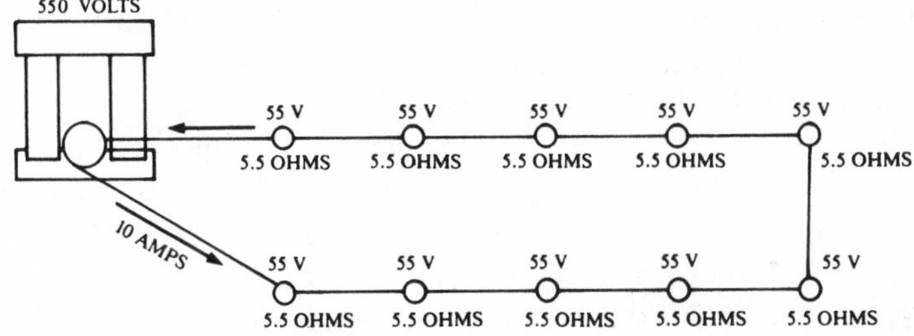

Electricity is transformed into heat, or power, according to Joule's Law, by multiplying the square of the current (amps) by the resistance (ohms) and the length of time the current is applied (in seconds).

Using one of the lamps in the arc-light installation, for example:

$$10^2 \text{ amps} \times 5.5 \text{ ohms} = 550 \text{ joules per second}$$

One joule of heat continued for 1 second is a watt, so the lamp would have a rating of 550 watts.

In contrast to a series installation, where voltage is the principal factor, in parallel circuits amperage plays the primary role. This is true because the current divides as it is fed to each individual lamp, so that while the pressure remains the same, the total current in the mains is equal to the sum of the individual currents. Furthermore, since each lamp offers an additional path for the current, the resistance of the total circuit *decreases* with each lamp that is added: the greater the number of lights, the less the resistance. An installation of ten 550-watt lamps, each drawing 10 amps at a resistance of 5.5 ohms under pressure of 55 volts, would be diagrammed as follows:

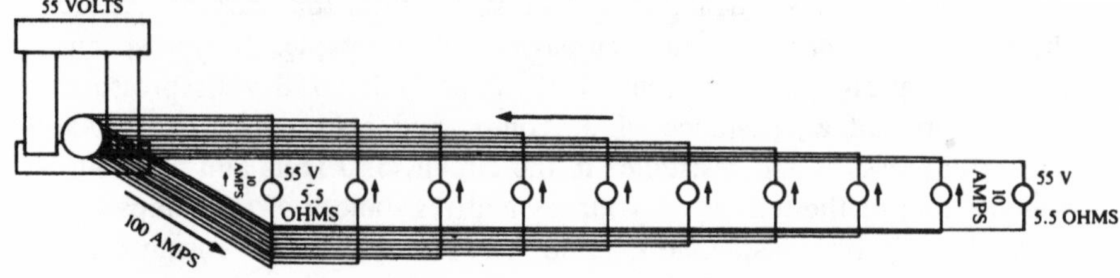

According to Ohm's Law:

$$55 \text{ volts} = 100 \text{ amps} \times .55 \text{ ohms}$$

Because each lamp reduces the resistance, the total resistance of the circuit would be only 0.55 ohms; hence the current generated would be 100 amps. But to transmit 100 amps would require a main of sizable proportion, and in a commercial installation, Edison would be dealing not with ten lamps but with thousands. Mains of such heroic diameter would be needed that there would not be enough copper in the world for the construction of even a modest system of parallel lighting.

Edison was undeterred. The mathematics of electricity was beyond him. "I do not depend on figures at all," he declared. "I try an experiment and reason out the result, somehow, by methods which I could not explain." Deduction could lead to the same conclusions as mathematics, albeit not with the same precision, and he decided that by increasing the resistance of the lamps, he could decrease the current needed to obtain the same amount of energy. Instead of using lamps with low resistance, he would develop a lamp with *high* resistance. If he could achieve, for example, a resistance of 1000 ohms, he would need a current of only 0.1 amp to obtain an output of 100 watts from each light ($0.1^2 \times 1000 = 100$). The diagram of an installation with one hundred lights would be:

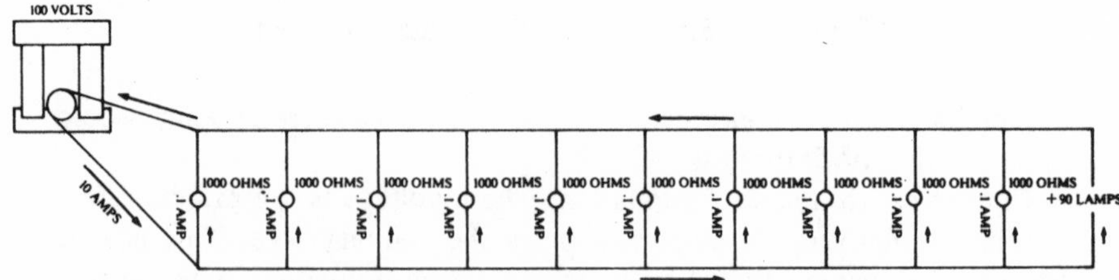

The current in the mains would be 10 amps, the same as in an arc-light system.

No one, however, had achieved an electric light except by means of a low-resistance "burner." (The resistivity of a conductor depends on its length, its cross section, and the substance it is composed of.) The longer and thinner a wire, the greater its resistance. Copper has one of the lower resistances, 1.7 millionths of an ohm per centimeter, and is a good conductor. The resistance of platinum is 10.6 millionths of an ohm per centimeter; that of carbon, 3500 millionths of an ohm. Carbon was the substance used in arc lights; but it was well established that it would burn up and disintegrate, and Edison early in his thinking dismissed it as a possible material.

None of the essential elements—the theory, the materials, the technology—for the development of a system of incandescent lighting existed. The generation of the vast amounts of electricity Edison would need from the small and primitive dynamos being manufactured was comparable to expecting the Great Lakes to furnish water for all of the Midwest by employment of a one-horsepower motor. To anyone conversant with the mathematics of physics and electricity, Edison's project was a pipe dream. In England, a Parliamentary committee of inquiry summoned the leading physicists and scientists; and they, after intensive investigation, concluded that the commercial subdivision of incandescent light was an impossibility.

Edison, however, was not hampered by excess knowledge of mathematics or physics. He was an explorer, who damned the theories and sailed on into uncharted seas; he was an inventor, who rushed in where scientists were reluctant to tread. He determined his goal, and only then applied himself to devising the technology and machinery to move from point A to point B—he did not first ask whether the technology and machinery existed or could be developed to make the goal feasible. Even if he never reached his goal, surely he would discover something else of equal or perhaps greater value; such had always been his experience. The experiments on underwater telegraph transmission had led, eventually, to the carbon-button telephone transmitter; the trials with the automatic telegraph had resulted in the electromotograph; the attempts to devise an autographic telegraph and a telephone transmitter had produced the phonograph.

Edison was convinced that, if gas could be subdivided to produce house lighting, electricity could be also—to think otherwise was illogical. He was of the firm opinion that whether or not something would work could not be determined by theory and mathematics—to answer the question, one built a model, then tested it.

"At the time I experimented on the incandescent lamp I did not understand Ohm's law," he revealed years later. "Moreover, I do not want to understand Ohm's law. It would prevent me from experimenting."

Batchelor was cast from the same mold as Edison. His math was little better, and he was, in actuality, spending most of his time working on the telephone.

Thus matters stood on Friday, the thirteenth of December, when

Francis Upton dragged his valise off the train, climbed the stairs up from the railroad tracks, and trudged along the wooden walk toward the lab atop the hill. A tornado and torrential rain had swept across Menlo Park two days before, and Upton with a downcast heart considered it one of the dreariest places he had ever encountered. Reaching the building at the end of the boardwalk, he inquired for "Mr. Edison," and was directed to a nondescript individual who looked like one of the laborers putting the finishing touches on the new buildings. A battered, broad-brimmed Kossuth hat atop his head and a soiled silk handkerchief around his neck, Edison had on a shapeless jacket, flannel shirt, and coarse pants blotched by acid, whitened by alkali, and spotted with the grease of countless midnight suppers.

It was not the image of the famous man Upton had had in mind. As for Edison, he spent only a few minutes with the reserved Upton, saw his Boston suit and heard his Boston accent, and then nicknamed him "Culture."

Yet there was here one of those unique conjunctions of talent from which great enterprises stem. Edison was only five years older than Upton; but Upton, who still depended on family advice in reaching decisions, was awed by him. The world's greatest men came to him as suppliants. He had already accomplished the achievements of a lifetime. Upton was well versed in science; but Edison had the ability to make people believe that science was not an absolute, but an infinitely expanding universe, in which to seek was to find.

Though Upton was to become closer to Edison than anyone but Batchelor, he never overcame his diffidence. The relationship remained always one of master and savant.

Upton, nevertheless, filled a missing link vital to Edison's success. It was all very well for Edison to disparage mathematics and depend on hunches and empirical methods. But never to know beforehand whether a thing would work, or even the probability of its working, was fiercely wasteful. Upton, trained in higher mathematics and abstract reasoning, was able to devise the formulas that indicated direction and probability of success. While Batchelor constructed a miniature "magneto-electric machine" (dynamo) to test the efficiency of various types of windings after the Wallace generator failed to come up to expectations, Upton accomplished the same task in a tenth of the time by means of pencil and paper.

Thus it was that within a few days Upton not only became a key member of the experimental team but assumed the role of Edison's information system and calculator. He was fascinated with the economic aspects of the scheme; from the very first day he was at Menlo Park, columns of figures danced across his notebook. Fluent in French and German, he had during the past month familiarized himself with everything done in the field of lighting. He brought the happy news that neither in America nor in Europe had patent applications anticipating Edison's work been filed.*

Edison had, in fact, been floundering in the same general darkness of ignorance that surrounded the whole field of electric lighting. Though his hypothesis that the lights would have to have high resistance to be linked in parallel was a critical insight, technically he had made virtually no progress. He was unable even to gauge his progress, for he lacked the instruments with which to do so.

It fell to Upton to add order and direction to the experimentation. He fashioned a Bunsen photometer to determine how much candle power a light emitted, and a Weber dynamometer to measure the strength of the current. He inaugurated the practice of systematic testing.

Whenever Edison did not understand a process through which Upton was moving or a theory he was employing, Edison put Culture through an inquisition. Upton was forced to probe deeper; and Edison, paying close attention, gained an education. In such fashion the two men, whose backgrounds and personalities were light waves apart on the American spectrum, challenged and drove each other on.

Nine days after Upton arrived, the executive committee of the Edison Electric Light Company once more ventured to Menlo Park. They discovered Edison in the midst of transferring machinery and furniture to his new brick buildings.

"The general dilapidation, ruin, and havoc of moving caused the electric light to look very small, and it looked rather as if you were getting ready to have an auction," Lowrey said, trying to be jocular about the investors' anxiety.

In reality, the directors were fearful that the course of the Light Company would follow that of the Phonograph Company. Although Edison wrote Painter he would soon tackle a new phonograph—"Batchelor has the drawings ready and its going to be a big success for dictating"—Edison was totally engaged on the electric light, and Batchelor had little time for anything but the telephone. The men who had invested in the phonograph on the assurance that Edison would produce a commercial model had all but given up hope. Gardiner Hubbard and Hilborne Roosevelt asked to be released from their commitment to raise $50,000. Edison responded favorably—providing he didn't have to return the $10,000 advanced for experimenting; a sum he had already spent, albeit not on the phonograph. Sales of the machines had reached a peak of fifty-one during May, right after the Washington demonstration. By 1880, however, they were down to two or three a month. The instru-

* Siegfried Marcus, a German Jew residing in Vienna, had, however, just filed an application for a patent on incandescent lighting using a parallel system of distribution. He ranks as a strong contender for the world's unluckiest inventor. Two years before, he had constructed the first automobile, but had had the contraption banned from the streets and had been unable to interest anyone in its development.

ments cost $95 to $164 each, depending on the model, and were available only on special order. In its arrested state of development, the phonograph had no general interest or utility.

Only with the telephone were Edison and Batchelor making favorable progress. The array of kerosene lanterns smoking night and day in the carbonizing shed were scarcely able to keep up with the demand for carbon buttons. Since, however, Edison and Western Union were infringing on the Bell magneto receiver (even as the Bell companies were pirating the Edison transmitter), Edward H. Johnson suggested to Edison that he try to convert the musical electromotograph receiver into a true speaking telephone instrument.

In September Edison had turned the task over to his eighteen-year-old nephew, Charley, and Charley very nearly had the problem whipped. Charley, Pitt's only son, was like Edison the offspring of mismatched parents—an earnest, religious, protective mother and a financially irresponsible, cussing, drinking, irreligious father. Edison had strong ties to his immediate family, and, remembering his own boyhood, he felt a sense of responsibility for Charley. Beginning in 1872, he had Charley visit and help out with experiments. Batchelor prepared lessons for the boy. Edison sent him educational books, such as *The Playbook of Science,* in Port Huron. Charley was an apt pupil, intelligent and dexterous. He matured rapidly. A few months after Edison moved to Menlo Park, Charley arrived as a more or less permanent assistant. He learned photography, and became Edison's official picture taker. Edison soon trusted him more than Mary. When he went away on trips, he gave Charley responsibility for taking care of the mail, because Mary had a habit of pawing through and reading his letters.

Mary, naturally, was furious at both Charley and her husband. But while Edison had no taste for confrontations, Charley was ever ready to take on a bear. As impetuous as he was quick-witted, he had "a will which would have ruled an army [and] an ungovernable temper which when aroused requires about two weeks to become calmed down." His cussing was a match for Edison's, Adams's, or any other man's. He had mechanical skills and an inventive knack that in precocity exceeded Edison's. During a stay in Port Huron in the winter of 1878, he won the bid for installing a fire-alarm telegraph in the town, and in three weeks had the system working. He told his mother he intended soon to prove he was no less a genius than his famous uncle.

To bear out his assertion, Charley worked assiduously on the electromotograph. The cylinder was transformed into a thumb-sized chalk (precipitated carbonate of strontium) drum. Pressed against it by a spring was a platinum or palladium-tipped stylus connected to a diaphragm. While Bell's receiver still produced only a mousy squeak, the electromotograph could be heard a hundred or more feet away.

Laboring uninterruptedly with Edison and Batchelor over Christmas and New Year's, Charley celebrated in his notebook: "Happy New Year! . . . Midnight! . . . middle of the night. . . . Now is the winter of our discontent."

Some of the discontent was attenuated by Edison's receipt of $10,000 for French telephone rights and his payment of back wages. So far as the electric light, however, was concerned, Edison was so dissatisfied that he considered dispensing with electricity entirely in incandescent lighting. Instead, he devised a lamp that used steam passed through copper pipes to heat a polished copper reflector plated with gold or platinum. The heat rays generated from the disk were, on the principle of the Archimedean mirror, focused on a small point of iridium, zircon, magnesium, or lime so as to bring it to "vivid incandescence." Ordering $3000 worth of copper, Edison hammered out the lamp—then, after trying it, hammered it in again.

Abandoning the eccentric light, Edison returned to developing a platinum lamp. But when Upton calculated the cost, with platinum selling for twelve dollars an ounce, Edison was staggered. Far from being more economical than gas, the system would cost at the most optimistic four times as much.

Yet that was not the worst of it. In technical books platinum was described as an infusible metal. Yet Edison now discovered that platinum did indeed fuse, or melt, at slightly more than 3000° F.—a temperature readily reached in incandescent lamps. It seemed as if the whole system would have to be scrapped. Edison cursed, coughed, and, with the January weather chronically uncongenial to his bronchial system, took to bed. He was a lifelong believer that "I can sleep anything away," be it physical illness or psychological malaise.

When in a few days he threw off the covers, he was full of enthusiasm. He ordered Kruesi to have the shop force make 480 spirals of iron. These he hooked up in one circuit, and managed to heat to a faint glow. "We have had 480 lamps in one circuit," he wrote one of his business associates in Paris. "All that remains now is a practical lamp. We have almost reached that point and now have a lamp that would last several months, but there is a defect which requires time to remedy." Speaking to a *New York Herald* reporter he expressed no doubts: "The electric light is an accomplished fact—just as soon as I decide upon the form of generators and lamps. We will have it here within a year."

But Edison had proclaimed "success!" too often. "As day after day, week after week, and month after month passes and Mr. Edison does not illuminate Menlo Park as he has so often promised to do, doubts as to the practicality and value of his widely advertised and much-lauded invention begin to be entertained in the public mind," the *New York Graphic* warned. "When the phonograph was invented and the telephone was paraded before the admiring public, promises of magical results were lavishly made. How signally they have failed of perfection everyone now knows."

In England, where the essence of the first of Edison's patent applications was becoming known, there was snickering. Edison had included in his application an adaptation of the tuning-fork motor he had made for sewing machines, a device the *English Mechanic* called "The very worst magneto-electric machine ever made. All anxiety concerning the Edison light may be put to one side. It is certainly not going to take the place of gas."

The American humor magazine, *Puck,* suggested: "Edison is not a humbug. He is a type of man common enough in this country—a smart, persevering, sanguine, ignorant, show-off American. He can do a great deal and he thinks he can do everything."

Frank Pope sniped bitterly: "I know of no one here who has any confidence in the practical success of Edison's scheme. The way that the world stands agape waiting for the Edisonian mountain to bring forth its mouse is really absurd."

The skepticism was contagious. At a gathering of J. P. Morgan and the other bankers in New York, Lowrey jokingly was asked if he knew of anyone gullible enough on whom the Light-Company stock could be unloaded. Relaying the comment to Edison, Lowrey thought the time had come to speak frankly about the problems: "You may feel that it would be prejudicial to let us see how great the difficulties are, [but I] would like to have a talk with you right down to the bottom of everything, and would like to have Mr. Morgan join in it."

But even to think of the possibility of failure was to Edison the equivalent of admitting defeat. And to admit defeat would be like thrusting a knife into his own heart. Edison's response to Lowrey was to return with ever more intensity to the experiments. On the second floor of the laboratory behind the organ was a glass case filled with a supply of all the world's rare metals—aluminum, boron, chromium, gold, iridium, platinum, ruthenium, silver, titanium, tungsten, et al.* Edison made spirals of every promising substance—platinum-iridium alloy, tantalum, nickel—and hooked them to the battery. Staring at a platinum spiral in the bright light of the new arc furnace he had bought for working with exotic metals, he was suspended in the brilliance for hour after hour—finally the light burst like a meteor in his head.

He could see nothing, but his eyes were like two balls of fire and his head felt torn apart. From ten that night till four in the morning he endured "the pains of hell." Eventually, with the aid of a large dose of morphine, he fell asleep.

* Some of these elements, of course, are abundant on the earth's surface, but were "rare" in the nineteenth century because means of extracting them cheaply had not yet been developed.

CHAPTER 12

A New Light to the World

It was a winter when everyone in Edison's family was sick. Dr. Ward shuttled between Newark and Menlo Park. In January, Tommy came down with such pulmonary problems that Ward warned Edison he had better get the boy out of the New Jersey climate. Edison sent him to St. Augustine with Alice. Mary had never recovered from the birth of Will. Both physically and mentally she was a wreck. Miserably lonely, she traveled back and forth between Menlo Park and Brooklyn, where her family now lived, and between Menlo Park and Newark; then, exhausted, she fell sick again. In mid-February Edison put her and the other two children on the train to join Tommy and Alice.

Unburdened of domestic difficulties, Edison seldom bothered to go home, but crawled into his dog coop beneath the stairs. Other lab workers, waiting for a globe to be exhausted or a part to be machined, stretched out on benches or coiled themselves between bottles on a table. It was not unusual for the laboratory to look like a flophouse.

In place of the noon whistle Edison had the midnight organ. Going over to the instrument against the far wall, he would attack it vigorously with two fingers, play the only tune he knew over and over, or make up some composition on the spot. Supper would be brought in and steaming coffee poured. Edison passed out cigars, and for an hour the air was redolent with smoke and stories. In Upton Edison had found his intellectual equal, and they discussed such subjects as magnetism, gravity, astronomy, philosophy. "It is unreasonable," Edison thought, "for men today to be afraid that they cannot find out any more. That all has been found."

During the past few months Edison had concentrated on raising the temperature of the wire to the point at which it would emit a good light.

He had concluded that the wire would have to be shaped in the form of a spiral or coil. But the discovery that the wire melted created another dilemma. The thrust of the experiments would have to be redirected —the oxygen would have to be removed from within the globe.

The concept of a vacuum to keep the burner from disintegrating was nothing new. A number of inventors had tried it. But, curiously, most of the experiments had taken place before 1865, when Hermann Sprengel had designed the first efficient vacuum pump. There were still only a few Sprengel pumps in the United States, and in the last week of January Edison, in his usual fashion, set off on a telegraphic hunt. He was unable to find a Sprengel pump he could obtain. Upton, however, knew of a mechanical vacuum pump at Princeton, so Edison put him aboard a train to borrow it.

Since the return train late that evening did not stop at Menlo Park, Upton, alighting at Metuchen, had to trek two and a half miles back to the lab. He arrived hungry, cold, and out of breath. Yet if he had been frozen stiff Edison would have put him in a pot of hot water and spurred him to work—at such a time Edison was an explosive mixture of enthusiasm and impatience.

A platinum-iridium spiral was sealed in the bulb. After a few minutes, however, the heat from the wire cracked the glass.

Two days later Edison managed to keep the wire incandescent for ten minutes. On February 12 he coated a spiral with acetate of magnesium to insulate and protect it. Although the spiral produced a brilliant light, both the glass and the wire disintegrated quickly. In another test at ten-thirty that night, Edison and Upton noticed a thin black deposit on the glass except where the pillar holding the wires screened it away. Without having the faintest intimation of what it meant, they had discovered the basic principle that would lead a quarter century later to development of the radio tube.

Edison and Batchelor observed that, as the wire was alternately heated and cooled in the vacuum, gases were driven from the pores, the metal was hardened, and the refracting power of the platinum increased. Encouraged by this discovery, Edison designed an entirely new lamp consisting of two cylindrical globes, one within the other. The outer chamber contained air, whose expansion when heated pressed against a diaphragm to interrupt the current. The inner chamber, exhausted of air, held a thumb-sized bobbin of compressed lime, resembling the chalk drum of the telephone receiver. Around the bobbin, thirty feet of platinum wire was to be coiled, providing a resistance of 750 ohms. As the heat increased, the bobbin was to incandesce, thus keeping the lamp aglow and reducing the flickering that would otherwise occur due to the on-off action of the current.

Such a design was all quite well in theory; but in practice Edison found it impossible to obtain a good vacuum in the inner globe, and the action of the heated air on the diaphragm was erratic and difficult to control. Most important, there was no way to put three feet, much less thirty feet, of wire on the spool, because the magnesium acetate melted and the wire shorted out.

Pressure, nevertheless, was building from the stockholders for another demonstration. Edison worked round the clock to improve the lights. "We was all night bringing up 12 lamps in vacuum worked all day Sunday all night Sunday night all day Monday," he recorded on March 16.

Since he had no hope of exhibiting a lamp with a bobbin, Edison decided to improvise and show instead a light produced from a short platinum spiral. Each lamp, however, had a resistance of only ⅓ ohm—a mockery of Edison's intent to produce a high-resistance light. It was impossible to work such lamps entirely in parallel even on a laboratory circuit, so Edison designed a mixed series-parallel layout consisting of four parallel sets, each with four lights in series. The total resistance of the circuit was only ⅓ ohm, completely impractical from a commercial standpoint.

On Saturday evening, March 22, Lowrey and other members of the company alighted once more at the rickety Menlo Park station. From there they were taken to Edison's elegant new library, then led to the machine shop. The gaslights were turned off. Edison ordered the resistance boxes cut out one by one. As the current increased, the sixteen lamps mounted above the workmen's benches glowed red . . . yellow . . . white. Each lamp produced 22 candlepower—compared to the 16 candlepower obtained from a gas jet—and the visitors exclaimed with pleasure as the machines were etched sharply like three-dimensional drawings on white paper.*

Within a few seconds, however, the lights commenced flickering. Sparks shot from the contact points of one lamp, then another, as the circuit was opened by the thermal regulators. The flashing increased whenever the current was interrupted on two lamps simultaneously. The stockholders, taken aback, flinched and shielded their eyes.

Suddenly, one light grew in intensity. A sharp crack resounded through the room. The glass shattered. A row of lamps died out. Batchelor jumped, and inserted a resistance coil, which simulated the burned-out light in the circuit. After a minute or two, another globe burst. One by one the number of lamps was reduced. Soon the visitors had seen enough.

Edison explained the trouble was not with the lights but with the

* In 1879 measurements were so crude and the mechanics of measurement so different than those employed in 1979 that it is impossible to give a precise comparison between the lights Edison was working with and the illumination provided by modern bulbs. Today, a 100-watt incandescent tungsten bulb emits between 120 and 135 candlepower. A 16-candlepower gas jet, therefore, would have been roughly equivalent to a 12-watt bulb.

generator, and would be corrected when he completed his own machine. Despite his assertion, the directors were dismayed. One of them, Robert Cutting, remarked that he thought Edison would have been better off to spend a few dollars for Starr's thirty-year-old book, "and to begin where he left off, rather than spend 50,000 dollars coming independently to the same stopping point."

A few days later several members of the syndicate went to Sawyer's shop in lower Manhattan. Sawyer demonstrated how he treated the thick, short carbons in olive oil, then inserted them into tube-shaped lamps filled with nitrogen. He had a half-dozen lights which incandesced like balls of fire, and could be turned off and on individually. He professed that a few thousand dollars more would bring the realization of the light. In actuality, though none of his visitors knew it, he had reached the end of Albon Man's financial backing, and was about to pack the lamps away. They were inefficient, the nitrogen leaked out, the carbons deteriorated, and none lasted more than four or five hours.

But while Sawyer was reduced to despair, Edison had learned to accept setbacks philosophically: "It has been just so in all of my inventions. The first step is an intuition, and comes with a burst, then difficulties arise—this thing goes out, then that—and months of intense watching, study, and labor are requisite before commercial success—or failure—is certainly achieved."

Upton, for his part, told his father, "I have not the least lost my faith in him, for I see how wonderful the power he has for invention."

As the experiments continued, it became evident that the thermal lamp was impractical. The regulator was not sensitive enough to function properly. The uneven cutting in and out of the lights created unacceptable peaks and valleys in the electrical load.

Edison, therefore, decided to replace the thermal device with an automatic, magnet-controlled regulator. This was designed to allow the current to reach the spiral only one third of the time. He reasoned that, according to probability, one third of the lamps would be drawing current at any given moment, so that the flow of electricity would be kept fairly steady.

For the generator, Edison conceived a machine radically different from those in existence. And Upton, performing the calculations, told him if his ideas appeared feasible. Because batteries operated best when the external resistance was equal to the internal, the makers of generators were following the same design. This resulted in the greatest theoretical output of electricity. Edison, however, concluded that it was a waste of energy, and told Upton to devise a generator with the lowest workable internal resistance. The total output of such a machine would be less, but the usable portion would be greatly increased.

Edison determined, furthermore, that to reduce the sparking resulting from intermittent operation of the lamps, the magnets of the generator would have to be removed from the circuit. The machine would have to be *excited* by means of another dynamo.*

Upton, recruiting everyone, including Edison, for winding bees, made such good progress with the generator that by mid-April Edison's optimism was restored. The Light Company's newest investor, Henry Villard,† the president of the Oregon Steamship and Navigation Company, was interested in obtaining an incandescent lighting system for his latest ship, the S.S. *Columbia*, due off the ways about Christmas. Villard, who was to play a prominent role in Edison's career, sent his chief engineer, J. C. Henderson, to Menlo Park.

Henderson, after spending a few hours at the laboratory, fell under the inventor's spell so completely that he not only reported, "[Edison's] machine and lamps will be in a finished state in about six weeks," but told Villard that Edison had convinced him that power could be transmitted directly from heated coal to the propeller. It would soon be possible, he asserted, "to do away with the steam engine altogether. You may be inclined to think that he has imbued me with some of his impractable notions but I am confident it is only a question of time until the steam engine will be behind the age."

In the thrall of the same influence, the *New York Herald* featured "The Triumph of the Electric Light." The *Herald*'s publisher, James Gordon Bennett, who believed in creating his own news,‡ was outfitting an expedition to the North Pole. For this purpose he had bought a British yacht built for sailing in Arctic waters, renamed her the *Jeannette,* and deeded her to the U.S. Navy. A young explorer, Lieutenant George De Long, was to command the vessel, which was being refitted at the Mare Island Naval Yards in San Francisco.

On April 21, De Long asked Edison if he would provide incandescent lights for the ship, due to depart the middle of June. Following up, a *Herald* reporter went to Menlo Park two weeks later, then wrote De Long: "Edison will give us his own generator (no joke) which is equal

* There is an intimate relationship between magnetism and electricity. Lines of force, easily demonstrable by means of iron filings, exist around the poles of every magnet. Electricity is generated when a conductor, such as a loop of wire, is rotated so as to cut the lines of force. A simple generator consists of an armature, wound with many loops of wire, rotating between the poles of a magnet. *Self-exciting* generators, such as those in operation in Edison's time, employed part of the current they produced to activate their magnets. Edison, however, found this design impractical for his purpose.

† Villard was the scion of a well-connected German family. Twenty years before, in youthful rebellion against his father, he had sailed to America and joined relatives in Illinois. During the Civil War he had been an outstanding war correspondent, and afterward had continued his newspaper career. When the Kansas-Pacific Railroad had been forced into receivership, the bondholders in Germany had asked him to look after their interests. As their representative he had stood up to Jay Gould—who was in the process of taking the railroad over from the Palmer-Reiff group—and so gained accolades and prominence. Villard had now developed into a full-fledged, flamboyant financier himself.

‡ In 1869 Bennett had sent Stanley to find Dr. Livingston in Africa.

to 2½ horse power. [It is] a beauty and gives enormous results."

On May 24 Upton noted: "The machine for lighting the North Pole has been sent away." Edison included two phonographs and a set of telephones with the generator, which was to provide power for a Jablochkoff arc light.

Unfortunately, the generator, which featured a large armature revolving between two pillarlike magnets, turned out to be a dud.* Upton spent June completely redesigning the machine. By the time he was finished, he had created a dynamo having little similarity to any extant. Four and a half feet tall, it consisted of a relatively small wooden armature revolving between powerful field magnets consisting of two massive, upright cores of iron, each weighing 1100 pounds.

From the very start, the machine's performance was immensely encouraging. The most efficient dynamo in production, the Siemens, converted 55 percent of the horsepower applied into usable electricity. Edison's gave more than 80 percent. Since the two cores reminded the men of a reclining female with her legs sticking up, the workers nicknamed the generator "the Long-legged Mary Ann," in honor of a girl who occasionally visited the lab. (This designation was subsequently altered to preserve Victorian sensibilities, and the machine acquired the somewhat mystifying appellation "the Long-*waisted* Mary Ann.")

But while Upton was completing the generator, Edison and Batchelor were experiencing only frustrations with the lamp. When magnesium acetate failed as insulation for the wire, Batchelor decided that oxide of zirconium would be better. Zirconium, however, was difficult to obtain, and the oxide was entirely unavailable, so the experimenters had to learn to make their own—a slow and laborious process.

Edison would have liked to get away from platinum and iridium (costing twelve and fifteen dollars an ounce) altogether. He thought of tungsten, but one of the leading rare-metal supply stores in the world had only thirteen ounces in stock. He tried aluminum; but, he recorded, "This would not come up at all above a whitish yellow, when it seemed to get weak in the knees." In truth, platinum, with a coefficient of expansion approximately equal to that of glass, was the only metal which did not crack the globes.

That being the case, platinum would have to be found. He would have to prove that it was not so rare a metal as everyone believed. He

* The machine failed to produce even the faintest spark when it was tried out in the Arctic. Little else on the expedition proceeded according to plans, either. De Long, who made Edison look like a pessimist, headed directly for the pole on no other basis than a hunch that he would find a break in the ice pack north of the Bering Strait. Within twelve hours of reaching the ice, he was frozen fast. For nearly two years the ship drifted westward with the ice off the north coast of Siberia. When the vessel finally broke up and sank, in the summer of 1881, the men struggled over the ice and water for two months to reach the barren, nearly uninhabited shore. About half the men in the expedition, including De Long, perished.

sent 2000 inquiries to prospectors, miners, and telegraphers all over North and South America. He ordered $127 worth of maps of mining regions from New Jersey to Bolivia. He hired Frank McLaughlin, a telegrapher friend and part-time prospector, to investigate a promising deposit in Canada. He was obsessed with platinum. On June 26 an all-night session of telephone experiments broke up when Batchelor and Griffin departed at 4:00 A.M. An hour later, McLaughlin, who had not had signal success in Canada, tried to leave. He was intercepted at the door by Edison, who had picked up an old milk pan and a spade with a curled edge.

"Come, Mac, let's go out prospecting," Edison said, and led the way through the damp grass to the abandoned copper mine a mile away. Handing McLaughlin the shovel, he had him fill the pan with dirt, then waded into the brook and shook the pan so vigorously spray flew in all directions. He peered intently at the pan. Nothing was on the bottom. "Dig away, Mac! We'll get it yet! We'll get it yet!" he urged on his companion. At last, after two hours, he detected a thin residue of black sand on the pan's bottom. "We've got it now!" he murmured joyfully. Gingerly he carried it to Dr. Alfred Haid, the chemist at the lab, and anxiously awaited the result of the assay.

"As fine a specimen of Jersey mud as I have ever seen," Haid reported.

For the moment, actually, the incandescent light did not have primary importance at the lab. By the end of April Edison had used up the allotment of $50,000 from the electric-company investors. Unpaid bills were piling up, his account was overdrawn, and the workers were, as customary, having part of their pay deferred. Edison's hopes now rested on the telephone. For several months Gouraud had been negotiating with British capitalists to form a London company, but the deal had been hanging fire because of the Bell Company's threat to sue for infringement of its receiver.

By February, however, Charley had the chalk receiver working well enough for a public demonstration, and Edison decided to send several models to England. Edison's relationship with Charley was both close and delicate. On the one hand, he could see in Charley an image of his own talent. On the other, Charley had a wild and irresponsible streak inherited from his father and grandfather, a streak that Edison had largely suppressed in himself. Edison, mindful of Charley's explosive temperament, had learned to treat him gingerly. A year before, Charley had pleaded with George Bliss, purveyor of the electric pen, to send him to Europe as the company's representative, but Edison had told Bliss: "You need not say it above a whisper, but I do not think that Charley would do."

Charley, however, was the expert on the chalk receiver. No one knew more about its intricacies and its eccentricities. On February 27 Edison, resisting the entreaties of Sam to go along as a chaperon,* dispatched Charley to England with six instruments.

Charley's arrival put Jim Adams, who had gone to England as Edison's telephone expert the year before, in a pique. Though only thirty-three years old, Adams was dying from tuberculosis and heart disease. He had oath-filled battles with Gouraud, who thought him "the most difficult individual to get on with that it has been my lot to be thrown with. He is really dangerously ill and utterly unfit for the purpose for which he is here." A month after Charley's arrival, Adams took off to join the Edison European Telephone Company in Paris. Charley assumed the principal burden of putting the system together for its demonstration before the British Royal Society on May 10.

The superiority of Edison's transmitter and the volume of his receiver generated considerable opinion Edison would drive Bell completely out of the telephone business. The Edison operation had a hell-on-wheels atmosphere reminiscent of a Western railroad camp. Twenty-three-year-old George Bernard Shaw, who was hired as a trainee, found "Their language was frightful even to an Irishman. They worked with a ferocious energy which was all out of proportion to the result achieved. Indomitably resolved to assert their republican manhood by taking no orders from a tall-hatted Englishman . . . they insisted on being slave-driven with genuine American oaths by a genuine free and equal American foreman. . . . They were free-souled creatures, excellent company; sensitive, cheerful, and profane; liars, braggarts, and hustlers; with an air of making slow old England hum."

Then, on Sunday May 4, Adams died in Paris. Charley, who had everything prepared for the demonstration, decided to rush to France to comfort Adams's widow and, incidentally, see what Paris was like. Gouraud was absent, but his deputy forbade Charley to go—which was the one way of making certain Charley went. Charley introduced him to a few expressions culled from Great Lakes sailors, and disappeared. When, four days later, he returned, he was barred from handling the equipment at the exhibition and told he had been separated from the company. Charley protested to Edison he had to contend with "a man with whom I am confident you would not have any business as his style is the exact reverse to yours. He is a domineering, overbearing, self-conceited Lardy duck, a swell, parts his hair in the middle and don't know a telephone from a Dutch clock."

All that, of course, was immaterial, since a huge sum of money was at stake. Edison wired Charley: "Return home immediately." But Edison wasn't about to get Charley back to Menlo Park after he had seen Paris. Charley took ship not for America, but for Calais.

* Sam was certain that a fortune awaited the Edisons in their ancestral land, Holland, if only he could get there to uncover it.

Even without him, the demonstration went brilliantly. Present were many Lords and Members of Parliament, the Crown Prince and Princess of Denmark, and the Prince and Princess of Wales. The heir to the British throne sent Edison congratulations. On May 15 the London company was formed, and a few days after that Edison had $25,000 in hand.

In high spirits, Edison lent Gouraud $2500, paid Batchelor his $2500 share, gave Mary $1000 for a present, and ordered five hundred books.

Then he turned to considering just what kind of a telephone system he would send to England. For what the Londoners had seen were really nothing more than laboratory devices. Under the artistic tinkering of Edison, Batchelor, Adams, and Charley they had worked—but then again, at times they hadn't. The chalk receiver still needed a good deal of perfecting. As for the speaker, Edison and Batchelor weren't sure but that the earlier "inertia telephone" would be more suited than the carbon-button transmitter.

Not only was the system technically immature, but there were many organizational problems. Edison needed a manager–chief engineer for the London company. The job would pay the impressive salary of $500 a month, but there was scarcely anyone who combined the requisite technical knowledge and managerial ability. Johnson had given telephone exhibitions for two years, but his business record was not one to inspire confidence.

Since, however, no one else was available, Edison broached the subject to him. Johnson was reluctant. During the past decade he had gone from the Kansas-Pacific Railroad to the Automatic Telegraph Company, thence to the Atlantic & Pacific Telegraph Company, the Electric Pen Company, the musical telephone, and now the phonograph. "I have been always seeing the bushy tail of the fox around the corner while another fox has drawn me off in another direction," he complained. But a few days later he had to put up his own and borrowed phonograph stock to bail himself out of bankruptcy, and England looked more attractive.

The last two weeks in June Edison and Batchelor, their shirts discarded in the sweltering lab, worked round the clock to prepare fifty telephones for Johnson to take to England. Day and night kerosene lamps smoked to deposit lampblack on the glass. The deposits were scraped off and pressed into transmitter buttons. The last day of June Edison held a rousing telephone exhibition and party for two score journalists and friends at Menlo Park. Two days later Johnson sailed for England with the task of setting up in London a telephone exchange similar to the one operating in New York.

Dishearteningly, he took with him only four telephones—the rest had been botched in the attempt to rush them out. So for Edison it was back to smoking lamps, capricious chalks, and crumbling carbon buttons. A receiver which did not deteriorate in the course of forty-eight hours caused Edison to exclaim: "This is simply wonderful!" But one day later it was only half as loud. On July 21 Edison had a new group of telephones ready. Half were carbon and half inertia transmitters. Johnson could try both until he made up his mind which he wanted.

Letting Batchelor continue with the telephone, Edison returned to the electric light. By early June he had finally made enough zirconium oxide to coat the wires, and constructed a number of magnet-regulated lamps containing platinum-wired lime bobbins in a vacuum.

On July 7 Edison wrote Tracy Edson: "Everything looks bright. The standard dynamo machines get beyond our most sanguine expectations." He was ready to demonstrate his system. He would hook thirty lights to the first dynamo, and add another thirty when he had a second.

The day arrived. The lamps were linked in parallel. The dynamo was belted to the steam engine. The valves were opened. The resistances in the circuit were removed one by one. The platinum lamps glowed, and turned into bright incandescence. Then suddenly a huge spark shot from one lamp. A second from another. Before the fascinated and despairing eyes of Edison and his crew, the system turned into a Roman candle.

To other men it would have been a disaster. To Edison it was a setback. It seemed almost irrational the way he kept attacking, the dogged determination with which he probed his way through nature's maze, convinced somewhere there had to be an exit. He told Upton he would either get an incandescent light or prove it was impossible. Since the measures he had taken to reduce the sparking had proved inadequate, he determined to take others. He revamped and expanded the control mechanism until it became by far the dominant aspect of the lamp. Affixed to a continuously revolving governor were two revolving magnets, complemented by two stationary magnets. Since sparking decreased according to the square of the number of points at which the circuit was broken, Edison added a device to interrupt the current in four places simultaneously, thereby reducing the sparks to one sixteenth of their previous strength.

Yet each new mechanism also encumbered the light more. It was beginning to look like the anticipation of a Rube Goldberg sketch. Each night when Edison or Upton touched a match to a gas burner and watched the simple flame come to life, they had to stop and wonder.

Maddeningly, they were not even able to try out the regulator and the lamp together with the generator, for Batchelor was unable to successfully insulate the platinum wire.* Throughout August, Batchelor experimented with one coating after another: titanium oxide, aluminum oxide,

* Nothing demonstrates the primitive state of the art more graphically than the fact that Edison was trying to heat the wire to its maximum degree and simultaneously insulate it.

silica thoria, tungsten hydrate, and, of course, zirconium oxide. Then, toward the end of the month, he discovered that zirconium oxide attacked the wire at high heat, and so was worse than useless.

The last day of August, Upton and Edison went to Saratoga for the meeting of the American Association for the Advancement of Science. Summarizing the results of the experiments, Upton read a paper he had written but attributed to Edison, "On the Phenomenon of Heating Metals in Vacuo by Means of an Electric Current." The scientists, impressed, applauded Edison liberally.

Upton, who was to be married in a few weeks, had earlier broached the question of his twelve-dollar-a-week salary to Edison. Edison agreed he had earned an adjustment. Why not work without salary? Batchelor had a 10 percent share of the future income of the electric light, and Edison offered Upton 5 percent and thirty-seven shares of stock.

Upton thought that with some pertinacity he might get 7½ percent. "Yet as it is pure generosity on his part I think it is not becoming to try to jew him," he told his father.

His father believed he was foolish. He should insist upon the pay, and let the profit go.

Upton, in truth, could well have claimed that he had become the linchpin of the operation. The twelve-horsepower generator, in the construction of which he had played the leading role, was at this juncture the only workable element in the system. Edison was prepared to introduce it into commerce even if he did not have a light. By reversing the action, and applying electricity to the armature, he could convert the generator into a motor. Deciding to hook the Long-legged Mary Ann to a sewing machine, Edison wrote sewing-machine manufacturers for samples of their wares. He sketched plans to establish twenty-four electric stations in New York for the purpose of running motors. Moreover, he was convinced that the Long-legged Mary Ann would shortly solve the platinum problem.

Upton had calculated that an electric lighting system, at the current price of platinum, would cost at least ninety-eight dollars per lamp, a prohibitive figure. Edison, however, concluded that he could easily bring the price of platinum down from twelve dollars to one dollar an ounce. "Contrary to the statements and sneers of mining experts and mineralogists and scientific frauds I have succeeded in finding vast quantities of platinum," he declared.

He based his assertion on the black sand of Pacific Ocean beaches and the tailings from hydraulic gold mines he had seen during his trip west. These, he maintained, were full of platinum and gold. If the precious metals had not in the past been profitably extracted, it was because the Long-legged Mary Ann had not yet been born. He, however, would use the dynamo to excite a huge magnet, and the magnet would draw aside the particles of gold and platinum as the sand and tailings were poured through a hopper. Proposing to establish a concentrating works at Menlo Park, he suggested that, as a modest beginning, a company capitalized at $10 million should be formed.

The question of how to obtain a more realistic sum to continue the electric-light experiments was becoming increasingly demanding. Lowrey brought some of the directors out. Edison demonstrated an electrically operated sewing machine, explained his plans for the stations in New York, and suggested $100,000 as the amount required to perfect his system. The directors, unswayed, were of the opinion that they should not be asked to contribute additional money unless they were given a larger share in the company. Lowrey opposed the directors' proposal, because his and Edison's interests would be reduced. He suggested to Edison that he try to continue on his own awhile longer.

Edison had sold the rights to his Polyform concoction for $5000. Any larger sum would have to come from further telephone sales in England. Negotiations were close to being completed for an additional $50,000 advance payment against 20 percent royalties. But there was considerable unhappiness among the British investors at the slow rate of delivery. The instruments were handcrafted under the supervision of Batchelor. He tested each on its completion. Number 142 was a "Belcher but bad transmitter," and 143 was just the opposite: a good transmitter but "Not a belcher."

To install and operate the system in England, Edison recruited telephone repairmen, called "inspectors." Ten telephones and an exchange were set up in the laboratory. Edison would short-circuit one instrument, cut the wires in another, and put dirt between the electrodes of a third. The inspectors were then given five minutes to discover the trouble. Ten men at a time, shouting "Hello! Hello! Hello!" hustled about the laboratory, fraying the nerves of everyone except Edison, who, quite unperturbed, went about his experimentation.

Edison had found that the carbons used in Jablochkoff and Wallace arc lights also worked well in the inertia telephone, but many were impure and the quality varied widely. Edison had just about decided that inertias were superior to the carbon-button transmitters when Johnson wrote from England to stop sending them. When they were thrown the slightest degree out of adjustment they became, in fact, totally inert. Then Johnson wired the even more crushing news that the receivers were arriving in England with the chalks crystallized and would not work at all.

Thus, in what was an increasingly supercharged atmosphere, labor on the telephone and light continued side by side. After the disastrous July test of the lamps, Upton expressed the opinion that the mechanical pump they were working with was better suited to pump water than to exhaust air, and persuaded Edison that a Sprengel pump must be obtained. A Sprengel pump was actually not a pump at all in the common connotation of the word, but an arrangement of long glass tubes, set up vertically. The globe to be evacuated of air was fused to the top of the tubing. Mercury was dripped down slowly from above. Each globule drove out some of the air as it fell to the bottom. It was a tedious process, taking the better part of a day, to achieve a vacuum. But the end result was infinitely better than that obtained from a mechanical pump.

To construct the arrangement of glass tubing that composed the pump, Edison hired a New York artisan named Boles; and Boles subsequently came to Menlo Park from the city whenever the globes mounted on the pump had been exhausted of air and required sealing. But the delays encountered in waiting for Boles led Edison to suggest to Upton that he learn glassblowing.

Upton gave it an exhausting try. Puffing through the long pipe projecting through his thick beard, he succeeded in producing nothing more than grotesque blobs. Edison finally gave up, and placed an ad for a glassblower in the *New York Herald*.

The youth who responded on August 18 was gangling and fuzzy-cheeked. His name was Ludwig Böhm. He spoke English with a tony Rhineland accent. His forehead supported the bill of his little red German student cap. He looked comical and hardly confidence-inspiring. Yet he had worked with the noted scientific instrument maker Heinrich Geissler at Heidelberg. And though he had a hefty ego, it was supported by his talent.

Böhm's arrival created a new dimension, and greatly speeded up the process of exhausting and sealing bulbs. Böhm made a Geissler pump which, similar in operation to the Sprengel, worked faster though not with as much precision. Edison, after returning from Saratoga, initiated a series of experiments combining the two pumps for best effect.

Evacuation of the globes was always started on the Geissler and completed on the Sprengel. Lowrey's former office boy, nineteen-year-old Francis Jehl, was the pump man—he raised and lowered the mercury bottle on the Geissler, and dripped mercury down from above into the Sprengel. Bobbins were fashioned of every conceivable substance—lime, zirconium, powdered quartz, lanthanium—and the platinum wires coated with every kind of oxide known. Thirty different combinations were prepared. The bobbins were inserted into the globes and glass was melted around the wires. Hours blended into days, days into weeks. Lamps cracked. The pump gave out and was repaired. Spools glowed spectacularly and eerily—red, yellow, blue. Lime melted to green phosphorescence. Upton took a long weekend and went to Massachusetts to get married. He brought his bride back with him, and settled her in

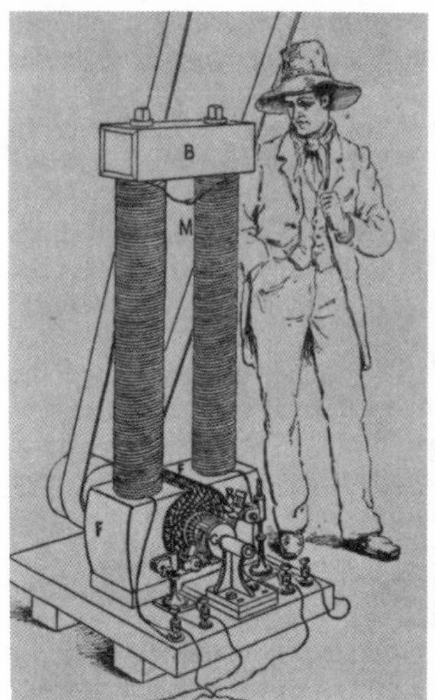

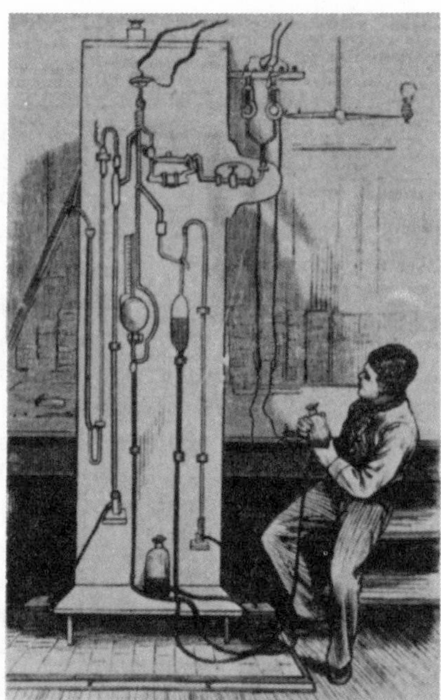

Key Steps in the development of the incandescent light were the production of the Long-waisted Mary Ann generator, being inspected by Edison (upper left), and the design of a unique new Sprengel-Geissler vacuum pump (upper right). Francis Jehl is working the Geissler pump, consisting of the two tubes at right. The two tubes at the far left compose the Sprengel pump. In the center is the McLeod gauge, which leads up to the spark gauge, from the top of which two wires protrude. The sausagelike horizontal device is a drying chamber to remove the moisture. Two globes to be exhausted of air are mounted at the upper right.

the village's only remaining substantial house. He left her late each morning and did not return until the following dawn.

With each new adjustment, the pumps became better and the vacuum improved. On Friday, September 26, Edison inserted a two-and-a-half-inch spiral made out of platinum-iridium wire coated with magnesium oxide into the globe. The spiral offered only 3 ohms resistance, yet Edison thought they had gotten as much as 8 candlepower (the equivalent of a 6-watt bulb) out of it. Fascinated, they had kept it glowing for thirteen hours and thirty-eight minutes.

"There is millions in it!" the normally reserved Upton enthused.

On the first of October, however, the pump broke. It had been producing the finest vacuum ever achieved in the world, but Edison thought it could be made even better. The pump had a spark-gauge attachment that enabled Edison and Upton to determine when a satisfactory vacuum was reached. Now, Edison proposed to add a McLeod gauge, which would measure the precise amount of air left in the apparatus.

Upton and his fellow Princetonian, Samuel D. Mott, designed a highly complex and sophisticated new pump. While Edison and Batchelor were busy on the telephone, Upton, Mott, Böhm, and Jehl put the apparatus together and mounted it on a door-sized board. When they tested it on Monday, the sixth of October, they found they were able to obtain an almost perfect vacuum—only one millionth of an atmosphere of air remained in the tubes. No one had ever attained anything like it before.

It was evident now that the wire could be maintained in the vacuum. But the platinum continued to short out on the bobbin. All known oxides had been tried as coatings. At the temperature of incandescent platinum all either melted or became conductors. "The trouble," Upton stated on October 8, "is to get an insulation for the platinum wire."

More than thirteen months had passed since Edison had initiated his quest. He had developed a generator—which had not been included in his original plans—but the realization of a commercial incandescent light seemed as distant as ever. On Saturday, October 11, Edison—himself beginning to have doubts that a high-resistance lamp was attainable—asked Upton to apply himself to the problem of making a 10-ohm lamp behave like a 100-ohm lamp in a circuit. Edison proposed to achieve this phenomenon by placing in the cellar of every house a regulator that would direct current to each lamp only one tenth of the time.*

Yet even as Edison pulled at his eyebrow and groped for a solution, the answer was about to manifest itself—once again the parallel experiments, improvements in technology, and step-by-step process of elimination were to be fused into an astonishing discovery.

Batchelor, working on the telephones, noticed "Some of the Wallace carbons were perfect insulators, probably due to one of the pieces of silicon being in contact with platina."

Edison, whose work place was next to Batchelor's on a raised table by the head of the stairs, decided to try silicon for insulation. He coated a short spiral of platinum with the material and, as was now the practice in testing insulations, inserted the spiral into an evacuated bulb. Silicon, a major ingredient in quartz and (as silicon dioxide) in glass, withstood heat well, and was both an insulator and semiconductor.

* Since each lamp in a parallel circuit reduces the total resistance, ten 100-ohm lamps drawing current simultaneously have a combined resistance of 10 ohms. Edison figured he could achieve the same result by having ten 10-ohm lamps drawing current only one tenth of the time. How he intended the lamps to remain lit during the nine tenths of the time they were disconnected is not clear.

When the first tests proved promising, Edison applied even thicker coatings of silicon. On Monday, October 13, he obtained a light of 1 candlepower from a spiral with 2 ohms resistance.

Upton, concurrently, was working with Batchelor on resistance coils. The coils were used both on telephone and electrical circuits, and Upton employed them to test the performance of the generator. Many of the resistance devices were made out of carbon, which was closely related to silicon.

Though Edison had dismissed carbon as a possible "burner" for the incandescent light, he had read in the July 12 issue of *Scientific American* an article describing the renewed experiments of Joseph Swan in England. Swan, who had abandoned his attempt to produce an incandescent light in 1860, had kept a carbon rod, one twenty-fifth of an inch in diameter, lit brilliantly for several minutes in a vacuum globe a few months before.

Swan's result, and the behavior of the silicon-covered spiral, piqued Edison's curiosity. The same day that he talked to Upton about making a 10-ohm lamp equal a 100-ohm lamp, he had Upton test an eight-inch-long piece of carbon one seventeenth of an inch in diameter. On Monday a similar stick heated in a vacuum remained dark in the center but emitted light from both poles.

Upton, measuring the resistance, found that it decreased as the carbon incandesced. This was entirely unexpected, for metals *increased* their resistance as they grew hotter. Edison thought that pure carbon would increase in resistance as it heated, and that the decrease in resistance had been caused by impurities in the rod.

Yet *if* carbon indeed decreased in resistance as it grew hotter, and *if* they were able to maintain it in a vacuum, then might not carbon, which would withstand temperatures of more than 5600°, contain its own built-in balancing mechanism, so that the current would not have to be interrupted, and all the regulatory gadgets could be jettisoned? But could, then, a *high-resistance* burner be made from carbon? Edison's head reeled with speculation.

Writing Johnson in England, Edison related he had worked on the telephones and the electric light for forty hours with only three hours' sleep. "That chalk catastrophe has sent a little leaden pain right in the pit of my stomach. I'll make those chalks last for 10 years before I get through or I'll eat a ton of them. We are striking it big in the electric light, better than my vivid imagination first conceived. Where this thing is going to stop Lord only knows!"

It stopped, for the moment, with another cracking of the pump. During the four days that were required to fashion a new one, the laboratory was almost suspended in time. Upton tested materials that might be combined with carbon to produce a high-resistance mixture. Edison speculated. The air was imbued with an explosive stillness reminiscent of the eve of battle. Friday night the new pump, containing in addition to the various gauges a drying chamber to take the residual water vapor out of the tubing, was ready. The experiments with coated spirals and with carbon sticks resumed. Edison was animated by anticipation. Then a telegram from Paris shook him to the core.

When Charley had gone to France in May, he had had no trouble establishing himself. Cornelius Herz, who held Edison's quadruplex rights for France and Belgium, hired him for $300 a month (an enormous sum in Europe) to operate a Paris-Brussels link. Charley had been Edison's photographer at Menlo Park, and in Paris he met an American artist and photographer, D. Murray, who fell in love with him. Charley took the apartment next to Murray's, and converted one room into a laboratory. The connecting doors between the two apartments were thrown open. From then on, Murray said, "We not only lived and pleasured together, but also worked together." In Port Huron, Nelly Edison could not understand what was going on and was wild with worry about her son. She wrote Edison inquiring if Charley was in trouble.

"Charley is not acting right. If possible I would advise that you get him to come home," Edison replied, then added warily: "But do not mention that I said so."

Charley had none of Edison's inhibitions, and he took to Paris as if he had been born in Montmartre. Murray was a sadist who confessed: "I had never had such an intimacy before." He guided Charley to all the historic places and unhistoric dives. Then they went on a three-week trip down the Rhine. Charley spent the $500 Edison had given him during his last week in London as if they were sous, and accumulated bills all over Paris. On Wednesday, October 8, he and Murray returned to the apartment in the early-morning hours, and at seven o'clock Charley complained of terrible pains in his abdomen.

Murray poulticed him and gave him laudanum, but by that evening Charley was in agony. Murray was unable to get a doctor till two o'clock in the morning. The doctor put twelve leeches on Charley's stomach and injected morphine. During the day Charley was better, but the following night he vomited blood. The doctor diagnosed general peritonitis. It was extremely rare in men, he said, but had been caused by "Inflammation or a cold in the bowel." There was no possibility of Charley's surviving.

Charley, however, refused to die. He shouted, cursed, struggled, and hallucinated. He yelled that Murray had put him between the poles of a battery, and told him to take off the wires and put on more resistance. For five days he was kept on morphine, but then he grew so weak the injections had to be stopped. Murray and the doctor kept popping him into baths. And finally, when he seemed suspended between living and dying, they poured a huge dose of purgative into him. That settled the issue. He never stopped eliminating and vomiting until he died, at six o'clock Saturday morning, October 18.

The first cable, with the news of Charley's critical condition, reached Edison on Friday. The next, announcing his death, arrived on Sunday. For more than forty hours in between, Edison and his co-workers conducted a "death watch." They were a close-knit group—the entire Menlo Park force included only two dozen men, and of these Batchelor, Upton, Jehl, and two or three others made up the handful of intimates who worked with Edison on the second floor. Charley had been one of them; and his dying at nineteen, so soon after the death of Adams, filled the shadow-sketched laboratory with a spectral gloom.

Pitt and Nelly received the news at the same time as Edison. Nelly was heartbroken. Pitt importuned his brother to come to Port Huron. But Edison, who felt awkward in close, personal relationships, could not face tragedy. He bore all the expense of shipping the body home and paid Charley's debts—nearly $3000 all told—but he himself would not go near Port Huron.

One of the executives of the Edison Telephone Company in Paris covered up Charley's tracks so as to avoid a scandal. Murray, loving to the last, snapped pictures of the embalmed body. "These portraits I took from the most favorable view, and will please you I think," he wrote Edison. "Please let me know that he has arrived all safe."

Adams and Charley were dead, and so many of the telephone men Edison was dispatching to England were meeting a similar fate that Batchelor grimly wrote Johnson: "Edison facetiously remarks that if you men are going to die off so, it would be better to have the large [telephone] boxes made a little longer, so that you could send back the corpses in them, and previous to our sending the inspectors we will see if they fit the boxes."

Edison's stomach churned, and the tobacco juice flowed more freely. But he had learned in childhood that expressions of emotion were construed as weakness. No matter how much something hurt, he would not let it get to him.

"As a cure for worrying, work is better than whiskey. Much better," he philosophized.

Sunday evening everyone was back in the laboratory. Upton placed a stick of carbon in the globe to be exhausted. Jehl, climbing onto the ladder, started dripping mercury. When, after a time, this produced a metallic clicking, Edison took a Bunsen burner and slowly heated the globe to drive out the air. The clicking increased in violence until it threatened to shatter the glass. (Sometimes it did.) Attaching one wire from the bichromate battery to the lamp terminal, Edison took the other wire and touched it for an instant to a wire protruding from the lamp, so completing the circuit. As the vacuum in the lamp was suddenly depressed, air bubbles appeared once more in the long glass tubes.

Jehl continued dripping mercury and Edison applying current from time to time until all occluded gases were driven out. Less than 1/500,000 atmosphere of air remained in the apparatus. Edison cut out resistances. The current flowed through the carbon stick.

As Edison, Upton, Batchelor, and Jehl looked on, the carbon grew brighter and brighter. The light was reflected from the myriad bottles, and imbued their contents with a gemlike glow. Measuring the intensity of the illumination, Upton found it to be 40 candlepower. No platinum lamp had incandesced so brightly. Yet the carbon rod, its temperature uncontrolled by any mechanism, did not overheat.

Because of grease in the pump and a broken gauge, the vacuum was lost. Yet it was evident that in achieving the nearly perfect vacuum they had created a new medium. Carbon, which volatilized in the open air, could be maintained in a stable state. The question of whether a *high-resistance* lamp could be made out of carbon nevertheless remained, since the resistance of the rod had been only about 2 ohms.

To raise the resistance, it would be necessary to increase the length and thinness of the burner. Edison set Batchelor to work fashioning a spiral of carbon similar to the spirals of platinum on which insulation had been tested. Batchelor made a mold from which he was able to squeeze Wallace carbons out like spaghetti. But when they were shaped into spirals, fastened to the platinum wires, and baked, contraction always caused the carbon to break at its junction with the platinum. After Batchelor failed for two days to produce a spiral, Edison suggested they might as well try some plain thread. On Tuesday night, the twenty-first, Batchelor carbonized a piece of thread 13/1000 of an inch in diameter attached to platinum wires. At nine o'clock Wednesday morning, the thread, which had perforce taken on a horseshoe shape as it was bent between the lead-in wires, was inserted into a globe. When the globe was evacuated and current applied, the thread emitted a light of only ½ candlepower, and soon flashed out.

The resistance of the thread, however, was 113 ohms! Thus, in separate tests, carbon had proved that in an almost-perfect vacuum it was stable, it emitted a good light, and, if the burner was made thin enough, it provided high resistance. *All the conditions theoretically necessary for a parallel system of incandescent lamps had been fulfilled.*

Edison plunged in to test filaments (as the burners were renamed) of every potential substance: fish line, cotton, cardboard, tar, architects' drawing paper, celluloid, coconut hair, wood shavings, cork—even visiting cards! He danced across the lab, and went up and down the stairs with the spring of a mountain lion. He never went home, but stole naps on and off while waiting for a globe to be exhausted. The sun rose; the sun went down—he was oblivious, and lost all track of the

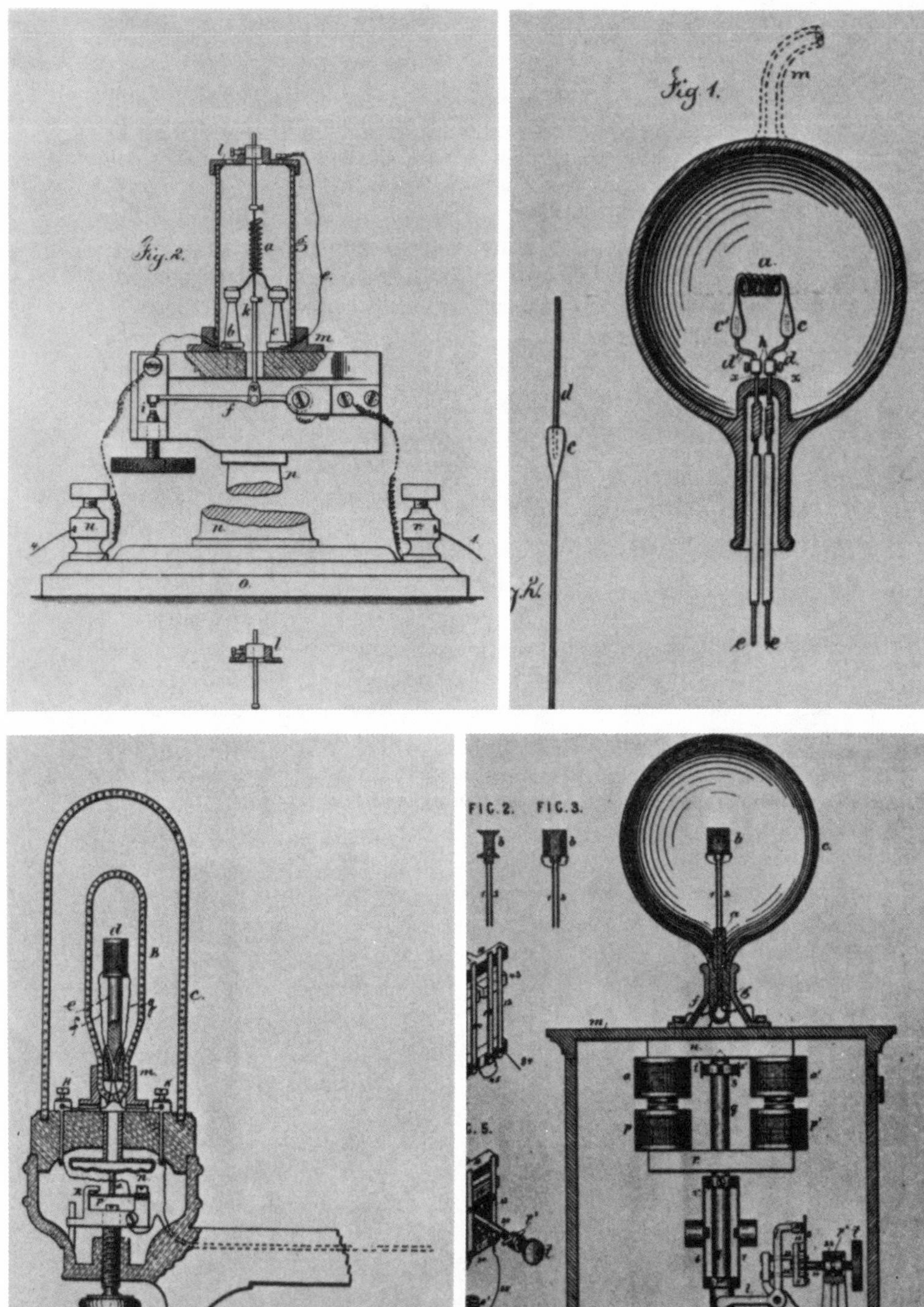

Fig. 2.
Fig 1.
FIG. 2.
FIG. 3.

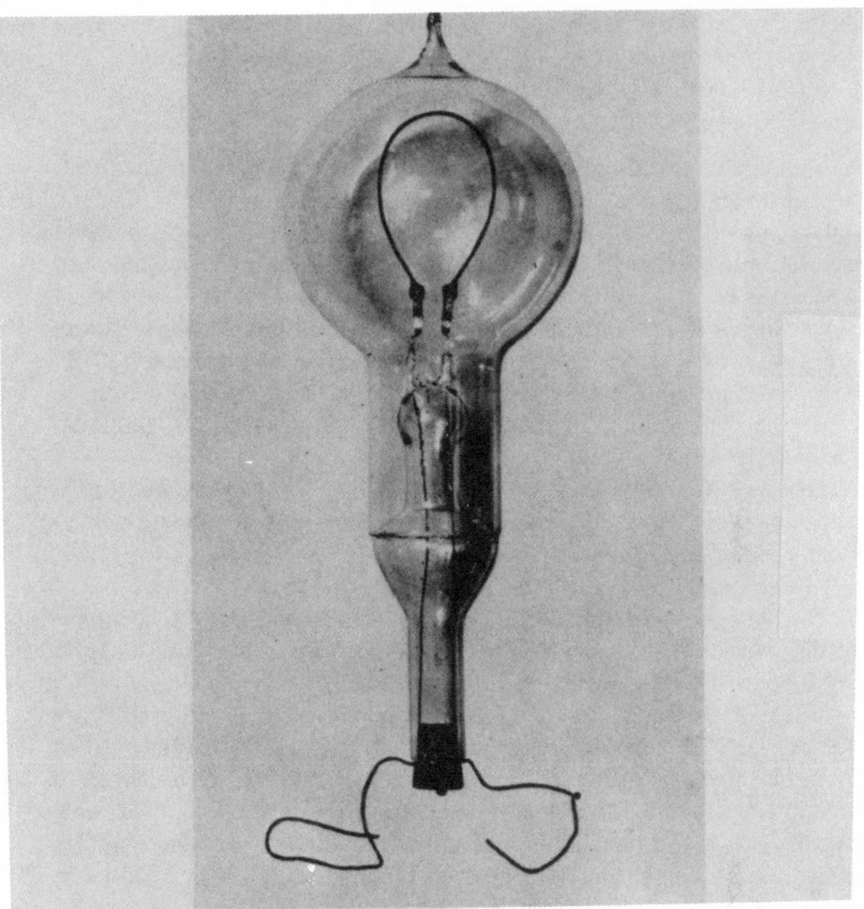

Development of the Incandescent Lamp. One of Edison's first designs in the early fall of 1878 envisioned a lamp with a heat-activated cutoff but no vacuum (facing page, upper left). As a rod (k) *was heated by the wire spiral* (a)*, it expanded and pushed against the lever* (f)*, cutting the flow of electricity. After Edison realized in the winter of 1879 that a vacuum was required, he devised the lamp at lower left. A bobbin* (d) *containing thirty feet of platinum wire was mounted in an exhausted inner globe* (b)*. The air in the outer globe* (c) *expanded and pressed against a diaphragm* (n)*, interrupting the current. This lamp was plagued by such sparking that Edison switched to a magnet-controlled lamp (facing page, bottom right), which became as complicated as a Rube Goldberg sketch. The bobbin* (b) *continued to be placed in a vacuum, but the current was cut in and out by two rotating and two permanent magnets* (o *and* p)*. Edison had committed himself to this design when he discovered, in mid-October, 1879, that carbon remained stable in a nearly perfect vacuum. He patented the spiral lamp (facing page, upper right), which he was unable to bring to fruition because of the difficulty of baking a carbon wire in spiral form. When Batchelor instead shaped the wire into a horseshoe, the first viable incandescent lamp (above) was realized.*

days. Toward the end of the week a lamp was placed on a current at one-thirty in the morning and glowed at ½ candlepower until three o'clock the next afternoon, when the glass cracked.

Immense delicacy and patience were required in the making and testing of the lamps. Frequently, hours were spent in preparing a filament, only to have it break. The twenty-four-hour days and the feeling that success lay just beyond reach affected everyone's nerves. The weather was hot, the grass dry and crackling, the laboratory with its pervasive smell of chemicals uncomfortable and throat-parching. Böhm, working with the blazing fire in the glassblower's oven, croaked repeatedly: *"Ein Königreich für ein Stein!"* Passionately fond of beer, he paid a boy "to chase the duck" and bring back a bucketful from Davis's Lighthouse, the local tavern.

Tuesday, the twenty-eighth, Batchelor made a filament, inserted it into the globe, watched the air exhausted, then saw the glass crack as Böhm sealed it.

"Shitt! Busted by Böhm!" he exploded.

The same day Edison wrote a new patent application: "The object of this invention is to produce carbon conductors of high resistance to produce the Electric Light by incandescence."

Much of this application, nevertheless, consisted of Edisonian hypotheses, rather than experimental results. On the horseshoe-shaped filaments the light flickered because of the pulsations of the dynamo; and Edison continued to believe the problem could be resolved only by shaping the filament into a spiral, so that it would retain heat. He furthermore thought the system would require a lamp with a resistance of several hundred ohms, not 100 ohms; to obtain that resistance he would have to make the filament longer; and to make it longer and fit it into a globe he would, again, have to shape it into a spiral. So despite the fact that he had yet to produce a lamp with a workable spiral filament, he specified a spiral or coiled form in the application, and even mentioned the possibility of mounting the filament on a bobbin.*

Pursuing the elusive spirals, Batchelor made filament after filament of interlaced strands of thread. Some he coated with various combinations of lampblack and tar before carbonizing them. But they all either broke at the contacts or, not being uniform, quickly developed hot spots where the resistance concentrated, and thus burned out.

The first week of November was a time of teeth-grinding trials. Hopeful eyes in a haggard face, Edison paced up and down. Painting visions of an electrical cornucopia that would spill the riches of the earth at the feet of all those in the laboratory, Edison coaxed, humored, and

* A decade later, when technology had improved sufficiently, it was demonstrated that it was possible to make a spiral lamp, and today General Electric produces an intriguing replica with a tungsten spiral filament. In 1879, however, the spiral filament was a wish, not a fulfillment.

drove his workers on. He told Upton that as soon as the check for British telephone rights arrived he would build "a new chemical laboratory, as fine a one as there is in the world." He talked of forming a new company with $3 million to $5 million capital floated by Drexel-Morgan. That could be done, he assured Upton, as soon as the light was exhibited. To prepare for the showing, he ordered, on November 4, 500 pounds of glass, and requested Western Union president Norvin Green, who had succeeded Orton, to send out two linemen to erect poles and string wires.

"Yet he is always sanguine, and his valuations are on his hopes more than on the realities," Upton told his family. The following Sunday, November 9, Upton grumbled that the electric light was "a continual trouble. For a year we cannot make what we want and see the untold millions roll in."

The next day, however, a bulb with a new filament was ready. Batchelor had convinced Edison to abandon the spirals, at least momentarily, and had taken a piece of cardboard used for mounting minerals, and cut it in the horseshoe shape of the threads. The cardboard horseshoe was boiled in sugar and alcohol, then carbonized. On Wednesday a fine lamp equal to a gas jet was brought to incandescence, and before it gave out, at noon the next Monday, it had burned, off and on, for sixteen hours. The combination of the cardboard with improvements in the generator eliminated most of the flickering, and Edison reconciled himself to working with a resistance of 100 ohms—it would necessitate greater outlay for conductors, but it was within reason.

Excitement replaced irritation. Impossible as it would have seemed only a month before, they might yet place a system aboard the S.S. *Columbia* when the ship was launched. On November 22 Upton expressed himself "Elated at the prospect of the electric light. During the past week Mr. Edison has succeeded in obtaining the first lamp that answers the purpose we have wished it for."

The next two weeks raced by with every advance tempered by setbacks. Attaching the carbonized filament to the platinum wires remained a major problem. The filaments from some cardboard were good, but others failed completely. Edison had still not yet reached the point of completely committing himself to carbon, so that in the specifications for the British patent he coupled the generator and distributing system with the platinum lamp, and made only a vague reference to "carbon sticks." He signed and mailed the application on November 25.

Three days later, Batchelor fashioned a steel cutting frame that enabled him to shape the horseshoes much more evenly. The same day, lamps number 89 and 94 were made from bristol board, which had fewer impurities than the stock tried before. They provided illumination of approximately 16 candlepower.

"My light is perfected," Edison wrote.

Leading out from the windows of the machine shop, the bare wires were strung to Edison's and Upton's houses. There they were attached to the gas pipes, and the bulbs placed atop the fixtures. The women decorated the lights with ribbons, flowers, and papier-mâché. The first display to members of the syndicate was set for Wednesday, December 3. The train arrived. Edison ordered the telegraph key that controlled the current at the machine shop switched on, and went to meet the guests.

At Edison's house, Mary, Alice, the children, and the servants watched fascinated as the globes turned to bright incandescence. Then there was a sharp crack, a bulb burst, the wires crossed, sparks flew. Suddenly the decorations in the parlor were ablaze. As they tumbled to the floor, puffs of flame shot up from the elegant rose-embroidered white carpet.

Alice rushed for a blanket to smother the flames. The financiers were expected any moment. Alice and Mary thought feverishly. What were they to do? Mary closed off the parlor. Edison arrived with the visitors. As his brows knitted in puzzlement, Mary apologized that she could not show the guests into the parlor—remodeling was under way.

The men were treated to a few lights mounted elsewhere in the house, but no more than half the directors had come. Edison had cried "Light!" in the darkness too many times before. Less than three weeks earlier, Lowrey had been unable to stir up any enthusiasm among board members to provide additional funds for Edison. One by one Edison's supporters had slipped away, while his critics chortled. A scientist had responded to a *Scientific American* article describing Edison's generator and distributing system by suggesting it would be "almost a public calamity if Mr. Edison should employ his great talent on such a puerility." Edward Weston, a young Newark arc-light and electrical-equipment manufacturer, called Edison's claims "so manifestly absurd as to indicate a positive want of knowledge of the electric circuit and the principles governing the construction and operation of electric machines." Edison's declaration he would run motors on the same circuit as lights brought only derision.

Yet here was the needle of a sewing machine powered into a streak of steel and operating on the same line with the lights. Edison had stood up against the scientific world, he had met nature head on—and he had won!

Everything, Upton told Lowrey, was "clear and perfect." The electric light was "an immense success." Rumors twirled like small tornadoes over the country. Edison sold twenty shares of the Electric Light Company stock for $10,000—a month later he could have gotten $70,000. Lowrey talked to J. P. Morgan, and the board voted Edison $75 a month for personal expenses so he could buy a decent suit, pay off his small accounts, and not be an embarrassment to Wall Street.

Money was coming in from all directions. After agonizing over the telephone chalks while Johnson bombarded him with letters that his repeated failure to keep promises "had placed the confidence of the English people at zero," Edison had solved the problem by shipping the chalks dry and having Johnson brush them with distilled water after they arrived. On December 2 Johnson reported "excellent results" —the flow of British royalties would be resumed. From the French telephone company Edison received the balance of $25,000 due. Even the platinum diversion would be made to pay dividends. With some of the Light Company investors, who now swore that the synonym for Edison was "riches," he incorporated the Edison Ore Milling Company. The magnetic separator would be put to work to sift the precious metal from the waste. "We are the boys to invent processes for any kind of ores," Edison crowed.

"A golden future dawning," the *Weekly Mercury* rhapsodized in California.

The prediction made in the *London Weekly Dispatch* more than a year before was being fulfilled: "All the existing conditions of life will be turned upside down. In the boundless spirit of enterprise which characterizes the undertakings of every true son of Columbia, this ingenious gentleman would seem to have entered into a stupendous contract with Nature to be permitted for a term of years to do exactly as he pleases with electricity."

Edison had no time for reflection. He slept on the couch in the office, and Mary seldom saw him the whole month. In the machine shop half a dozen generators were in various stages of production. The laboratory force could now make ten to fifteen globes a day, though not more than a third of these survived. New problems cropped up all the time. The glass was too thick and cracked under heat—Böhm had to learn to blow it thinner. On December 16 the 228th lamp was made; on the nineteenth the 260th. Not till that day did Edison file his caveat for the horseshoe lamp and the system to go with it—in England the patent office had just recorded the receipt of his papers on the magnet-controlled platinum lamp. Caveats and patents on different lamps and systems were tumbling over one another in the American and European patent offices. Hastily phrased, sometimes contradictory, they were to lead to nearly endless disputes and suits.

Throughout the fifteen months, Edwin Fox had been the writer most closely following the incandescent trail. He still owed Edison $125 of the $150 he had borrowed on the western trip, he had been gifted with several shares of Light Company stock, and he had an agreement with a New York publisher to write a book on the development of the light. Every few weeks he had gone to Menlo Park and taken notes. As soon as the horseshoe caveat was filed, he wrote the story. Momentarily he had difficulty convincing the editor at the *Herald* to run it—

after all, the paper had embarrassed itself in April with "The Triumph of the Electric Light." But the *Herald* was a newspaper that would rather be sorry than scooped. On Sunday, December 21, "The Great Inventor's Triumph in Electric Illumination" covered a whole page. It was "a bright, beautiful light, like the mellow sunset of an Italian autumn. A light that is a little globe of sunshine, a veritable Aladdin's lamp." Fox detailed the whole, grueling course that Edison had followed. But most of the article—in keeping with the history of Edison's pursuit—dealt with platinum. Fox did not know how the carbon filament had come about. It popped up near the end, almost as an afterthought.

All through the month the speed of events at Menlo Park had been accelerating toward the frenetic climax. Some sixty surviving bulbs and three Long-waisted Mary Anns—Fox's new designation—were ready. Two of the dynamos were connected to the wires; the third served to excite their field magnets. Lights had been installed in the laboratory, the railroad station, and the half-dozen houses in the village. A few globes were placed on poles along connecting ways. Such frills as switches were still to be developed. The whole system was controlled by the telegraph key in the machine shop.

Within a day of the *Herald* publication, sightseers were drifting out. On Christmas night the whole system was tested for the first time. In other places arc lights had transformed night into a garish dusk, and yellow gas lamps cast mysterious shadows. But never had there been a light like this: pale yellow-white, unflickering in the wind, a light that seemed to turn the countryside into a mirror of the sky.

Each night the number of visitors increased, while elsewhere experts, tangled in the *Herald* story, still scoffed in disbelief. Platinum lamps and carbon burners had all been tried before and failed. The crucial elements, the perfect vacuum and the slender filament whose existence depended on it, were not perceived. The prestigious Du Moncel could only shake his head. "One must have lost all recollection of *American hoaxes* to accept such claims. The sorcerer of Menlo Park appears not to be acquainted with the subtleties of the electrical science. Mr. Edison takes us backwards." At nearby Stevens Institute, Professor Henry Morton, who had been with Edison in Wyoming, felt compelled "to protest in behalf of true science." The results of Edison's experiments, he asserted, were "a conspicuous failure, trumpeted as a wonderful success. A fraud upon the public."

And every night Edison added a few more lamps, a few more people came to gaze and wonder, and leave enchanted by the sorcerer. On New Year's Eve, when the first "public exhibition" was held, the platforms of the Pennsylvania Railroad in Newark and Jersey City, Trenton and Philadelphia were jammed. The railroad added extra trains. Three thousand people poured into the one-store village. They might have turned the place into a carnival. But most had a sense of historic wit-

Jubilation is demonstrated by this laboratory-notebook sketch at the end of 1879.

ness. They came in wonder, they walked in awe, they left touched by the magic. Even the cynic *Puck* was muted: "The Northern Lights pack valise and go on. The moon goes into mourning. A New Light to the World."

CHAPTER 13

A Wide-Awake Wizard

Except for a dozen lamps* broken or plucked from their sockets, the installation had come through unscathed. After the first of January Edison announced the lab was closed to visitors. But, actually, anyone with a pretext or perseverance was admitted. Sometimes they stayed on for the midnight songfest at the organ:

> "I am the wizard of the electric light
> And a wide-awake wizard, too,
> Quadruplex, telegraph, or funny phonograph
> It's all the same to me.
> With ideas I evolve and problems that I solve
> I'm never, never stumped, you see."

The making of the lamps was a laborious and uncertain process. Much depended on luck, for the slightest imperfection in the carbon or the seal caused the lamp to burn out. On January 3 the lamp at the head of the stairs leading to the railroad station held the longevity record of 40 hours. Within three weeks the record had been extended to 550 hours. The morale and confidence of Edison's workers and associates was typified by Upton, who jotted down:

"Francis R. Upton *Immortal.*"

A dozen young men who were to have important roles in the development of the electric industry were drawn by the excitement, and joined the lab force without pay or for very little pay—terms on which Edison would hire anyone. Even the university-trained among them knew little about the properties of electricity. Frequently they short-circuited themselves as well as the equipment.

* The term "light bulb," subsequently shortened to "bulb," did not come into use until many years later.

Henry Villard, who a few months before had formed the Oregon Railway & Navigation Company and was riding the crest of a financial surge, was the newest member of the Light Company's executive committee. The first week in February Villard wired one of his German banker associates: "Will you join me in purchase of half of Edison patent rights for Germany, Austria, Russia, France, Italy, Spain for about $450,000? Invention much perfected in these few days and practical success, notwithstanding published unfavorable reports. Believe this extraordinary opportunity for quick enormous profits."

The banker replied dourly that he might be willing to take a $25,000 interest, "Whenever it is proved that the invention is practicable." After the first explosion of enthusiasm, skepticism was reasserting itself.

Upton was regretting he had not sold some of his shares at the end of December, when the price was $3500 bid and $5000 asked—only a month later the price was down to $1500.

William Sawyer was unrelenting in his attack on what he called Edison's "horse-hair lamps." Sitting in a Manhattan saloon, he drunkenly dictated letter after letter to the newspapers. Believing that the only novelty in Edison's lamp was the horseshoe shape of the filament, Sawyer went to the *New York Tribune* and exhibited a lamp with a similar-shaped carbon burner he had made two years before. Such a lamp, however, Sawyer asserted, was worthless: "I challenge him: First—to maintain a vacuum in his lamps. Second—to run his carbonized paper lamp three hours. In practice, in a perfect vacuum, it will last twenty minutes."

Fully convinced that whatever results Edison was obtaining were due to trickery, Sawyer held exhibits of his own nitrogen-filled lamp with a pin-sized amber burner. So far as the observers could see, the burners were permanent. In actuality, Sawyer unfastened the globe and inserted a new burner between every demonstration.

Yet for all his drunkenness and buffoonery, Sawyer was a respected electrician, a citable authority for Edison's detractors. The hooting was loudest in England. There, although the *London Times*'s American correspondent vouched for the success of the lamp, the *Saturday Review* remarked:

"Three times within the short space of eighteen months he has had the glory of finally and triumphantly solving a problem of worldwide interest. It is true that each time the problem has been the same [but] there is no reason why he should not for the next twenty years completely solve the problem of the electric light twice a year without in any way interfering with the interest or novelty. . . . There is a strong flavor of humbug about the whole matter. . . . Mr. Edison's efforts in electric lighting seem cursed with a total absence of originality. . . . He is an inventor who is absolutely intoxicated with his own reputation. . . . Successes seem to have completely turned his head."

The hounds were led by William Preece, the Post Office's electrician, who was intent on making Edison pay dearly for the attack on his integrity. Edison's British patent application on the telephone had, as so often was the case with his papers, been drawn up hastily and haphazardly. Exploiting Edison's carelessness, Preece charged that "Mr. Edison has laid aside his peculiar form of transmitter" and replaced it with "a mere form of microphone." The telephone, Preece asserted, was merely a variation of the telegraph, and therefore should fall under the telegraph monopoly granted to the British Postal Authority.

Edison's London solicitor, Theodore Waterhouse (of the firm of Price-Waterhouse), thought Edison should come to England. Edison, however, replied that such a journey "at the present stage of my Electric Light Developments would be paramount to an abandonment of the stronghold in the face of an eager and aggressive enemy."

The affairs of Edison's British telephone concerns were, in fact, in a terrible tangle. In London, Bell had 800 subscribers, Edison only 200. Both Johnson and Gouraud were promoters, not administrators or technical men. "This Co. is like a ship without a commander, it would take the bank of England to keep it going at the rate it is going now," one of the men Edison sent over reported. In May, agreement was reached to fuse the British Bell and Edison companies, the Bell interests receiving 40,000 and the Edison 23,000 shares in the new company.

In America, the competition between Bell and Western Union had been resolved some six months earlier. From January to September, 1879, Bell and Western Union had been pitted against each other in court. No fewer than ten inventors claimed all or significant portions of the telephone as their handiwork. Western Union controlled the patents of the principal contenders against Bell, and both Elisha Gray and Tufts University Professor Amos E. Dolbear had well-founded claims that Bell had appropriated substantial portions of their conceptions for himself. But there was no gainsaying that Bell had been the first to construct a workable apparatus. (Wilber was not to make his affidavit of the assistance he had given Bell for another seven years.) Western Union magnate Hamilton McK. Twombly could foresee only endless litigation and costs—in fact, before the cases of the individual inventors against Bell and one another were settled, the testimony filled 149 volumes over a span of eighteen years.* On

* The Speaking Telephone interference cases were as follows: *W. L. Voelker* v. *Dolbear* v. *Gray* v. *Edison* v. *Bell; Voelker* v. *Gray* v. *Edison* v. *Bell; Gray* v. *Edison* v. *George Richmond* v. *Bell; Gray* v. *Edison; Edison* v. *Gray* v. *Dolbear* v. *Bell; Dolbear* v. *Gray* v. *Bell; Edison* v. *Gray* v. *Dolbear* v. *J. W. McDonough* v. *Bell; Gray* v. *Edison* v. *Bell; Edison* v. *Emile Berliner; Gray* v. *Berliner* v. *Richmond* v. *Dolbear* v. *A. G. Holcombe* v. *Bell.*

The oddest and perhaps most interesting suit was that of Daniel Drawbaugh against Bell. Drawbaugh, a small-town mechanic, lived near Harrisburg, Pennsyl-

November 10, 1879, Western Union agreed to transfer its telephone patents and business to the Bell Telephone Company. Bell, in turn, gave Western Union 40 percent of the stock of the New York and Chicago telephone companies and assigned 20 percent of earnings to Western Union for seventeen years (the life of the patents).

Edison's name was stamped on each telephone transmitter installed by Bell, and he was anxious "to be *perfectly satisfied* that the W. U. must continue the payment of $6,000 yearly." In March he began shipping carbon buttons to the Bell company. (The carbon continued to be produced by smoking kerosene lanterns. The secret lay in keeping the flame down, so that the deposit was made at a low temperature.) Theodore Vail, president of Bell, complained of the price, eight dollars per hundred, especially since the buttons frequently crumbled and had to be replaced.

"A carbon button is *never* broken in a properly constructed transmitter," Edison replied assuringly. "They will last a century."

Edison's feat in producing an incandescent light dramatically reduced anyone's inclination to dispute with him. No longer was he, the inventor, at the mercy of the capitalists. He had achieved a power above and beyond them. On January 2 Western Union settled the quadruplex dispute by paying him and George Prescott $100,000. Several months before, Edison had disposed of American rights to the electromotograph telephone receiver to Western Union for a royalty of $1 per instrument, though the total he was to receive annually was limited to $3000.* He now pressed Western Union to buy the basic electromotograph patent, which he had received two years before signing the contract with the telegraph company in 1877. On February 23 he told President Norvin Green (who was also president of the Edison Electric Light Company) he wanted $100,000 for the patent, and expected an answer within ten days.

Edison's virtual ultimatum caused consternation among Green and

---

vania. Supported by scores of witnesses, he claimed that in 1872 he had constructed an apparatus with which he was able to hold conversations over a wire one hundred yards long, but had given up the experiments because his wife had nagged him that he would have the whole family in the poorhouse if he kept working on his "talking machine." On March 19, 1888, the United States Supreme Court, in a 4–3 decision, upheld Bell's patent against Drawbaugh, the minority ruling that Drawbaugh had preceded Bell and that fraud had been involved in the obtaining of Bell's patent. So, by a one-vote margin, it remained the American Bell Telephone Company, instead of, perhaps, becoming the American Drawbaugh Telephone Company.

* This was an almost unbelievably ambiguous contract. Western Union obviously intended to place a low ceiling on the total amount it would have to pay Edison. But the construction of the language was such that Edison might have accumulated hundreds of thousands of dollars in royalties, but been able to draw only $3000 a year.

the company's executive committee. In June of 1879, Jay Gould, dissatisfied with his revenue-pooling agreement with Western Union, had organized another competitor, the American Union Telegraph Company, and this was growing into a far bigger threat to Western Union than the Atlantic & Pacific ever had been. When Green asked Edison what value the basic electromotograph patent had without the subsequent improvements to which Western Union already held title, Edison replied:

"It has immense negative value if not positive value."

Green knew quite well what he meant. If Gould were to obtain control of the basic patent, all the succeeding patents owned by Western Union would be worthless.

Green suggested that, "after due experiment, for which the contract allows three months time, royalties should be agreed upon by reference as provided for in the contract."

But Edison, no longer the suppliant, retorted: "If any such damned nonsense is to be attempted on me it will not work."

"The existing contract under which you are operating with this company appears to be a liberal one on our part," Green pleaded. "There should be no hesitancy in awarding us cheerfully all the patents you produce."

"Liberality is not in the W. U. dictionary," Edison snorted.

He had learned from the toughest, most amoral all's-fair-in-stocks-and-money operators who ever walked Wall Street, and he was not loath to play Gould and the Vanderbilts off against each other. "If the company thinks the price too high, well and good, that settles it," Edison advised Green, and initiated negotiations with Gould.

Five days later, on March 22, the Western Union executive committee took a collective hard swallow and capitulated. Edison would get his $100,000—$16,000 down and the remainder in monthly installments spread over twelve years.

For British telephone rights he was to receive $100,000 during the course of the year, raising the total to $175,000, and Canadian payments brought in another $14,000. During 1879, even while frequently acting like a pauper, he had steadily increased his balance at Drexel-Morgan from $13,000 to $45,000. At the age of thirty-three he was worth some half-million dollars. He was one of the richest men in America, and what the incandescent light would bring only Midas might reckon.

The first incandescent installation was to go on Villard's new 3200-ton steamer, the *Columbia*, and Upton was placed in charge of the project. In mid-April the ship sailed from Chester, Pennsylvania, where it had been built, to New York. There it was anchored at the foot of Wall Street, near John Roach's ironworks on Goerck Street, to be fitted out. Sigmund Bergmann's shop in New York was given a rush order to make four Long-waisted Mary Ann dynamos to power the 150-lamp installation. (Each dynamo produced enough current to operate sixty 16-candlepower lamps. The fourth machine served to excite the field magnets of the other three.) The wires, wrapped in cotton and treated with hot paraffin, were tacked along the corridors. Switches were located in locked metal boxes outside the rooms, and had to be turned on and off by the stewards. Lamps were placed loosely in crude wooden plugs, and of course had to stand upright, for there was as yet no means of fastening them in.

At the laboratory, Batchelor, Upton, Böhm, and Jehl slaved to produce the delicate carbon horseshoe bulbs. Upton personally supervised their transportation to the ship, for the slightest jar could break the filament. On April 28 the system was tested for the first time. New Yorkers were treated to a spectacle of glowing lamps reflected by the dark water. The next evening a gala supper and dance were held aboard.

Ten days later, on May 9, the vessel left for Portland. The first commercial incandescent installation in the world sailed out onto the Atlantic Ocean. Circumnavigating the tip of South America, the ship stopped at Rio de Janeiro, Valparaiso, and San Francisco. Everywhere the floating display was greeted with exclamations of amazement. By the time the steamer reached its destination, on July 26, half the lamps had given out, and Edison had to rush a shipment to the West Coast. But the practicality of the incandescent light had been demonstrated to the world.*

Edison applied himself with equal intensity to adapting the Long-waisted Mary Ann to uses as a motor. In the summer of 1879 Werner von Siemens had demonstrated a lilliputian electric railroad at the Berlin Industrial Exhibition. But, with the operator sitting astride the locomotive as if it were a hobby horse and the track only 1000 feet in length, it had been nothing more than an amusing novelty. Edison, possessing a vastly more powerful dynamo, decided to build the model of a full-scale road.

Purchasing some light, secondhand horsecar tracks, he had his men cut down trees for ties, and laid the rails over an ungraded roadbed a third of a mile long. He and Batchelor mounted a twelve-horsepower Long-waisted Mary Ann on a truck with solid iron wheels, and attached brushes to the wheels to pick up the current from the rails. Resistance

* The *Columbia* remained in service until it collided with the *San Pedro* and sank—with the loss of eighty-eight lives—off the West Coast in July, 1907. One of the remarkably well-constructed dynamos had previously been removed and is now at the Smithsonian Institution in Washington, D.C.

boxes surrounding the operator's seat were cut in and out to control the speed. A long-handled wagon brake brought the machine to a halt.

On May 13 the engine, pulling a horsecar with twenty passengers, received its formal trial. Ensconced on an upside-down beer crate, Batchelor assumed the role of engineer. Pulling out resistance boxes, he accelerated the one-car train to twenty miles per hour—a terrifying speed, at which it humped and swayed over the rough tracks like a berserk camel.

Edison guaranteed the riders the railroad was perfectly safe—so long, at least, as they were acrobatic. One day in early June Kruesi was at the controls when the engine jumped the track, and he and another man were catapulted into a thicket. They emerged, after a moment of suspense, picking brambles from their hands and faces.

"Oh yes! Pairfeckly safe!" Kruesi muttered in his German accent; while Edison, jumping and laughing, declared the accident to have been "A daisy!"

Although there were such occasional derailings, and the leather belt that connected the motor to the wheels frequently broke, within a few months the speed of the engine was doubled. The president of the Rockaway Railroad expressed interest in having Edison electrify the tracks and build engines for operation of the railroad between Rockaway and Perth Amboy, New Jersey. Villard, Butler, Lowrey, Fabbri, and other Light Company investors proposed the formation of an Edison Electric Railway Construction Company. Colonel Gouraud, arriving from England, advanced Edison $1800 for a European option, thus partially recompensing him for the $16,500 he had expended on the engine and the tracks.*

Edison had, at this juncture, a lead on other inventors in the development of an electric railroad. But it was a major project, one for which he did not have the time or personnel as he wrestled with the problems of composing the world's first incandescent-lighting system. For more than a year, Edison let the railroad lie fallow; and it eventually turned out to be an opportunity on which he failed to capitalize.

* Gouraud was not only interested in the railroad, but was fascinated by Edison's revelation that he had employed the vacuum pump to preserve food in glass vessels exhausted of air. Gouraud, who could have stepped straight into a Gilbert and Sullivan operetta, gave Böhm a large steak and had him seal it into a huge jar, from which it took five hours to exhaust the air. Gouraud carried the jar back to England, where, opening it, he consumed the steak. It had acquired a fermented, alcoholic flavor, which Gouraud, who liked his whiskey, did not particularly mind.

"Conclusively there's something in it," he wrote to Edison. He was willing to invest a thousand dollars in the process, and wanted it patented—a request Edison complied with. He asked Edison to send him a few grouse and prairie chickens—some with the feathers on and some with the feathers off. "I want this thing followed up—set somebody at it without delay," he wrote. In the summer of 1881 Edison packed peaches, and shipped steaks and chops, but they arrived in such advanced states of fermented putrefaction that Edison labeled the process "not yet perfected" and dropped the project.

CHAPTER 14

The Giver of Light

By the summer of 1880, Edison was unquestionably the world's greatest inventor. Yet he had shepherded few of his inventions on to commercial success, and many skeptics continued to predict the incandescent light eventually would also be abandoned.

The durability and efficiency of the bristol-board filament were poor. So many lamps were broken during production that Edison could, at best, turn out three a day. In all the lamps there was a rapid deposit of carbon from the filament onto the inside of the glass—within a few score hours the amount of light emitted was sharply reduced, and before long the bulb was totally black. One horsepower could be converted into only 76 candlepower, equal to appproximately five of Edison's standard bulbs.* In arc lamps, 1 horsepower produced as much as 1200 candlepower.

To increase production capacity, Edison, on April 25, paid $825 for the building along the railroad tracks that had housed the electric pen factory of Robert Gilliland. To improve the quality of the lamp, Edison and Batchelor began testing every conceivable substance for a filament: Southern moss, palmetto, monkey grass, Mexican hemp, jute, bamboo, coconut palm, and manila fiber were dipped in rock-candy syrup and carbonized. Strands were plucked from Kruesi's and Mac Kenzie's beards, and bets placed whether Kruesi's black hair or Mac Kenzie's red would prove the longer lasting. (Mac Kenzie won a disputed decision.) A thread snipped from a spiderweb turned beautiful light pink and produced green phosphorescence.

Edison was using hot mercury, with which he could obtain a better

* Since 1 horsepower is equal to 746 watts, it took 9.8 watts to produce 1 candlepower. Today 1 watt produces about 1.3 candlepower—in other words, incandescent lights are approximately twelve times as efficient.

vacuum, instead of cold, but this greatly increased the amount inhaled by the men. Everyone associated with lamp production was developing severe salivation. Undeterred, Edison prescribed iodide as an antidote, and made the workers take large doses daily. But the men were as opposed to swallowing the nauseating iodide as they were to continuing to work in the poisonous atmosphere.

In late June, Batchelor collapsed from excessive exposure to mercury and years of overwork. A month later, when he had recovered enough to travel, he took his family on an extended vacation to Europe. Although he returned to Menlo Park in the fall, Europe was to be his primary area of operation for the next four years, and the intimate, shoulder-to-shoulder working relationship with Edison was never re-established.

With Batchelor's incapacity, Edison pressed twenty-year-old Francis Jehl into service as his principal assistant in the filament experiments. On Friday, July 9, Edison absentmindedly plucked at a bamboo fan frequently employed in the laboratory to evaporate liquids in dishes—despite the electrical pioneering and experimenting, air was still stirred about by a vigorous back-and-forth movement of the hand. Edison had tried bamboo before without obtaining particularly encouraging results. But, examining a fiber from the top of the fan under a microscope, he noticed it to be of slightly different composition, and told Jehl to carbonize it.

The filament was sealed in test lamp number 1253. Electricity was applied Saturday, then, as was the custom, turned on and off intermittently during the next few days. While other lamps made of palmetto and paper burned out, the bamboo lamp continued to glow.

Six weeks of further experimentation convinced Edison bamboo was the material he was seeking. The problem was, where could he find bamboo similar to or better than that contained in the fan? He wrote the American consuls in Puerto Rico, Paraguay, Jamaica, and Panama for information. He sent one man to Japan, two to the Amazon, and asked Murray if he would go to India. He dispatched an adventurer named John Segredor to the Florida swamps. From there Segredor wrote:

"What makes this job extremely interesting is the strong probability of getting bitten by a snake."

A month later he was ordered on to Cuba by Edison, and on October 27 he was dead of yellow fever.

"Bury him at my expense," Edison wired.

By the end of the year one of the 16-candlepower bamboo lamps had lasted 700 hours and an 8-candlepower lamp 2450 hours. Tests of a shipment of bamboo from Japan were so successful that on March 3, 1881, Upton ordered 50,000 fibers. Yet because bamboo was slightly more expensive, Edison hesitated about abandoning paper for the filament. On May 4 Upton remonstrated with him:

"The difference between a paper loop and a bamboo loop is about 2 cents in cost. As the life of a lamp depends entirely on the carbon, if the bamboo is 10 per cent better we have saved the 2 cents. Bamboo is far more than 10 per cent better than paper."

What settled the question beyond doubt was that the lamps Edison shipped to the patent office all arrived with the paper filaments broken —a disaster resulting from the slight jarring occurring when the packing cases had been nailed shut.

Batchelor's departure and the absence of Mary and the children, who often visited relatives and spent August at Asbury Park on the Atlantic, made the summer of 1880 a lonely one for Edison. Menlo Park provided few diversions, one of the notable ones being the occasional visits of Upton's sister, Sadie, whose appearances had the men burning their fingers, increasing their intake of beer and cider, and writing desperate notes: "My dear Sadie: Will you meet me tonight at . . ."

After the laboratory closed Saturday evenings, most of the men hustled off to New York or to their families in nearby cities. Edison customarily took an early-evening nap, and then returned to the lab. Frequently he found Jehl, who slept in the attic, still there. Sometimes Edison invited his assistant to accompany him to Manhattan. Off they would go on the train, usually not reaching the city until after 11:00 P.M.

The ferry landed at Cortlandt Street, and Edison habitually walked Jehl past the old familiar haunts: the Western Union building on Broadway, the Astor House near City Hall, Printing House Square with its newspaper offices, and the Bowery with its nightlife and throngs of people. Edison liked to lose himself in the crowds, and watch sleight-of-hand artists take suckers, and quacks sell medicines. He got a kick out of the noisome tenderloin gymnasium, where slum kids from Five Points and Mulberry Bend pummeled each other for a square meal while sweaty, cigar-puffing men placed bets with gold dollars. He frequented the theaters: Wallack's, Union Square, and the Old Bowery, with its blood-and-thunder dramas.

One Saturday midnight, with Jehl tagging along, Edison stopped at a three-story house just off the Bowery and banged on the door. A head appeared in a third-story window, and a German-accented voice called out:

"Where's the fire?"

"Get dressed and come down. I want to speak to you," Edison shouted back.

A few minutes later Sigmund Bergmann, in good humor despite the late arousal, appeared in the doorway. He was a handsome, spade-bearded man who had arrived in the United States from Germany in 1869, when he was eighteen years old, and had worked for Edison at the Newark shop. When Edison had had his blowup with George Harrington, Bergmann had come to New York and opened a one-man, one-helper establishment on Wooster Street. He was a sharp bargainer, a good businessman, and an excellent craftsman. He made many of Edison's telegraphic and telephonic devices for Western Union, and was the exclusive manufacturer of the phonographs.

Edison took his companions to a German restaurant, and there discussed business with Bergmann. Edison was planning the installation of his first incandescent-lighting system in New York, and wanted Bergmann to machine some of the dynamos, fixtures, and accessories he would need. Mains and feeders, insulators, sockets, fixtures, meters, regulators—all had to be developed from scratch.

To construct the model for the system, Edison laid out across the fields of Menlo Park a grid of imaginary streets and houses. Upton calculated the size of the conductors. To compensate for the drop in electrical pressure as the mains led farther and farther from the station, he devised a feeder system to boost the voltage at intermediate points.

The size of the mains—whose diameter would have to equal the sum of the individual wires leading to the houses—made it impossible to suspend them from poles. They would, therefore, have to be buried like gas pipes.

Edison thought wood a good insulator and predicted that, because of the low voltage of the current, there would be no trouble with leakage. A couple of old telegraphers shook their heads, but he was not to be swayed. Shallow trenches were dug, wooden moldings laid in them, and the wires placed in grooves. The trenches were then covered with six inches of earth. From the main, wires led up to hundreds of wooden posts, which jiggled like loose teeth and were emblazoned with six-inch-diameter lamps looking like hussar helmets. On July 20 everything was ready. Edison waited for a wet day to conduct a full-scale test.

The rains came. Electricity went in all directions. Even the trees were charged. The lamps, like dying fireflies, produced only the faintest flicker.

The entire system was uncovered. An attempt was made to insulate the wires by pouring coal tar into the grooves. It failed because of the acidity of the tar. Muslin was wrapped around the wires, but produced no improvement. One thing after another was tried, until Edison, finally tiring of random experiments, told Wilson S. Howell to go into the library, read everything on insulation, and report back in two weeks.

Working with the information supplied by Howell, Jehl started testing various compounds and combinations: three layers of tar and three of rubber cloth; thick muslin with hot linseed oil; three thicknesses of rubber cloth treated with rubber cement; three layers of white rubber cloth combined with two servings of pine tar.

The compound ultimately settled on was Trinidad asphalt mixed with linseed oil and small amounts of paraffin and beeswax. The mixture was cooked up in 250-gallon kettles in the first-floor chemistry lab, and permeated the building with a nauseating stench that drove everyone out. Boys dipped two-and-a-half-inch-wide strips of muslin into the hot compound. The wires were lifted from the trenches onto sawhorses, and three layers of the muslin were wrapped around them. It was tedious, filthy work that paid seven cents an hour. It had one fringe benefit: the compound turned out to be quite chewy, and soon chewing insulation was the rage at the lab.

The tar also attracted land turtles, of which there were thousands in the vicinity. Some days as many as twenty turtles were pulled from a single two-foot-square insulation-filled junction box.

On November 1 the first portion of the reconstituted system was ready for testing. The next day was the Presidential election. "If Garfield is elected, light up the circuit. If not, don't!" Edison ordered. That evening Edison and some of the men gathered around the telegraph key on the second floor. As the returns came in, they were tallied. When it was evident that Garfield had won, the wheel was turned. The current stayed in the wires. The lights winked on across the fields.

By Monday of the following week the electricity had reached Edison's house. By the end of the month Edison had 400 lamps connected.

Eleven dynamos were in operation. Edison was using so much water for the steam engine that the well ran dry. One of the men was rushed to New York to buy a mile of pipe so that water could be pumped from the brook to the powerhouse. On November 20 Edison sent Tracy Edson, who was chairman of the Light Company committee negotiating with the city council for the right to put in street mains, his plans for laying the wires. Two weeks later he asked the company's executive committee to authorize the hiring of a mathematician for mapping and calculating the size of the conductors.

Edison employed more than a hundred men in the lampworks along the railroad tracks. Bulbs were purchased in semihemispheres from Corning Glass, and fused together after the insertion of the filament. Charles Clarke, an engineering classmate of Upton's who had come to work for twelve dollars a week the first of February, devised a ground-glass stopper for sealing the lamps. It was an invention that proved a key to large-scale production.

The lamp-factory workers were paid only a dollar a day, and were worth every penny of it. Drunkards were numerous. The alcohol used for sealing the bulbs was exhausted so quickly that Upton had the liquid tinted green; and green-lipped men were dismissed.

One man used to ask for alcohol to rub his feet, but stopped after he was given a cup of wood alcohol instead of grain alcohol. Returning a few days later, he rasped: "That alcohol you gave me last did not do my feet any good."

The person who had most difficulty adjusting to the new atmosphere of the mini boom town was Ludwig Böhm. He was sensitive, artistic, and egotistic. At midnight suppers in the lab he played the zither and yodeled. He liked children, and made toy glass figures, including trick swans that squirted a fine spray of water into one's face when one blew into their tails. He had started working side by side with Upton in the laboratory, then been moved into the adjacent, small photographic and blueprint shed. (This, thereupon, acquired the name of the Little Glass House.) He now spent much of his time at the lamp factory. There he came into conflict with William Holzer, a balding, handlebar-moustachioed Philadelphia glassblower who had dropped off the train while on his way to the Corning Glass Works and been hired in February. While Böhm prided himself on his craftsmanship, Holzer stressed productivity, and made iron molds to increase output. Holzer was as convivial as Böhm was introspective. A great storyteller and humorist, Holzer became friendly with Mary's sister Alice, and initiated a courtship. He and James Hipple, a lamp-factory worker who was the husband of another of Mary's sisters, were soon close pals.

It was no contest. Böhm, who looked like a gawky toy soldier with his lorgnette and funny little red hat, was the butt of the rowdy Americans' jokes. He moved out of Sarah Jordan's boardinghouse, where men were doubling and tripling up in the rooms, and put up a bed in the attic of the Glass House. There the men proceeded to drive him into a frenzy by tossing rocks onto the roof at night, tapping on the windows with a remote-controlled device, and running the "corpse reviver," a ratchet that sounded like a derailing freight train, along the walls of the shed.

One night, in retaliation, Böhm unleashed a shotgun blast over the heads of his tormentors. The next day when he sat down at Sarah Jordan's boardinghouse to eat, every man slammed a club, knife, or pistol onto the table. Niel Van Cleve, another of Edison's brothers-in-law, waved a huge army revolver and imperiously commanded: "Böhm! Pass the butter!"

Böhm bolted. Aggrieved, he complained to Edison in mid-October that he had received "a kind of treatment which no man with any sense of honor can bear. I do not want to be bossed by people that understand less than I. Things have gone too far." He resigned.

Edison told Böhm he could come back and work in the laboratory instead of the lampworks. But Edison had little understanding of personal relationships, and lacked the touch for handling men. Böhm left.*

He applied for a position to Hiram S. Maxim, who was emerging as

* After some time he returned to Germany, and in 1886 received his Ph.D. from the University of Freiburg. He proudly wrote Edison of his accomplishment, but Edison refused to see him when he returned to the United States.

Edison's most serious rival in America.* Maxim, of course, welcomed Böhm, and Böhm designed a filament in the shape of a Maltese cross for his new employer. Maxim had considerable experience in arc lighting, and had filed an application for a low-resistance incandescent lamp using a carbon strip in October, 1878. He had followed this up with another application on December 22, 1879, the day after the *New York Herald* story appeared. Three months later he had come to Menlo Park, and Edison had spent several hours explaining the process of making filaments and lamps to him.

Shortly thereafter the investors in Sawyer's lamp had asked Maxim to take over Sawyer's work. (Sawyer thereupon accused Maxim of piracy.) A few years before, Maxim had invented a machine for manufacturing gas from light hydrocarbons. Working with the Sawyer lamp, Maxim devised a means of "flashing" the filament—the bulb was filled with hydrocarbon vapor, which was then combusted. The process deposited a hard finish on the fragile carbon, strengthening it greatly.

Edison was aware of Maxim's technique, since Sir Henry Bessemer, the inventor of the Bessemer furnace, had written him on January 12, immediately upon hearing of Edison's success, to suggest use of carbureted hydrogen gas for hardening the filaments. The process could make any substance "crisp and hard and immensely stronger," Sir Henry had noted.

Edison, however, was instinctively prejudiced against any process or apparatus devised by someone else. "It is a want of knowledge on the part of Maxim that makes him use carbon for the decomposition of a hydrocarbon," he rationalized. Instead, he tried coating the filament electrically in a vacuum, but gave up after the apparatus disappeared in a spectacular explosion of fulminate of mercury.

The chemist, Alexander Haid, was allowed free rein to see what he could do, and treated filaments with hydrocarbon and other gases. After a few months, Upton reported: "Dr. Haid has just sent down 75 of the finest bamboos I have ever seen treated."

Edison, nevertheless, ignored Upton's recommendation, and spent year after year agonizing to produce filaments perfectly uniform—the only condition in which they were durable for any length of time without flashing.

Maxim, to the contrary, adopted all of Edison's processes, and applied flashing to the filament. By October, 1880, he had twenty men working in the rooms of the United States Electric Light Company, an arc-light concern formed two years before. The facility was in the

* Maxim was a prolific inventor, whose best-known invention is the machine gun. He narrowly missed preceding the Wright brothers in developing the airplane, his steam-powered machine having too heavy an engine to get off the ground. He dropped out of the lighting field, and did not return to the United States after attending the Paris Exposition in 1881, but settled in England.

Equitable Building on Broadway, and there a demonstration lighting system was set up in the basement room of the Mercantile Safe Deposit Company. Sixty lamps were lit the first week in November, and remained burning night and day.

Maxim, of course, except for the flashing, had done nothing Edison had not accomplished the previous December. But he did it in New York, where hundreds of people saw the lights every day. Edison, meanwhile, was uncharacteristically out of the public eye as he strove to put together a whole system.

Maxim sent Barker and Draper samples of his lamp. But when the two professors asked Edison for some of his lamps for comparison, Edison refused to supply them. Barker and Draper burned one of Maxim's lamps for twenty-four hours at 650 candlepower without any apparent deterioration.

"I have never seen such results as these in your laboratory," Barker advised Edison, and told a reporter it was "the most remarkable performance of an incandescent lamp ever made. Edison has a good generator, but his lamp was old 20 years ago."

The *Illustrated Science News* predicted: "In connection with electric illumination, [Maxim's] name will be remembered long after that of his boastful rival is forgotten."

Lowrey, in retort, punned: "It is a good Maxim not to crow until you are out of the woods." But it was obviously important to get Edison's system blazing across the pages of the New York papers.

Robert Cutting, a Light Company director, was a friend of Sarah Bernhardt's impresario. The "incomparable Sarah" was invited to behold the system's first public demonstration after her final performance in New York, on the afternoon of Saturday, December 4. She was delayed when the frenzied audience cheered her on to twenty-nine curtain calls, and thousands of people jammed the streets outside the theater. It was a cold, slushy evening of damp snowfall, and as the hour grew late Upton, skeptical from the beginning, was sure she would not come.

Finally, a train with a special car pulled into the station. There she was, the world's most renowned actress, wrapped in furs, beautiful and pixyish, slender with a classic profile—the antithesis of the busty Gibson Girls that were the male ideal. Edison's carriage drove her to the laboratory, where, taking her extended hand, he bowed deeply, as was his custom when meeting important people.

At Edison's command, Jehl turned the wheel of the rheostat that was now used to control the current. All across the fields the lights sprang up like budding flowers, the snowflakes glistening iridescent in their glow. The actress clapped her hands and, enraptured, took the wheel herself. Edison led her from room to room, and on to the powerhouse. There he demonstrated the magnetic power of the Long-waisted Mary

Anns, and showed her how a piece of iron wire laid across the mains instantly melted. To the actress, spurred by a boundless imagination, it seemed as if she were "crossing bridges suspended in the air above veritable furnaces. . . . Light then burst forth on all sides, sometimes in sputtering, greenish jets, sometimes in quick flashes or in serpentine trails like streams of fire."

When they went outside to return to the lab, she impulsively and animatedly took his arm. The shy inventor and the dramatic Sarah looked as if they had been friends for years. On the top floor of the laboratory he showed her an electromotograph telephone, and she said she must have one for her house in Paris.

They went from table to table until they reached the phonograph. Edison talked and sang into it, and when she heard his voice repeated, she laughed, clapped her hands in glee, and jubilantly recited into it herself. She demanded he let her take the machine with her, and relented only when he promised to have one made for her immediately. When she returned to France she painted two landscapes for Edison, and addressed them to "The giver of light."

CHAPTER 15

Edison Asked Only for the Earth

Eleven days later the Edison Illuminating Company of New York was incorporated by the Light Company directors as a separate entity for the installation of the first incandescent-lighting system in the world. Edison was on the board of the new company as well as of the old, but offered to resign. He was not interested in lengthy business discussions or administrative matters, and never attended meetings.

To impress New York officials with the practicality and advantages of the Edison system, a gala demonstration was arranged for the evening of December 20. Shepherded by Light Company directors, the New Yorkers arrived on the late-afternoon train. A number of commissioners but only eight aldermen—fewer than half the total—came. Mayor Cooper sent his regrets. Edison repeated the Bernhardt exhibition for them, and demonstrated two sewing machines and a lathe run by a one-horsepower motor. He plunged a bulb into a globe filled with green-tinted water to show how the light could be used by divers and night fishermen.

A feast catered by swallowtailed, white-gloved waiters from Delmonico's was served on horseshoe-arranged tables in front of the organ. Edison displayed a huge map, covering an area of more than a square mile from Canal to Wall Street, on which every potential electric light and power customer was plotted. He spoke of the danger of fire from the existing tangle of wires. He emphasized the safety of the underground conductors, compared to the high-tension arc lights which had just been installed along a three-quarter-mile stretch of Broadway. He pointed out that the incandescent lamp would end the rash of inadvertent suicides by hotel guests who, accustomed to candles and kerosene lamps, blew out but did not turn off the gaslights, and then were discovered extinguished themselves in the morning.

J. C. Henderson, Villard's chief engineer, gave an account of the success of the *Columbia*'s system. Lowrey made a forceful and witty speech.

Illuminated by electricity, brandy, and champagne, the aldermen clouded the air with the smoke of Edison's cigars. Wilber, who had come up from Washington to assure the New Yorkers of the strength of Edison's patents, suddenly called for "Three times three for Mr. Edison." The hip-hip-hurrahs reverberated through the lab. At 9:30 P.M., belching and bubbling, the New Yorkers left on the last train back to the city.

A week later Jay Gould and other New York financiers came to look at the system. The hill was covered with a canopy of white, against which the lights glowed like topaz. Edison reckoned his expenses on the electric light at over $100,000, and he had received only $50,000 from the Light Company. Economizing was a concept with which he lacked patience. Two months before, the company's secretary, Calvin Goddard, had sent out a proposal on his behalf to borrow $60 per share from the stockholders, but this had been coolly received. Gould's appearance at Menlo Park, and Edison's decision to parallel Western Union wires into the lab with American Union telegraph lines, made Twombly apprehensive and brought frowns to Lowrey's face.

In fact, the six-year-old telegraphic rivalry between Gould and the Vanderbilts that Edison had exploited was about to come to an end. Gould had acquired such a large interest in Western Union that William Vanderbilt had lost working control. On January 19, 1881, Western Union bought the American Union Telegraph Company for 150,000 shares and the Atlantic & Pacific Telegraph Company for 84,000 shares of Western Union stock. Gould became the dominant stockholder in the restored monopoly, and Vanderbilt retired from the board. Eckert finally achieved the presidency of Western Union, and Edison lost interest in doing business with the telegraph company.

A week before Gould's triumph, the Light Company directors came to a new, complicated agreement with Edison. The company's stock was increased from $300,000 to $480,000. Edison subscribed to $50,000 of the new shares, but was not required to pay in any money. Instead, the Light Company credited the payment to him in return for 52 percent of the shares of the newly formed Edison Electric Railway Company. Edison retained a quarter interest in the railroad—which had not run since the end of summer because rain and snow shorted out the system—and was to receive $100,000 on its successful completion.

On January 8 the New York Board of Aldermen granted Edison a permit to lay the wires, but attached conditions that he and the board of directors found unacceptable. The company was to pay $1000 per mile to lay the wires, and after five years turn over 3 percent of gross receipts to the city—in contrast, the gaslight companies were charged nothing, and paid only property tax.

Since the Light Company had some of the wealthiest and most influential men in the city as investors, the aldermen did not persevere in their stand. Within three months they reduced the total charge to 1 cent per linear foot—$52.80 per mile. When the mayor vetoed the enabling bill, the aldermen overrode him 19–2. New Yorkers quipped that "Some of the electric companies wanted all the air, others had use for all the water, Edison asked only for the earth." The Edison Illuminating Company of New York offered $750,000 worth of stock to the shareholders of the Electric Light Company (only 5 percent of which was required to be paid in immediately), and Edison was ready to begin.

For the company's New York office and showplace he leased a four-story brownstone (known as the Bishop Mansion) at 65 Fifth Avenue for $6500 a year. Fifth Avenue north of Washington Square was the city's most prestigious residential area, lined with the townhouses of millionaires, each of whom strove to outdo the others. The nearest cross street was Fourteenth Street, the east-west artery of the fashionable business district. Edison could not have chosen a more strategic or attention-focusing location. A gold-lettered sign announcing the Edison Electric Light Co. went up on the façade.

At Menlo Park the inventor keyed his whole force up to complete and make a final test of the system. Edison contended a man should be able to work three days and nights with only occasional catnaps. But it was a pace no one, including himself, could maintain. He habitually took a snooze following supper. Then, returning to work in the middle of the evening, he knocked off again for the ritual midnight snack.

Even on the stormiest, windswept nights one of the apprentices was sent to the Woodward farmhouse, nearly a mile away, to return with hot coffee, pie, and a steaming kettle of food. Everyone told tales. One night Edison claimed one of the men would have to be 140 years old to have accomplished all he claimed; but he himself was no slouch at relating "whoppers."

Afterward, some of the men might perform a clog dance, or don boxing gloves for an impromptu match. One man played the organ, Mac Kenzie strummed his guitar, and all joined in singing "Over the Hill to the Poor House," "Good Night, Ladies," and "We Won't Go Home Until Morning."

More often than not, they went home soon after supper. But in the frenetic days of January and February, 1881, work continued around the clock. Reporters were again dogging the lab. Frequently they missed the 9:30 train back to New York and spent the whole night. Edison would put on a show for them, and when they staggered off bleary-eyed on the morning train they were ready to write pages about Edison's Herculean endurance and the lab that never slept. It was a legend Edison never tired of perpetuating.

Upton's friend Clarke, a superb engineer, was given the task of improving the efficiency of the system and designing a new, bipolar dynamo many times larger than the Long-waisted Mary Anns. Charles Dean, the shop foreman, worked seventy-two hours without interruption to solder the ends of the armature, made of solid copper bars instead of the wire windings all previous dynamos had featured. Batchelor, who had returned from Europe after being absent for a few months, sacrificed a five-dollar gold piece to plate the tips of the bars and keep them from oxidizing. On Friday, January 21, Clarke started a full-scale economic test of the six miles of underground mains, to which 600 lamps were now connected. The dynamos ran uninterrupted for twelve hours until nine-thirty Saturday morning. Clarke took a week to complete his calculations, and reported the results to Edison February 7. The efficiency of the system compared favorably with that of gas lighting.

A few days later, Edison, bubbling with enthusiasm, returned from New York and told Clarke to pack his bags: "The company has made you chief engineer." The week of February 14 Clarke moved into a bare, second-floor office at 65 Fifth Avenue. Edison reassured him: "The furniture will be here this afternoon." He was to be a live-in engineer, with his bedroom on the third floor of the mansion.

Clarke was back at Menlo Park on February 24 for the test of the 100-horsepower bipolar dynamo. The generator was shafted directly to a high-speed engine designed by Charles Porter, one of America's foremost steam engineers. The direct coupling was intended to eliminate the slippage of the belt connections and multiply efficiency. The engine was to operate at twice the pressure and more than ten times the sixty revolutions per minute of the average steam engine in existence.

The boilers were fired; the pressure rose; the speed increased. The foundations were none too solid, and by the time the shaft reached 300 revolutions per minute the hill was trembling. The valves were opened wider; the shaft spun faster—500, 600, 640 revolutions per minute. "Every time the connecting rod went up she tried to lift the whole hill with her," Edison recounted. The racket was drum-bursting, the building was being shaken like a can on a paint mixer, and the machinery appeared momentarily ready to take flight. Men scattered and slid out into the snow.

After the engine was brought under control and shut off, a vigorous argument ensued between Edison and Porter as to what had gone wrong and who was at fault. It was evident the generator could not be run at more than half its designed speed. It could be viewed as a limited success or a successful failure.

In any case, drinks appeared in order. Edison led the way to Davis's Lighthouse. The place was closed, but old man Davis was roused. Mac Kenzie, who was temporarily lodging upstairs, came down. There were many toasts and much clinking of glasses. Not till long after midnight did the men totter to their beds.

They did not know it, but it was a last hurrah for Menlo Park as the center of Edison's activities. He needed the machinery to prepare the New York installation. Dismantling the entire machine shop, he shipped everything to the city, and there took over the Roach Iron Works on Goerck Street. From then on, Edison, Kruesi, and other key personnel spent most of their time in Manhattan, and the Menlo Park laboratory quickly declined in importance.

At first it appeared the lamp factory might remain in the village. Upton, however, was uneasy. "We must move. We never can make the lamps cheap enough until we can have plenty of boys and girls at low wages," he advised Edison on March 7. Occasionally Upton was able to recruit whole families trudging along the railroad tracks, and put all the members to work. But in an era when child labor was universal, he needed masses of children, and these "can never be had except where there are other manufacturing establishments employing men. I think a little outside of Newark will be just right."

The next day Edison muscled the board of directors of the Electric Light Company into orally granting the Edison Lamp Company exclusive rights to manufacture incandescent lamps. (The Lamp Company, formed with a handshake by Edison, Upton, Batchelor, and Johnson, represented an initial investment of $17,500, most of it contributed by Edison and Upton.) In return for the exclusive license, Edison agreed to supply all the lamps the Light Company could use for 35 cents each, a price at which he projected a profit of 3 cents. (The price charged for the lamps was soon renegotiated, and never fell that low.)

In East Newark (now Harrison), Edison purchased the fire-damaged factory of an oilcloth manufacturer who had gone out of business. The plant consisted of three four-story buildings and the ruins of another, plus offices and stables, and had ten times the capacity of Menlo Park. Edison obtained the $136,000 plant for $52,000—only the first of a number of times he was to buy a factory at a bargain price.

Production commenced during the summer. But for the moment the new facility was a liability rather than an asset. Demand for the lamps was small. During all of 1881 fewer than 35,000 were sold, and at the end of the year Upton had an inventory of 90,000.

In February Edison had remarked: "Don't think we owe $1000, hope nobody will ever give us credit." But by December the partnership had put in $105,000. (The total was to reach $163,000 by June, 1882, when the final assessment was made.) The company was having difficulty meeting its payments.

Although Edison was receiving thousands of inquiries for lighting installations, he resisted putting up "isolated" plants for companies and individuals, because he considered such business "demoralizing." He was intent on concentrating on municipal systems to rival the gas companies. The groundwork had been laid in New York; Villard indicated readiness to build a system in Portland; and a Detroit syndicate had been formed. But it was a task that was going to take years and require huge amounts of capital.

To promote his system in New York, Edison installed a lighting plant for the printing firm of Hinds and Ketcham in January, 1881. Consisting of 240 8-candlepower (B-type) lamps, it was the world's first commercial land-based lighting installation. Three months later, 65 Fifth Avenue became the first building lighted exclusively by electricity.

To put in the wires, Edison had to pull up the floors. The only people with experience in this type of wiring were burglar-alarm installers. And with the general lack of knowledge, numerous mistakes occurred. Sparks and small fires kept erupting underfoot; everyone treaded lightly. Two Otto gas engines were set up in the cellar to power the dynamos. But after several disconcerting explosions, one of which blew Edison down along with the doors, a steam engine was substituted. An April 4 the 100 lights went on. The building became a magnet for New Yorkers, and every week drew hundreds of visitors.

Not until a short time before the move from Menlo Park was the problem of improving the sockets solved. At Menlo Park the lamps all stood upright, and even a slight jar could knock them loose. Then, one night, Edison was cleaning his hands with kerosene from the can that always stood atop his worktable in the lab. One of the men, watching him unscrew the top, suggested the lamps be fastened into the sockets in the same manner. So the screw socket was born. Displayed at 65 Fifth Avenue along with other accessories, such as "safety catches" (fuses), the socket added a new dimension to the advantages of electric over gas lighting. Edison hired Luther Stieringer, a noted gaslight engineer, to design ceiling boxes, drop cords, and chandeliers, which were called "electroliers."

The fascination and success of the incandescent lighting on Fifth Avenue increased the pressure on Edison to install isolated plants. With the Lamp Company languishing, and individual installations offering quick returns, he bowed to the demand. During the summer and fall a steadily increasing number of places were wired: hotels, machine shops, cotton, wool, and silk mills, railroad and locomotive works, flour mills, piano and organ factories, meat packers, printers and engravers, sugar refineries, and steamships. James Gordon Bennett had his newspaper plant and his yacht outfitted—and of course if Bennett's yacht had electric lights, Gould's had to have them, too. By November 11, 1881, the business had grown so extensive that the Edison Company for Isolated Lighting, capitalized at $500,000, was organized and licensed by the parent Edison Electric Light Company.

Much of Edison's attention and efforts were, however, directed across the Atlantic, where negotiations had been dragging on since the beginning of the year for formation of an Edison Electric Light Company of Europe. The world's first electrical exposition was to be held in Paris during the summer and fall, and was to be followed by another in London's Crystal Palace during the winter and spring of 1882. Europe was still the world's financial and industrial center. It was of paramount importance to display the lighting system in Paris and London. In early February, 1881, Edison dispatched Batchelor to France, and during the spring and summer Edison men kept the transatlantic-steamship lines in business.

Edison's display at the Paris Lighting Exposition completely eclipsed Swan's and Maxim's. Prospective purchasers of isolated plants were knocking on the door. Batchelor importuned Edison to provide him with the organization and means to get started.

A French financier had advanced 150,000 francs ($30,000) for the exhibit, but was threatening to impound the machinery and the lights unless progress was made toward a European manufacturing concern. On September 13, Edison, speaking as always in grandiose terms, proposed the formation of a syndicate of various financiers, including Eggisto Fabbri (Morgan's partner), "to form a Patent Company for operating the light on the Continent of Europe, and to prevent the new company from being a purely speculative one shall pay up one million of dollars for the purpose of forming a large manufacturing company."

At the same time, he advised the directors of the Light Company that he intended to ship the giant new twenty-seven-ton generator, which he had built and been experimenting with for the New York station, to Paris. He nevertheless billed the Light Company $6200 of the $12,000 cost. When the directors protested they should not have to pay for something they were not receiving, he responded that the experiments had shown the machine would be useless for the New York station; in shipping it to Europe and charging them only half the cost of the experiment, he was saving them $6000. This outing in Edisonian finance brought a sharp communication from Calvin Goddard, the Light Company's secretary, that in the future all requisitions would have to be approved by a company officer, and no bills would be paid unless okayed beforehand. It was the opening round of a three-year struggle between Edison and the board.

On October 9 the giant generator underwent what was intended to be its final test. As the engine speed was increased, the crankshaft snapped, hurled across the room, and smashed a hole in the opposite wall.

Only three days remained before the machine was scheduled to be loaded aboard ship. A new crankshaft was installed, but another test was not completed until four hours before the ship was to sail. Edison had a host of men with written instructions clambering over the generator to knock it down for shipping. Tammany Hall was paid off, and the police were furnished with a barrel of beer. Sudsy officers cleared the streets and shepherded the convoy of horse-drawn trucks, led by a loudly clanging fire bell, as the generator was whipped down to the docks.

Barely an hour before the S.S. *Assyrian Monarch* was to cast off, the machinery was loaded aboard. Since the ship had brought P. T. Barnum's huge elephant, "Jumbo," to the United States, the generator became associated with the elephant, and likewise acquired the name Jumbo. Jumbo arrived too late for the Paris exhibit. But the jury had been sufficiently impressed without it to award Edison five gold medals and the diploma of honor—the top award.

Edison's latest triumph brought about an invitation to Batchelor from the Paris Opéra to install incandescent lighting experimentally for two months, an offer that carried with it prestige almost equal to the diploma of honor. Since the city did not permit the installation of large steam engines in buildings, the only alternative seemed to be to couple a number of small engines together. But Batchelor, harking back to the New York experience, was not sanguine about such an arrangement. Then M. Garnier, the Opéra's architect, suggested running the system with the new Faure storage battery, claimed to be of great power and durability. When the claims turned out to be much exaggerated, Edison told a *New York World* reporter:

"Whenever a man begins to talk storage battery, it brings out all his latent capacity for lying."

In fact, the Edison generator, with which the storage batteries were to have been charged, proved useless for the purpose. It could safely produce a maximum current of only 24 amps. When its output was stepped up to 80 amps, the strength some of the Edison men in Europe had been advertising, the armature became so hot the insulation on the wires caught fire. The project of employing the batteries to light the opera house was abandoned.

The off-and-on negotiations for formation of a European light company had, on the other hand, taken on vigor. Before the end of the year not one but three firms were established. The Edison Continental Company, capitalized at $200,000, was to be a patent-holding company on the line of the Electric Light Company in the United States; the Société Electrique Edison, headed by Batchelor, was to be the manufacturing concern; the Société Industrielle & Commerciale would install isolated lighting plants. With money advanced by the Continental Company, Batchelor in January, 1882, purchased a huge factory, belonging to an organ manufacturer, at Ivry-sur-Seine in the suburbs of

Paris. For the moment, Edison's European production capacity outstripped his American.

Orders for the Edison system came from all over central Europe, and Batchelor was pulled first this direction, then that. In Vienna he arranged for the lighting of the opera house; in Brünn the theater; in Antwerp a sugar refinery; in Lorraine the world's largest porcelain factory; in Budapest the central telegraph agency; in Milan the Galleria, a glassed-over shopping arcade. The isolated plant in Milan was so successful that it led to the erection in that city of the world's second permanent central station.

The demands of the European operation resulted in an outflow of the limited number of Edison-trained men to the Continent. Several of them, like Jehl, were to spend a large part of their lives away from the United States.

The spotlight, after the end of the Paris exposition, shifted to London. There Johnson, after a visit to Edison in September, 1881, was given the task of supervising the Edison exhibit at the Crystal Palace and constructing a complete model station. Since it was, in Johnson's words, "A city that lives and breathes foul Gas polluted air in the daytime made artificially night by the London fog," it seemed an ideal place for a showcase. Johnson searched for a location within a half mile of the Lord Mayor's residence. He discovered, however, that although the Metropolitan Board of Works had control over street lighting and had allowed Siemens, Gramme, and Jablochkoff to dig freely in putting up arc lighting, "the Gas Companies have the exclusive right to 'open the streets' for private lighting." No less than an act of Parliament would be required to confer the right "to scratch a cobblestone" on anyone else.

So far as Edison's chances with Parliament were concerned, he might as well have been Irish. Not only had he the festering feud with Preece and a bad press to contend with, but one of the members of Parliament was J. H. Puleston, who had lost $10,000 in the aborted Automatic Telegraph venture. Puleston had met Edison in the United States in 1879, and Edison had told him he did not like to see Puleston and his friends lose money, and would try to help them get it back. With the success of the electric light, Puleston thought, "You may now be able without difficulty to pay us fully."

Edison, however, contended he had spoken only in general terms, with the intent of appealing to Harrington and Reiff. He replied: "You labor under a mistake if you think I promised a settlement. I shall not pay that which I do not morally or legally owe, and that for a speculation from which I have suffered as much as yourselves."

Puleston reacted with incredulity: "You most certainly do owe every penny of it morally and legally."

But Edison refused even to see Puleston when he made another trip to the United States.

Johnson, laboring under the burden of this acrimonious heritage, was told confidentially that there was one London street "peculiarly adapted" to his needs. This was Holborn Viaduct, an arch that crossed over Farringdon Street and a set of railroad tracks in a strategic section of central London. Nearby were two well-known churches (St. Andrew's and the City Temple), the General Post Office, Old Bailey, the Central Market, and Fleet Street, the focus of the press. Underneath the street ran tunnels that carried the gas and water supply for the houses and street lamps on the viaduct. From the tunnels a hole opened to each house and lamppost. An extension of the viaduct went down to Fleet Street.

"Thus you see," Johnson wrote Edison, "with one station in a Building on the viaduct we could run to the right 2 blocks & to the left 5 or 6 = Light all the street lamps—a Handsome Bridge and 4 Buildings called the Bridge Towers—this making a magnificent street display—and then run into as many private shops as we chose—two Railway Stations, two Hotels, a church etc etc—& never dig out a brick or come to the surface at all = thus saving in Labor and time—saving the necessity for applying to the City or Parliament." Johnson had only to petition "the City Sewer Commission for permission to put any wires temporarily in this viaduct. Could anything be more lovely?"

By not requesting permits for a permanent installation, Johnson made his task very much easier.* Financed by Drexel-Morgan, which owned the British Edison rights, he exploited his subterranean coup. To obtain permission to light the Post Office, Johnson courted Preece, and worked on Edison to make up. Edison wrote a conciliatory letter that Preece received with "great pleasure." Preece was glad "to let bygones be bygones."

Favored by such warming relations, the Edison exhibit on the Holborn Viaduct and at the Crystal Palace opened in mid-January. Holborn Street was initiated with only a few hundred lamps. But over the months more and more were added—400 at the Post Office alone—until finally nearly 3000, fed by two Jumbos, were connected. Johnson and Gouraud hosted a dinner for 125 peers, press, and important personages at the Crystal Palace. Unfortunately, early in the evening a safety plug blew out on the steam engine, so the dinner was consumed by gaslight. But at its conclusion power was restored, and the guests were treated to a brilliant display of lighting. Its architect was William J. Hammer, a twenty-three-year-old Newark resident who had popped up at Menlo Park in December, 1879, and was the most imaginative and artistic of the Edison men. Hammer built a hand-operated sign flashing EDISON letter by letter. He designed an electric fountain, and an enormous gilt, floral-

* He had, for the boilers, to construct a 7000-gallon water tank and a 90-foot chimney. Holborn Street has, in the past, sometimes been mistakenly referred to as the first permanent central station.

shaped chandelier, with colored (painted) bulbs depicting the flowers. A number of appliances were shown, including an electric cigar lighter. Although there were four competing electric companies in England and America, none came close to challenging Edison.

"If a combination of all other inventors and their product could be effected they would still be unable to do what Mr. Edison is now doing," Johnson trumpeted. "His system is already so complete in every detail that were his patents to offer him no protection he would still be absolutely alone in the field of real competition with gas. No combination or aggregation of Mediocrity ever did or ever can equal in fertility of resource this single brain of genius. He is not to be overtaken so long as he lives."

"There is but one Edison and Johnson is his prophet," the *London Standard* noted.

CHAPTER 16

The Magic Touch

While Johnson was boosting Edison in Europe, Johnson's protégé, twenty-two-year-old Samuel Insull, was turning into the inventor's alter ego in New York.

Sammy, as Edison called him, was one of eight children of a poor, middle-class British family. At the age of fourteen he had gone to work as an office boy. In 1879, after spending his nights studying shorthand and bookkeeping, he had become Gouraud's private secretary. A few months later Johnson had taken him over. When Stockton L. Griffin quit as Edison's secretary early in 1881, Johnson, who was then in the United States, sent for Insull. Insull arrived in America on March 1, just as Edison was moving into 65 Fifth Avenue.

Insull was slight, below average height, nearsighted, with slicked-down prematurely thinning hair and long sideburns. His voice was nasal and high-pitched. He had the demeanor of a shop clerk. Edison, not favorably impressed, put him to the test immediately by entering into an all-night discourse with him and Johnson about British affairs. Edison needed money for his manufacturing enterprises, for the New York station, and for the preparation of the Paris and London exhibits. He questioned Insull about the financial status of the English telephone company, and was anxious to know what the stock would bring and whether he could expect the payments due him to be made on time. He found Insull to be responsive, knowledgeable, and mature.

Insull, in turn, was adjusting to his surprise at Edison's visage—not at all what he was accustomed to in an English or even an American gentleman. Edison's speech still had burrs on it. His hair needed cutting. His Prince Albert coat, dark trousers, silk neckerchief, and wrinkled white shirt all looked as if they had been slept in—which, as Insull would soon discover, they had.

Edison moved Insull into an office and a bedroom at 65 Fifth Avenue, introduced him to Menlo Park (where Insull took a spin on a railroad handcart and crashed), and within a few days was letting him handle all the routine affairs of both his business and his personal life. Insull learned which bills to pay and which to put off, and how to juggle accounts so as to keep three out of five continually in the air. He took care of the correspondence, which Edison usually let pile up for days before going through it in one burst. He separated the typical Edisonian "No Ans" from the letters on which Edison scrawled a cryptic comment, which Insull then converted into a polite response. He spent his nights with Edison, and his days transacting Edison's business.

Edison was devoting the greatest portion of his time to the New York station. He had started out overoptimistically, as usual, intending to cover one square mile and predicting he would light all of New York in two and a half years. In May he wired the new uptown mansions of three Vanderbilts, William, Cornelius, and W. K., so that they would be ready to receive current from a central station. He estimated the cost of putting in the first district at $160,000. But that turned out to be the figure asked for nothing more than two buildings in one of the city's worst slums. By early summer Edison was having doubts that electric lighting, because of the great initial investment required, could offer dollars-and-cents competition to gas. He reduced his plans for the first district to fifty square blocks, about one sixth of a square mile. Bordered by the East River and Wall, Nassau, and Spruce streets, the location was not a promising residential area. But it embraced a large part of the financial district, had considerable commerce and manufacturing, and bordered City Hall and Printing House Square, with its newspaper offices. Just to the north were the high, massive towers of the nearly completed Brooklyn Bridge, more than ten years in the building. Fifteen hundred gaslight customers were within the boundaries. The key factor was that the area had a potential for industrial installation of 750 electric motors, which would enable Edison to sell power in the daytime as well as at night.

The property he purchased was a three-story double building at 255–257 Pearl Street, to which the wind from the East River carried the pungent smell of the Fulton Fish Market. The interior of the structure had to be rebuilt and reinforced to support the heavy machinery, a task that continued on into the winter of 1882. The boilers and steam engines were manufactured by Charles Porter's ironworks in Philadelphia. At Goerck Street Edison was crafting a dozen dynamos.

In August, 1881, canvassers were hired to make a survey of every gaslight in the district. Six dozen "wire runners"—most of them telegraph, telephone, and burglar-alarm linemen—were taken on to wire the premises of every potential customer, so that only the connections to a junction box need be made when electricity was ordered. The wires were run along the gas pipes to the fixtures, which were left in place, so that a customer could try both means of lighting and make his choice. Since gas did not require bulbs, the Electric Illuminating Company decided to supply and replace these free of charge—a policy that was to be followed for many years by incandescent-lighting companies.

Edison had not only to install the wiring but convince a skeptical public of its safety. On Maiden Lane, which ran through the heart of the district, a telephone lineman was electrocuted when a thinly insulated Maxim high-tension wire came in contact with the telephone wire. The Board of Underwriters passed a resolution that all buildings with electric wiring were to be rated "extremely hazardous." Edison met with the underwriters to demonstrate the insulation and the "safety valves" he had developed. Consequently, the board established a category of approved "Underwriters' Wire."

(The Edison men promptly referred to it irreverently as "undertakers' wire.")

The mains presented the greatest problem, and the design that had been tested at Menlo Park required considerable modification before it could be used in New York. The mains were constructed in twenty-foot lengths of solid copper bars in half-moon shape. Two of these—one the outgoing and the other the return conductor—were placed face to face, separated by insulation, in an iron tube. The tubes were packed with hot, liquid asphalt, laid in trenches, and connected by junction boxes, which served also as outlets for the lines running into buildings.

To calculate the size of the mains and the amount of power that would be required, Edison hired a mathematician, Dr. Herman Claudius, late of the Imperial Austrian Telegraph Agency, for two dollars a day. Claudius constructed a fifteen-by-twelve-foot model of the district, complete with wiring. Potential customers were represented by spools of German silver. Claudius was thus able to tell Edison what the total resistance of the wiring and the potential consumption of the district would be, and what capacity would be required.

Not until December, 1881, was Edison able to start laying the mains, though with typical euphoria he told Johnson: "The men are easily broken in at laying tubes. Putting in this central station will be very much easier than I at first imagined." To prepare the mains, Edison established a tubeworks in a twenty-foot-wide building on Washington Street. There Kruesi presided over kettles of asphalt and linseed oil. Pipes under treatment hung out the windows, and the whole neighborhood suffocated in the stench. Edison's "gang" consisted of ten men tearing up paving stones and another ten digging the trench. Behind them came the pipe layers, then the aligners and connectors. Another group broke through the cellar walls and ran the lines from the junction boxes into the buildings. There were, of course, bugs. The first batches of pipe had air holes in the asphalt through which electricity leaked. The mains had to be dug up and replaced.

The city allowed the work to be performed only between 8:00 P.M.

and 4:00 A.M., which suited Edison fine. Placing an engine, dynamo, and arc light on a wagon, he trucked them from street to street. Often he was on the scene himself, jumping into freezing ditches to inspect the work. Across the way from him the New York Steam Heating Company was digging simultaneously, and he talked with the engineer about their mutual problems. One night a leak developed in a steam pipe and three tons of lampblack being used for insulation sprayed up and drifted down like black snow. When Edison and his gang emerged from the ditch they looked as if they had been digging in a coal mine.

Edison was enjoying himself. In February, however, he left the construction site and New York for two months because of Mary's condition.

Mary had never been truly well again after the birth of Will. For a few weeks she would seem to be better. Then she would suffer a relapse. Her physical state was atrocious. She weighed more than 200 pounds. She invited friends to Menlo Park, and together they drank port by the bottle and ate candy by the pound. Sometimes, instead, Mary hosted parties in Newark. Edison, of course, never attended.

Mary had charge accounts at fashionable New York stores. Her dresses were made of the most expensive materials—one red brocade gown was decorated with stuffed birds. Edison gave her diamonds and other jewelry. In addition to indulging her, he attempted to allay her anxiety by making her independently wealthy—the Menlo Park house was hers, and he put $50,000 in United States bonds in her name. Still, he was parsimonious with what she wanted most: his attention. She had periods of intense anxiety alternating with severe despondency. When Menlo Park boomed and men roamed about at night, her fears increased. Edison turned over to her a .38 Smith & Wesson revolver given to him as a present after his western trip, and she slept with it under her pillow. Once Edison, returning to the house after midnight without his key, clambered onto the porch roof and pushed open the bedroom window. Mary screamed, frantically dragged out the revolver, and nearly altered the history of electric development.

The children suffered from the uncertain health of their mother. The youngster, Will, spent most of his time with his black nanny, and was more attached to her than to either of his parents. Tom was introverted, loved to be read to, and clung to his mother, who dressed him in long Little Lord Fauntleroy kilts. Edison, like any proud parent, told his father: "Tom is a *smart* one, and make no mistake." Yet the boy was thin and sickly.

All three children were spoiled and careless. Will and Tom often left their expensive toys, which were the envy of the farm boys, scattered about the fields.

The queen of the household was Dot. Vivacious and blonde, she had the favor of her father. She bossed the young assistants around, and had them fetch water to her from a half mile away. She made them eat the meals she "cooked" for them, which put them in as much pain as Edison, who would bolt the large piece of pie that was always atop the lunch basket but leave the rest of the food untouched and complain about his indigestion. Dot drove the surrey hitched to the pair of ponies, and delighted in dancing on the table for the men, but refused to learn sewing from her mother. In the fall of 1881, when she was eight, Edison enrolled her and her aunt, Jenny Stilwell, in Bordentown College, a girls' academy near Trenton.

Both Dot and Mary were much attached to Alice, who was sister, surrogate mother, nurse, and confidante. On the surface, Mary was delighted when Alice married William Holzer, the Philadelphia glassblower who had become the foreman at the lamp factory, on December 2, 1881. A week later she staged a grand reception for them at the house in Menlo Park. But afterward her condition deteriorated so rapidly that Dr. Ward advised Edison:

"I have noticed during the past few weeks that she seems very nervous and despondent and think that she will never recover. I think that an entire change would be of benefit and if you could take or send her to Europe for a few months she might return improved in health and be better pleased with her surroundings here. She seems so changed physically and mentally of late that something ought to be done and I can suggest nothing better."

Mary liked Florida, where she always seemed to get better. Edison took her and the children to Green Cove, near St. Augustine, and did not return to New York until April.

Edison's absence and bad weather retarded progress on the New York installation. Not until early summer was the work of laying the fifteen miles of mains completed or the first of three Jumbos and their engine installed at Pearl Street. Motor-driven fans cooled the giant armatures, which often became so overheated they had to be packed in ice. The first engine was started up on July 5. When it operated satisfactorily, a full-scale test followed three days later. A bank of 1000 lamps was arranged around the walls of the station to carry the load. With one engine running, everything was fine. "Then we started another engine," Edison reported, "and threw them in parallel. Of all the circuses since Adam was born, we had the worst. One engine would stop and the other would run up to a thousand revolutions. And then they seesawed. The trouble was with the governors. When the circus commenced, the gang that was standing around ran out precipitately, and I guess some of them kept running for a block or two." Vivid colored sparks and flames lashed from the commutators. The racket was nightmarish. Edison grabbed the throttle of one engine, and Johnson, who was on a visit from London, took hold of the other. Together they shut off the engines.

Through the summer, Edison sweltered at Pearl Street and in the Goerck Street works, where the temperature often exceeded a hundred degrees. The plant was so overcrowded that he paid off the Tammany leader so he could run lathes on the sidewalk. He ate in an all-night dive where, he contended, the same clam was used for seasoning in each day's chowder and there were seven flies for each pie. Every evening after supper he lay down on the floor or on a pile of pipes to take a nap. One night the men got the idea to set the clocks ahead six hours, and when Edison woke up the hands pointed to 4:00 A.M. Puzzled but believing, he told the men to knock off, and headed for his hotel. Not until he reached the theater district and saw the crowds emerging from the shows did he realize a hoax had been played on him.

He had expended $480,000. That was three times his original estimate, and twice the amount the contract between the Edison Electric Light Company and the Illuminating Company specified. As Edison failed to meet one deadline after another, investors grumbled, the stock of the Light Company fell to $600 a share, the newspapers were once more skeptical, and Maxim and other competitors made snide remarks.

It seemed prudent to get the station in operation, even if only at a small fraction of its planned 33,000-lamp capacity.

At 3:00 P.M. on Monday, September 4, the switch was thrown on a single dynamo. In all, 800 lamps were connected at the Drexel-Morgan Building, the *New York Times* office, and twoscore other establishments. Standing in front of the station in a collarless shirt and a high-crowned derby, Edison told the reporters:

"I have accomplished all that I promised."

Two months later six dynamos were in place. But the problem with the governors had not yet been licked, so only one dynamo could be run at a time. This, in fact, sufficed, for although 1700 to 1800 lamps were now hooked up, the number in use at any one time never exceeded 1300. Because Edison was having difficulty fashioning a meter, the customers were receiving their electricity free. Not until six months after the beginning of service was the electrolytic meter he had devised ready. Customers sometimes objected because they could not themselves "read" the meter, and suspected meter men were guessing. Occasionally the device registered backward. In winter it had a tendency to freeze. The instruments were in use about a decade, then gradually replaced by a mechanical meter invented by Elihu Thomson.

The questionable condition of the New York station and the large capital required made investors in other cities shy away. Edison traveled to Boston and Philadelphia to meet with financiers, but three years after the great discovery in Menlo Park, the gaslight companies were as entrenched as ever in the metropolitan areas.

Yet, while development of big-city lighting was stalled, Edison was in fierce competition with the United States Electric Light Company and its Maxim lamp in the installation of isolated lighting plants. The United States Company advertised cleverly, and generally underbid Edison. It won the contracts for lighting the Philadelphia, Chicago, and St. Louis post offices, and, after a bitter struggle, the New York capitol in Albany. Edison was especially galled when the Pennsylvania Railroad succumbed to "the lies of these infamous shysters" and picked the United States Company to illuminate its eight Hudson River ferries.

But even as the United States Company was coming out on top in the more publicized battles, Edison was winning the war. Marshall Field and J. P. Morgan were well pleased with the plants at their mansions, though Morgan had the common mishap of having his walls and carpets scorched, and the steam engine, located well away from the house in his extensive gardens, kept his neighbors upset with its smoke and clanking. In mills, hotels, distilleries, ironworks, banks, meatpacking plants, canneries, and theaters Edison was installing more than five times as many systems as his rival. He put plants into the Iowa State Prison and the Government Printing Office in Washington. In Chicago he lighted the Rand, McNally Company, Marshall Field's store, and Cyrus McCormick's machine works. The University of Missouri's president, Dr. Samuel S. Laws, who had been Edison's initial employer in New York, ordered a plant, and the university became the first in the world to be illuminated by incandescent lighting. In 1882 Edison had 130 plants producing electricity for 22,000 lamps. In another twelve months the number increased to 246 plants with 61,000 lamps. The Isolated Company was making a 50 percent profit on its $500,000 investment.

The finances of the United States Electric Light Company, in contrast, were in a parlous state. In September, 1882, the company put its six-story, $170,000 factory at Avenue B and Seventeenth Street on sale. Edison perceived another opportunity to make a bargain acquisition. He proposed to Sigmund Bergmann, whose establishment was making machinery and accessories for the lighting system, that Bergmann purchase the United States Company's plant; Edison himself would put up half the money. Since the United States Electric Light Company would never sell to anyone connected with Edison, Bergmann initiated negotiations through a dummy, who pretended he wanted the building for a cigar factory. A price of $77,000 was agreed on. When the U.S. Company found out who the real purchaser was, they offered Bergmann $10,000 to cancel the deal. Bergmann, of course, refused the offer. Shortly after the opening of the Pearl Street station, he trucked his equipment from his Wooster Street shop to Avenue B.

Edison established his laboratory on the top floor of the new factory.

Now thoroughly committed to New York, he moved his family out of the Clarendon Hotel and took a two-year, $400-a-month lease on a mansion at 25 Gramercy Park. It was one of New York's finest residential areas, and only ten blocks from 65 Fifth Avenue. Mary and the children, who had been spending the summer at Menlo Park and the Long Island seashore, moved in.*

"Do you not think this looks very much as if he will never go back to Menlo Park again?" Sherbourne Eaton, who had replaced Norvin Green as president of the Electric Light Company, wrote to Batchelor. "He told me it was in consequence of the necessity of his being close to the central station. In the next breath he said he would never come near the city if it was not for the women continually bothering him to do so. Johnson and myself are of the opinion that he wanted to come in just as much as the women do."

Indeed, Edison had a fine time at 65 Fifth Avenue. He transferred most of his library from Menlo Park to the top floor. The building was usually crowded with people until late in the evening. Often Edison presided over the equivalent of the Menlo Park suppers. Ed Remènyi, a popular violinist who had once worked as a telegrapher alongside Edison, came and played after concerts. Henry Dixey, a famous actor, competed with Robert Lincoln (Abraham Lincoln's son) and Edison himself in telling tales. At midnight, the gatherers did not have to satisfy themselves with farm fare, but went to Delmonico's, the Hoffman House (noted for its painting of nude nymphs), or Theiss' Beer Garden (a fashionable gathering place despite its plebeian name), where Remènyi sometimes gave concerts.

Only a few workers remained at Menlo Park. Upton urged Edison to shut down the laboratory entirely. "There is a decided tendency to picnic showing itself there," he contended. "The expense account at Menlo Park should be a watchman." He himself had phased out the Menlo Park lampworks in April, and offered to sell Edison the building for $500.

At the electric-light factory in East Newark, Upton was able to hire all the girls he needed at fifty to seventy cents a day. Everyone was paid on a piecework basis, the price set according to the output of the fastest and the most skilled—though the efficiency engineer was not to become fashionable for another generation, Upton was one in everything but name. Demand, however, continued to be comparatively low, so that the factory was losing eight cents on each of the 200,000 lamps it turned out annually.

* Rummaging through some of the owner's stored furniture in the garret, Edison discovered a portion of Samuel Morse's journal. A quarter century later he donated the diary to the American Institute of Electrical Engineers.

The biggest problem in selling the lamps was that, although the company advertised them as good for 600 hours, they quickly became clouded by carbon deposits. At first the light was refracted into myriad colors, and the lamp sparkled prettily. But, often within 20 to 40 hours, the lamps took on the smoked look of kerosene chimneys, and soon they were useless. In France, Batchelor completely "lost the art" of carbonizing the filaments. After spending an enormous amount of money and nearly bankrupting the company, he decided he would have to resort to flashing. Edison's brother-in-law James Hipple, who set up a plant in Berlin, had the same experience.

Upton was rescued only because Edison was at hand. It was Edison's opinion that "The process is entirely dependent upon judgment [and] a most illusive one. We have at times carried on the process concurrently with regular observations as to the various atmospheric conditions of temperature, moisture, pressure and direction and velocity of the wind." Nothing seemed to have any relationship or effect. All depended on intuition. In late July, 1882, when Upton was frantic after several weeks of turning out horrendous lamps, Edison took time off from Pearl Street to set things aright in the plant. His touch was truly magical. He left Upton shaking his head:

"You were here Thursday. There was no change made in Thursday's lamps, yet they were the best we have made for several weeks. I have read that the Rothschilds believe in luck, and appoint for their agents men that are reported lucky. Your luck is almost a proverb, otherwise it seems strange that your mere presence can change the run of lamps. I will give you a testimonial with particulars recommending you to the Rothschilds."

CHAPTER 17

A Jackknife and a Bean Pot

In two projects, however, luck and intuition were not quite enough.

Even while he was struggling to put together the model lighting system at Menlo Park, Edison had contemplated arriving at his fortune through a "golden future." The stockholders of the Edison Ore Milling Company, incorporated in the euphoric days of January, 1880, were no less confident and anxious than the investors in the Light Company—in fact, the capitalization of the Ore Milling Company was $50,000 greater than that of the Light Company. Throughout 1880 an immense amount of activity was conducted at Menlo Park relating to ore milling. Edison obtained samples of sand and ore from all over the United States and Canada, and had them assayed by Dr. Haid. One day the painfully honest Kruesi, who always wound up playing straight man to Edison, brought in a thermopile polished to gleaming gold. Edison took the bar, weighed it, and said to Upton: "We have it this time. Our worries will soon be over." He continued in this vein until Kruesi could stand it no longer, and asked:

"What is it, Mr. Edison?"

"Why, it's solid gold! We can make it by the ton. We don't need the light now!"

Outside the laboratory, Edison constructed an ore separator. It was a simple machine. The black sand containing gold and magnetic iron ore was poured into a hopper, from which it sifted down past a large magnet excited by a Long-waisted Mary Ann. The magnet pulled the stream of iron particles in its direction, while the gold and sand fell straight. Though this worked quite well in separating the 20 percent of iron, it left the tiny quantities of gold still mixed with the 80 percent of sand, so not much had been accomplished.

There was, however, a great shortage of high-quality iron ore in the eastern United States. Despite the vast deposits of the Lake Superior region, it was cheaper to import ore than to ship it from the Midwest. The ore separated from the sand was almost pure iron, and seemed ideally suited to enrich the impure ores found east of the Alleghenies. If the magnetic separator was impractical for taking out the gold, might it not be employed profitably for obtaining iron?

By the summer of 1881 Edison had located iron-bearing sand deposits on two remote stretches of beach at Quonocontaug, Rhode Island, and Quogue, Long Island. From Rhode Island the ore, after being separated, had to be taken off by ship. One vessel went aground and broke apart on the rocks. Another started sinking when the ore was put aboard, and had to be hurriedly unloaded. There was as yet no market for the ore. Edison nevertheless advised the stockholders that the company could expect to make a profit of four dollars per ton. He was building five separators at Goerck Street, and told a newspaper reporter he expected to launch a search along the entire seashore from the St. Lawrence River to the Gulf of Mexico. The ore would be as fine as that obtainable from Sweden.

While the main operation was at Quonocontaug, Edison had seven men and two carts hauling sand to the separator on Long Island. The next summer, while taking a few days off from the Pearl Street station, Edison increased the Quogue force to twenty-six men. The deposit actually lay offshore on a spit of land that was little more than a glorified sandbar. On Sunday, September 3, the day before the Pearl Street station opened, Edison went with a banker friend to Quogue. Late in the morning a storm came up. Rain and waves lashed the beach. A few hours later, when the gale abated, Edison returned. The sand, instead of black, was gleaming white. The iron had all disappeared into the sea.

It did not seem too great a loss, for the company had failed to develop any customers. One ironworks had contracted for 200 tons a month, but closed up before the first shipment. Another wanted its money back because Edison had represented the ore as being absolutely pure but in actuality it contained 4 percent foreign matter. "With 4 percent impurity it is unfit for any direct process for making steel," the manufacturer complained. "If it was *really* what it is represented to be, a pure magnetic iron, *large* quantities could I have *not the slightest doubt* be sold at $9 per ton in this district [Pittsburgh]."

Having spent $4000, Edison let the operation slip into dormancy. A few men had been fleeced when they paid $300 a share on reports that the company owned $50 trillion in tailings. But, in general, no great damage had been done.

Edison had little more success with the electric railroad, to whose development he returned in the late summer of 1881.

For a year the tracks had lain idle. But then Henry Villard, the financier and Light Company director, became very much interested in the railroad's potential.

Villard, exercising financial legerdemain that startled Wall Street, had taken less than two years to parlay his holdings in the Oregon Railway & Navigation Company into formation of the Oregon & Transcontinental Company and, through this, into control of the Northern Pacific Railway, which was under construction from Chicago to the Northwest. Early in September, 1881, Villard was named president of the Northern Pacific, and on the fourteenth he came to an agreement with Edison. To make the Northern Pacific a paying proposition, Villard badly needed narrow-gauge, low-cost spurs to haul wheat to the main line. Edison was to attempt to meet the need by constructing a two-and-a-half-mile experimental electric railroad. He was to build two locomotives, one for carrying freight and the other for passengers, with a speed of sixty miles per hour and a hauling capacity of ten tons. Villard agreed to lend Edison $12,000, for which Edison was to put up one hundred shares of the Electric Light Company as security. If, by July 2, 1882, Villard's engineer J. C. Henderson certified the railroad a success, the loan would be written off, and Villard would purchase at least fifty miles of electric railroad and appropriate rolling stock. Edison, who had flunked out as a locomotive engineer, applied himself to building a railroad.

He adopted the Siemens method of using a third rail to carry the current. A four-and-a-half-ton engine, simulating the appearance of a steam locomotive, was constructed for the two-and-a-half-foot-wide track. The ties were tarred and japanned to insulate them against wet weather.

The road ran north to Mine Gully, then west to a local landmark called Pumptown. It crossed the land of four neighboring farmers. Cows frequently stepped onto the third rail, and charged off with their tails waving in the air. Horses reared and threw their riders, who had to learn to rein carefully in crossing the tracks. Chickens and turkeys roosted in the roadbed, were swept up by the cowcatcher, and occasionally, with feathers flying, landed on the engineer.

By the end of June, 1882, the railroad was still not completed. When Henderson came out to inspect the road, Charles Hughes, a former telegrapher and railroad conductor who was now in charge of the project, took him aboard. Hughes unplugged resistance boxes and off they tore, up the hill, then down, around sharp curves and through the woods, where branches flicked at them along the tunnellike passage. An assistant kept clanging a large cowbell, which added to the sense of peril. When they crossed a trestle built across an eighteen-foot-deep gorge, the engine swayed so much the bell rang of its own accord. Henderson reached the end of the line with perspiration dripping down his paste-white face. "When we go back I will walk," he told Hughes, "and if there is any more of this kind of running I won't be in it."

The engineer was not impressed. Edison had fallen far short of the specifications. The fifteen-horsepower locomotive could attain a speed of thirty miles per hour on a straightaway and forty miles per hour downhill, but not much more than half that with three loaded cars. Mechanically, the improvements over the crude contrivance of 1880 were minimal. The locomotive was nowhere near being a commercial model.

Edison was not unique in failing to meet deadlines, and pioneer projects seldom come up to expectations. In the normal run of events he would probably have been able to string Villard along. But the Northern Pacific was as much behind schedule as Edison, and far from completion. Villard was expending more than $2 million a month, and, despite the efforts of his friend Senator Conkling, was unable to pry the expected help out of the Republican administration in Washington. The electric railroad had been reduced to chaff in Villard's scheme. Needing every penny, Villard held Edison to the contract. He wanted the $12,000 returned, or else he would sell the Light Company shares.

The stock, of course, would bring five times that amount, so Edison scrambled to raise the cash and redeem the shares. The general interest in the railroad was so great that Villard's dropping out scarcely seemed significant. The Manhattan Railway and the Minneapolis & Minnetonka Railway, both of which were under enormous pressure to abandon steam locomotives in metropolitan areas, virtually hung on to Edison's Prince Albert coattails. The Denver & Rio Grande wanted to try out the locomotive. The president of the Long Island Railroad offered his tracks for a test, and the president of the Staten Island Railroad went to Menlo Park to see a demonstration. Inquiries came from San Francisco, where the Market Street railroad was preparing to abandon horsecars.

Looking at the nearly completed Brooklyn Bridge, Edison proposed to a New York politician, Judge William Kelly, the placing of an electric railroad on the bridge to carry traffic between Brooklyn and Manhattan. Edison, Kelly, and a few cronies walked back and forth across the span, and deemed it eminently satisfactory for the purpose. Judge Kelly managed to sneak a bill, whose purpose was disguised, through the legislature during the adjournment rush. Governor Cornell, however, failed to sign it (or a host of similar last-minute legislation).

"Its true object he never discovered," Judge Kelly was pleased to inform Edison. "The strongest influence was brought on him without avail. Meantime let us carefully guard our secret for the scheme is a most excellent one."

Interest in Europe was as active as in the United States. Drexel-Morgan provided funds with the view of producing an electric locomotive for England. Then, on September 1, 1882, Edison concluded an agreement to build an electric railroad in Geneva, and sent several men to Switzerland to prepare the groundwork.

Most of Edison's time and money, however, were taken up by Pearl Street and the problems of central lighting. He gave the railroad only secondary attention. The building of a locomotive and the transmission of the current presented problems of immense complexity. Where the tracks were exposed to the elements and not secured against people, the use of a third rail was not a practical approach. Edward Fox,* a young New York engineer who was to have a notable career as an electrical inventor, offered to sell Edison a system of delivery based on overhead wires and trolleys for $500. Edison asked the board of directors of the Light Company to put up the money, they declined, and he did not pursue the matter.

The brushes on both dynamos and motors gave dreadful trouble. Henry Draper, who had purchased one of Edison's isolated plants for his house, told Edison he was using carbon in contact with the commutator to stop the sparking. Because of the high resistance of carbon, it seemed as outlandish an idea as the carbon filament, and Edison did not follow it up. Yet six years later another electrical experimenter, Charles Van De Poele, substituted carbon brushes for metal, and the innovation turned out to be a key to the success of the electric streetcar.

At Goerck Street Edison began construction of a forty-ton, 250-horsepower locomotive designed to haul a full-sized train. He wrote out a long examination form for electric-locomotive engineers, and tried to interest Frank Thomson, president of the Pennsylvania Railroad, in financing the engine. But Thomson, after taking a ride at Menlo Park, was unencouraging. He doubted the Pennsylvania Railroad would ever have use for an electric locomotive.

Johnson, who had now returned to the United States permanently, tried to persuade Edison to turn the project over to Frank Sprague. Sprague was a brilliant graduate of the United States Naval Academy who in 1875, as an eighteen-year-old freshman, had written a highly knowledgeable dissertation, "Electricity, its Theory, Source, and Application." After graduating from the academy, Sprague had come to Menlo Park and talked to Edison about an idea he had for improving the carbon telephone. During the succeeding years he had cultivated the acquaintance of Farmer, Wallace, Draper, and other leading electrical experts. At the time of the Crystal Palace exhibition his ship had been in Europe, and he had obtained a leave of several months to serve as secretary to the jury, later writing a report on the electrical displays for the navy. In London he made the acquaintance of Johnson and other Edison men. He took pains to be useful, and read a paper he had written on Edison's system of distribution before the British Association for the Advancement of Science. Johnson convinced Sprague without much difficulty that an electrical engineer had a better future with

* Not to be confused with Edwin Fox, Edison's article-writing friend.

Edison than with the navy. In the spring of 1883 Sprague resigned and came to New York.

Edison, who had spent $38,000 on the railroad over the span of three years, was not enthusiastic about investing more.* The Light Company directors were anxious to have something to show at the Chicago Railway Exposition in the summer of 1883. Next to Edison, the most important experimenter in the field at this time was Edison's acquaintance Stephen Field, who had taken out a number of key patents.† Without these, it appeared it would be difficult for Edison to continue. Edison's young Newark rival, Edward Weston, had a dynamo better suited for propelling the locomotive than Edison's. Against Edison's will, the Light Company directors, who had the controlling votes, decided to merge the Edison railroad company with Field's. On April 20, 1883, the United States Electric Railway Company was organized. Field was the engineer of the new concern. But "Judge," the lilliputian engine he designed, operated at a speed of only nine miles per hour on the short track at Chicago.

The merger marked the end not only of the Menlo Park railroad but of laboratory activity there as well. A few men were still employed in the late spring and early summer of 1883, but they spent much of their time loafing and gathering wild strawberries. By the end of August, only Nicholas Stilwell, Mary's father, remained, as caretaker.

The rails stayed in place for another year. Then, in the midst of a financial crisis, Edison pulled them up and sold them. Like a small plot of earth transported through a time machine into another century and then pushed back, Menlo Park returned to kerosene lamps for two more generations.

Sprague, instead of going to work on the railroad, applied his mathematical knowledge to Edison's operations. Sprague was part of a new generation of academically trained men already emerging in the electric industry, a generation that would, in another decade, become dominant. He was appalled when he saw the laborious methods Dr. Claudius used to work out the wiring requirements, and soon devised a formula that shortened what had been a task of weeks into one of a few hours.

Of even greater importance, Sprague brought with him the knowledge he had gained in London of the ingenious new three-wire system developed by John Hopkinson, a British electrical engineer who had served as consultant to Edison's Holborn Viaduct station and Crystal Palace exhibit.

* If, however, the 520 shares Edison had received from the Light Company at the time of the formation of the Edison Electric Railway Company are taken into account, Edison was considerably on the plus side of the ledger.

† Field and his partner, it may be recalled, had obtained telephone and quadruplex rights for several European countries in 1877.

Hopkinson, by applying some of the series principles to a parallel hookup, succeeded in reducing the amount of copper required for the mains by two thirds. On July 27, 1882, he filed the provisional specifications for his system. Edison, making some small changes, applied for a patent on a similar system in the United States five months later. Sprague gave Edison extensive advice on the system, and Hopkinson, after some thought, agreed to accept a royalty and not contest Edison's patent.

The three-wire system was ideally suited for small-town installations, to which Edison now gave his attention, since construction of central stations in metropolitan areas remained stalled—by the end of 1883 New York and Milan were still the only two cities with incandescent-lighting systems. Though Villard's brother-in-law, W. L. Garrison, and another financier agreed to erect a station in Boston, banks were unwilling to lend money—the experience with Pearl Street made it questionable whether electric lighting could compete economically with gas. (Electric wires, for example, cost three to four times as much as gas pipes. Even twenty years later, electric rates were still generally 50 percent higher than gas, and gas remained the poor man's lighting well into the twentieth century.) Edison, furthermore, continued to experience difficulty synchronizing the Porter-Allen steam engines with the direct-coupled dynamos. He enlisted the services of Gardiner Sims, the superintendent of the Armington & Sims Engine Company of Providence, and when Sims, anxious to please Edison, promised to build better engines at cheaper prices, Edison chortled to Insull:

"You see, we got the young man by the balls."

At Sunbury, an east-central Pennsylvania mining town of a few thousand, Sprague and William S. Andrews began installing the first three-wire system in May of 1883. An engine and two dynamos were placed in a shack at the edge of town. The mains, about the size of a fingertip, were suspended from poles. Connections were made to twenty-five buildings using a total of 500 lamps. The lightly insulated wires were stapled to the walls and ceilings. The sockets were attached to the gas fixtures.

The $14,500 station was intended, like most of the subsequent small-town installations, to operate only from sunset to sunrise. The grand opening was set for July 4. Edison arrived the night before in the picturesque Susquehanna River town, and there was a great excitement. On the afternoon of the fourth, however, when the dynamos were tested, they refused to pick up. Edison, Sprague, and Andrews scrambled about. "Lighting up" began to look doubtful. Finally, the wires were discovered crossed outside the station. When evening came, cannon boomed, fireworks traced across the sky, and the downtown area turned radiant with light.

Everything worked reasonably satisfactorily for a few days, until a thunderstorm burst over Sunbury. Then, suddenly, a fireworks display rivaling that of the fourth of July started up in the city hotel. A hysterical messenger summoned Andrews, who found all the inhabitants of the place standing outside in the torrent. Spectacular sparks were shooting from the gas fixtures—the lightning was streaking in along the poorly insulated wires, and exploding from the pipes. The proprietor demanded the removal of the lights. Andrews, straining his conscience, calmed him by explaining that the hotel had been struck by lightning and would have burned down had not the wires acted as lightning rods.

Within a year seven small-town stations were in operation in east-central Pennsylvania, four in Ohio, three in Massachusetts, and three elsewhere. The Appleton, Wisconsin, plant became the first to be worked by water power.

Luther Stieringer, the noted gaslight engineer, designed an insulating joint to prevent repetition of the lightning spectacular. But there were a host of other problems. Meters and regulating devices were primitive. (The watt was not devised as a unit of measure until the end of the decade.) The only guide to the proper voltage was the brightness of a lamp placed in a socket on the headboard of the dynamo. The single bulb was so unsatisfactory an indicator that Edison introduced a red-blue, magnet-controlled device. When the voltage rose above normal the electromagnet was charged and attracted a bar, closing the circuit and lighting a red lamp. When the voltage fell below normal, a spring drew the bar back and a blue lamp came on. In either event, a bell rang to alert the engineer. But the continual ringing of the bell and the red-blue flashing increased the liquor consumption of the tenders, and the regulation of the dynamos improved little.

Drunkenness seemed almost to be an occupational concomitant. Added to the unreliability of the indicators, the poor quality of some of the equipment, the penny-pinching, and the more-than-occasional friction between Edison and the local entrepreneurs, it came nigh to producing havoc. Edison was constantly after Sprague "to cheapen these stations down." He thought $3 a day was far too much to pay engineers: $2 would be plenty.

Less than two years after its construction, the Sunbury plant reached an advanced state of dilapidation and needed extensive repairs. "We are all disgusted and want to get out," one of the owners wrote. W. L. Garrison threatened to sue if the mistakes in the Brockton station, in which he had invested, were not corrected. The agent for the Ohio Edison Electric Company told Insull: "Middletown and Piqua is the only hope of the Edison light in Ohio, and that is very slim. I never had the *Blues* half so bad in all my life."

Edison had a way of giving a man his head but then looking over his shoulder, nitpicking and making none-too-subtle suggestions as to how things might be done better. After only a year, Sprague was chafing. On April 24, 1884, he indicated to Edison that unless he was given more freedom, permitted more initiative, and allowed to do more of his own work, he might resign.

"I think it would be the better plan for you to resign," Edison told him.

Sprague followed Edison's advice. Within a few months after leaving, he had patented the first true electric motor, one which maintained a constant speed no matter what the load. In November, 1884, he, Edward H. Johnson, and William Hammer, whom he had gotten to know well in London, organized the Sprague Electric Railway and Motor Company. This was a largely paper concern, of which Johnson was the president and Sprague the treasurer and electrician.* Six months later, since Edison himself had failed to develop a practical electric motor, the Edison Electric Light Company directors decided to adopt and manufacture the Sprague motor.

Edison was disappointed, and considered it "a galling thorn [to have] a stranger like Sprague in a late stage of the art thrust in upon what was previously a happy family." He thought Johnson, Sprague, and Hammer demonstrated ingratitude when they did not make him a gift of stock in the new concern. Nevertheless, he, together with Batchelor and Bergmann, invested in the company.

The Sprague Company, with its executive and financial ties to Edison, now became a competitor of the United States Electric Railway Company (formed by the merger of the Edison and Field interests) and several other small firms in the development of electric traction. In the summer and fall of 1885, the Edison-Field company ran test cars on the New York elevated tracks in competition with Leo Daft, an English-born Jersey City inventor. Since Stephen Field's uncle, Cyrus, was one of the principal investors in the elevated railroad, the Edison-Field forces had every advantage, and Calvin Goddard, former secretary of the Light Company, even headed a syndicate promoting a subway. But neither Edison-Field nor Daft had developed their motors or the mechanical art of delivering power to the wheels to the point of commercial application. Edison was occupied with a variety of ventures, and when Cyrus Field, Jay Gould, and banker Sidney Dillon came to him to discuss electrifying the elevated railroad, he greeted them, then left them waiting for three hours while he pursued an experiment. Afterward he wondered why they had departed so quickly.

While incidents of engines crashing through brick walls and Gould jumping off a short-circuited car caught the popular imagination, it was technical shortcomings—made glaring by the onset of winter weather—and the frustration of trying to do business with Edison that retarded

* Johnson's primary connection, however, continued to be with the Edison Light Company.

the electrification of New York's elevated system until the twentieth century.*

Sprague, on the other hand, concentrated on motors, and in May, 1887, signed a contract to electrify the Richmond (Virginia) Street Railway. It was a display of nerve that would have done justice to Edison, for Sprague had never seen the city, nor did he have a working model of a car. When he came face to face with Richmond's steep hills, his optimism skidded. It was a task of a magnitude and technical difficulty never before contemplated in electric traction.

By placing the motor beneath the car, and protecting it from dirt and dampness, Sprague changed the whole pattern of streetcar development and set the example for the future. Despite many difficulties, the line went into experimental operation in February, 1888. A few months later it had forty-one cars running on twelve miles of track—the largest and first comprehensive system in America. Unfortunately, Sprague was no more of a financial manager than Edison, and had the same inventive quirks and failings. In developing the streetcar system, he was experimenting on production models, only to discover that some elements did not work and he had to start all over again. Although Edison committed the same errors, he derided them in Sprague as "due to too much mathematics," and refused to help him and Johnson out of the financial bind they got themselves into when they lost $40,000 on the Richmond project.

The line, nevertheless, redounded in great prestige for Sprague. Within a year of its completion, he had twenty-six additional systems, encompassing 133 miles and 261 cars, built or under construction. The stationary-motor business was very profitable, and by the end of the decade company sales were nearing a million dollars a year.

The course of Edison's relations with Sprague was to a large degree representative of his dealings with other talented men who were drawn into his orbit. As great as was his instinct for probing the secrets of the inanimate world, as sorely deficient was his ability to handle men. He lacked the capacity for supervising—whether it was men in a machine shop, or inventors and scientists attempting to solve a problem. He himself always had to be a part of what was going on; he himself had to be the heart pumping blood into the project. Although his ideas and plans were often boundless, he worked best in intimate surroundings, where a few steps could take him from Batchelor experimenting with the telephone to Upton wrestling with the incandescent light.

What he could not understand, he was not interested in. Despite the

* Edison and Field eventually parted ways, and the company went into receivership. Field's electric-railroad patents wound up with the Westinghouse Electric Company, and Edison's with General Electric.

camaraderie of the laboratory and the pleasure he obtained from nocturnal socializing, his deafness acted like an auditory veil—his intake of knowledge was overwhelmingly visual. He had difficulty participating in lengthy conversations or absorbing involved explanations.

Essentially, his was a lonely striving, within a social setting. It was alien to his nature to consider that success might depend on the strength and capabilities of his retainers. He was uneasy with men who were too independent, or whose knowledge challenged his. He was indifferent and even antagonistic toward men who wanted to pursue their own ideas under his aegis. He preferred men of limited talent, like Edward H. Johnson, who he knew were dependent on him.

Edison was on the threshold of creating a new world. But who were to be his lieutenants? Men like Sprague came to him, stayed awhile, then moved on.

Samuel D. Mott was a versatile draftsman and engineer who worked on the incandescent light and electric railroad, and had taken out several patents. At Menlo Park he had had a dispute with Edison over the payment of two dollars, which led him to comment: "Small economics wrongly applied are fatal to a great enterprise." In 1882 he was sent to work on the electric railroad in Geneva, but shortly left Edison's service amid great acrimony. Thereafter, he was a successful inventor for thirty years.

Edward L. Nichols was a friendly, communicative twenty-six-year-old Ph.D. in chemistry and physics whom Upton persuaded in 1880 to come to Menlo Park. Edison placed him in charge of the testing department, but he remained only a year before accepting the professorship of physics and chemistry at the University of Kentucky. From there, he went on to Kansas and Cornell universities. When he retired, the heads of the physics departments at thirty-five universities had been his pupils.

Edward G. Acheson was a thin twenty-four-year-old who arrived at Menlo Park in September, 1880. Experienced in iron making, surveying, and railroad engineering, he was promised $100 by Edison if he succeeded in pressing a filament from graphite. One day, after trying to produce a coated filament electrolytically, Acheson found crystals that Nichols thought might be carbon in a diamond state. Acheson asked Edison to buy a $2500 platinum tube so he could experiment with making artificial diamonds. Edison smiled, said he didn't have that kind of money, wasn't interested in diamonds, and suggested:

"No man can become an inventor unless he can do everything with a jack knife and a bean pot."

Acheson went to Europe with Batchelor, worked on the Paris and London exhibits, then left Edison's employ. After he returned to the United States he spent several fruitless, hard-up years trying to establish himself as an inventor. In 1891, however, he produced carborundum, almost equal in hardness to diamonds, and one of the most important industrial compounds ever discovered.

Nikola Tesla was a tall, gaunt, stiff-necked, humorless, somewhat mad Croatian touched with genius. He arrived at 65 Fifth Avenue in 1882 with an introduction from Batchelor. Batchelor had discovered him working for the Edison telephone company in Budapest, and invited him to come to Paris. Tesla carried in his head one of the great theoretical discoveries in the history of electricity—the rotating magnetic field. From this he was to develop the alternating-current motor, a dynamo, and a transformer. He was a skilled engineer who proved his adeptness working with Edison's dynamos and motors. After a year, Tesla asked to have his salary raised from $18 to $25 a week. Edison, influenced by Batchelor, who thought "The woods are full of men like him," turned Tesla down. Tesla, who had an ego that reduced Edison's to a mite, thereupon suggested Edison buy all of his inventions for $50,000. Edison jokingly declined. He considered Tesla "the poet of science. His ideas are splendid, but they are utterly impractical."

Tesla walked out in a huff. Five years later his inventions were to provide the most important support for Edison's principal competitor, George Westinghouse.

Charles Dean was an excellent, tough machinist, one of the few men remaining from Edison's Newark days. Although addicted to head-racking blowouts on Saturday nights, he left little to be desired as a worker, and Edison made him superintendent of the machine shop on Goerck Street. Unfortunately, the massively built Dean incurred the enmity of the frail Insull, who plotted his downfall. Insull put Jim Russell, a detective formerly with Gould but now employed by Edison, on Dean's tail. Russell could come up with nothing positive against Dean, except that he gave contract work to boon companions at high prices—a practice almost universal in industry at the time. Insull nevertheless managed to cook up a flimsy case of stealing scrap, worth perhaps $200, against him. By threatening him with jail, Insull succeeded in frightening him into signing a release and resigning—thereby saving Edison $9000 in accumulated bonus money he had promised Dean.

So they went. Edison believed men made their own rewards. Occasionally he would give key people like Batchelor, Adams, and Upton a percentage of inventions they had helped develop. He did not object to employees patenting their own inventions. But so far as offering a man pay commensurate with his talent and effort—that was not the way Edison operated.

Upton, for example, by October, 1883, had worked for five years without pay. He had lent the Lamp Company $17,000. During the last six months he had pushed the factory into the profit column, and made $25,000. Considering all that, he thought it was time Edison started paying him a salary, and suggested $6000 a year.

Edison talked him down to $2750, then gave him a twelve-month note for the sum. When the note became due, Edison expressed his regrets—he was a little hard up and didn't have the cash to pay it.

Insull, describing himself as a "busted Britisher," thought it "hardly probable that I shall make money out of the electric light." To flesh out his salary and satisfy his elegant tastes, he did a thriving little business as a broker of securities in the Edison companies, and was always on the lookout for new schemes. Edison did not mind—he liked hustlers.

Yet depending on men who had no particular loyalty to him, men who would take a buck from anyone, also had its drawbacks.

In Washington, Wilber had left the patent office some years before, and been retained by Edison as his patent solicitor. As Wilber's debts mounted in step with his alcohol consumption, he began drinking up the money Sherbourne Eaton sent him to file Edison's patents. The unfiled patent applications stacked up on his desk. In due time Eaton became concerned about why the patents were not coming through, and discovered Wilber had embezzled $1300, the filing fees for some threescore patents.

No lasting damage was done. Wilber was dismissed. But when Colonel George Dyer, who replaced Wilber, spread word about the embezzlement in Washington, Wilber turned mean. He threatened that, unless Dyer stopped talking, he would start. He had in his possession blank affidavits which Edison had executed by the score, and Wilber had let him file at leisure. Should these be brought to the attention of the patent commissioner, hundreds of Edison's patents might be invalidated. He would, furthermore, reveal "other damaging facts" if everyone did not shut up.

It was but one more example of a patent system that tended to corrupt everyone, of an era when graft and bribery in government and business were as American as apple pie, when honest men were considered dolts to be fleeced by the dishonest, and when outsharking the other guy was part of the art of survival. While in the light of the ethics expected from business today Edison's behavior was often below standard, he was merely practicing what he had learned from Gould, Orton, and other financiers. In truth, while he could be as sharp and hard-nosed as any Wall Streeter when making a deal, he lacked the inclination to apply himself to business, and frequently lost out in the long run.

CHAPTER 18

A Time of Crises

Edison's impact on the development of electric lighting was enormous. From 1865 until April 22, 1879, when Edison was issued his first patent, a total of 31 patents had been granted in the field. Edison alone surpassed that total in two years. By 1883, the patents pooled in a holding company by eight inventors numbered 321, and of these 147 were Edison's.*

Among the directors of the Edison Electric Light Company there was controversy as to how to react to the apparent infringement of Edison's patents by the United States Electric Light Company and Maxim. At first it seemed clear-cut that suit should be filed. Lowrey set the case in motion with the taking of depositions. In France court action was initiated against Maxim, and in England against Swan. But, to the surprise of the directors, Edison opposed filing suit in the United States. "My views are very strongly in favor of not suing either Swan or Maxim," he declared. The effort was not worth the time or money. The competition should be left "to their own destruction."

"The wisdom of this course was at first disputed by his associates," Johnson elaborated in a lengthy broadside in September, 1881, "but is now fully recognized. By preserving secrecy Mr. Edison has rapidly entrenched himself in an absolute monopoly without educating others. Were Edison to appear in court at any time prior to the issue of his final Patents he would be compelled to show his hand fully. This would be a great disadvantage as there is no education in scientific work equal to a legal contest over an invention. When Mr. Edison does enter the courts it will be to show an invulnerability never before attained."

* Other patent holders were Farmer, 10; Weston, 53; Brush, 35; Jablochkoff, 4; Maxim, 39; Fuller, 7; and Thomson-Houston, 26.

By the fall of 1881 Lowrey, who had a better grasp than the others of what had happened at Menlo Park in 1879, was as hesitant as Edison about filing suit. The applications Edison had submitted to the patent office were so slipshod and chaotically drawn that Lowrey had great doubts whether they would stand up. Edison's application of November 4, 1879, which embodied the principal portions of the system, referred only to carbon spirals, which had never worked. His application of December 11, 1879, was for the horseshoe filament, but did not specify it be made of cardboard or parchment. In fact, since the cardboard had not proven practical and Edison had switched to bamboo, he did not file an application for cardboard or paper until May 27, 1881. That was more than a year after Swan's application on a cardboard horseshoe—Swan and Maxim meanwhile having made the paper filament workable by subjecting it to flashing. Between 1880 and 1882 Edison filed numerous amendments to his patents, which were hung up in interference with other inventors' applications in the patent office, so as to try to pin down the patentable portions. But the language only became more confused and contradictory.

In the Light Company, the directors did much indecisive tacking. The Gramme Electric Company, primarily an arc-light concern, organized a patent pool in which the Light Company participated, but that was a temporary and expedient grouping that soon fell apart. Eaton held a number of meetings with the president of the United States Electric Light Company to discuss a merger, which the U.S. Company favored. Edison was not opposed so long as the new company carried his name only, a demand he made clear when a combination was proposed in England with Swan:

"The company shall be called the Edison Electric Light Company, Ltd., or at least shall be distinguished by my name without the name of any other inventor in its title. I am bound by pride of reputation, by pride and interest in my work. You will hardly expect me to remain interested, to continue working to build up my new inventions and improvements for a business in which my identity has been lost."

Exigencies finally forced him to settle for the Edison and Swan United Electric Company, but in America he was more unyielding. Yet when the Light Company issued one warning after another that it would sue, but failed to take action, the competition grew bolder. Edison agents complained they were being damaged. In June of 1882 the Brush Electric Company acquired rights to the Swan lamp in the United States and loomed as a new rival. The Light Company's board of directors instructed the company's president, Sherbourne B. Eaton, to sue. Edison told him not to. In the first week of November the bedeviled Eaton tried to get a decision:

"The question to be decided is whether we shall sue Swan and Maxim, or either of them. If we do not, we shall lose prestige and possibly business. I am loth any longer even to threaten suit for infringement unless we really intend to sue."

Once more Edison blocked action. Then, on October 8, 1883, the patent commissioner applied the crusher. William Sawyer, he ruled, had preceded Edison. The patent for an incandescent lamp with a carbon burner was granted to Sawyer and Man.

The Light Company, of course, appealed and announced it would go to court. But Edison, who unquestionably had developed the first incandescent lighting system in the world, was now the underdog so far as his claim to invention of the incandescent lamp itself was concerned.

The per-share price of the Electric Light Company plummeted to $130. The stock of the European company was unsalable; and even that of the Isolated Company fell by more than half.

The patent commissioner's decision was but one of several crises with which Edison was forced to cope in 1883–84—crises that were to bring about substantial changes in his life.

During Edison's financial twistings and turnings in 1874, he had obtained approximately $6600 by giving two notes to George Harrington, notes that Harrington had then endorsed over to William Seyfert of the Philadelphia financial firm of Seyfert, McManus & Company, which was one of the principal investors in the Automatic Telegraph Company. Seyfert was supposed to receive 320 shares of the Atlantic & Pacific Company from Jay Gould to liquidate the notes—but, of course, never did.

Seyfert derived his wealth from his wife, Lucy, a banking heiress, whose business he diligently mismanaged. By 1880 he had wasted $42,000 of her inheritance, and they were on bad terms. Among the assets he turned over to her was one of Edison's notes for $3351.70, due five years before. In the spring of 1880 William Seyfert threatened Edison with legal action if the note were not paid.

Edison, however, maintained that responsibility for the note was not his, but Harrington's; that Harrington had promised the notes would be written off as part of the deal with Gould. In November, 1880, Lucy Seyfert, receiving no satisfaction, filed suit against Edison.

Edison managed to keep the case from coming to trial until December, 1882. Then the jury ruled that, whatever his arrangements with Harrington, the note was his. With interest, he owed Lucy Seyfert $5065.84. Edison appealed. But in February, 1884, the New Jersey Supreme Court upheld the verdict.

It might reasonably be expected that a man who could now be numbered among America's millionaires would pay the judgment. But Edison owed Drexel-Morgan $43,000, money he had borrowed to buy the United States Electric Light Company factory; it was no time to sell depressed electric-company stocks; his bonds were in Mary's name; most of all, verdict or no verdict, he felt he was not responsible. He would make it as difficult as possible for Lucy Seyfert to collect. It was one more manifestation of the Edison obstinacy.

The Menlo Park house and some of the land were already owned by Mary. Grosvenor P. Lowrey and John Tomlinson, Edison's attorneys, advised him that if he were to transfer the rest of his possessions in Menlo Park to her or the Light Company, he would have no assets in New Jersey. There would be nothing to attach to satisfy the judgment.

Mary, unfortunately, was in no condition to be involved in a legal dispute. Despite periods of seeming improvement, her health was progressively deteriorating. She could no longer care for herself or for a large house. In the fall of 1883 Edison sublet the Gramercy Park place to a German baron. When he and his family moved out, Mary had to be accompanied by a doctor. In February and March, 1884, Edison took her to St. Augustine again. When they returned to the house in Menlo Park, her father was critically ill. A few days later Mary became violently irrational. Edison wired Insull:

"Send trained man nurse who is not afraid of person out of her mind —send as soon as possible."

Four days later Mary's father died. Within a week or two she recovered enough to come into New York. But at the end of April she complained:

"I am so awfully sick. My head is nearly splitting and my throat is very sore."

Two weeks later, on May 13, Sheriff Andrew Disbrow of Middlesex County went to Menlo Park and took inventory of everything on the property, including "ice in ice house, two plows, three wagons, one stove pipe, one gray horse, two sleighs, one tent, two wagon dusters, one looking glass, lot of old planks and boards, one dog house, two pigs, 17 fowls, one Alderney cow, one brindle cow, one bull, lot of manure."

In response to Edison's plea, "Duck, please sign your name," Mary swore that everything at Menlo Park belonged to her, not to her husband.

A sheriff's sale was, nevertheless, scheduled for Menlo Park on July 22, then postponed until August 12. Usually Mary and the children went to the seashore during the hottest weather; but she was now too ill to go anywhere. Concern over the impending auction aggravated her mental condition. Her headaches increased in intensity. Edison retained two additional doctors to supplement Dr. Ward. But they were unable to make more than a general diagnosis, and could do nothing to relieve Mary.

Only the trains passing by sporadically on the Pennsylvania Railroad disturbed the quiet of the farming village to which Menlo Park had returned. On the hill, the laboratory buildings were like ghostly relics, empty except for a few pieces of machinery and the remnants of experiments. In the house, the light fixtures were still attached to the chandeliers. But they were lusterless, devoid of electricity.

At two o'clock on the morning of Saturday, August 9, with the crickets chirping outside the open windows amid the fragrance of a summer night, Mary, twenty-nine years old, died of "congestion of the brain"—apparently a tumor.

The following Tuesday she was buried in Newark. Edison's friends rallied about him, and Lowrey, who had regarded Mary almost as a daughter and had often had her stay with him in New York, wrote Edison a beautifully sympathetic letter.

It had not been a happy marriage. Yet Edison had provided liberally for her; she had vacationed in Florida in winter and along the seashore in summer. If the time he had spent with her had been limited, she would probably have seen an ordinary workingman husband, on the job ten to eleven hours daily, little more. During her last years he had gone to Florida with her for months at a time, even though the electric light system languished for lack of his attention.

In the end, Edison was as much stricken by the cause of Mary's death as by the death itself—not only she but his mother had succumbed to a mind-affecting disease. He regarded it almost like a blot on the family, and when Dot asked him eleven years later what her mother had died of, he told her, "Typhoid."

Since Edison was Mary's heir, her death frustrated the plan to place the property beyond the reach of the sheriff. A new date, November 10, was set for the auction. Batchelor was at the sale to protect Edison—when the bidding was over, he had bought the Menlo Park property for $2750. Tucking the deed away, he forgot about it until he discovered it lying in the bottom of an old trunk seven years later and returned it to Edison. Lucy Seyfert received the money, but (with interest and penalties) the sum was still $2900 short of the judgment—so the matter was not ended.

Even while coping with the dying Mary, and refusing to yield to Lucy Seyfert, Edison was involved in "a question of unusual delicacy and difficulty," as Eaton termed it.

The Light Company had been created as a patent-holding company, which was to derive its earnings from the sales of franchises and participation in the profits of the local illuminating companies. It had seemed a way of realizing almost limitless profits on a fairly small investment.

Major-city development of electric lighting systems was stalled, however. Small-town plants brought small profits. The Isolated Lighting Company was successful, but only half owned by the Electric Light Company. The principal profit makers were turning out to be the Edison manufacturing companies, of which Edison, Upton, Batchelor, and Johnson were the exclusive owners. The Edison Machine Works did well from the start. The Lamp Company, responding to Upton's Draconian economics, realized a profit of five cents per bulb in 1883, and this grew to seventeen cents (on a sales price of forty-two cents) in another two years. The profits of the manufacturing companies,* however, siphoned off the earnings of the utility companies, in whose profits the Light Company shared.

The entire relationship between the Light Company and the manufacturing companies was built on good faith; but the faith was faltering. Edison and Upton wanted a contract, confirming them as the exclusive manufacturers for the Light Company; without a contract the Light Company, as owner of the patents, could at any time go to court to shut down the plants. The Light Company directors, on the other hand, charged that "The manufacturing companies' interests are antagonistic to ours. It is only human nature for them to strive for the largest profit practicable, leaving the smallest to us." They clamored for a merger between the patent company and the manufacturing companies so as to end the detour of profits. To this, Upton—who had invested half his inheritance in the Lamp Company, which he now valued at a quarter million dollars—was even more opposed than Edison.

"Now that we have risked $200,000 in this place and pointed out the way, they wish to reap the profits," he angrily told Edison. "I think that Major Eaton and Major Lowrey are leading us a wild goose chase and that they never intend to tie the Light Company's hands to any one place to get lamps."

Finally, in the spring of 1884, Edison and Upton proposed a settlement that would limit the profits of the manufacturing plants and give the Light Company a percentage of gross sales. The board of directors appointed a committee of three, including Lowrey, to study the offer and propose a settlement.

Lowrey was in a difficult position: he had too many masters to serve. He was the attorney for the Vanderbilts, he had been the driving force in putting the Light Company together, and he was Edison's counsel. For a time all interests had seemed to come together, and he had been not only lawyer but friend to Edison. But now his clients were pulling apart. Edison was growing distant. The stockholders were in a bilious temper. One asserted: "The results of the past show either imbecility or worse. No man connected with the Edison Company has reason to feel a pride in the work." The opinion of a director was "We are drifting toward destruction." Villard declared himself "not at all discouraged. But our general situation is certainly anything but satisfactory."

* In addition to the Machine Works and the Lamp Company, Edison had organized a tubeworks, which made the pipes that were laid in the streets, and a shafting company.

Other issues aside from the conflict over the manufacturing companies were involved. The Light Company's five-year contract with Edison, giving it rights to all his electric-light patents, was about to expire. The board of directors wanted the contract renewed for at least three years, but Edison was not inclined to accommodate them unless concessions were made to him. It was a key point, for as a patent company the Light Company's value depended on its continuing acquisitions of patents.

There was chronic quibbling about expenses. Sixty-five Fifth Avenue was shared by the Electric Light Company, the Isolated Company, the New York Illuminating Company, and (for a time) Edison's Construction Department, which installed the isolated lighting plants. Eaton objected to the Light Company's bearing nearly half the $11,000 annual administrative costs. He questioned the amount of charges made against the Light Company for Edison's lab on the top floor of Bergmann's factory.

Edison retorted: "Your language would indicate indigestion," and countered with a claim of $133,000 for expenses and a half million dollars for patent rights.

On June 18 Lowrey's committee, trying to find a compromise, recommended that Edison, Upton, Batchelor, and Johnson sell 40 percent of their interest in the manufacturing companies, receiving in exchange stock of the Light Company and the Isolated Company. Edison objected that he did not have confidence in the present management of the Light Company and that Eaton was not aggressive enough in developing business. The lines of dispute hardened.

Four factions were represented on the board. One was made up of Edison and his adherents, another of Lowrey, Twombly, and the Vanderbilts. A Boston group was led by Villard and his brother-in-law, W. L. Garrison. J. P. Morgan and Anthony Drexel formed a New York–Philadelphia financial axis.

Because of Edison's increasingly bad relations with the Vanderbilts, Norvin Green had been shunted out of the Light Company presidency, and Sherbourne Eaton, who had marched through Georgia with Sherman and distinguished himself at the Battle of Atlanta, had been the Morgan-Villard choice to replace him. He was a fine lawyer,* but he had grown bald and paunchy, and was hard put to cope with Edison and the hard-driving, scheming, sarcastic Insull, who sat in an office across the hall from him at 65 Fifth Avenue. Insull and Edison wanted Eaton out, and the Boston faction had been undermined by the financial collapse of the Northern Pacific Railway and the withdrawal of Villard to Germany.

Although Edison was still the single largest stockholder in the Light Company, his position had been eroded by two further stock issues that

* One of the attorneys in the firm of Carter and Eaton was Charles Evans Hughes, who was to become Chief Justice of the United States Supreme Court and 1916 Republican Presidential nominee.

had raised the total to $1,080,000, and by secret sales of his own holdings.

"I have lied a damn hell of a lot," a broker complained, "in saying it was not Mr. Edison's stock I was selling. Everyone asks the question."

To win the election at the annual meeting on October 28, Edison needed proxies. He solicited his friends, and made overtures to the dissatisfied shareholders. He charged that the problems stemmed from the company's management, and if he were in control he would put better men in. Reiff, now a stockbroker, helped round up votes. A day or two before the meeting, Edison could count 3000 of the requisite 3500 shares needed to control the company.

The key holdings were those of J. P. Morgan and his associates. Morgan was heavily involved in railroad financing with the Vanderbilts, but his own interests naturally came first, and it was always better to support a winner. Without Edison, the company would be worth very little, so if Edison lost, there would be no winner. Edison, conversely, courted Morgan, and placed him at the head of a proposed slate of directors that included several Morgan men. As a further inducement, Edison secretly presented Morgan and J. Hood Wright, one of Morgan's partners, with 155 shares of Machine Works stock each.

Lowrey tried to head Edison off by warning: "If you or persons under you manage the Light Company as you do the shops, then both sides of that important business will be in one control with the natural risk that it will be carried on so as to pay the largest profit to the party in control. The law not only frowns upon but forbids it."

Nevertheless, when the votes were counted, Edison, Batchelor, Upton, and Johnson, together with Morgan and his adherents, had been elected and controlled the board. Lowrey, Eaton, and most of the Vanderbilt men were out.

Eaton, in acknowledgment of his legal ability, was retained as general counsel, but Insull could crow over his fallen rival:

"There are times when revenge is sweeter than money, and I have got mine at last."

Johnson, Edison's handpicked choice, replaced Eaton in the presidency. Edison's personal attorney, John Tomlinson, the new and rising favorite, asked Edison about Johnson. He was understandably startled by the reply:

"EHJ is a telegraph operator, near sighted and generally of no account."

Fifteen years before, the penurious Edison had hovered about the floor of the New York Gold Exchange and watched men frenziedly strive for fortune. Now he himself had bested New York's most powerful family. He had risen from obscurity to the pinnacle of the inventive world and to power on Wall Street. A man who had little interest in administration and no patience with finance, he was to be an industrialist.

PART II

THE INDUSTRIALIST

CHAPTER 19

An Inventive Obstetrician

In the last half-dozen years Edison had lost his lean and hungry look. He had gained some twenty-five or thirty pounds, his face was full and his body stocky. He appeared prosperous. Mary's mother, Margaret—Grammach, as the children called her—kept house for him and cared for the children at the comfortable apartment he rented on Eighteenth Street. He spent little time with the two boys, who, when they were not with Grammach, were sent on visits to Pitt or to Edison's sister, Marion, in Ohio.

Dot, however, was his constant companion. He took her to the theater and the opera, of which he was passionately fond. By sitting in one of the front rows and cupping his hand to his ear he could pick up much of the sound. The more spectacular the show, the better he liked it. A humorous scene or a good joke reduced him to paroxysms of laughter.

After the final curtain, he would take Dot to the men's dining room at Delmonico's, a sanctuary she would have been barred from had she been older. There, amid the haze of cigars and pipes, he searched out friends, and kept his daughter up until 12:00 or 1:00 A.M. Consequently, her attendance at Madam Mears's Madison Avenue French Academy was spotty.

Dot, mature for her age, assuaged his loneliness. He accompanied her to the dentist to get her teeth straightened, then had her join him for diversion at the Light Company board meeting. Sometimes they went for short holidays to Menlo Park, where the house was occupied by Mary's sister, Alice, and her husband, William Holzer.

Edison still kept a horse, as well as a deaf-and-dumb parrot, for Dot at Menlo Park. The parrot, Edison remarked, "has the taciturnity of a statue and the dirt producing capacity of a drove of buffalo." Dot took Edison on drives—he could no more handle a horse than a locomotive—and when they passed fields of red raspberries he punned that it seemed like "a berrying ground." She asked him to play ball with her. So, for the first time in his life, he tried to catch a ball. But it nearly broke his finger. She announced she was planning to write a novel on the theme of "marriage under duress." He replied she should "put in bucketfuls of misery. This would make it realistic."

Holzer wanted to use the land for farming and the laboratory for hatching chickens artificially with an electric incubator. Edison thought he would succeed: "Everything succeeded in that old laboratory." But since Holzer was a scientific man with no farm experience, Edison "explained the necessity of having a rooster. He saw the force of this suggestion at once. Just think, electricity employed to cheat a poor hen out of the pleasures of maternity. What is home without a mother?"

He was an omnivorous reader who maintained a charge account at Brentano's and bought books by the shelf. He mixed Goethe with Thomas Aldrich, Hawthorne with Darwin, Disraeli with Mark Twain, Longfellow with Virgil, and Madame Récamier's memoirs with the *Police Gazette*. The *Encyclopœdia Britannica* served to steady his nerves.

Plagued by dandruff, he asked rhetorically: "What is this damnable material? Perhaps its the dust from the dry literary material I've crowded into my noodle lately. Perhaps dandruff is the excretion of the mind—the quantity of this material being directly proportional to the amount of reading one indulges in."

He smoked so much that, looking into a mirror, he observed, "holding a heavy cigar constantly in my mouth has deformed my upper lip, it has a sort of Havana curl." Often he had no appetite: "Stomach too nicotinny. The root of tobacco plants must go clear through to hell. Satan's principal agent Dyspepsia must have charge of this branch of the vegetable kingdom."

Tortured by his stomach, Edison suffered not only because of his eating habits and excess smoking and tobacco chewing; he had a knack—almost, it seemed, a necessity—for creating problems and anxieties for himself. With his Jules Vernian imagination, he might have been a writer—and he wrote articulately and facilely, though without discipline. Fate, however, had given him an unparalleled opportunity to materialize his visions. The difficulty was that his visions were limitless. There was no way he could engage in everything that interested him, or accomplish all that he proposed. He had little patience with detail, or interest in the maturation and improvement of inventions after they were born—he was like an inventive obstetrician who, once the baby is delivered, wants to get on to the next pregnancy.

Most of all, he could not discipline himself to work within his resources. No matter how much money he had, he would always spend more. No matter how large or how small the bills, he chronically failed to pay them all.

Two months after Mary's death Dr. Ward complained bitterly that for ten years he had attended the family at all hours of day and night. His remuneration had been two shares of Edison Light Company stock (for which he had allowed $1200), a few shares in other companies, but only $195 in cash. "To have the last bill ignored is more than I expected."

Upton had left more than $6300 due the Lamp Company in Edison's hands "on the express agreement that you would meet a note for $2750 that would come to your office for payment on Nov. 3. [This was the note Edison had given Upton for his salary.] Today we received a formal note saying that you would not fulfill your obligations. This compels us to pass a large portion of our payroll and to destroy the credit we are trying to build up at the bank."

Edison seemed incapable of budgeting. He had no sense of financial responsibility, and appeared oblivious of the difficulties he caused his creditors. When money was owed him, however, he fastened on to his debtors like a bulldog.

In January, 1885, he filed a claim of $100,000 against the Isolated Lighting Company for that company's share of expenses he had incurred during the last four years for the Edison Electric Light Company. Included were the money expended on railroad experiments; $2750 paid to Sprague for his inventions; $6200 to C. E. Chinnock, the superintendent of the Pearl Street station, "in consideration of the extraordinary efforts he made to bring the station from a state of absolute failure to one of perfect success"; $1753 spent in lobbying and payoffs to New Jersey legislators for a bill allowing illuminating companies to do business in the state; $388 "in working up an agitation in the daily press having in view the injury of the gas interests"; and $2175 to detective Jim Russell for doing the dirty work.

Ultimately Edison's claim was settled for $67,000, to be paid out of the profits of the company. But it was only part of the overall financial tangle in which the Light Company, the Isolated Company, and the New York Illuminating Company were involved.

Few of the directors would have agreed with Edison's assertion that the Pearl Street station was now a "perfect success." On June 4, 1885, a committee established to work out a settlement between the three companies reported: "The capacity of the station even now is 20 percent below what was originally the minimum contemplated, its present capacity being only about 8,000 lamps burning at one time."

When the New York Illuminating Company had been organized, the Light Company, taking Edison's figures, had estimated the cost of the district at $250,000. But, the committee charged, "The Light Company assumed a knowledge which it had little ground, beyond wild enthusiasm, to believe that it possessed." So far, $750,000 had been expended, "thereby absorbing the entire cash resource of the Illuminating Co. and leaving it with a debt which well nigh proved its ruin and which has absorbed its entire net earnings." The committee recommended that, in order to get the construction of two further New York stations, as well as installations in other major cities, under way, the Light Company reduce its royalty from 35 to 20 percent, wipe the existing debt of the Illuminating Company off the books, and contribute $170,000 of Illuminating Company stock to finance expansion.

Edison was quite willing to accept the committee's recommendations. He considered his victory at the Light Company board meeting in October not as a beginning that would have to be consolidated and exploited, but as an end in itself. He left the running of the electric-light business to Johnson, Upton, Bergmann, and Batchelor. Johnson was the president of the Light Company; Upton was the manager of the lamp factory; Bergmann was the proprietor of his own establishment, in which Edison held a one-third share; and Batchelor, who had returned from Europe in early 1884, was the general manager of the Machine Works (where Kruesi continued to be the supervisor of production).

Edison was thus left free to shift his focus back to the telephone and a new form of the telegraph.

He was upset with the Bell Telephone Company because it refused to market the electromotograph receiver. The receiver, despite its ingenuity, was impractical and unwieldy. Because of its loudness, one could not listen to a conversation privately. One had to keep turning the crank while operating it. And the water moistening the chalk cylinder had to be replenished every few days. Edison, however, believed the receiver was not being offered because Bell did not want to pay royalties. Now that the telephone was coming into increasing use, Edison was much dissatisfied with the contract that paid him only $6000 a year for the employment of his transmitter. Bell President Theodore Vail refused to renegotiate that contract. In October, 1884, however, he agreed to pay Edison an additional $6000 a year for five years in return for an option on Edison's future telephonic inventions.

The contract with Bell brought Edison in touch again with Ezra Gilliland, recently named head of the Bell Telephone Company's experimental department in Boston. Gilliland, after the unsuccessful venture with Edison's electric pen, had returned to Ohio. In 1878 he had had the Edison phonograph agency in Cincinnati. Later he had worked for Western Electric in Indianapolis, where his father held the telephone franchise. During the past few years Gilliland had invented a number of telephone devices, the most important of which was a switchboard. Gilliland was jolly, portly, and, like Edison, given to punning. The renewed friendship grew apace. Soon they were nicknamed Damon (Gilliland) and Pythias (Edison).

In Indianapolis, Gilliland had come to know W. Wiley Smith, the Western manager for the Bell system. Both the telephone and the telegraph wires were plagued by induction. Transmission would jump from one wire to a parallel wire, so that strangers would suddenly find themselves in intimate conversation. Smith puzzled over how the phenomenon might be put to use, and thought of applying it to telegraphing from trains in motion.

When Gilliland broached the subject to Edison, it revived in Edison's mind his experiments with etheric force and the possibility of telegraphing without wires. In 1878 Edison had put up a pair of kites at Menlo Park and attempted to send an electric charge between them. Now, in the spring and early summer of 1885, he rigged up two hot-air balloons at the old laboratory site. Each balloon carried aloft a condenser. In the transmitter, the condenser was connected by wire to a battery and induction coil on the ground. In the receiver, the condenser was wired to an electromotograph telephone. Edison found he could transmit only on line-of-sight, and had to take the curvature of the earth into account. On land, because of the terrain and the electricity-absorbing effect of obstacles, the maximum distance he could send was two to three miles. At sea, he believed, he would be able to communicate ten times as far.*

Gilliland convinced Edison to join with Smith, and the three men organized the United States Railway Telegraph & Telephone Company. On each train wires were run along the roofs of the cars so that they paralleled the telegraph lines along the tracks. The initial test took place on the Staten Island Railroad February 1, 1886, and Edison sent a message to Dot at the Hotel Normandie. Paradoxically, the system worked when the train was traveling in one direction, but failed when it was going in the other. One quipster suggested placing the island on a turntable, so that the train would always proceed in the same direction.

Though the Grasshopper system, as it came to be called, worked well in theory and sometimes in practice, problems plagued it when it was put in service. Where the telegraph lines did not precisely parallel the tracks—as for instance when the train went through a cut but the lines were atop the embankment—induction failed. Bad weather reduced transmission to static. In the cars, as simple a thing as a stovepipe could ground the system. It was expensive to equip the cars, and several connected together were required for the system to work, thus limiting flexibility and use of the cars. The Grasshopper was a luxury. In April, 1887, it was merged with a rival, the Phelps Induction Railway Telegraph; but the expense failed to justify the limited utility.

* Because transmission was better at night, Edison sometimes sent up the balloons after dark, with lights to mark their location. This led to reports of the appearance of an "Edison star," rumors of which continued for decades.

Edison had more success with a cousin of the Grasshopper, the phonoplex. The basic principle of the phonoplex, a duplexing of the wire to enable telegraph and telephone messages to be transmitted simultaneously, had been worked out by a Belgian, Van Rysselberghe. Edison made some improvements and alterations, and employed the electromotograph as a receiver. He began selling the system to the railroads in 1885, and it provided him with a steady income of several thousand dollars a year until its discontinuance, in the early 1900s.

Edison hoped to dispose of the phonoplex patent and several new telephone ideas, including a platinum-contact transmitter and a tellurium receiver, to Western Union and Bell. Neither company, however, was enthusiastic. On May 9, 1885, Vail commented that he saw nothing in the patents he wanted to buy, and rejected Edison's offer. Edison brooded over the rejection for a while, then told a friend:

"Well, David! They can't shut up my factory."

David asked what he meant.

Edison tapped his head.

A short time later, on July 20, he wrote Insull from Gilliland's beachfront home near Winthrop, north of Boston:

"I am going to bust the undulatory theory."

What Edison intended to do was disprove the theory on which Alexander Graham Bell's patent hinged, that telephonic transmission depended on the use of an undulatory current. Edison was of the opinion he could employ a "make-and-break" current.* If he were correct, the Bell monopoly would be destroyed. Edison based his belief on experiments he and Bergmann had conducted at the New York lab in November, 1883. There, by inserting the platinum contacts in oil, they had been able to transmit articulated speech with a make-and-break current.

The World Industrial and Cotton Centennial Exposition was being held in New Orleans from December, 1884, to May, 1885, and the Bell Telephone Company had some of Edison's devices on exhibit. In early February Edison and Gilliland decided to attend the exposition and then continue on to Florida for a hunting trip.

They arrived in New Orleans the last week of February. Gilliland was accompanied by his wife, Lillian, a ninety-pound brunette with a small waist, but bulging hips and bust. She had married Gilliland in Indianapolis, and was a good friend of the Lewis Miller family of Akron, Ohio.

Miller, fifty-five, the son of a prosperous Ohio farmer, had also come to New Orleans for the exposition. Having early acquired a distaste for farming, he had gone into business with his brothers and stepbrothers to manufacture farm machinery. In 1855 he had invented the first

* An undulatory current, it may be recalled, is one consisting of a wavelike flow of electricity modified by the human voice. A make-and-break current is one activated by the voice.

efficient grass mower, and since had grown wealthy as a manufacturer of mowers, reapers, and other implements. A fervent Methodist, he had helped organize the Sunday-school movement. He liked children—which was just as well, since he had fathered eleven.

The seventh of these children was Mina, who was attending finishing school in Boston preparatory to entering Wellesley. Mina had silken black hair, great dazzling eyes, and a pronounced Roman nose set in a piquant face. Her figure was slender. She was gay and intelligent. She was called "one of the most royal girls on the face of the earth," and "The maid of Chautauqua."*

Her father had brought her along on the trip to New Orleans, and Lillian Gilliland, of course, knew her well. One day the Millers, the Gillilands, and Edison met at the exposition. It was a fleeting encounter. But it left Edison with a yearning for another, more extended one.

From New Orleans, Edison and Gilliland went to St. Augustine, then backtracked on a Toonerville Trolley train across the peninsula to Cedar Key on the Gulf Coast. The 125-mile trip (during which the train ran off the track three times) took two days. From Cedar Key the only way to continue was by boat, so they hired a fishing sloop to sail down the coast to Punta Rassa, the terminus of the Cuban telegraph cable. At Punta Rassa they learned of a small community named Fort Myers a few miles up the broad Caloosahatchee River.

Most of Florida was as wild as Wyoming. Fort Myers, with about fifty inhabitants, was a cattle town populated by cowboys and dotted with saloons. Unlike Wyoming, it had a balmy climate and an immense variety of tropical game, birds, and fish. Deer, alligators, and flamingoes wandered about at the edge of town. Orange trees grew thirty feet tall. Edison became enamored of the place, and badgered Gilliland to build a compound with him. They would each have a house, and share a laboratory where they could work during the winter. When the pair left near the end of March, Edison had an option to buy thirteen acres, including 400 feet of river frontage, for $3000.

Edison's business with Bell took him to Boston a number of times. In early June, Lillian Gilliland arranged a dinner so he could see the intriguing Mina again. Mina, accustomed to the constant hubbub and companionship of Oak Place, the Miller home in Akron, found Boston lonely. Although her father was wealthy, the demands of his large family and several philanthropies left little cash to spare. He financed European tours, which he considered cultural necessities, for all the girls. But at school, Mina, who liked pretty things, periodically ran short of cash.

* Chautauqua was the lake where a summer school for continuing education had been established by Lewis Miller and the Reverend John Heyl Vincent. The Reverend Vincent had invited Edison there in 1878, but Edison had not shown up.

Her affectionate and fussy mother continued to keep her wing over her daughter at long distance: "I would not commence going out with Arch if you do there will not any other one come and he is not quite the right one for you," she admonished. Mina must ask her teacher's opinion on everything, including boy friends.

Mary Valinda Miller believed it was a wife's principal, and perhaps only, purpose in life to serve and obey her husband, and the family was clearly dominated by Lewis. He was an intelligent, honest, and gentle man whom all the girls loved and thought of as the ideal husband. So when Mina sat down at the table with Edison, he was not necessarily at a disadvantage because he had gray hair and was more than eighteen years older than she. He was one of the most renowned men in the world, and she had read a sketch about him in a book, *How Success is Won,* which included profiles of her father and John Heyl Vincent. Yet he had few pretensions, and exercised a courtesy born of an ineradicable bashfulness.

"Ask me nothing about women. I do not understand them. I do not try to," he asserted. He had, nevertheless, a reservoir of interesting stories. His odd expressions were amusing. And Mina had been conditioned to his Midwestern brand of humor. When she sang and played the piano for the group, he was as captivated by her self-assurance as by her charms.

"I could not help being interested immediately in anyone who would play and sing without hesitation, when they did it as bad as that," he later recalled.

School was ending, so the Gillilands invited Mina to spend a week at Woodside, their beachside house. "I do hope my Dear Mina will have a very nice and prophetable time at the seaside with her friendes," her mother wrote.

One of the friends "unexpectedly" was Edison. The Gilliland house was large. Both Ezra and Lillian liked company, and they were partial to young women. Parlor games, conversations about subjects in fashion, music, rides on the bay in Gilliland's steam launch, and philosophical and prophylactic discussions on metaphysics and love filled the days and evenings.

Edison, wearing Gilliland's nightshirt, was usually the last to go to bed and the last to rise. He slept "as sound as a bug in a barrel of morphine," and speculated he had been "inoculated with insomniac bacilli when a baby." He had splendid dreams that would have kept Freud in delight for a season. Edison and Mina could talk and brush sleeves; but, surrounded by people, their conversations could hardly be personal.

Then, in a few days, the idyll was over. Mina was to attend the Yale graduation of twenty-one-year-old George Vincent, the son of the Reverend John Heyl Vincent, and afterward visit with the Vincent family. George was handsome, rugged-looking, and a scholar. The

Vincent family assumed that, when he married, Mina would be his bride. It was an assumption Mina did not necessarily share, but her mother advised: "It is rather a hard place for you I think but just try and go along as though there was nothing thought of."

Edison, however, in the weeks that followed could think of nothing but Mina. She had promised to send him her picture; when, after a month, it had not yet arrived, he lamented: "Oh dear, this celestial mud ball has made another revolution and no photograph yet received from the Chataquain Paragon of Perfection. How much longer will Hope dance on my intellect?"

Edison made trips to New York and Menlo Park, but kept getting drawn back to Woodside. "If I stay there much longer," he worried, "Mrs. G. will think me a bore." Other females came and went, and he invited the twenty-five-year-old Insull, who was in such obvious and desperate need of feminine succor that all his acquaintances were trying to find a match for him, to spend the Fourth of July at the Gillilands'. "There is lots pretty girls," Edison enticed.

Edison's mind was on only one, and to such an extent that he was nearly run over by a horsecar. "If Mina interferes much more, will have to take out an accident policy," he noted to himself. He attempted vainly to ease the turmoil in his stomach by chewing gum and taking long walks. He constructed "a mental kaleidoscope, and tried to improve upon Mina by discarding some and adding certain features borrowed from Daisy and Mama Lillian G. . . . Sort of Raphaelized beauty."

Pining away for Mina in July weather so hot he speculated, "Hell will get up a reputation as a summer resort," Edison was like a Jekyll and Hyde. In the parlor he sat around with the women, played silly games, helped put paper dresses on daisies, showed the girls how to make shadow pictures out of crumpled paper, and was badly beaten at checkers by one of the girls' watchful aunts. Alone late at night with Gilliland, he plotted the downfall of the Bell Telephone Company.

Gilliland had invented a new carbon transmitter that Edison thought was superior to any previously developed, including his own. In Kenosha, Wisconsin, where the Gilliland family had a skate factory, a businessman, Zalmon G. Simmons, was anxious to establish a telephone system outside of the Bell monopoly. If Gilliland's transmitter, Edison's new receiver, and a make-and-break current could be successfully worked together in a system, the monopoly would be demolished. Gilliland and Edison entered into negotiations with Simmons.

Two obstacles presented themselves: Gilliland was an employee of Bell, and Edison was under contract to Bell for his telephonic inventions. In the latter part of July, however, Edison informed Insull: "I have committed myself too deep to withdraw." Tomlinson would have to devise a legal zigzag around the obstacles.

Gilliland would, of necessity, have to leave Bell. Edison signed a contract, appointing him his "confidential agent" and guaranteeing him

Woodside Villa July 17 1885-

Slept so sound that even Mina didn't bother me as It
would stagger the mind of Raphael in a dream to
imagine a being comparable to the Maid of Chatoqua
so I must have slept very sound— as usual I
was the last one up, this is because Im so deaf
—found evrbody smiling and happy— Read more of
Miss Clevelands book, think she is a smart woman
—relatively— Damons diary progressing finely—
Patricks went to city get tickets for Opera of
Polly, we can comparrot with Sullivans— We are
going out with the ladies in yacht to sail perchance
to fish, The lines will be baited at both ends,
Constantly talking about Mina who me an Damon
use as a sort of yardstick for measuring perfection
makes Dot jealous, she threatens to become an
incipient Lucretia Borgia, Hottest day of season
— Hell must have sprung aleak, at two
oclock went out on yacht— cooler on the water,
sailed out to the Rock-buoy— this is the point
where Damon goes to change his mind, he circles

A page from Edison's diary shows how he became enamored of Mina Miller.

an income of $5000 a year. In early October Gilliland moved to New York. Simmons agreed to furnish Edison and Gilliland with $5000 to conduct the experiments, which Edison estimated would take four or five months. But since Edison could be connected with the operation only covertly, the agreement with Simmons named merely Gilliland, Bergmann, and Johnson. If Simmons exercised his option on the system, Gilliland would receive $40,000, Bergmann and Johnson $36,000, and Edison $24,000, plus a substantial amount of stock.

Through the fall of 1885, Edison and Gilliland worked in the fifth-floor laboratory atop Bergmann's factory. Much of the equipment from

Menlo Park was there—the organ, the books, the test tubes, the spiderweb of wires. Edison usually arrived about two o'clock in the afternoon and stayed until midnight. The building was illuminated by gas—six years after the great discovery at Menlo Park, neither the factory where the fixtures were made nor Edison's workshop was lit by electricity. At night the place was dark, deserted, and spooky. Edison, looking like a meditating alchemist, worked at a large table, haloed in the glow of two or three gas jets. Sometimes there were visitors. If Edison was in a jocular mood, he would hold mock fencing matches with them or dance around the benches.

In the dark of one night, Edison's faithful employee John Ott stepped into the elevator—which wasn't there—and plunged five stories down the shaft. Miraculously, he escaped with only a bruised back, but as the years progressed he became more and more incapacitated and eventually lost the ability to walk.

In November, Edison's father, Sam, appeared. At the age of eighty-one he shunned the elevator, and bounded up the steps two at a time. Edison had repeatedly offered to support him, and in 1879 had told him: "I will have everything fixed for you and you will have a nice easy job. You can have anything that I have. I'm not poor by any means. If you wish I will give you money enough to go to Florida, and your salary will be sufficient to keep you in fine style."

Sam, however, was of far too independent a spirit, and was not to be lured away from Mary Sharlow and Port Huron so easily. Together with a chemist, he started up the Fidelity Chemical Works, which produced soap, bleaching powder, and flavoring extract. Toward the end of 1884 he made a comic and pathetic attempt to collect from the Grand Trunk Railroad of Canada the twenty-eight dollars in wages Edison had left behind while hightailing it twenty-one years before.

Edison thought the old man needed a vacation. For years Sam had talked about the great fortune the Edison family was entitled to in Holland, if only he could get there to dig up the documents. Edison decided to finance a grand tour of Europe for him and his sidekick, Jim Symington. They spent two months traveling over the Continent, and ended by being invited by the mayor of Liverpool to the dedication of the Liverpool-Birkenhead tunnel. When Sam came back he reported he and the Prince of Wales had been the guests of honor (approximately in that order), and appeared ready for another go at the prime of life.

The attack on the undulatory current, meanwhile, was bogging down. Fortunately for Edison, his thesis that a telephone could be worked with a make-and-break current was wrong. (Had Edison succeeded, the lawsuit that would surely have followed would have made the quadruplex case seem like a legal finger exercise.) His dudgeon against the Bell Telephone Company was abating, and he was diverted by other interests.

CHAPTER 20

The Advent of Thomas Edison

Ever since he had installed the lights on Bennett's and Gould's yachts two years before and taken a trip up the Hudson on Villard's steamer, Edison had had an itch to acquire a yacht of his own. The collapse of the Electric Light Company stock had cut short his initial search. But the outings on Gilliland's launch, and the information that Lewis Miller owned a forty-passenger yacht, revived the urge. By acquiring a yacht he could include Mina in a party for a leisurely sail up and down the coast. He advertised in newspapers to buy or charter a sixty-five- to eighty-five-foot yacht, giving the return address of "PLEASURE." The least expensive boat he could find cost $1000 a month, with a minimum charter of three months. At the end of July he directed Insull, "Let up on the yachts. I have changed my mind, the expense is too great when you get the proper yacht."

Instead, he decided to follow up on John Heyl Vincent's invitation of seven years before and attend the meeting of the Literary and Scientific Circle at Chautauqua, there to contest George Vincent for the hand of Mina.

In the eleven years since the inception of the institution, a thriving village had sprung up along the shore of the lake, nestled in the western corner of New York. A shrinelike Hall of Philosophy had been erected in 1879. A huge amphitheater with a great organ accommodated the lectures. What had started out as a Sunday-school instruction camp had become the fashionable resort for Eastern and Midwestern Protestants, where religion was blended with education, social graces, crafts, athletics, and romance. There were classes in fancy dancing, bookbinding, music, pottery making, philosophy, and a score of other subjects. Athletics were supervised by the Physical Institute at Yale. At the Feast of

Lanterns the whole shoreline and a one-mile regatta on the lake were illumined by lights. The Recognition Day procession, led by white-gowned girls carrying flower baskets, was as solemn as a coronation.

Lewis Miller had invested a great deal of money in his project. He was pleased when the Gillilands brought Edison—the attendance of well-known men increased the stature of the institute. There was no intimation that Edison was a suitor for Mina, and his meetings with the Miller family were casual. Despite the romantic setting, there was little opportunity to be alone. Edison spent the late evenings not with Mina, but with Mina's older brother Ira, and Ira's friend Walter Mallory. Mallory was an iron manufacturer; and, sitting on the hotel porch until well after midnight, he and Edison discussed not conjunction but separation. Edison revived his idea for the magnetic ore separator, an invention that had lain dormant for three years, and Mallory was a receptive listener.

After four days, Edison interrupted his stay to attend the first national meeting of the Edison Illuminating companies. When he returned, the Gillilands formed a party, including Mina, to visit Mount Washington. The highest peak in the Eastern United States, it was ascended by a cog railroad, the first in America. Edison was invited along.

From the twenty-first to the twenty-fourth of August, the group lodged in Bethlehem, New Hampshire, in the heart of the White Mountains. With its peaks, narrow valleys, and cradled meadows, the region was one of the most beautiful in the country. Already some of the leaves were coloring, and the nights were chilly—it was pleasant to gather around the fireplace in the evenings. Edison turned his deafness to advantage, and suggested to Mina she learn Morse code so she would not have to shout into his ear. Mina assented. Edison was provided with the opportunity to tap upon her hand, and she was an apt pupil. On the last evening of their stay, when they were exhilarated from the day's riding and hiking, Mina placed her hand on his knee so that they could communicate. Edison tapped out a paean to her beauty, accompanied by a Morse-code sigh. She blushed, but did not withdraw her hand.

Then they went their separate ways—the Gillilands to Boston, Mina to Akron, and Edison to New Orleans, where he sketched the plans for a central station.

When Edison was obsessed with an idea, he could not rest until he followed it to success—and he was now obsessed with Mina. Since, he confessed, "a postoffice courtship is a novelty to me," he did not correspond with her. But he could scarcely wait until she returned to Boston.

Gilliland let him know she would be back on September 23. Edison arrived the next day and laid siege. Before the week was out he had won her permission to write to her father.

Edison returned to New York, where he composed a formal letter at his desk in the laboratory. He loved Mina, and had asked her to entrust her happiness into his keeping, he wrote Lewis Miller. "I trust you will not accuse me of egotism when I say that my life and history and standing are so well known as to call for no statement concerning myself. My reputation is so far made that I recognize I must be judged by it for good or ill."

Edison's letter was received with considerable shock in Akron. There seemed to be a peculiar affinity between Miller girls and older men. The upright, very-much-married fifty-three-year-old John Heyl Vincent, who was soon to be a bishop of the Methodist Church, was so desperately in love with Mina's twenty-nine-year-old sister, Jenny, that for fifteen years she put off another rockfast suitor, Richard Marvin, the brother-in-law of B. F. Goodrich. (When, finally, she married Marvin, she was sick with rheumatic heart disease and had but six years to live.) The feelings Jenny and Vincent had for each other were, of course, illicit and unmentionable. But they exacerbated the reaction to the prospect of one of America's most notable religious families gaining a gray-haired, nearly middle-aged nonbeliever for a son-in-law.

Lewis Miller invited Edison to be a guest at Oak Place when Mina returned from Boston before Christmas. Miller placed high value on education. Even before he had been the moving spirit for the Chautauqua program of continuing education he had been a generous contributor to Mount Union College in Alliance, Ohio. As the president of the board, he had advocated equal education for women, and the college had been the first in the United States to initiate a women's program equal to the men's. He was progressive in his labor relations, and instituted a nine-hour day in his plants. Although he had been a moderate drinker in his youth, he had become a teetotaler, and the organizational meeting of the Women's Christian Temperance Union had taken place at the Miller cottage in Chautauqua.

Edison's attitude toward education was indifferent, and he lacked Miller's social progressiveness. Still, in many ways the two men had similar backgrounds. Both were of middle-class, Middle West, Protestant background. Both were ingenious, self-made, and successful inventors. Both read a great deal. They talked on the same plane and were interested in some of the same subjects. Miller had great respect for Edison as an inventor, and they established a rapport and liking for each other.

Mary Valinda had a harder time adjusting to the thought of Edison as Mina's husband. For her, religion was the sine qua non of life—every morning the family gathered for prayers. Edison, in diametric contrast, averred: "My conscience must be incrusted with a sort of irreligious tartr." Mary Valinda regarded her earthly existence as rather insignificant, and instructed Mina: "When this life ends, you will enter a brighter and happier life, one that will never end." How, Mary Valinda wondered, could one be comfortable with a man who did not think of their present acquaintance as merely a prelude to everlasting relations in a divine hereafter?

During the mid-December days at Oak Place Edison was at his most charming and generous. Again, he had little opportunity to be alone with Mina, if for no other reason than the fact that Mina's two younger brothers, twelve-year-old John and ten-year-old Theodore, followed them like persistent fleas.

When Edison departed, a few days before Christmas, he went without a definite reply from Miller. After returning to New York, he sent the boys a telescope, equipment to construct a telegraph line, and an induction coil, with copious instructions for its operation: "The wheel should be turned about 200 times per minute for a black cat and 199½ for a cat with a sanguine temperament. The coil is *very powerful.* I tried it on a Dutch Carpenter today and it knocked him down instantly. Hoping you will have a Merry Christmas and not watch me and Mina so closely when I come again."

Edison had won over most of the family. Jenny, however, who had always been Mina's guide and confidante, supported Mary Valinda's doubts. But their hesitation only increased Mina's determination. She had made up her mind, a process her father had always maintained a woman had a right to. After a few days, Lewis Miller informed Edison of his consent to the marriage.

Edison had bought the Fort Myers, Florida, property in September. He and Gilliland established a joint $27,000 account (which, predictably, proved inadequate) to erect the buildings and equip the laboratory. The structures were to be prefabricated in Maine, then shipped to Florida. Edison persuaded Mina that Fort Myers would be a fine place for a honeymoon.

For a permanent residence, a New York real estate firm offered Edison a mansion on an estate in West Orange, New Jersey. The place had been built by Henry C. Pedder, a trusted employee of Arnold, Constable, Inc., New York's large and fashionable department store. When Pedder felt he was not being paid in keeping with his responsibilities, he, together with two other employees, began embezzling. A large part of the money that he diverted he poured into his dream house. He kept adding rooms, bought exotic furnishings, and accented the house with elaborate stained-glass windows, paneled and bronzed walls, and frescoed ceilings. Before he was discovered, he spent $300,000 to $400,000. The estate was on the market for $235,000.

Glenmont, as it came to be called, was surrounded by eleven wooded acres and sat atop a knoll that commanded a view of the Orange

Valley, with New York in the distance. Included were $40,000 worth of greenhouses and a $15,000 stable. The list of furnishings ran thirty-five pages. In keeping with its unplanned expansion, the place had all sorts of odd corners, as well as eccentricities, including a toilet one ascended to like a throne. But, built in the Queen Anne style at a time when Victorian turretry and embroidery were the vogue, it had a remarkably modern appearance.

Edison liked to do things in grand fashion. Here was the chance to make Mina mistress of a house comparable to the one over which her mother presided. Selling securities, he came up with $150,000, and took out an $85,000 mortgage for the rest.

There was one problem: if he occupied the house, he would be arrested.

After the sale of the Menlo Park property, Edison had still owed Lucy Seyfert $2900, which, with interest, had grown to $3200. The New Jersey Supreme Court ordered Edison to appear for a discovery hearing, to determine whether he held assets enabling him to pay off the remainder of the note. When Edison ignored the order, a writ was issued against him for contempt of court. He paid no attention to that, either, until he decided to buy Glenmont. Then, finally giving in, he left it to Tomlinson and a New Jersey attorney to settle the case.

On Saturday, February 20, 1886, Edison's friends gave him a ribald widower's dinner at Delmonico's. (The proper Upton and the un-jocose Kruesi were left out.) Three days later they all departed on a special car to Akron for the wedding. The event was scheduled for Oak Place at 3:00 P.M. on the twenty-fourth. A long red carpet was spread from the entryway several hundred feet down the knoll. Flowers from Miller's greenhouses filled the mansion. The couple stood beneath an arch of roses. Edison was in his familiar Prince Albert coat and black tie. Mina had on a white silk dress with duchess lace. His wedding gift to her was a necklace of pearls with a descending crescent of diamonds. After the ceremony, a huge dinner was served by fifteen waiters for the throng of guests.* The mammoth wedding cake was accompanied by harlequin ice cream. But before the meal was finished the bride and groom rushed off—they were to catch the 6:18 train for Cincinnati. The train was late, so, surrounded by reporters, they spent an hour in the waiting room.

Making nightly stops in hotels along the way, the couple reached Fort Myers via the roundabout land-and-sea route ten days later. The Gillilands and Dot were already there—Edison had yanked his daughter out of school so precipitately Madam Mears was at a loss to know where she had gone. Dot, who at times displayed the maturity of a sixteen-

* George Vincent, understandably, did not attend. He recovered from his disappointment and became a leading educator, attaining the presidencies of first the University of Minnesota and then the Rockefeller Foundation.

year-old, had been delighted to have her father for an escort, and was not at all pleased with developments. Whenever Edison had mentioned Mina's name, her nose had become dislocated.

"She threatens to become an incipient Lucretia Borgia," Edison had mused to himself.

The Edisons' arrival at Fort Myers was anything but propitious. One of the vessels carrying the goods had sunk. Only one of the buildings was advanced far enough to even give hope of completion. For the first weeks, the eccentric honeymoon party crowded into the town's frontierlike hotel. Afterward, they all moved together into the one house that could be roofed over. They were completely isolated. Mail from New York took two to three weeks, and was routed via Tampa. But the climate was lovely, the surroundings exotic, and Mina had the resilience of youth. Mary Valinda worried that, since Edison didn't sleep, Mina certainly wouldn't get much either. But Edison, cuddled in the fashionable silk nightshirt he enjoyed luxuriating in when he didn't go to bed in his clothes, slept long and late. After a few days at Fort Myers, Mina ventured to her younger sister, Mary Emily, that she was not finding married life a bore. Mr. Edison made "a nice roommate."

The honeymooners remained at Fort Myers till the end of April, then returned to New York via Akron. Grammach and the two boys, meanwhile, had moved into Glenmont. Edison assumed that since Mina had lived in a stately home she knew everything about running an estate. The house and the household, including the love-starved ten-year-old Tom and the rambunctious seven-year-old Will—not to speak of the resentful Dot—were dropped into her lap.

Mina was a resourceful young woman; but she really did not know her husband at all. She knew she was marrying a great man, and he had seemed good and generous, complaisant and charming besides. But she had never seen him in his natural surroundings. She went on her honeymoon with Thomas A. Edison; and when they returned to Glenmont she discovered Al.

Al was undignified, unpredictable, undisciplined, and often an enigma. Nothing in her staid Methodist and Boston upbringing had prepared her for his character. He nicknamed her Billy, and played prankish word games that left her sisters perplexed and gasping.

"I don't know but what he was calling you the worst names on earth so I wish he would translate for me," Mary Emily requested of Mina. "I don't know whether he was swearing or saying the sweetest words imaginable."

Al did not fit Mina's conception of her husband at all. She refused to accept him. Edison was "Thomas." She would dedicate her life to burnishing his dignity and sculpting a new image. To all the relations and close friends who had known him prior to his second marriage, Edison was and remained Al.* To those who came after, he was Thomas; or, occasionally, Tom.

Mina was immensely proud of her house—though, understandably, her tastes differed somewhat from the H. C. Pedder interiors she inherited—and invited all her friends to visit. But she disliked housework, and was not prepared to cope with the children. She knew her husband was rich—his wealth exceeded her father's several times over—and so she set about assembling a staff commensurate with the estate. For the children she obtained a governess. For the home and grounds she hired a cook, an upstairs maid, a parlor maid, a laundress, a gardener, a groundskeeper, a stablehand, and a coachman. Grapes and a vegetable garden, containing every variety that would grow in the climate—thirty-odd in all—were planted. The stables were filled with cows as well as horses.

In his courtship, Edison had been as obsessed with Mina as with any of his experiments. But once the goal had been attained, he was ready to go on to other things. His interests were always diverse. Mina had a great capacity for love, and was at a loss to know what had happened to the persistent suitor. She worried that she was not pleasing her husband. She was unsure of his love. She was lonely in the big house out in the country, where she had no friends.

Mary Emily urged: "Love Thomas with all your soul and gizzard and everything will be all right." But Mina, the desired and lighthearted, was given to melancholy.

* For example, Fred Ott, who worked for Edison all his life, named his son Alva Edison, not Thomas Alva or Thomas Edison Ott.

CHAPTER 21

A Laboratory of Grand Design

In fact, on his return from the South, Edison's time was demanded by a multitude of problems that had accumulated during his courtship and honeymoon. They dealt not only with the electric light, but also with the phonograph, which suddenly came very much to life.

It was a difficult situation, for he was removed by nearly two hours of rail, ferry, and carriage travel from his laboratory and office in New York. The lamp factory in East Newark was the closest of his facilities to Glenmont, yet even that was about an hour away.

On May 1, 1886, the union at the Machine Works presented Batchelor with a set of demands: a nine-hour day in place of ten hours, without reduction in pay; time and a half overtime pay till 8:00 P.M. and double time after that; no more piecework; one man for one machine (thirteen planers were being run by two men and six lathes by one man); and a union shop. Edison and Batchelor talked the matter over. They would grant the nine-hour day, but none of the other demands. With the inventory on hand, they could shut down for three months without serious loss. On the seventeenth, 350 men walked out at Goerck Street and a second, Brooklyn shop, to which the Machine Works had expanded.

Edison had bought land in Brooklyn on which to erect a factory; but the strike caused him to reconsider. In New York, wages were high —some men earned as much as thirty to forty dollars a week—and the unions were strong.

"When we open up again it may be in some other place away from this city," Batchelor told a reporter. A real estate agent told him of a factory for sale in Schenectady. On the twenty-second Batchelor traveled up the Hudson to take a look.

The McQueen Locomotive Works consisted of two brick buildings, one 350 by 80 feet and the other 325 by 120 feet. Walter McQueen had intended to go into competition with the Schenectady Locomotive Works, but his funds had run out and the property was to be sold at auction. Batchelor thought the place was a decided bargain. Best of all, there were no unions, and skilled mechanics could be had for $1.75 to $2.25 a day.

Edison entered into negotiations to acquire the property.

On May 31, before they were completed, the strike at the Machine Works collapsed. The men returned to work on Edison's terms. On June 6, nevertheless, Edison purchased the Schenectady property for $45,000.*

Work began immediately to equip the Schenectady plant. To raise the $100,000 in capital needed, Edison sold all his stock in the Edison Electric Illuminating Company of New York and lent the proceeds to the Machine Works. Between Christmas and New Year's the move was completed. At the onset of 1887 operations commenced at the new factory.

Kruesi remained supervisor and chief engineer. Edison, who badly missed Batchelor in the laboratory, exhorted him to return to their partnership in experimentation. Batchelor, however, replied such an action "would be fatal at present" to the fortunes of the Machine Works. He would continue as general manager, but keep his office in New York.

Insull was unhappy in his position. He had learned from Edison how to juggle money, and his finances were in as much disarray as those of his employer. He had his hand in a number of businesses, none of them very profitable. He owed $1175 to the Light Company for nineteen shares of stock. He was in debt to Kruesi, Bergmann, and several stores. The sheriff was knocking on his door to collect a court judgment. He was supercilious and arbitrary, and had a talent for making himself unpopular. But he was hard-driving, efficient, ruthless, and an *operator*—qualities Batchelor and Edison considered desirable for the running of the Machine Works. On December 3 Insull was named secretary and treasurer of the Works.

Edison, however, had grown so dependent on Insull's handling of his finances and personal affairs that—even though Insull relocated in Schenectady—he was to continue as Edison's private secretary.

Moving in to care for Edison's correspondence and day-to-day

* The story that Edison offered $37,500 and the merchants of Schenectady chipped in with the remainder when it appeared negotiations might break down is spurious. Edison originally offered $42,500, and paid the full, final sum with a check drawn on Drexel-Morgan.

matters in New York was Alfred O. Tate, Insull's friend and protégé. A Canadian, Tate had been in charge of experimentation with the phonoplex system, and together with Edison's old telegrapher companion, Milt Adams (whom Edison employed briefly during the 1880s),* had been sent out to market an arc light that Edison tried, but failed, to develop.

Edison owned 75 percent of the stock of the Machine Works and 60 percent of the Lamp Company. Their combined assets were now close to $1.5 million, their net profits $100,000 and soaring. Production of lamps continued to be beset by troubles, however. Complaints were legion. In many instances lamps were lasting only fifty to sixty hours. Before going on his honeymoon, Edison had written a sarcastic memorandum to Upton:

"How would it do for you to personally learn the lamp business ie the Carbonization—you are a scientific man like myself Batch etc. It seems to me if I was running the Lamp factory there wouldn't occur any such thing as losing the art. I suggest you do like the rest of us learn the business thoroughly and not be dependent on others = you are degenerating into a mere business man—money isnt the only thing in this mud ball of ours."

Then, shortly after Edison returned, Upton took his doctor's recommendation and departed on a vacation for Europe. He left behind such a mess in the carbonization department that in June Edison moved into the lamp factory to apply his intuitive skill once more to the making of filaments.

It was there, a few months later, that Edison returned to work on the phonograph.

While Batchelor had been abroad, Edison had let the invention lie idle. In the early 1880s he told Insull: "Sammy, they never will try to steal the phonograph. It is not of any commercial value."

The investors in the phonograph company that Edison had left

* Adams led an extraordinarily diverse though financially unproductive life. Every time he got drunk, which was frequently, he headed off in a new direction. In 1869, while Edison plugged away in Boston, he had taken off for San Francisco on the first transcontinental train, and once there made his living by selling patent medicines. He then became a lion tamer, and afterward took charge of telegraph construction on the Oroya Railroad in the Andes. Bouncing through Peru, Chile, Bolivia, Argentina, and Brazil, he wound up running up such a large bill at a Pernambuco hotel that the proprietor figured the only way he was going to get his money back was to sell the place to him. Before departing for Africa, he was in charge of a skating rink, a bull pen, and telegraph construction in Rio de Janeiro. He rambled through the gold and diamond fields of South Africa, and made stops in various European countries. After returning to the United States, he was ordered by one company for whom he worked to report to Lewis and Clark in Helena, Montana, but discovered on his arrival that he was a few decades too late. For a time he lectured, then went back to selling patent medicines. He died in 1910 after the amputation of one of his legs in a Pittsburgh hospital.

hanging in limbo in 1879 had thought otherwise. One of the men was Gardiner Hubbard, whose son-in-law, Alexander Graham Bell, received the French Volta prize of $20,000 in 1880. Bell used the money to establish the Volta Laboratory Associates, consisting of himself, his cousin Chichester Bell, and Charles Tainter.

Tainter, the most active of the Volta associates, had started as an apprentice in Charles Williams's Boston shop in 1870, when he was sixteen. He had come to work for Bell in Washington at the beginning of 1881. Edison's seeming disinterest in the further development of the phonograph piqued Tainter's curiosity. Since tinfoil was obviously not a satisfactory recording material, Tainter turned to wax, another substance Edison had suggested. After many months of experimentation, Tainter and Chichester Bell (Alexander Graham Bell was away most of the time) produced a cardboard-backed wax cylinder on which the recording was made by cutting, or incising, rather than by indenting, as specified by Edison. The machine they used was in almost every element the same as Edison's. To distinguish it, they reversed the syllables and called it the graphophone. Since they could not market it without infringing on Edison's patents, they did not pursue the experiments further, but on October 20, 1881, deposited a sealed box in the Smithsonian Institution with the records of what they had done and a cylinder containing the words: "Grr—Grr—There are more things in heaven and earth Horatio than are dreamed of in our philosophy—Grr —I am a graphophone and my mother was a phonograph."

At this point the Bells lost most of their interest in the experiment. Tainter, however, plugged on. One of the great difficulties with the phonograph was that the reproducer was carried by a stationary arm which was very difficult to adjust laterally to the track of the record. In February, 1882, Tainter achieved a breakthrough by devising a floating reproducer that adjusted itself to the groove. Thus matters rested until 1885, when Edison failed to renew his British patent on the phonograph, and it lapsed. According to court decisions then prevalent, the lapsing of a foreign patent automatically ended the American patent protection also. On June 27 the Bells and Tainter filed five patent applications. Between August and October they had Bergmann make six graphophones for them.

Edison, of course, learned of the development, but did not seem particularly perturbed. The idea of producing a machine for stenographic use had been stored in a compartment of his mind, and in December, 1885, he pulled it out and told a *New York World* reporter about it.

The next month the Bells and Tainter incorporated the Volta Graphophone Company, and in May, 1886, their patents were granted. Another few months passed. Then Edison finally came to life. On October 5 he pulled Gilliland off the telephone project, on which he was making no progress, and assigned him the task of designing a small phonograph for office use. It was to be driven by a motor, preferably electric. The cylinders were to be interchangeable—the diaphragms and needles, therefore, were to be attached to them, so that they could be moved from machine to machine as a unit. Shellac, gum, or wax was to form the recording surface. The user would paint the material onto the cylinder himself. Then, when he was done with a recording, he would dissolve it off.

Edison himself set to work at his auxiliary lab at the lamp factory. To test the recordings, he recruited singers, one of whom became famous for his "wolf howling."

Late one night in the bitter, damp cold of the last week in December, Edison became engrossed in a conversation in the factory yard. He got started telling stories, and before he left for home went through a half-dozen—which was at least five too many. The next day he had a fever and a bad cold. The cold developed into pneumonia and pleurisy. As January, 1887, dawned he lay desperately ill at Glenmont. Though Mina ministered to him night and day, he hovered on the threshold of death. Batchelor was so shocked he made out his own will.

Edison's enormously resilient constitution pulled him through, but his recovery was to take months. Yet as he lay in bed during the first days after the crisis, he came to a decision: he would build a new laboratory. Not only would it be convenient to his home, it would be an inventor's dream, a laboratory of grand design, such as the world had never seen before. It would be so complete as to contain everything required for any experiment in chemistry, electricity, or physics. Through Tomlinson, Edison entered into negotiations to buy a parcel of land amid the scattered farms and orchards near the foot of the hill on which Glenmont was situated.

Early in February, Edison, accompanied by Mina and a male nurse, left for Fort Myers and a recuperation of three months. (Pitt was summoned to take charge of Glenmont and the children during Edison's absence.) The first week in March, Batchelor made the eight-day trip to Florida to confer with Edison about the laboratory. Three days after Batchelor left to return to New York, Edison developed abscesses below his ear. Erysipelas, a potentially fatal streptococcus infection, was feared. In the backwoods of Florida, Edison underwent a delicate operation. His recovery was retarded, but he came through without complications.

He had periodic dizzy spells and was still feeling poorly when he, Mina, the Gillilands, the Millers, and a group of friends who had come to visit Fort Myers started north at the end of April. They decided to interrupt their journey so Edison could bathe in the curative waters of Bartow Springs. The whole group was crossing a causeway when the termite-eaten timbers gave way. As the women screamed and the men gallantly tried to save them, all except Edison and Lillian Gilliland were plunged into the four-foot-deep water. While they floundered about, Edison gleefully bounced up and down—nothing could further his recuperation so much as a good pratfall. Afterward, Miller had a cartoon drawn: "Women and children first!"

Edison returned to Glenmont with the abscesses still unhealed and his impatience swelling. Batchelor had prepared the rough plans for the laboratory. Edison wanted an architect to go to work immediately. Hudson Holly, who had designed Glenmont, was selected. The three-story red-brick building with an attached powerhouse was to be 250 feet long, 50 feet wide, and 40 feet high. Most of the first two floors were to be taken up by the machine shop. The third floor was to be occupied by the phonographic and photographic departments, as well as by experimenters.

The entire west end was devoted to a huge two-and-a-half-story library. With its alcoves, open center, and three tiers of galleries, it was an acrophobe's nightmare.

The building shell was to cost $38,000. Before Edison was satisfied, however, his total investment came to over $150,000. It was a laboratory far beyond the scope of any ever contemplated—in truth, it seemed more like a large machine shop or a small factory. Even Edison was somewhat overwhelmed, and shook his head:

"The Lord only knows where I am to get the shekels—Laboratory is going to be an awful pull on me."

He would have been more accurate to substitute gold doubloons for shekels—seldom would anyone pay as high a price as Edison for that laboratory.

The foundations were laid in late May and early June, but the course of construction did not run smooth. Both Edison and Batchelor demanded perfection. Edison was at the site daily. "The foundations are not deep enough," he complained to Holly. "The cement being used is of inferior quality." He thought Holly should personally supervise the construction, and when Holly demurred, Edison replaced him with another architect, Joseph Taft.

But with Edison on Taft's shoulders, and Taft on the artisans' backs, matters did not improve. The piers between the second-story windows were as much as an inch out of plumb, and when Taft complained to the contractor, the man took a sledgehammer and drove them back into place.

"Instead of improving the quality of the work," Taft complained to the contractor, "it is growing worse all the time, so much so that Mr. Edison is thoroughly disgusted and never will be satisfied with the building,"

In mid-September Batchelor took personal charge of the construction. Although the building contained nearly 40,000 square feet of floor space, Edison was dissatisfied with its scope, and in midsummer decided to add four ancillary structures: one for metallurgy, one for chemistry, one for woodworking, and one for galvanometer testings. These 100-by-25-foot buildings were placed parallel to each other and perpendicular to the main lab. No iron was used in the galvanometer building (the nails were copper), so that no magnetic action could throw the delicate instruments off. One element, unfortunately, was left out of account, and when streetcars started running just outside the laboratory two years later, all the instruments in the galvanometer building were thrown out of adjustment.

By the end of September the foundations, roofs, walls, and flooring of the main building were complete. But the boiler, steam engines, shafting, machinery, electric wiring, and a dozen other things had yet to be installed. Edison was making a steady but slow recovery, and his workday averaged about six hours. He devoted most of his time to the lab; and after the visit of the Duke of Marlborough in late October, he refused to see any of the scores of visitors who flocked to West Orange. By the end of the year the lab was operational, but as late as the following April Edison would not talk to the editor of the *Scientific American*, who wanted to do an article on it, because it was not yet finished.

"An experimenter never knows five minutes ahead what he does want," Edison opined, and proceeded to lay in every conceivable substance. He bought $6300 worth of chemicals and a supply of every metal in existence, including the entire stock of the American Nickel Works. He ordered hog bristles, porcupine quills, tanned walrus hide, skins of every known animal, a pound of peacock tails, five pounds of hops, fifty pounds of rice, every kind of grain, a dozen bulls' horns, a dozen walrus tusks, twenty-five pounds of marlin—"everything from an elephant's hide to the eyeballs of a United States senator." The purchase orders filled volumes and volumes.

From the Bergmann lab, the lamp factory, and other places, Edison reassembled the things he had had at Menlo Park. One chemist was given the task of identifying the contents of hundreds of bottles without labels.

Edison envisioned the laboratory as a mill that would transform ideas into commercial products. New things would pour out in an endless stream. He would operate, in effect, a factory of creativity. For a start, he compiled a list of more than a hundred projects—some old, some borrowed, some new. "A cotton picker to do for cotton what the Mowing Machine has done for serials." An apparatus for deaf people. (Edison still received hundreds of letters yearly asking what had happened to his ear trumpet.) An improved battery. Artificial silk, artificial ivory, and artificial mother-of-pearl. A snow-pressing machine for cleaning streets. A cheap India ink, and an ink for the blind. Butter direct from milk, and electricity direct from coal. An electric piano. An electromotograph mirror. Heat-activated photography. A miner's lamp. And a molecular telephone.

Edison was in an optimistic mood. The nation was in the midst of a long period of prosperity. Some of his old inventions, which had been limping along, were suddenly catching on—Edison bought the rights to the mimeograph back from George Bliss, then sold them to A. B. Dick, a Midwest lumber manufacturer. Dick marketed the machines aggressively, and with paperwork increasing rapidly and typewriters coming into general use, mimeographs became standard equipment in large offices—within five years Dick was registering more than $60,000 in net profits annually.

Edison foresaw a host of products like the mimeograph that, as they emerged from the laboratory, would be manufactured in the huge industrial complex he intended to establish nearby. In his usual stream-of-consciousness jottings, he filled page after page with his "Prospectus and Plan of the Edison Industrial Co." Accompanied by Tate, he went riding in the Bloomfield–Silver Lake area, three miles northeast of West Orange.

"Tate! See that valley!" He pointed.

"Yes, it's a beautiful valley," his secretary replied.

"Well, I'm going to make it more beautiful. I'm going to dot it with factories."

Searching for a means to finance his proposed complex, Edison poured out his industrial soul to W. L. Garrison: "I expect to turn out a vast number of useful inventions and appliances in industry. I propose putting up a factory to manufacture, making only such articles that can be sold through jobbers, and yield large profit. [Do] you think Boston parties would put money in such a factory? I should want to start small at first, say for buildings and land $20,000; $15,000 for machinery to start on 2 or 3 things I already have. In time I think it would grow into a great industrial works with thousands of men. You know Mr. G. that I have already 1300 men under me. The works at Schenectady, of which I own 75 per cent, employ 900 men. I am proud of those works, and I believe them to be the finest in the country, and any expert will say they are *well* managed. I do not manage shops myself as I am incompetent for that class of work, but I do know how to select the right kind of men to do it for me. My ambition is to build up gradually and surely a great Industrial Works in the Orange Valley. With my Laboratory and skilled men as the creator of highly profitable specialties, but not big cumbersome things like a system of Electric Lighting. Now Mr. G. what do you think, would people invest in it?"

"No," Garrison replied, as gently as he could.

He thought the amount of capital Edison requested was "too modest"; but even so it was not easy at present to interest financiers in new enterprises. He himself had $13,000 tied up in Edison's Brockton station, and in four years had not received a cent of return. He had persuaded men to invest in Edison enterprises in the past, but "Disappointment and expectations make me more circumspect and less enthusiastic in urging new enterprises on friends. I believe in you and your star, although I have had occasion to question the quality and the conscience of some of your lieutenants."

In Edison's life, nothing ever proceeded quite as planned. He professed he could devote himself to new inventions and enterprises because the electric light no longer required his active attention—a declaration astonishing to Johnson, Upton, and Insull, who were pleading with him to develop a new system and new devices. Thus it came to be, in an ironic twist of circumstances, that as the principal justification for not responding to their pleas, Edison made one of the laboratory's first major projects the development of an instrument of death:

The electric chair.

CHAPTER 22

The Electric Chair

As far back as 1880, Edison had realized that the long-distance transmission of power required high voltage. In a caveat for a high-voltage dynamo he wrote that to transmit "electricity for great distances without the erection of costly conducting wires requires that the current should be of a very high electromotive force, as much as 2000 to 3000 volts."

Edison was not alone in this thought. The difficulty lay in how to step down and divide high voltage for practical employment. Almost fifty years before, Michael Faraday and Joseph Henry had demonstrated that voltage could be increased or decreased by transferring electricity from one coil to another, one with many turns of wire and the other with few. This was the basic principle behind Edison's old joy, the induction coil. For the device to work continuously, however, the current had to be alternating, not direct. Since, in the first decades of electricity, batteries were the sole source of power, all current was direct, and the transformer remained a theoretical curiosity.

The development of the generator, and Edison's subsequent work, stirred interest in the alternating current. In 1883 Lucien Gaulard of France and John D. Gibbs of England patented an alternating-current transformer. It had serious defects. But three Hungarians, Karl Zipernowsky, Otto Blathy, and Max Deri, improved upon it. When their system was demonstrated at the Bucharest Exposition in 1885, it drew a whirlwind of interest. The Milan Edison station bought the ZBD system to transmit power to a theater a half mile beyond the range of the Edison system. In September, 1886, Blathy came to the United States and talked to Edison. When Upton went to Europe a short time later, he inspected the operation of the system, and immediately recommended that Edison acquire rights to it, so as to keep "a dangerous enemy" out of the field. Edison paid $5000 for an option.

The combination of an alternating current with a transformer was as significant a breakthrough in electric transmission as Edison's development of the Long-waisted Mary Ann and the carbon-filament lamp had been in incandescent lighting. It obviated the need for the huge initial investment in mains, for, as Edison had pointed out in 1880, the tremendous pressure behind the electricity made possible the employment of "a surprisingly small wire."

And there, precisely, was the rub, so far as Edison was concerned. His entire experience had been with direct current. He could visualize the action of a direct current, but the action of an alternating current was something that he could not comprehend—water could not be made to go back and forth in a pipe; and, even if it could, it would be useless for performing work. Because of the orientation of his mind, he had allowed the development of a transformer to slip by him while he struggled with less radical solutions to the problem of transmission. With control of the ZBD system, the first operational for long-distance delivery of power, Edison was in position to rout his weaker rivals. But the system was not *his*. He immediately set Batchelor to work on development of a direct-current high-voltage system, and commenced a campaign of rationalization why the Edison companies should not employ the ZBD system.

(In theory, DC is more efficient than AC, but the invention of a DC transformer involved enormous technical problems that were not solved until well into the twentieth century.)

George Westinghouse, a latecomer to electrical development, did not have Edison's ego involvement in the field. A year older than Edison, he had at the age of twenty-two invented the railroad air brake, and subsequently established large works in Schenectady and Pittsburgh. In 1880 he had visited Menlo Park, and he and Edison had gotten along well. But a year later, when he wanted to go into competition with Armington & Sims on production of a high-speed, direct-coupled steam engine, Edison had replied: "Tell Westinghouse to stick to air brakes. He knows all about them. He don't know anything about engines." This had not gone down well with Westinghouse, who had proceeded to make not only steam engines but dynamos, and entered into competition with Edison.

In diametric contrast to Edison, Westinghouse built his electric business by purchasing existing patents. The ZBD system was beyond his reach; but in Great Barrington, Massachusetts, William Stanley had built upon Gaulard's and Gibbs's work. In March, 1886, he transmitted current three quarters of a mile by stepping 500 volts up to 3000, then down again to 500. Stanley first offered his transformer to Edison. When Edison naturally turned it down, he sold it to the eager Westinghouse. At approximately the same time, Westinghouse purchased the flickering United States Electric Light Company, and became a full-fledged rival to Edison.

Edison, responding to the expressions of anxiety that poured in from his sales agents, in November, 1886, wrote a thirteen-page memo to Johnson. He pointed out that power loss from various causes for an AC system could be reckoned at about a third, so that 130 volts would be required to deliver 100 volts to customers. Electric motors could not be run on AC.* He would head Westinghouse off by developing his own system. "Do you know that the Dynamo Batch made for transforming is a very perfect contrivance, in fact it is perfection. The more I study our converter business, the more I am satisfied that we shall be able to give Westinghouse all the law he wants on this particular subject or any other. None of his plans worry me in the least, only one thing that disturbs me is the fact that Westinghouse is a great man for flooding the country with agents. He is ubiquitous and will form innumerable companies before we know anything about it. When it comes to dollars and cents nothing that anyone else could possibly do could touch us in the least. The moment capital has confidence and will furnish unlimited capital if they can make 8 per cent we will think no more of putting in $20,000 feeders than the P.R.R. [Pennsylvania Railroad] does of spending $100,000 to straighten a curve."

By the time he was finished Edison was, almost self-hypnotically, convinced he had "an infinitely better and the ultimate system to beat competitors." When the DC transformer proved illusory, Edison proposed to elaborate the three-wire system by combining two three-wire systems into a five-wire system. But the people in the field groaned they were being enmeshed in a tangle of wires.

During the first two years of operation, Westinghouse installed AC systems in 130 towns and cities. Pressure from Edison agents to introduce a competitive system grew intense. William S. Andrews, now chief engineer in New Orleans, wrote: "We could get at least 2000 or 3000 more lights in at very profitable rates in a nice residential district from one to two miles away if only we had an alternating or continuous current system to support our regular three-wire system. It is idle to scoff at the Westinghouse people—they are hard and persistent workers."

Edison lost the contracts for Portland, Oregon, and Tacoma, even though the purchasers would have preferred an Edison system, had he offered one. Westinghouse's AC system could not only transmit much farther but was far cheaper than the three-wire system. Edison's most productive agent in the West demanded: "Are we going to sit still and be called old fashioned, fossils, etc. and let the other fellows get a lot of the very best paying business?"

* This defect was remedied two years later, when Nikola Tesla, rescued by the advent of alternating current from a New York ditch-digging job, patented an AC motor. He sold this patent, as well as others, to Westinghouse.

Edison agents deserted to Westinghouse. At the very time that the bottleneck in installation of Edison central stations was finally coming uncorked, business in small and medium towns collapsed. Edison blamed Johnson's business methods. Johnson replied: "The three-wire system is the only thing you have given me to work with. Give me a system that will *enable* me to compete [or] accept the fact that we will do no small town business, or even much headway in cities of minor size. A development of our system can only come from you. Is it fair to hold us accountable for the absence of this development?"

The stress and friction shook the Edison organization. Edison, in his memorandum to Johnson, emphasized the dangers of a high-voltage alternating current: "1200 volts continuous current will never do greater harm than blister the flesh, and I'll bet any amount that 1000 volts alternating current will kill certain. Why Zip uses 2000 volts alternating—this gives a difference of 4000 volts (!) (HOLY MOSES) and as it is not continuous he has to get a mean which gives *6000* volts dif. The first man that touches a wire in a wet place is a dead man. Just as certain as death Westinghouse will kill a customer within six months after he puts in a system of any size."

The killing power of electricity was not lost on other people. As early as 1878, electric shocks had been used as a method of punishment at the Ohio State Penitentiary, where prisoners were made to sit naked in three inches of water and suffer the effects of an induction coil. Shortly thereafter the *Scientific American* had carried an article titled "Electricity as an Executioner." In the fall of 1887, the New York State Legislature, searching for a more humane method of capital punishment than hanging, established a three-man commission. On November 8 one of the commissioners, A. P. Southwick, wrote Edison asking his opinion on how execution by electricity could be accomplished.

Edison replied he wanted nothing to do with it. He questioned the right of law to kill, thought society could protect itself in some other way, and declared he would "join heartily in an effort to abolish capital punishment."

Southwick, however, kept after Edison. "Science and civilization demand some more humane method than the rope," Southwick declared. "The rope is a relic of barbarism."

Then, early in 1888, Harold P. Brown, a New York engineer, perceived an opportunity to capitalize on the issue and associate his name with Edison's. Although he had never met the inventor, he came to the lab, said he was alarmed at the number of accidents resulting from the use of alternating current, and offered his service to simultaneously demonstrate the dangers of AC and investigate the applicability of alternating current for execution. Edison, seeing a chance to dramatize his stand against AC without involving himself personally, made the laboratory available to Brown.

The dynamo room was reserved for experiments, which took place mostly at night, when the power was not otherwise employed. Bare wires carrying 1200 volts dangled loose overhead, and occasionally brushed against a man, knocking him out more surely than a blow from a crowbar. One dog subjected to the direct current froze for a minute and a half before toppling over. Another trotted off after receiving 1400 volts and was nicknamed Ajax, for having survived the thunderbolt. But when Batchelor was careless enough to nudge one reluctant dog onto the electrified plate after the current had been turned on, he felt "body and soul wrenched asunder," and did not recover for several days.

Arthur E. Kennelly, who was, in effect, operations manager of the lab, became Brown's associate in the experiments. The son of the harbor master of Bombay, Kennelly was an exceptional mathematician and theoretician who, after he left Edison in 1894, was (ironically) to crystallize the theories and applications of alternating current. Kennelly took the electrical resistances of every man in the lab and found they varied from 6100 to 9900 ohms. Alternating current proved more efficient as a killer—$\frac{1}{10}$ second of exposure to AC was enough to kill a dog.*

After the killing of some fifty dogs and cats, Brown pronounced AC a perfect medium for execution, and launched a campaign against Westinghouse as a merchant of death. Pointing out that accidental deaths from electricity had risen from 10 to 42 in two years, he sent pamphlets to the mayors, leading businessmen, and insurance agents of every town of over 5000 population. In them he dramatized the dangers, and urged the limitation of alternating current to 300 volts. When Westinghouse protested that in his system the current was stepped down to 50 volts by the time it reached the consumer, Brown challenged him to a duel: Brown and Westinghouse would sit side by side, Brown to be subjected to 1400 volts DC and Westinghouse to 160 volts AC, and they would see who died first.

By the end of 1888, New York State was ready to adopt electricity for execution, but critics complained that all the animals dispatched had been small. So Brown and Kennelly led a 125-pound calf and then a 1230-pound horse into the dynamo room, and knocked them off in thirty seconds. If that was not convincing enough, they proposed to follow up with an elephant.

Brown devised "an electrical cap and shoes," and sold them, together with three Westinghouse AC dynamos, to the state for $8000. Competition commenced to name the new method of execution. Suggestions included "electromort," "dynamort," and "electricide," while Edison held out for "Westinghoused." In August of 1890, William Kemmler,

* The Society for Prevention of Cruelty to Animals asked Edison if he could recommend a method to replace drowning for disposing of animals at the pound, and Edison suggested a small AC generator.

convicted of murder, became the first man to be electrocuted, an execution that was bungled and turned into "an awful spectacle, far worse than hanging."

"They could have done it better with an axe," Westinghouse said.

So far as Edison's battle with his rival was concerned, the experiments had accomplished little except to enable Edison to use them as justification for his failure to introduce an alternating-current system. (The company that owned the ZBD patent sued Edison in 1890 to pry him loose from his option.) Edison fell back to the position that "It is in the lamp that I hope to make it positively impossible for them to exist in Central station work to say nothing of Isolated."

Since the state of lamp production was anything but satisfactory, and the crux of the problem was the filament, Edison kept experimenting. Among the substances he tried were turnips, pumpkins, squash, eggplants, and apples (as well as bran for a battery).

Once more he sent out a half-dozen men to search the world for bamboo. "I want to get a fibre that is hard and dense like a wire," he instructed them.

James Ricalton, a forty-five-year-old New Jersey schoolteacher, spent a year circumnavigating the globe via India, Ceylon, Nepal, Bhutan, Assam, Burma, China, and Japan. Frank McGowan, a feisty, incredibly tough bantamweight, started on an epic exploration across the Andes from Peru to Ecuador, and then Colombia. Seeking a giant gramina (a type of grass) that was supposed to grow eighty feet tall, he whirled down rapids, was attacked by mosquitoes the size of wasps, had his skin ripped to the bone, and was nearly drowned in Amazonian deluges. After an eight-month journey he arrived in July, 1888, at the Colombian town of Cali, and from there shipped 1181 pounds of gramina. Unfortunately, the shipment was exposed to alternating rain and searing sun, then steamed in the hold of an iron vessel, so that by the time the gramina reached West Orange it was a rotting mass.*

In fact, while his explorers were out, Edison had been diverted by news from Europe of the development of a filament made from squirted

* For the next few months, McGowan, waiting for instructions in Cali, heard nothing from Edison. "What is the cause of this Rip Van Winkleism?" he inquired of Edison in November. "Has a blizzard cut off New Jersey from all contact with the outside world?" More than half a year passed before Edison finally rescued McGowan from the backwoods of Colombia. Several months after his return, McGowan, carrying on him $2000 of the money he had been paid for his explorations, kept a luncheon appointment in New York, walked out into the Manhattan jungle, and disappeared. Although Edison engaged detective William J. Burns, subsequently the first head of the FBI, to search for him, he was never seen again. McGowan's letters from South America are classics of exploration.

cellulose. (This consisted of treated cotton pressed through a die, and made possible the production of a filament much more even in composition than those derived from natural fibers. During the next twenty years the squirted filament gradually displaced bamboo.) One day Edison returned to the lab from Glenmont carrying the skin from the top of the rice pudding he had had for dinner, and assigned an experimenter to develop a squirted filament from it. The man worked on it for months without producing anything, but Edison told him to keep at it—one never knew when there would be a breakthrough.

Edison's laboratory staff had mushroomed to 120 employees—a veritable cuckoo's nest of learned men, cranks, enthusiasts, ambitious youth, plain muckers, and quite insane people. If Edison believed in a project, he would never let go of it: Patrick Kenny, who had started on the autographic telegraph ten years before in Menlo Park, was still plugging away. One man was working on closed tubes that had a habit of exploding—Edison told him to try his experiments on suicide in a separate room. (There he promptly blew a hole through the ceiling.) A boy washing bottles in a room where hydrogen was being used to increase the size of soap bubbles decided to determine whether hydrogen, flammable, or soap, nonflammable, would prevail when a match was lit above the tub. A veritable explosion of soap bubbles, floating off in the wind through the shattered windows, provided the answer.

Several "muckers," as Edison referred to the experimenters, were assigned to produce highly poisonous chlorine gas and pentachloride of antimony. This often seeped out and was inhaled by numerous men. Since chloroform was a partial antidote, workers lolling about sniffing the anesthetic gave the laboratory the atmosphere of an opium den.

Edison had thought that by creating a bigger and better Menlo Park he would be able to develop more and better products more rapidly. Matters did not turn out that way. For while at Menlo Park the operation had been small enough for Edison to be the axle around which the experimental spokes revolved, West Orange was too large, impersonal, and diversified to lend itself to a re-creation.

Edison assigned projects to people, and tried to make the rounds of the experimental rooms like a hospital physician. But often he became interested in what was happening in one room, and spent the whole week there. Since he disliked employing people with the intelligence and initiative to pursue experiments independently, he left the men in other rooms, who were deprived of his direction, floundering about.

He continued to measure employees more by their salaries than by their output. He dismissed chemists, but hired hordes of boys with little education on the premise that it was cheaper to train his own workers.

"Our present staff of juveniles are excessively stupid," one of the supervisors complained to Tate. "All of them combined have not as much common sense as would be required to keep a ton of pig iron from floating out to sea in a calm."

In an atmosphere where decision and direction were hard to come by, years passed and problems went unsolved. The development of improved insulation was a critical project, and Edison had several men working on it; but Insull was convinced Edison couldn't be aware of what they were doing. "Water tests show your wire does not insulate at all," Insull criticized. "We have spent so much money on experimental work on insulation that we ought to be getting some results out of it."*

More and more it would become evident that only in those projects in which Edison himself was personally involved was reasonable progress made. Almost never were more than one or two major projects sustainable simultaneously. Edison's view of the laboratory as a creative cornucopia was impracticable.

* In Milan the Edison tubes were pulled up because of shorts and leakages. In the Pearl Street district the tubes were deteriorating. In a celebrated incident that took place on Nassau Street in 1892, a hearse was passing over a spot in the street when the horse suddenly started performing like a rodeo bronco. An expressman's horse came next and bolted. A large crowd gathered on both sides of the street, and eagerly anticipated the appearance of each wagon. When an electrician finally arrived, he stated: "There was no danger. The beasts will likely be all the better for it stirring up their blood."

CHAPTER 23

The Perfected Phonograph

The primary activity at the laboratory, an activity that was to continue for decades, was the improvement of the phonograph.

While Edison was sick in Florida, Gilliland had attempted to carry out his ideas, but the concepts had not proved workable. In the first week of May, 1887, a few days after Edison returned, Bell and Tainter placed a treadle-powered graphophone on exhibit at a New York hotel. A week or two later, Tainter and Chichester Bell visited Edison. They offered to turn over all their work, let the graphophone slide into oblivion, and provide the capital for the revival, further experimentation on, and marketing of the phonograph. Edison would retain all the honors. In return, they wanted a half interest in the enterprise. Edison, suspecting they were not dealing from a position of strength, offered them a fourth. Tainter turned Edison down; and the two men parted acrimonious competitors.

To get around Tainter's patent for incising the recordings, Edison proposed mounting a cutting tool to make grooves in the cylinder ahead of the recording needle, which would continue to operate on the indenting principle. Batchelor and Gilliland were urged by Edison to go on a crash schedule. Batchelor sketched three or four different concepts, a couple with distinctly Martian appearances. One substance after another was tested for the diaphragm, needle, and pickup. Various mixes of resin, kaolin, beeswax, and paraffin were tried for the cylinder, with tinfoil continuing to be superimposed on top of them. None worked well. The reproducer was still plagued by the old scratching noises.

Gilliland was the first to give way under the stress, and became seriously ill the last week in May. When he recovered, he departed on an extended vacation. Batchelor, now worth more than a third of a million dollars, was no longer willing to subject himself to the heroic exertions of a decade before, and took off all of July for a Canadian vacation with his family. Edison was involved with the laboratory construction, and was on a limited work schedule. In Washington, Tainter was laboring like a bee (and fuming that Alexander Graham Bell was receiving credit for the invention of the graphophone). Pressure kept building.

From England, Colonel Gouraud wrote that he was involved in the formation of a graphophone company. He was attempting to protect Edison's interests, and hoped to have *Edison's name given equal prominence with Alexander Graham Bell's!*

Edison exploded: "Under no circumstance will I have anything to do with Graham Bell with his phonograph prounced *backward* graphophone. I have a much better apparatus and am already building the factory to manufacture and I not only propose to flood England with them at *factory prices* but I shall come out with a strong letter the moment they attempt to float the Co. They are a bunch of pirates!"

Gouraud pretended puzzlement as to why Edison was annoyed: "Bell and those associated with him having worked out your idea more or less perfectly." He was, however, willing to drop Bell "like a hot potato" if Edison would give him European rights to the phonograph. Two months later, Gouraud arrived in the United States and manipulated Edison into bestowing on him all rights outside the American continent, a generosity Edison was soon to regret.

Despite Edison's assurances to Gouraud, he neither had a new instrument, nor was he building a factory. On October 1, however, he rented a building in Bloomfield, just north of West Orange. He was far too busy with the construction of the laboratory to involve himself directly, so Gilliland was placed in charge of manufacturing and appointed exclusive sales agent for North America, with a commission of 15 percent.

The recording medium of the new machine was to be tinfoil cushioned by a plumbago-stearite compound. Records were to be reproduced by suspending the cylinders in a vacuum and plating them electrically with gold—Edison filed a caveat with the patent office, but the process was rejected as unworkable. (More than fifteen years passed before Edison was able to perfect it.) The phonograph reappeared in newspaper headlines. Edison promised to have the first 500 machines out by January. He predicted "Phonogram" sheets, holding up to 4000 words each, would supplant letters.

Essentially, the machine Edison was preparing to market was the same model that had flowered and wilted in 1878–79. Indeed, Tainter's instrument, except for the use of wax and the addition of the flexible pickup, was also the same. Gilliland had no confidence he could sell such a phonograph. Without telling Edison, he set to work to improve the machine. He added a lever that permitted the repeating of sentences and the erasing of any part of the recording. The instrument could take duplicate copies, be halted during the recording, and be operated by any form of motor. On December 14, Gilliland, in the process of producing a model of the improved phonograph, dropped a hint to Edison about what he had been up to.

Edison was furious—it was as if Gilliland had turned Tainter in his own camp. "No man can invent and do business at the same time," he stormed, the irony of the declaration escaping him. Gilliland's job was to manufacture. He was to drop his model and proceed with business.

"Whatever harm has been done, has been done," Gilliland, swallowing his pride, tried to soothe Edison. "Surely I must be entitled to some consideration. The present form of machine that we are at work upon in Bloomfield would not compare favorably in any respect with the Graphophone, and I have never felt that you would put it upon the market in that condition. Certainly we are capable of producing a machine superior to the Graphophone. I have worked out a machine not only equal to the Graphophone but superior."

On reflection, Edison concluded it would be foolish to go back on Gilliland's work. With the laboratory ready for occupancy, he set up a phonograph section on the third floor and took over from Gilliland. Tinfoil was abandoned, and a "white wax" cylinder substituted for recording. On March 22, 1888, Edison believed he had a model ready, and decided the Bloomfield facility, which had a production capacity of thirty to thirty-five phonographs a day, would not be substantial enough. He would, therefore, build a large phonograph factory adjacent to the lab. Looking for capital, he invited the New York bankers Jesse and Theodore Seligman and several of their associates to West Orange to hear the phonograph. Edison read into the speaker, swiveled it aside, and put the reproducer in place. He switched the motor on. But the fixed-arm cylinder produced merely a derisive hiss. No matter what Edison did, the assembled group heard nothing but knocking, scraping, and humming.

(As an explanation for the failure, Edison resurrected the story he had told Harrington seventeen years before about a workman changing a part on the machine without his knowledge.)

So it was back to experimenting. Edison adopted Tainter's flexible pickup. A keen eye and a steady hand were required to operate the machine. In reply to one inquiry at the end of April, Edison noted that a skilled technician had to accompany each instrument. "With an expert it would be a big success, without one it would be doubtful if it wouldn't get a black eye and set it back for a long time."

On May 12 Gilliland and Edison demonstrated the phonograph at the New York Electric Club before a distinguished audience, including General William Tecumseh Sherman. Edison brought ten instruments, but to avoid the pitfall of the earlier demonstration, nine were set up only for reproducing the recorded cylinders of piano music, cornet solos, and humorous recitals. That same day Edison incorporated the Edison Phonograph Works to manufacture the machines, and two weeks later construction started on two large buildings (75 by 200 feet and 75 by 400 feet).

On May 24 Edison held a recording demonstration on the third floor of the lab. In one corner was an organ, in another a piano. Funnels, some round, some square, some short, and others up to fifty feet long, were arrayed around the room. H. H. Rosenfeld, a popular composer, recorded some of his selections, including "Kentucky Gallopade" and "Kutchy, Kutchy Coo."

Mina was present, rotund in the last week of pregnancy. She thought herself too young to have children and was so big she believed she might have twins. (The baby weighed more than ten pounds.) During the last few months Edison had spent more and more time at the lab, often not returning to Glenmont for meals. In March, when a great blizzard swept across the East, he had remained at the laboratory for days and not even realized it was snowing until Mina finally sent a sleigh to bring him home. Sometimes she took a picnic basket and stayed at the lab, uncomfortable and out of place. Yet it seemed to her it was the only way she could share her husband's life. He forgot their wedding anniversary (he even had to be reminded when Christmas came), and Mina, biting her lip, held in her disappointment until she burst into tears.

"Why, bless you," Edison said, uncomprehending. "Everything I have in the world is yours. Why should I make a point of giving you something special?"

Mina's sister Mary Emily was so impressed by Edison that she told Mina: "If I can make my future husband love me half as much I will be very, very happy." Yet she was sad to see Mina so depressed. "Ah, Mina, you must try and shake off all your cares that are not real and not borrow so many for the future. I so long to have you happy and light again as you were before you were married."

Yet the more Mina's pregnancy showed, the more Edison seemed to look right past her. After the child was born, on May 31, he absented himself as much as possible, as was his wont when babies were around. Even Mary Emily, with her tender spot for "the fraud," voiced her indignation that he had deserted Mina "just at the time you need the most sympathy and husbandly love. If I were there and got to see him—which would be doubtful—I should give him a piece of my mind. I hope you make it interesting for him and use your charms in making him stay at home."

The group of people present at the phonograph exhibit could not help noticing that he seemed to regard Mina as a stranger. With his hands thrust into his pockets, a half-chewed cigar in his mouth, and his hair in disarray, he might have been entirely alone, so oblivious was he of his surroundings.

"He invents all the while, even in his dreams," Mina remarked.

The exhibit was a success. But the phonograph was not yet ready for the market, Edison admitted. "There are little minor defects."

The defects, in reality, where anything but minor. Yet the mechanical condition of the phonograph was relatively good compared to its organizational state.

Gardiner Hubbard, Uriah Painter, Charles Cheever, and the other men who had formed the Edison Speaking Phonograph Company in 1878 now had hopes that their investment might turn out to be valuable. Hubbard, who was still the company's president, was also a large stockholder in his son-in-law's and Tainter's Graphophone Company. And Edison, who had sold or given away most of his stock, so that he now held only 1200 of the outstanding 25,000 shares, was fearful that the majority stockholders might favor a combination with Bell and Tainter. He demanded that the company be reorganized with himself as majority stockholder, and that he be granted money for his experimental expenses.

Hubbard and Painter saw no reason why they should dilute their equity. Nor, having previously given Edison $10,000 for experiments that had never taken place, were they prepared to put up more.

Edison responded by asserting that since the patents had lapsed in the United States as a consequence of not having been renewed in England, the old phonograph company was defunct. In October, 1887, at the time of equipping the Bloomfield factory, he incorporated a new concern, the Edison Phonograph Company, with 12,000 shares of stock. He offered to exchange 4000 of these shares for the stock of the Edison Speaking Phonograph Company, "Not because I was legally bound to give them a cent," but merely as an act of friendship.

The majority of stockholders declined Edison's proposition.

"Since Painter and Hubbard have refused my donation, the only thing for them to do is bring a suit to establish equities," Edison snapped. "Their rejection puts it out of the pale of friendship and honor and places it on a business basis." He informed Painter he considered himself "to have been treated outrageously," and told him to address future correspondence to Tomlinson, his attorney.

Painter protested, "If you will establish any specific outrage that I have committed against you I will present you with a majority of the stock in the old company. I have no desire to see Mr. Tomlinson or to make his acquaintance any further as it has neither been pleasant nor profitable to see what I do know of him," a comment that was to take on additional meaning in a few months. "The question of the validity of the Phonograph patents is a question that can be tested and if necessary an act of Congress can be had to revive them."

Even Johnson, who had always stood by Edison and was not noted for outspokenness, thought: "I am in duty bound to try and show you the other side of a case which so greatly affects your reputation for fairness." Johnson detailed the history of the relationship and the points in dispute. "Have they in point of fact ever broken the letter or spirit of their agreement with you? If *not,* then is it material whether the patent is valid? You are not so far lifted above the obligations of ordinary men to respect those factors as to be able to ignore them with impunity; and knowing this as I do, I say here what no money consideration could induce me to say—recognize the error committed in so ignominiously setting aside those who have had Pride, Hope and Faith locked up in the Phonograph for the past ten years, and consent to meeting them here at an early date, to ascertain if some amicable adjustment may not be had."

Edison, however, refused to see Hubbard or Painter, or to attend any of the meetings of the board of directors, which consisted of himself, Hubbard, Batchelor, Johnson, Reiff, Painter, and Cheever. (Roosevelt had died in August, 1887.) Hubbard resigned the presidency. The other directors and stockholders squabbled among themselves.

"Boyish attempts were made to pull me into the fight but I refused to participate," Edison wrote, and told Reiff: "I don't want to mix in it. I have had enough of U. H. Painter and his methods."

While Edison contended for control of the phonograph, the Volta Associates joined with two United States Supreme Court reporters to reorganize themselves into a new firm, the American Graphophone Company. In the spring of 1888, the Graphophone people were approached by Jesse Lippincott, a middle-aged glass manufacturer who was coming into a great deal of cash from the sale of the Rochester Tumbler Company of Pittsburgh. Lippincott was a gay blade who staged lavish parties, backed Broadway musicals, and knew many politicians. He desired to establish a recording monopoly. As a first step he agreed to pay a half million dollars to the Graphophone group for marketing rights to their machine.

Shortly afterward, in mid-May, Lippincott obtained an interview with Gilliland. Gilliland had lost enthusiasm for his position vis-à-vis Edison, and the snarl with Painter and the Edison Speaking Phonograph Company threatened to make his agency contract worthless. He was swamped by inquiries resulting from the spate of publicity generated by Edison, and was weeks behind trying to explain why he had no phonographs to sell. His health was breaking down. He welcomed Lippincott's approach as a possible deliverance from his imbroglio.

Lippincott, despite his success as a glass manufacturer, was such a gull that if he had been interested in cheese Gilliland could have sold him the moon. In the Graphophone he had an instrument that would prove nearly unsalable, and the phonograph he was negotiating for still required years of work. Lippincott offered Gilliland a half million dollars for the phonograph rights, the same sum he had contracted to pay for the graphophone. Gilliland said he would try to persuade Edison to accept the amount, but Edison must not be told about the graphophone deal or he would demand much more. A settlement with the Edison Speaking Phonograph Company would have to be part of any contract; and Lippincott agreed to pay the stockholders $133,000.

Finally, Gilliland brought up the contract he himself held as exclusive sales agent. Lippincott professed himself ready to buy it. He would give Gilliland and Edison's attorney, John Tomlinson, who now joined Gilliland in the negotiations, $250,000 in the stock of his new North American Phonograph Company. Tomlinson replied this was a nebulous sum—the stock could eventually be worth more, or it might turn out to be worthless. Knowing intimately the state of the phonograph and the financial history of past Edison inventions, Tomlinson and Gilliland had no great faith that it would make their fortunes. Lippincott guaranteed that, if they wanted, they could sell the stock back to him at par.

Thus, with the Machiavellian scheming under way, Gilliland and Tomlinson met with Edison at Glenmont on June 3. Gilliland had apprised Edison of the negotiations, but Edison was reluctant. Why should he take $500,000 when, he estimated, at $150 per share,* the phonograph company was worth $1.8 million? Gilliland reminded him that he had offered one third of the shares to the stockholders of the old company, he had sold 150 shares, and he had already given away 1650 of the remainder. (Gilliland held 300 and Tomlinson 150.) At par, that would leave $620,000 worth of stock in Edison's hands; and Lippincott was willing to sweeten the pot. He would provide $160,000 for experimental expenses over 10 years, give Edison 5 percent royalty and 20 percent manufacturing profit on the phonograph, and assign him exclusive manufacturing rights to the graphophone. The package was worth more than $100 a share; it would get Edison out from under the dispute with Painter; it would net him a lump sum of a half million dollars, five times as much as he had received for any previous invention —and *that* for a product he had maintained not too long ago was of no commercial value!

Furthermore, Edison badly needed the money. The masons were ready to start laying the bricks for the phonograph buildings, the rent on the Bloomfield factory hadn't been paid since March, and Insull was complaining from Schenectady: "I am told that the Phonograph people do not pay their accounts. Of course I invariably reply that this is simply a matter of neglect and that it is a blank, blank shame. This failure to pay accounts affects our credit. If you people at Orange are going to abuse your credit you will cripple us."

* Edison had sold 150 shares to a Boston philanthropist, Mary Hemenway, for $22,000.

On June 12 Gilliland and Tomlinson took Lippincott out to see Edison, and Edison agreed to the deal. But when Lippincott went to Washington, where he attempted to get Bell and Tainter to approve his proposed contract with Edison, Tainter considered it far too generous. Since Edison was using wax recordings, Tainter demanded Lippincott pay a 10 percent royalty to the Graphophone Company on all phonographs sold. Lippincott agreed. The Graphophone Company was to be given a five-year option to purchase the phonograph patents. Lippincott agreed. Lippincott would have to buy a minimum of 5000 graphophones yearly from Tainter and his associates; Edison could bid on the manufacture of the remainder. Lippincott agreed. Edison was not to receive any royalty, and the sum for experimenting was to be cut in half. Lippincott agreed. On the twenty-first Lippincott returned to Edison and explained the less generous terms demanded by his "associates"—who remained unidentified. Edison reluctantly accepted them, providing he was "satisfied that the parties are biz men and the scheme isnt to be a stock speculation run by a lot of lunkhead directors."

On the afternoon of the twenty-eighth, Edison, Gilliland, and Tomlinson went to Lippincott's office in New York. Edison was to receive $500,000, a fourth within two months, and the remainder within four months. Edison asked Gilliland what he was to get.

"Only some stock in the company," Gilliland replied.

"How much?"

"Two hundred and fifty thousand dollars [at par]."

"What's it worth?" Edison inquired of Tomlinson.

"Fifty thousand to seventy-five thousand dollars," Tomlinson estimated.

Edison didn't really care. It was of no particular interest to him what Gilliland had hustled for himself. He remained entirely in the dark as to Lippincott's graphophone deal—Gilliland had told him Lippincott had paid nothing to Bell and Tainter for marketing rights. The agreements between Lippincott and Edison and between Lippincott and Gilliland were signed. Everyone departed. An hour later Tomlinson came back and had Lippincott sign the secret codicil: he and Gilliland could return the shares and receive $250,000 in cash over a period of five months, starting August 1.

Three weeks later, Tomlinson presented Lippincott with a claim for the first $50,000. Tomlinson kept $15,000 and gave the remainder to Gilliland, who bought Bergmann's yacht. (Gilliland and Tomlinson had agreed to split 70–30 between themselves.) Tomlinson also convinced Edison to pay him $7000—representing long-accumulated legal fees—out of the first installment Edison received on the sale.

Gilliland and Tomlinson requested another $50,000 from Lippincott on September 1. It was clear they were going to cash in all of their stock—a development that Lippincott had not foreseen. Painter, Hubbard, and the other stockholders in the Edison Speaking Phonograph Company meanwhile refused to accept $133,000 for their shares, but asked for $250,000.

The bewildered manufacturer did not have the ready money to meet all of the demands. After Gilliland and Tomlinson put in for their second installment, Lippincott notified Edison he did not have the cash to make the second $125,000 payment due, and asked for a delay of twenty-five days.

Edison sniffed molding cheese. "I consider it my duty to become informed fully on the present status of affairs before granting you the extension," he responded.

A week later Lippincott went to see Edison. Gradually most of the story rolled out. Edison exploded. He had been made a fool of. Not only were Gilliland and Tomlinson receiving half as much money as he himself, but their private bargain was threatening payment of the sum due him.

"I have this day abrogated your contract," Edison cabled Gilliland on September 11 in Europe, where Gilliland and Tomlinson were demonstrating and marketing the phonograph. "Since you have been so underhanded I shall demand refunding of all money paid you, and I do not desire you to exhibit the phonograph in Europe."

"You certainly are acting without knowledge of facts and are doing me a great injustice," Gilliland wired back.

Edison, however, never saw him again. Although Lewis Miller, for whom the break was an embarrassment, tried to effect a reconciliation, Edison forbade the very mention of Gilliland's name. Fifty years later Mina still respected his wish, as if "Gilliland" invoked some sort of curse. Edison even shunned Fort Myers until Gilliland was seriously ill and approaching death.

Edison decided Gilliland and Tomlinson could keep the money they had already received. But he wanted the remainder of the $250,000 used to pay off the old phonograph stockholders. If his two former associates refused to return the stock, he would sue. He sent Insull to "raid" his own offices on Wall Street and remove the records from Tomlinson's office there. Lippincott requested he take no action for six months, so as not to undermine confidence in the business and jeopardize sale of the stock. Edison consented, though commenting:

"I shall send over a civil engineer with a theodolite to see if your head is really level."

The scandal leaked out and broke in the press in January, 1889. Edison charged his former friend and his attorney with fraud and conspiracy. Suits and countersuits were filed in April, and depositions taken in May. The case attracted enormous attention from the press, and the coverage was all favorable to Edison. Sherbourne B. Eaton was restored to full honors as Edison's attorney, and kept urging the case's vigorous prosecution. Tomlinson, however, managed to obtain one postponement after another.

In truth, one party was reluctant and the other unwilling. The contract for the undoing of the Bell Telephone Company monopoly hung like an ethical sword of Damocles over both Gilliland's and Edison's heads. Edison could not look forward to having his handling of the old phonograph company dredged up, and Gilliland might reasonably claim that many of the ideas on the new phonograph were his. Thus, the case was allowed to drift from one date on the court calendar to another, until it finally disappeared with the demise of the star witness, Lippincott.

By suspending the payments to Gilliland and Tomlinson, Lippincott was able, by the end of October, 1888, to pay Edison $325,000. By the following July he had paid off all but $65,000 of the $500,000. As collateral for Lippincott's final note, Edison was given 6100 shares of the North American Phonograph Company; and these ultimately enabled him to gain control of the organization.

During the same months that Edison was entangled in the organizational thicket, he was striving to bring out a salable machine. From a personal standpoint it was, perhaps, the most difficult task he ever faced; and if he was irritable, testy, and pushed his friends' loyalty to the limit, it should be understood that the strain on him was enormous.

Edison always worked best when he had a partner—a co-inventor on whom he could try out his ideas, a talented technician who would follow directions and shape the things that Edison conceived. These partnerships had been all the more important in the development of the telephone and phonograph, where Edison depended on other men to tell him what they heard that he could not.

But Jim Adams and Charley were dead. Batchelor was general manager of the Machine Works, and available only for consultation. John Kruesi, the model maker par excellence, was in Schenectady. The men whom Edison had worked so closely with during the decade in Newark and Menlo Park were, for one reason or another, gone.

The phonograph was the one invention that Edison had felt was securely his; the invention that no one else would ever be able to lay claim to. When Tainter produced the graphophone, and Gilliland subsequently turned out his own version of the phonograph, the darts of challenge pierced Edison deeply. He had no one on whom he could depend, as he had in the past. He would, with his limited hearing, have to do the job himself.

After Edison agreed to Lippincott's proposal to buy the phonograph company on June 12, 1888, he went from Glenmont to the laboratory. Joseph Pulitzer, the publisher of the *New York World,* had sailed for

France the previous Saturday, and wanted to exhibit the phonograph in Europe—displays that would, of course, redound in great publicity. Edison had promised phonographs to Gouraud, who had been waiting for more than six months, and to Henry Villard and Werner von Siemens in Berlin. He decided to send one of his key assistants, H. de Coursey Hamilton, to Europe with two or three machines that he would operate for Pulitzer, Gouraud, and the others. An "expert" was still required.

Hamilton was scheduled to leave on a steamer the morning of Saturday, June 16. For three days and nights Edison, taking only periodic naps, remained at the laboratory, and exhibited once again the drive that established his reputation. The latest innovations were added to the phonograph. Revisions and adjustments were made. A number of cylinders were recorded. Edison had Mina bring the baby, Madeleine, down, and when Madeleine refused to cooperate he pinched her.

"The baby's articulation is quite loud enough, but a trifle indistinct," Edison explained her cries on the cylinder. "It can be improved, but is not bad for a first experiment."

Several pieces by Strauss, Gounod, Mendelssohn, and Chopin were recorded. "The Ravings of John McCullough," the great Shakespearean actor who had gone mad, were purportedly straight out of the asylum, but in actuality were the work of an impersonator, Con Nestor. Edison himself recited "Fair Bingen on the Rhine," and told a number of humorous stories. The Reverend Horatio Nelson Powers, Gouraud's brother-in-law, declaimed an ode he had written: "I seize the palpitating air. I hoard Music and Speech . . . I am a tomb, a Paradise, a throne. . . . In me are souls embalmed. I am an ear flawless as truth, and truth's own tongue am I . . . I am the latest-born of Edison."

At three o'clock on the morning of the sixteenth the group knocked off. Hamilton went home to pack. Gouraud's phonograph was not yet fully assembled, but Edison trusted Hamilton to put it together once he reached England. Numerous pictures were taken of Edison. One showed him propped slumped on his elbow at the table, a phonograph before him, a lock of hair falling Napoleon-like over his forehead—it ranks as one of the remarkable portraits in the history of photography. The publicity accompanying it proclaimed it to have been taken at the end of a five-day vigil during which Edison produced "the perfected phonograph."

When Hamilton reached Europe, he went with Gouraud to Sandringham Castle for a long evening with the Prince and Princess of Wales and Prime Minister Lord Salisbury. At various times, former Prime Minister William Gladstone, the Lord Mayor of London, Robert Browning, Arthur Sullivan, Henry Stanley, P. T. Barnum, and Florence Nightingale were recorded. The phonograph bacchanalia typically ended with tipsy notables of the realm shouting hip-hip-hurrahs into the machine. Gouraud made such a buffoon of himself that Edison wrote irritatedly:

"I am well aware of the value of the press, but it appears to be wholly unnecessary to make a parade after the fashion of Barnum and his White Elephant. Personally I have no desire for notoriety."

Hamilton continued on triumphantly across Europe with the phonograph. He recorded the Handel Festival at the Crystal Palace in London and Edward Strauss playing "The Blue Danube" in Vienna.* The trouble was that the machines were so eccentric no one but he could operate them. The most serious immediate problem was that the wax shaved off the cylinder clogged the mechanism.

Edison broke down from overwork and suffered serious intestinal difficulties during the summer. But, after recuperating at Chautauqua, he returned to the laboratory in September. On October 1 he sent Gouraud an optimistic message:

"Treadle machine perfect; new indenting material elegant. Hard, black, no chips to dirty machine, falls as fine powder into *closed chamber.* You will drop dead when you see it. Music now audible in large room and perfect. Tainter not practical man; dont know how to make cheap. Wont be ready for three months; meantime you can flood the country."†

As usual, Edison's pronouncements were ahead of his accomplishments. If Gouraud was to flood the country, it would have to be with water, for the factory was not yet producing machines. Alfred Tate, attempting to use an instrument for dictating, discovered that acids in the wax pitted the steel stylus and soon made it useless. In December Edison began searching for a jewel to replace the steel, and eventually settled on sapphire.

The underlying difficulty was that the wax was too soft to permit use of a recorder and reproducer of the same weight—the latter had to be 50 percent heavier than the former, so the two could not be combined into one element. Adjusting the reproducer to the grooves required the steady hand and precision eyesight of a diamond cutter (as Edison had discovered during the Seligman visit), and it was a practical impossibility to record on one machine and reproduce on another.

Edison hired a skilled chemist, Jonas W. Aylsworth, who was to be as important to the development of the phonograph cylinder as Upton had been to the incandescent bulb. In the winter and spring of 1889 Edison, Aylsworth, and a number of others slept on cots in the lab and spent days without emerging. Aylsworth made an infinite number of concoctions of wax, gum, stearites, resin, and paste. Edison listened hour

* In the United States, the extraordinary resurgence of interest in the phonograph led William Hammer to revive the telephone concerts of a decade before. The development of long-distance telephone lines made possible a phonograph concert piped direct to fourteen cities—the first "network" broadcast.

† The London Stereoscopic and Photographic Company, however, having bought British phonograph rights in 1878, claimed it, not Gouraud, held title.

after hour to recordings of his favorite tune, "I'll Take You Home Again, Kathleen," to the accompaniment of knocks, scratches, and weird noises. The phonograph had a terrible lisp, and refused absolutely to pronounce the hissing sibilants. For weeks and months Edison, like Professor Henry Higgins, tried to teach it to say "speaking." Every time it responded "peaking." "Say specie, damn it, say specie," he roared. And infuriatingly, it came back "pecie." Finally, on March 25, 1889, Edison, outfitted in a long laboratory robe and looking noticeably aged, declared:

"We've got it now. There has never been a perfect phonograph made till now, but there has always been a little bug in the machine that we could not catch. But now we have caught it and the result is a perfect phonograph. By a simple little device we have contrived an automatic spectacle [a magnifying glass on a hinge] which is applied to both recorder and reproducer."

Three weeks later a crucible of molten wax that Edison was working with exploded. The liquid poured over his face and head. He was seriously burned and his eyes were damaged. Swathed in bandages, he returned to the laboratory in a day or two and, looking half mummified, went back to work.

Complaints continued to pour in. The perfected "perfected phonograph" was a complex machine with a separate glass battery connected to an open electric motor which was belted to a wheel on the phonograph shaft. The array of instruments hovering above the cylinder looked like the superstructure of a ship. The entire combination of phonograph, motor, and battery weighed closed to a hundred pounds. The shavings continued to clog the mechanism, and the cylinders warped. The expensive battery lasted only ten hours. Lewis Miller sighed that his battery always gave out in the middle of a performance for friends, and wondered, "Could they not use a spring power such as is used in those music boxes?"

Edison, however, was adamant against developing a spring-motor machine. His entire career had revolved about electricity, and he was not going to go back on electricity now.

"The amount of work necessary to get the phonograph out has been stupendous, and we had no time to fool around on getting a proper battery," he told Batchelor in early May. But, with the phonograph improving, he was ready to turn his attention to a battery, and expected to produce one that would last thirty days. "If people will exercise a little more patience, every complaint and defect will be remedied—as rapidly as human flesh can stand it."

CHAPTER 24

Edison General Electric

Ten years before, Edison had abandoned the phonograph for the electric light. Now, as he came full cycle, the phonograph and the demands of the laboratory nudged him away from the lighting field.

In conceiving the laboratory, he trusted providence and Insull that money would be found not only to erect it but to run it.

"He of course wants a great deal more money than he at first anticipated," Insull unburdened himself to Tate, "but this is simply a repetition of what has occurred so frequently before. The trouble is that Mr. Edison does not seem to have anyone with him who urges him to curtail his expenses on his new laboratory. Exactly how I am going to carry out his wishes and give him what he requires, I don't know. Heretofore, when I have had to provide money, I have always had something to say about how much should be spent."

Not only the laboratory but his second marriage was bringing about an enormous increase in Edison's expenditures. Mina's household budget was $20,000 a year. Added to that was the $85,000 mortgage to be paid off, and the cost of maintaining the place at Fort Myers. Edison gave Insull power of attorney for corporate matters, including the voting of all Edison's shares, and hoped that by placing the responsibility on someone else he would not need to concern himself further about money. Matters were not improved by the fact that Tate at West Orange was handling Edison's personal finances, another secretary, Johnny Randolph, at the Wall Street office was caring for his business finances, and Insull in Schenectady was responsible for his income. Deposits and withdrawals flitted in and out of accounts with such rapidity that Insull could not keep up with them.*

* To handle the rapid-fire financial dealings, Insull devised a telegraphic code. "Do not fail to get darlings counter signature and endorsement delight notes and deposit two small notes in defray and the large one in defraud" was a typical wire to Edison.

In Edison's mind, Insull's positions as his financial manager and the manager of the Machine Works were not separate. Whatever money was required, Edison expected Insull to provide.

The demands became too much for Insull. "I have a letter from Tate, which is almost like a thunder clap," he wrote Edison. "He says you require $8,000 on Thursday [that day] or Friday; at the same time he asks me to arrange to pay the Lamp Company about $2,000, you having overdrawn your account $3,000 with them. Now I have no objection to struggling to finance for everybody, but I have $65,000 to meet within the next 15 working days. To suddenly have $10,000 put on the top of this, without even 24 hours notice, is a little more than I can stand without protesting. I am, as you well know, perfectly willing to run all around the country to hunt up money for you, and I can by hook or crook supply your wants, but I cannot do it unless you advise me ahead of your requirements."

Aside from the money raised by Insull, Edison had during 1887 an income of $148,000, earnings surpassed by only a handful of men in America. In the last six months of the year Insull provided an additional $55,000. When, early in January, 1888, Edison wanted another chunk, Insull wrote Tate:

"His demands on us begin to scare me. They are now growing so heavy that I must call a halt."

Upton complained to Tate that Newark's principal bank was refusing the Lamp Company further credit. "I feel sure that Mr. Edison's account is in a bad way. The bank has grown worse and worse ever since Edison's account went into it."

The average pay of the men at the laboratory was 25 cents an hour, and Edison reckoned the annual payroll at $72,000.* He decided that the Electric Light Company, the Machine Works, the Lamp Company, and Bergmann should bear the cost in return for the benefits they were deriving, and announced he would charge each $350 a week.

Upton was the most acquiescent, though even he demanded bills rendered "in a thoroughly comprehensive manner which can be easily verified by the records." Insull was more outspoken: "The amount you state as now being spent for our account simply appalls me. Our experimental accounts are going to be very, very much higher in the future than they have been during the last six years." Three days later Insull decided the Machine Works could not afford the money, and he would allow Edison only $250 a week. Bergmann knocked Edison's allotment down to $100 a week. Johnson demanded more definite information, assurances that the Light Company would be able to use the inventions produced without payment of royalty, and the appointment of a "receiver" who would be responsible for each party's interest and make sure each got its money's worth.

*The payroll was not a constant since the size of the work force frequently went up and down.

"Assurance from me that all inventions due to the money advanced by the Light Company shall be given without royalty puts it in a business light and it would be manifestly unfair that I should give my time and experience to devising apparatus without consideration other than being a 1/16 owner of the [Electric Light] Company," Edison replied. "Upon mature consideration I have concluded I shall not ask the Light Company to contribute toward the expenses of the Lab."

But that reduced the amount received for the running of the lab to $700 a week, half the requested sum. When the scheme got under way, Edison failed to provide a breakdown for or substantiation of the charges, which represented approximately twice the actual cost of the work.

"Some of the charges are simply outrageous," Insull stormed at Tate. "I propose to take the whole batch and see Mr. Edison personally about them. You charge us for material about double what we bill the same to you at. It will take more than the profits of an ordinary business to meet the charges received from Orange for the experimental account."

Edison was months behind in the payment of tens of thousands of dollars of accounts. He owed $1100 to the mason who had built the lab, a large sum to the Newark shop that had bound nine hundred books for him, $2570 to B. F. Goodrich, $543 to Merck Chemicals, and $3000 to a supplier of astronomical and engineering instruments.

Gouraud suggested Edison really should pay his subscriptions to important British journals. "You ought to speak to the party responsible for these accounts as these little things are calculated to prejudice you with people who are very good friends. I made all necessary apologies in your name."

Whenever a creditor camped on the doorstep or threatened court action, Edison directed Tate to send him to Insull.

"Are these instructions given out at your establishment?" Insull wanted to know. "If so, you ought to stop any such thing, as it does not do our credit any good to have people referred to us to meet the bills of another party."

Edison was forced to face up to the fact that, wealthy as he was, he didn't have the money to run the laboratory on the scale he had anticipated. If he were to continue, he needed an angel. The man with the most promising financial wings appeared to be Henry Villard.

Villard resembled Edison in that his designs always exceeded his resources, his vision knew no horizon, he was a romantic who behaved as if to desire was to conquer. He was indulgent and an opportunist, and, like Edison, in the grip of his imagination. He was the very opposite of the iron-fisted, down-to-earth business manager Edison desperately needed; and for that reason just about the worst person Edison could have picked on.

When Upton had gone to Europe in the fall of 1886, he had found the affairs of the Edison electric companies on the Continent in terrible shape. The Edison Electric Light Company of Europe had liabilities of $120,000. Its assets consisted of one safe worth $140 and $3.50 in cash. The French and German companies were at such odds with each other that they were on the verge of declaring an electric Franco-Prussian war. "If the Edison companies break, central station lighting will be done by outsiders and your name will be left out, and the light be known as Swan's and Siemens' light," Upton warned Edison.

In Berlin, Upton saw Villard, who, following the collapse of the Northern Pacific in 1882, was in the process of restoring his reputation in the financial community. Villard had access to two of Germany's most powerful bankers. He promised Upton that if Edison gave him power of attorney for European light affairs, he would not only unite the French and German companies but provide large amounts of capital for central-station lighting.

After Villard facilitated reorganization of the European concerns and returned to the United States, Edison initiated discussions with him in the winter of 1888. Edison proposed that Villard provide $42,000 of Edison's estimated $90,000 annual cost of running the lab and receive in return half interest in all inventions (except those applicable to electric lighting, the telegraph, the telephone, the phonograph, the mimeograph, and ore milling).

Villard, however, soon expanded the talks and turned them in a new direction. He admired the organization and efficiency of Germany's two largest electrical concerns, Siemens & Halske and General Electric of Berlin,* which were integrated from top to bottom. As chairman of the committee on manufacturing and organization of the Edison Electric Light Company, Villard thought the time had come to reexamine the position of the Light Company vis-à-vis the manufacturing concerns.

Edison's victory over the Vanderbilts four years before had suppressed the issue but not resolved it. The consequence of the battle was that J. P. Morgan had become the Light Company's leading stockholder—Edison himself held no more than $100,000 of the $1.5 million shares issued. Johnson had been Edison's choice for president; but power lay in the hands of the executive committee, largely controlled by Morgan, and Johnson knew whose bidding to heed. During the last two years the Light Company had become profitable. Prospects for the future were highly favorable. While the Edison system could not compete with alternating current in low- and medium-density areas of population, it was more economic in city cores. Two more districts had finally been added in New York, and negotiations were under way for a fourth. Ground was to be broken for a Brooklyn station, the largest to date, in the spring of

* General Electric of Berlin had been formed by Emil Rathenau, who had purchased German rights to Edison's patents at the Paris Exposition in 1881.

1889. Stations were in existence or in the planning stage in Des Moines, Harrisburg, Wilmington, Reading, Chester, Erie, Jackson (Michigan), Boston, Kearney (Nebraska), Little Rock, Minneapolis, Detroit, Atlantic City, Topeka, Rochester (New York), Lancaster, Altoona, Scranton, Wilkes-Barre, New Orleans, Cincinnati, St. Paul, Dallas, Dayton, Birmingham, Chicago, Columbus (Ohio), Washington, D.C., Grand Rapids, Kansas City, Denver, and Philadelphia. In the latter city the two largest dynamos in the world were installed. By the end of 1888 200 central stations and 1500 isolated plants were in operation. Only two of the central stations were losing money.

After inaugurating the talks with Villard, Edison became heavily involved in the phonograph, and left the initiative to Villard. Villard's original scheme collapsed, and the financier spent the first part of the summer in Europe. When he returned, Johnson, prompted by J. P. Morgan, persuaded him to take the matter up again.

"We shall speedily have the biggest Edison organization in the world with abundant capital when goodbye Westinghouse et al.," Johnson enthused to Edison.

Insull, however, told Edison that Johnson was preparing to feather his own nest at the cost of Edison's interest, and was planning to have the Light Company and the Sprague Motor Company brought into the new enterprise at high valuations while depreciating the manufacturing works. Edison, returning from Chautauqua at the end of August, met with Villard. He opposed including the Sprague Company—of which Johnson was still president—in the package. The difficulties of placing a valuation on each of the companies that was to be included in the new corporation caused acrimony between Edison, Bergmann, and Johnson, and for a time threatened their friendship.

An independent auditor established the value of the Machine Works at $1,400,000, the Lamp Company at $1,025,000, and the Bergmann Company at $900,000. The worth of the Electric Light Company was placed at $3.5 million. Villard put together a syndicate of Drexel-Morgan, Winslow-Lanier, the Deutsche Bank of Berlin, Jacob S. H. Stern of Frankfurt, Siemens & Halske, and the General Electric Company of Berlin to provide capital for formation of a new concern, the Edison General Electric Company, which Villard envisioned as only the first step in the creation of an international electrical giant. In early January, 1889, papers of incorporation were filed in New Jersey.

And then the whole thing blew apart.

In order to obtain Edison's consent for the combination, Villard promised him a half million dollars in stock beyond the shares he was entitled to. Edison was to offer the new company all his inventions in the lighting field through 1896. In return, the company would pay all costs of experimentation and allow him 20 percent royalty. J. P. Morgan and the other bankers objected to the terms as too generous to Edison, and as saddling the company with expenses for which benefits were nebulous.

From January to the middle of April the reorganization appeared stalemated. Finally, Villard broke the deadlock by taking the burden on himself. If the company would not agree to the contract with Edison, Villard would underwrite it himself. Conversely, Villard agreed to Morgan's demand that "A permanent board of directors will be constituted in a manner satisfactory to us."

In return for providing $3 million cash, the syndicate headed by Morgan got by far the better of the deal. The Machine Works alone had assets of $1.5 million and a net profit of $238,000 during 1888, compared to assets of $1.8 million and profits of $311,000 for the Light Company. Yet the new stock was issued on the basis of $3.5 million for the Light Company and $3.5 million for *all* the manufacturing concerns combined. The new capitalization was $12 million, of which $3.5 million in stock remained in the company treasury.

Edison had owned slightly over 6 percent of the Light Company, 75 percent of the Machine Works, 59 percent of the Lamp Company, and 33 percent of Bergmann. He wound up with 17½ percent of the stock of Edison General Electric, and the underwriting of a large part of his laboratory expenses. He could take comfort in the fact that the Vanderbilts had controlled Western Union holding 10 percent of the stock. Yet it was soon clear who was calling the shots. When one of Edison's favorites came into conflict with the new management, Edison refused to take his side, and declared: "Mr. Hix will have to stand on his own bottom," a remarkable feat of contortion that the man was unable to perform.

CHAPTER 25

The Pinnacle

While the negotiations had been in progress, Edison had told Villard it was of the utmost importance to push the consolidation as fast as possible—the court case on the validity of Edison's incandescent-light patent was nearing a decision, and Edison was afraid that, if it was favorable to him, "The stockholders of the Electric Light Company might get an inflated idea of the value of their property."*

In the three principal European countries, France, Germany, and England, the courts had upheld Edison as the inventor of "the filament of carbon," and ruled that the filament, sealed in a near-perfect vacuum, was the sine qua non of incandescent lighting. In the United States, to the contrary, Edison had been trapped by his gusher of patent applications, one superseding and sometimes contradicting the preceding. In the horseshoe-patent application, which he had delayed filing, he had declared: "It is not intended to claim anything herein shown and described except the incandescing conductor made of paper, as all such other matters of invention shown [are] incidental."

The patent commissioner, however, ruled in 1883 that Sawyer had established precedence to "an incandescent conductor for an electric lamp formed of carbonized paper." Edison thereupon sought to amend his application to claim all the features shown, but on June 27, 1885, the patent examiner turned him down. Edison filed notice of appeal to the patent commissioner. But since he had no new evidence to present, Richard Dyer, who had taken over as Edison's patent attorney, began playing a delaying game to prevent issuance of the patent to the United

* That is, since the Electric Light Company was a patent-holding company, a court decision sustaining the patent would have the effect of increasing the Light Company's value vis-à-vis the manufacturing concerns.

States Electric Light Company. Before the commissioner tired of the game, Dyer succeeded in obtaining twenty-three continuances. Finally, on February 27, 1888, the commissioner ordered the patent issued to Sawyer.

Westinghouse, who had obtained control of the Sawyer patent through the purchase of the United States Electric Light Company, filed suit. During the next twelve months feelers were put out to Edison regarding an amicable settlement or even a merger between the Westinghouse and Edison companies. Westinghouse wrote a friendly letter inviting Edison to visit him.

"The laboratory work consumes the whole of my time," Edison replied, "and precludes my participation in directing business policy." Several months later, on the second of February, 1889, Edward Adams, the second largest stockholder in the Edison Light Company and a good friend of Westinghouse, urged Edison to accompany him to Pittsburgh. Westinghouse was enthusiastic about a meeting.

Edison declined: "Am very well informed of all his resources and plans, and his methods of doing business lately are such that it cannot be accounted for on any other grounds than the man has gone crazy over sudden accession of wealth, or something unknown to me, and is flying a kite that will land him in the mud sooner or later."

A few days later the preparations for trial began. Westinghouse shrewdly had had the Consolidated Light Company, controlled by him, file suit against the Edison Company in McKeesport, just outside Pittsburgh. The case was scheduled for the United States Circuit Court in the Western District of Pennsylvania, where Westinghouse could hope to have the home-court advantage. The judge gave the McKeesport Light Company—which of course was to be represented by attorneys for the Edison Electric Light Company—only a month to collect evidence. Johnson thought Edison had better prepare for the worst and develop a nonfibrous filament in case the decision went against the company. (Edison was already experimenting with the squirted filament.) "I have seen you in many situations of emergency like this and never yet known you to be unequal to it," Johnson bolstered Edison.

The onset of the trial was preceded by two key court decisions. Some of Edison's light patents, like his phonograph patent, had lapsed in England (as well as Canada), and consequently their validity in the United States was brought into question. On January 25, 1889, however, in an unrelated case (*Bates Refrigeration Co.* v. *G. H. Hammond*), the United States Supreme Court ruled that the lapsing of a foreign patent had no effect on the American patent. Then, on February 18, 1889, the British appeals court upheld the lower-court ruling that Edison was the inventor of the carbon filament. In the lower court, Justice Butt had declared there was "one fact beyond contest, namely, that before the date of Edison's specification no good and efficient incandescent electric lamp was made or known." The opposing counsel had asserted that the specification of a "carbon filament" was not limited to a slender thread of high resistance, but could include a carbon pencil or rod, as used by Swan and others before Edison's feat. Justice Butt, however, ruled that "The carbon pencil or rod was a very different thing from the carbon filament." It was clear to Edison, Lowrey—who was to head the defensive team—and others involved that the crux of the case was to demonstrate the filament had no connection with any form of carbon rod.

And there was the rub: for the filament had, in fact, evolved from experimentation with carbon rods, Upton having tested resistance devices and Edison's attention having been drawn to Swan's work in England. On October 19, 1879, Upton had written in his notebook: "A stick of carbon brought up in a vacuum to 40 candles. Mr. B—— trying to make a spiral of carbon." The entry for October 21 was: "Stick of carbon about .020 and ½ inch long gave cold 4 ohms incandescent 2.3 ohms very good light." Batchelor's entry for that date in his notebook was: "Carbon Spiral—made enclosed tube for baking of spiral."

Those entries obviously would make the case for the Westinghouse lawyers. An impartial observer might conclude that, whatever the connection between the rod and the filament, the filament was a novel form, just as the automobile was a new form of carriage. In the heat of battle, however, Edison's problem was to prevent the court from discovering the link. Sawyer had lied like Baron Münchhausen in his patent application, and it seemed a question of which party could outfox the other. Since the Upton notebook would make the case for Westinghouse, it could not be submitted. The Batchelor notebook, if all entries from October 7 to October 21 were removed, would seem odd but provide evidence for Edison; and the manner in which the notebooks were bound —in sections—facilitated such a deletion without making it obvious that pages had been eliminated.

Neither in this case, nor a subsequent one in which the Edison interests went on the attack, was Edison a particularly effective witness. His memory was almost dateless. The carbon filaments, he said, had been made "either in October, November, or December, 1879." He fixed the date of adopting bamboo as January 30, 1880, or shortly thereafter, anticipating the actual discovery by six months. Under cross-examination he was evasive and antagonistic. To one question he responded: "I do not understand what you mean by electrical distribution. I am not well versed in the metaphysics of words and I do not like to answer the question because I think it might be too deep for me."

Westinghouse was able to call on a number of experts who had had dealings with Edison, chief among whom was Frank Pope. Pope not only backed Sawyer but, to give his views wider circulation, published a slim volume, *Evolution of the Electric Incandescent Lamp*, in which he depreciated everything Edison had done.

William Sawyer was long since dead from the effects of alcoholism, but Edison was able to put his brother, George, on the stand.* George described how William had lied to everyone, and had employed chicanery to make people believe he had a working lamp. Professor Charles Crosse of MIT testified that before Edison's incandescent bulb no lamp had had a resistance of more than five ohms, and no carbon rod of high resistance had practical utility. (Edison thereafter always had a warm spot in his heart for MIT.) Lowrey, who was a firm believer in the importance of the human element, had Mina and other family members sit in the front row of the courtroom. Handling the defense wittily and well, he charged in his summation that there was "not an honest *thread* in the plaintiff's case." By the time both sides had finished, at the end of May, Edison could look forward optimistically to a favorable verdict.

Edison was forty-two years old. At the midpoint of his life, he had a charming, loving young wife, a beautiful home, a unique laboratory, the attention of the world, and the esteem of his close associates and workers, who gave him a splendid birthday present.† During the course of a few months, he received more than a million dollars cash, and as the summer of 1889 approached he was, at least momentarily, free from financial worries.

Yet he still remembered his mother's tribulations over money, while his father schemed how to become wealthy; and he did not trust himself. He had made Mary financially secure, and he would do the same for Mina. Buying $200,000 worth of first-class railroad bonds, he gave them to Mina, then paid off the mortgage on the house, and transferred the deed to her name. No matter how he might dissipate his money, Mina need never worry—a few pen strokes had made her one of the wealthiest women in America.

In Europe, the great Paris Exposition was under way. The most popular exhibit was the phonograph, to which as many as 15,000 people flocked daily. William Hammer, a genius at promotion and display, built a 45-foot-tall tower covered with 20,000 lamps for Edison.‡

* Since George Sawyer was seriously ill and the Edison Electric Light Company was paying his medical bills, he was not an entirely disinterested witness. His testimony, nevertheless, was corroborated by other material and has the ring of truth.

† Since the completion of the laboratory, the great library with the huge, ornate built-in clock over the fireplace had contained only Edison's desk. The 3000 to 4000 volumes stood lonely in stacks designed to hold ten times that number. It was a gaunt and cheerless place. Batchelor and Kennelly led a drive to collect a birthday fund. With the total of $783 they bought three dozen monogrammed chairs, four tables, a reclining chair, glass cases, a gas log for the fireplace, Smyrna rugs, and a palm tree. They moved the furniture in secretly, so that when Edison arrived in midmorning everything was in place.

‡ Hammer's ingenuity was virtually limitless, and had he lived a half century later he might have out-Disneyed Disney. For New York's 1884 Columbus Day parade he designed a marching square of men, in the center of which was a steam engine and dynamo. Wires led from the dynamo to lamps set atop the men's helmets. The following New Year he held an astonishing party at his Newark home. When

Edison decided he would take Mina to Europe. Sixteen-year-old Dot, having been pulled out of the girls' academy she had been attending in Massachusetts early in February, was already there, traveling with Mary Emily and Grace, Mina's two younger sisters. Through much of July, Edison, his mind on the trip, merely puttered about the lab.

He and Mina left with Upton on August 3. Though Upton was not as close to Edison as most of his other old associates, Mina liked him very much—he, with his Boston background, was a person she could feel comfortable with.

Edison, though he liked to pretend he was a born sailor, suffered his usual mal de mer. (In his entire life he took only three trips to Europe, and none elsewhere on the high seas.) Word of his coming had reached France, but Edison had no inkling of the reception that awaited him. A decade after the development of the incandescent lamp the impact of electric lighting was beginning to transform industry and society. Whatever the controversy among scientists and lawyers, the public thought of Edison as the inventor of the light. The revitalized phonograph was creating a sensation. Edison's picture was displayed in shop windows all over the Continent. When he and Mina landed in France they were greeted by a storm of acclaim unparalleled since Napoleon's return from Austerlitz. A prominent Parisian declared the city "wished to receive Edison as it would wish to have received a king." He was treated like a head of state. At the luxurious Hôtel du Rhin huge bouquets of flowers were delivered daily. Calling cards piled up in snowy mounds. He was the honored guest at more than a half-dozen banquets given by the Municipality of Paris and electrical, telephonic, telegraphic, and scientific societies. *Le Figaro* put on a grand dinner for him attended by the most important literary and theatrical figures. Edison nearly stole the show at the fiftieth-anniversary dinner commemorating Louis Daguerre's development of photography. Louis Pasteur invited Edison to his institute; and Edison, who had responded to a Florida yellow fever epidemic by experimenting on killing microbes, went enthusiastically.*

---

a guest trod on the first step leading to the door, the number of the house appeared in small electric lights; when he trod on the second, the door bell rang; on the third, the door opened automatically. When the guest entered the hall, a revolving electric brush cleaned his shoes. In one room the ceiling was covered with electric stars, while the moon rose from behind clouds on one side and set in the other. When a person sat in one chair the gas went out; he moved to another and the gas went on: he sat in a third and a mysterious rapping was heard beneath the floor; in a fourth and drums started playing; in a fifth and the piano began tinkling. Hammer had devised an electric cigar lighter, coffee maker, toaster, and thermostat. At midnight a life-sized figure of Jupiter raised a glass to its lips, its nose lit up, and it began orating through a phonograph concealed in its innards. A luminous skeleton paraded about the room shouting, "Happy New Year! Happy New Year!" Inside the chimney a cannon went off, and fireworks shot into the sky electrically.

* As usual, Edison had jumped into the endeavor with more verve than deliberation. Once again his hunches had been simultaneously brilliant and outrageously

Nations entered into competition to honor him. King Humbert of Italy dispatched a special envoy to make him a Grand Officer of the Crown of Italy, conferring upon him the rank of a count. French President Sadi Carnot held a reception for him. At a dinner hosted by the American minister, Carnot decorated Edison with the Grand Cross of the Legion of Honor and elevated him to Commander of the Legion, the highest rank. The presidential box at the Paris Opéra, now lighted by an Edison system, was turned over to him. When Edison and his party entered the box, the orchestra launched into "Hail Columbia!" and the audience arose and accorded him an ovation. Alexandre Eiffel invited him to lunch in his apartment at the top of the tower. Afterward, composer Charles Gounod played the piano.

The press treated him as a sage and a seer. Edison, however, sounding as if he were wagging a finger at himself, said: "It is easy enough to invent wonderful things and set the newspapers talking, but the trouble comes when you try to perfect your inventions so as to give them a commercial value."

Having given himself over to that moment of sobriety, Edison relapsed into inventomania. He declared he was investigating the field of aerial navigation. "You've got to have a machine heavier than the air and then find something to lift it with. That's the trouble, though, to find the *something!*" He announced that he had made a model of a birdlike flying machine that would be propelled by the rising and falling of the wind.* He was in the process of developing a "far-sight machine," by means of which pictures could be transmitted, and expected soon to install the system between the laboratory and the phonograph works.

On a more mundane level, Edison transacted two pieces of business dealing with the phonograph. Gouraud so far had ordered only 1000 phonos, and wanted to space the order out over months, or perhaps even years. Edison sent Tate to England to investigate. Tate reported that Gouraud had no organization and no money. "The commercial success of the phonograph will never be worked out energetically until it leaves Gouraud's hands," Tate declared. The Seligmans and their banking partners were willing to form a million-dollar company to market the phonograph in Europe, but in order to accept their offer Edison had to get

---

off base. He believed "the fever microbe is parasitic, as it travels slowly along the ground and is known to have been stopped in some cases by the repaving of streets." He concocted a solution of caustic soda, gasoline, and rhigolene to exterminate the microbes; and, since there was evidence cold killed the microbes, he planned ways to produce cold. With his pronouncements he managed to get the medical profession in a state of agitation equaling the scientific in 1878.

* Six years later, however, he said: "We must abandon both the balloon and the airplane, neither will work. I would construct actual ships of the air—yachts, schooners, and brigantines—which would tack and sail and gybe before the wind. [The sails were to be made up of gas bags.] A balloon cannot carry an engine strong enough to drive it against the wind, and neither can an aeroplane."

the rights back from Gouraud. Gouraud proved a hard bargainer. When the contract was eventually signed, he and Edison each held 25 percent of the shares of the Edison United Phonograph Company, and Gouraud retained distribution rights for England.*

To provide better motive power for the phonograph, Edison on September 9 concluded an agreement for American rights to the French Lelande battery, at the time the most efficient in the world. The next day he and Mina took the train to Berlin, where they were the guests of Werner von Siemens. Siemens found him "a man of clear and sharp intellect, quick comprehension and far-reaching mind, who takes real delight in newly acquired knowledge. The fact that he does not work for gain only, but loves his inventions with an idealistic enthusiasm, makes him especially sympathetic to us."

The Germans, not to be outdone by the French, arranged for him to be received by the Kaiser, and he promised the emperor a phonograph. A week later he went to Heidelberg for a dinner—which lasted until 3:00 A.M.—with 1200 members of the German Association for the Advancement of Science. Afterward he was the guest of Friedrich von Krupp in Essen. When he presented Krupp with a phonograph, Krupp reciprocated, apologetically: "I could not send you a gun for your own private use [so I] put together an arrangement of writing table implements out of models of guns and projectiles."

Because of his disputations with the British—and fear, even, that a writ might be served on him—Edison had intended to avoid England. But Sir John Pender, who four years before had demanded $50,000 for settlement of the Automatic Telegraph matter, promised that all was forgiven. Pender, now the head of the Edison and Swan United Electric Company, offered Edison the use of his estate. Edison, exhausted, his stomach decimated, gratefully accepted, but had hardly landed in England when he was felled by the grippe. When the Lord Mayor of London sent him an invitation to dinner, he did not respond. When the invitation was followed by a telegram, he ignored it. It was an unforgivable breach of etiquette. Tate called on the mayor to apologize and explain that Edison had not received the messages. Instead of dinner, a less taxing private luncheon was arranged.

On September 29 the Edison party boarded the French liner *La Champagne*. They brought with them $3000 worth of goods bought in Paris. The largest item was a statue, *The Genius of Electricity*, a sexless figure of a winged nude holding a working electric light over its head. Edison, who paid $1850 for it, installed it as the centerpiece of the library.

A more intriguing member of the Edison party was an attractive

* Gouraud lived until 1912, but during his latter years was quite mad. In 1904 he announced himself as "The governor-general of the Sahara."

young Frenchwoman, who had been Dot's companion, and whom Mina invited to the United States in the hope of learning French. (Mina encouraged Edison to let Dot stay in Europe, and a female tutor and companion was engaged to travel with her.) As soon as the news reached Akron, the gossips were chattering. Mina's mother wrote concernedly:

"I wish you would tell me just how Madamsell came to come with you They say if it is as I tell it you had better be a little careful with your things Please excuse me for asking such a question for wanting you to tell me about Madamsell and tell me all the newes if you and Mr. Edison are on good termes you know what I mean."

Mina, proud, overwhelmed, only now, at the age of twenty-three, fully impressed with the world fame of the man she had married, was on very good terms with her husband—a month after their return she became pregnant again. If there was anything Edison was not on good terms with, it was his own anatomy.

"I barely escaped while abroad the complete disruption of my internal apparatus," he complained. "Eight days on the ocean liner have failed to repair the damage done to my digestion by a series of French dinners—I have returned a perfect wreck."

He had returned, also, a national hero. Insull, Batchelor, and other friends chartered a large launch, placed a band aboard, and met the *La Champagne* as it nosed into New York harbor on Sunday, October 6. With the musicians playing and the passengers aboard the liner crowding the rail and cheering, Edison and Mina were lifted off the ship into the launch. As they headed for Jersey City, Insull produced a tape measure and proposed to determine if Edison's head had swelled. To climax a triumphant trip, Edison was presented with the brightest jewel of all. The day before, Justice Bradley had rendered his decision in Pittsburgh. Edison was adjudged the developer of the high-resistance carbon filament. "But for this discovery electric lighting would never have become a factor. It is undoubtedly the great discovery in the art of practical lighting by electricity," the judge declared.

Never again would it be seriously questioned that Edison was the inventor of the incandescent bulb. He had attained the pinnacle of the inventive world, and reached the zenith of his life.

CHAPTER 26

Unprofitable Theories

Edison had never been a liquid millionaire before. He had always been dependent on someone's confidence in the profitability of his projects, he had scratched for money like a prospector for gold, he had never felt he had enough to operate on the scope to do his imagination justice. Now that he was financially independent, he had to convince only himself, just about the easiest task he had ever undertaken. And he was not about to change his philosophy that a dollar saved was a dollar wasted.

Two days after he arrived back in the United States, Edison went to Ogdensburg, in the mountainous, lake-rich region of northwest New Jersey. On a straight line, it was less than thirty miles from West Orange. But on the roundabout, herky-jerky train, it was nearly a four-hour journey. He was off on an adventure that promised to satisfy his need for giganticism, a project he had fretted over during the last two years while his time had been taken up with the laboratory, the phonograph, and the reorganization of the electric-light concerns.

His marriage to Mina had brought him into contact with men involved in the iron trade. During the 1880s the United States supplanted England as the leading iron and steel manufacturer in the world. Despite discoveries of immense deposits of iron ore in Michigan and Minnesota, industrial demand in a period of rising prosperity was such that the price of ore climbed to a historic high. In the old New York–New Jersey–Pennsylvania iron-mining region of the East, however, the veins had been depleted and the low-grade ores that remained could not be worked profitably in competition with the Western deposits. Because of freight charges, which made up a substantial portion of the cost of the ore, Eastern mills were at a disadvantage compared to Midwestern.

Two years before Edison's marriage to Mina, Lewis Miller had come into possession of an iron mine in Michigan's Upper Peninsula. When Edison had gone to Chautauqua in August, 1885, he had found two late-night companions in Lewis's son Ira and Ira's good friend Walter Mallory (a Chicago iron maker), to whom he had expounded upon the ore separator.

At the time, Edison's ore-milling company had "no assets, no cash, and no quotation price for its stock." But Edison was confident that he was on the right track. It was only a matter of eradicating the technical bugs. Once he had his new laboratory built, he would deal with these swiftly. In the summer of 1887 he constructed separators for both gold and magnetic iron ores. The ore-milling company was reorganized. Edison agreed to spend up to $25,000 in "experimenting upon separating the precious metals from the rebellious ores."

Another tenant of the building at 40 Wall Street, where Edison's office was located, was the firm of Witherbees, Sherman, & Co., the owner of mining property at Port Henry, New York. In the fall of 1887 a member of the firm asked Edison whether his separator could remove the relatively high percentage of phosphorus from the ore, so as to make the ore usable in Bessemer furnaces. Edison unhesitantly answered in the affirmative.

A few weeks later, one of the company's engineers, John Birkinbine, lectured on iron mining at the Franklin Institute. The Lake Superior region, where production was increasing exponentially, was responsible for nearly half the ore mined in the United States. At the same time, the depletion of mines in the Eastern United States had resulted in a quadrupling of ore imports in a decade. Magnetites, with an iron content of 72.4 percent, were the richest ores, and were abundant in Canada, New England, and parts of New Jersey and Pennsylvania. But because in a Bessemer furnace the phosphorus content could not be greater than 1 part in 1000, the ores were unusable unless the material could be crushed and the impurities separated.

In the spring of 1888, Edison constructed a separator for Witherbees, Sherman, and then hired Birkinbine as a consulting engineer. Through most of the year Edison was engaged with the phonograph, but in December he confidently declared:

"The ore separator is now perfected and leaves nothing further to be desired."

He planned to construct separators, license them to various iron-mining companies, and sell territorial rights throughout the United States. Mallory formed a company for Michigan, Wisconsin, and Minnesota, and Batchelor went to inspect the magnetite deposit owned by Miller in Michigan. Another company was formed for New York. Edison himself, in conjunction with Light Company investors Robert and Walter Cutting, organized the New Jersey and Pennsylvania Concentrating Works and obtained a lease on a marginal mine at Bechtelville, in eastern Pennsylvania. The Bechtelville ore, although containing considerable sulfur, was low in phosphorus, and results were encouraging.

The operation was far different from and more complex than that conducted on the beaches several years before. Since the ore had first to be reduced to fine powder, large Cornish rolls were used to crush it. The separator and magnets were built many times larger. The problems, however, increased in direct proportion to the equipment's size and complexity.

A man sent to demonstrate the machine in England disappeared in the clouds of dust that were generated, and the dust soon made the magnets useless. In Michigan, Mallory became bogged down. He was unable to obtain the 70 percent iron concentrate needed to make the ore usable. Three months later, Lewis Miller wrote: "It dont seem as though we had added anything to the proportionate amount of iron even though we have taken half of the other material away."

There was great skepticism in the iron trade that what Edison was attempting was practical. Indeed, the experience at Bechtelville soon demonstrated that Edison's process, while workable, was uneconomic. The vein carried only 14 percent iron, and the proportion would have to be twice that in order to compete with Southern ores. The experiment at Port Henry was not a success. Mine owners were reluctant to lease and try out the separators.

Edison, however, was not the least discouraged. "I have firmly decided to waste no time on proving the benefits of the process to mine owners," he declared before leaving for Europe. "I am buying mines myself. Have lately bought a half dozen and propose putting up mills myself. Some day they will come to trade with me without any parleying."

The magnetic-ore mines in Pennsylvania and New Jersey had very little value because of their low ore content and the high adulteration of other minerals, but Edison was sure he could make a fortune out of them. "My principal men and myself have a large amount of money to invest. Our intention is to buy all the magnetic ore mines in the state of Pennsylvania." Soon he added all the ore mines in New Jersey to his plans. He intended to put up sixteen concentrating plants a year until he reached a total of one hundred. To complement these, he would build blast furnaces to smelt the ore. He would eclipse Andrew Carnegie and Henry Frick to become the iron-and-steel baron of the East Coast.

"The money made in mining depends not so much on the richness of the ore as good management and *plenty* of ore," he asserted. Nor would he neglect the separation of precious metals. "There is as much gold in North Carolina and Georgia as there is in California."

To survey the properties, Edison retained Simeon O. "Sim" Edison, his father's half brother, who had had some success in iron making in Ohio. Within three or four years, Edison acquired mineral rights to or options on 145 old iron mines.

The Ogden mine, three miles from Ogdensburg, where some veins held as much as 27 percent iron, was one of the more promising. Edison intended to make it his pilot project. Rather than follow the veins down into the earth, he decided to strip-mine. It was a massive attack on a questionable property, and of a nature never attempted before, but others were carried along by his enthusiasm. The down-to-earth Insull reported to Tate:

"There is no doubt but that we are going to make a great deal of money in concentrating iron ores. Livor [the project's manager] and Edison are positively intoxicated by the business. Discounting what they have to say and estimating that the cost of production will be twice as much as they really estimate, I am sure we have got an extremely good thing."

An experienced mine owner, Jonathan Glidden, nevertheless expressed his doubts to Lewis Miller: "Mr. Edison proposes handling of an immense quantity of refuse that is neither good for man nor beast. Like all geniuses Mr. Edison will probably have to carry out his pet theories to a practical issue when he will be able to take new bearings and discard the unprofitable theories (if any) which he may find to exist."

Such criticism was not new to Edison—he had heard it all before when he was laboring on the incandescent light. Through the terrible, damp, windy, freezing winter of the New Jersey highlands Edison struggled to install his machinery and mill in the wilderness. Then, on New Year's Day, 1890, he was suddenly halted by a telegram reminiscent of a message that had arrived from Europe ten years before.

Even while contemplating the eradication of yellow fever and visiting Pasteur, Edison had not thought to have his daughter Dot vaccinated against smallpox. Now she was critically ill in Dresden, her survival an hour-to-hour proposition. It seemed a repetition of the two days when Charley had lain dying, and the next telegram might bring the dreaded news. Medicine, however, had made strides during the decade, and Dot received the best available care. After a week in crisis, she began to recover.

Wracked by the tensions of Dot's smallpox and the difficulty he was encountering at the mine, Edison came down with a serious case of his annual bronchial and pulmonary illness.

"Please show Upton details of process, company fears I might die," he directed two of his experts in the filament department at the lamp factory.

When Edison was sick, his personality changed. He loved to be babied by Mina, and stuck out his tongue and took the medicine at her command. She, in turn, was in her element as a nurse. Of course, once he recovered, he insisted he knew only one good doctor—himself—even though the family physician was E. R. Chadbourne, New York's most prestigious society doctor.

"I have studied myself and I can mend any little breaks in my system better than anybody else I ever met," Edison said. "When my liver or kidneys are out of order I know that my diet is at fault. I change my diet completely and then I am a new man again. I don't believe in doctors or drugs. Doctors' theories are never to be relied on, and most of the drugs aren't worth anything. I am a chemist and know the resulting action. Change of diet destroys any harmful bacteria. I believe heartily in the laws of common sense."

In his twenties Edison had been lean; in his middle thirties he turned stocky; in his forties he was becoming positively portly. Lacking any inclination for exercise, Edison put on pound after pound on Mina's hearty meals. When he attempted to take out a life insurance policy he was turned down by two companies because of elevated blood sugar and indications of incipient diabetes. Although the condition stabilized, his habits, mode of living, and general physical state were such that his life expectancy seemed considerably below average.

"I am pretty well broken down with overwork and am going down in the North Carolina mountains to freshen up," he wrote Henry Villard on February 8, when he declined the financier's invitation to involve himself in a stock-boosting scheme.

Villard had as great a penchant for overextending himself financially as Edison did inventively. Whenever Villard rocketed to prominence, it was certain he was simultaneously laying the foundation for an equally precipitous descent, for he inflated relatively small sums of money until they were entirely distended, and eventually the air was bound to rush out.

To Edison, the merger of the separate companies into Edison General Electric had been a matter of expediency as much as of necessity. It placed the enterprises and their frequently fractious heads under one authority. It promised to free him from all responsibility, as well as the necessity of dredging up money to operate the lab. The reorganization gave him a huge amount of cash that he could apply to anything he wished, and he gave priority to the ore-mining venture.

Villard, however, continued to view the formation of Edison General Electric as merely the opening battle in a campaign to establish a giant German-American electrical trust. Edison regarded such a trust, like all combinations, as unacceptable, since his own identity within the organization would be diminished. Although Villard telegraphed his intentions clearly, Edison preferred not to perceive them—throughout his life he blocked unpleasant facts from his mind.

In truth, Edison, whose triumphs were majestic, was concurrently committing two gargantuan errors: one was his association with Villard, the other his attempt to mine ore magnetically.

Villard in early 1890 needed three things: money; an electric traction system; and an alternating-current system. Edison had money, but he would not put it at Villard's disposal; he promised Villard electric traction, but he was unable to deliver; and he continued to be unalterably opposed to alternating current.

Villard, in addition to his presidencies of the Oregon & Transcontinental Company, the Northern Pacific, and Edison General Electric, held the lighting franchise for Cincinnati, and controlled or was in the process of buying the Milwaukee City and Badger (Wisconsin) Electric companies and the Milwaukee City and Cream City railways. For the city railroads, Villard needed streetcars immediately—a Milwaukee ordinance specified that the line was to be electrified by the end of the year, or the franchise could be revoked.

After the formation of Edison General Electric in the spring of 1889, Edison set Batchelor to work on a third-rail system utilizing such low voltage that there would be no danger to anyone who came in contact with the current. Electricity was to be fed to the rail every 100 feet from direct-current dynamos buried in the roadbed or installed in the basements of nearby houses. Tracks were run into the laboratory grounds from the Orange trolley-car line, and during the fall and winter of 1889–90 Edison's experimental car was driven in and out. Batchelor's design, however, met with little more success than Edison's previous attempts at electric traction. Villard later committed himself to provide another $250,000 for experiments,* but his attorney warned: "It might not be practical to wait for Mr. Edison to mature the inventions upon which he has been working."

Although Edison still owned a few electric-traction patents, he had fallen far behind in streetcar development. The leaders were the Sprague Electric Railway and Motor Company and Thomson-Houston, who divided the bulk of the patents and the business between them. Thomson-Houston's position stemmed from the firm's acquisition of the patents and services of Charles Van De Poele, a Belgian-born Chicago resident who had recently achieved a major improvement by substituting carbon brushes for metal on the motor—a substitution Edison had ignored when suggested to him by Henry Draper six years before.

In December, 1889, Villard assured himself a large part of his traction requirements by buying the Sprague Company and absorbing it in

* Of this sum, only $40,000 was actually advanced.

Edison General Electric. His monopolistic inclinations then led him to perceive that if he could come to an accommodation with Thomson-Houston, the two companies would have a stranglehold on the production of electric motors and traction systems in the United States. Early in 1890 Villard met with Charles Coffin, the president of Thomson-Houston.

The two men had mutual acquaintances among Boston bankers. They were similar in that, although they headed two of the three major electrical firms in the United States, they were businessmen and financiers, and scarcely knew a volt from an ohm. Coffin had been a Lynn, Massachusetts, shoe manufacturer when he organized a group that in 1882 had purchased Thomson-Houston, formed by two Philadelphia professors, Elihu Thomson and Edwin J. Houston. Coffin, like Villard, was an acquisitor, and in subsequent years had bought several other concerns, the most important being the Fort Wayne Electric Company and the Brush Electric Company.*

Villard and Coffin had little difficulty coming to an agreement to fix prices and split business 50–50 in the streetcar field. Shortly thereafter, Congress inopportunely passed the Sherman Anti-Trust Act, and the agreement had to be abrogated.

Villard's appetite for consolidation had merely been whetted, however. The chronic legal battles over patents damaged both Edison General Electric and Thomson-Houston. Edison was winning most of the suits, and was in a strong position as a result of the filament decision, but lawyer fees were over $100,000 a year. The cost to Thomson-Houston was even greater. Thomson's wife lamented: "I wish sometimes that Elihu was not in any such uncertain business as the Electric Light. Just as soon as it succeeds, the money all flows away in litigation."

Villard worked diligently on Edison to convince him of the wisdom of a merger. But if Villard was like the ocean, relentlessly attacking the shore, Edison was Gibraltar. He had already told Villard he would have nothing to do with Westinghouse: "You may see things differently from what I do; you may see them even through a telescope, while I see the subject through a microscope; still I am sure that if you enter into the slightest connection with him, it will be at the General Company's expense. We must all expect competition; if not from one person, then from another; but no one can ever convince me that a competitor whose system gives an average efficiency of only 47 per cent can ever prove a *permanent* competitor for large installations in cities against a system giving 79 to 80 per cent efficiency. But if for other reasons I am incorrect, then it is very clear my usefulness is gone—viewing it in this light you will see how impossible it is for me to spur on my mind under the

* With the Brush Electric Company, Thomson-Houston acquired the rights to Swan's incandescent lamp. Before that the company had marketed a pirated version of Edison's bulb.

shadow of future affiliations with competitors, to be entered into for financial reasons."

Nor was Edison less firm regarding any understanding with Thomson-Houston. "They speak of fierce competition, low prices etc. and assume if we make a coalition this will be stopped. The Thomson-Houston and Edison Companies can no more control the prices than the tides. The way to reduce expenses would be for the Thomson-Houston Company to turn over their incandescent business to us who know it and own it, and keep their arc light to themselves. The ruinous competition may be correct from their point of view, but I maintain that to do a great business in this country prices must be got down 50 to 75 per cent. I can only invent under powerful incentive. No competition means no invention."

Edison, in fact, saw no reason to disturb the situation in his electrical works.

"Mr. Edison could not wish for any better state of affairs than now exists in the General Edison Company," Insull remarked. He himself had all the business he could handle at Schenectady, workers were standing elbow to elbow, and he was in the midst of a $250,000 expansion. In New York, the company was building a seven-story headquarters on Broad Street.

No matter in which direction Villard turned, he and Edison were at loggerheads. Edison refused to help him out financially, because "I have been under a desperate strain for money for 22 years, and when I sold out, one of the greatest inducements was the sum of cash received, so as to free my mind from financial stress, and thus enable me to go ahead in the technical field. To put it back in the business is something I never contemplated. I had an income of $250,000 per year, from which I paid easily my Laboratory expenses. This income by the consolidation was reduced to $85,000, which is insufficient to run the Laboratory. I do about $20,000 worth of work for the General and local companies of such a nature that I cant charge for it, and devote about half my time to the same work. The fact that I am placed in such a position has produced absolute discouragement, and I feel it is about time to retire from the light business and to devote myself to things more pleasant, where the strain and worry is not so great."

The declaration was, of course, full of rationalization and hyperbole. Edison had scarcely been at work in the laboratory in six months, and when he was, his main concern was with ore milling and the phonograph. Even including Insull's extraordinary payments, he had never had an income of $250,000 from the manufacturing concerns. In complaining about the laboratory expenses, he inverted the true situation—a neat trick he employed more than once.

Villard was coming to the conclusion that he had miscalculated in his deal with Edison. His first move was to get out from under his private agreement for support of the laboratory; and he convinced the board of directors that Edison General Electric should pay the entire cost of running the lab in return for Edison's devoting half his time to developing inventions for electric lighting.

Next, as money became tighter and tighter and the threat of an international panic developed, Villard tied all his various concerns, except Edison General Electric, together into a holding corporation, the North American Company, so that he could float more stock and simultaneously have greater financial flexibility within the organization. He scraped up money here and there by such means as packaging a block of 5000 Edison General Electric shares for Hamilton McK. Twombly—now that Edison was no longer in sole control, the Vanderbilts were gradually reestablishing their holdings in the enterprise.

But that was patchwork financing. As the summer of 1890 neared its end, Baring Brothers, a giant British banking firm on which Villard depended heavily, tottered on the brink of insolvency. Villard, needing to raise $6 million, went back to Germany. Two million was due the Northern Pacific, and $4 million Decker-Howell, a brokerage firm that had sold to him so heavily on margin it was facing bankruptcy.

On October 4, 1890, North American's executive committee wired Villard in Germany: "Don't fail. Everything, in our opinion, hinges on it." They urged him to return to the United States immediately.

It was too late. On November 11, Decker-Howell failed. The overblown North American Company almost went down with the brokerage firm. North American's stock plummeted from fifty dollars to seven dollars a share.

Villard liquidated what he could. The most readily marketable securities were the shares of Edison General Electric. The North American Company held 5000 (valued at about $500,000) and had an option on another 5000. Villard put all 10,000 on the market, where they were snapped up by J. P. Morgan and the Vanderbilts. Through the first nine months of 1890 Villard managed to keep the North American Company afloat from crisis to crisis with loans from Berlin and Frankfurt bankers, as well as from Drexel-Morgan. But he was paying 9 percent interest—an enormous rate for the times—and it was clear he could not continue the financial juggling indefinitely.

Edison, meanwhile, proceeded blithely as if Villard's thrashing about were of no concern to him. Though telling Villard that he expected the cash he derived from the formation of Edison General Electric to deliver him from financial stress, he was in actuality pouring the money into his iron-mining complex in the New Jersey highlands.

As the ore was blasted loose, it was scraped by a huge steam shovel from the mountain, then carried by a system of conveyor belts to the crusher. Poured into a hopper at the top, it was ground ever finer as it dropped from one set of rolls to another. After passing through a screen, it fell past the magnets, which drew the stream of ore aside from the refuse.

To provide power for the rollers and the magnets, Edison erected a steam plant and large dynamos. The New Jersey Central Railroad built a spur, along which telegraph and telephone lines were put up, to the installation. On July 23, 1890, the first experimental run was made. The concentrate was not as good as expected, and when the tailings were passed through again in order to extract the remainder of the iron the results were "abominable." Edison, however, was not concerned:

"Carry out all my instructions, and then you say when you want me to start and I will come up and can right things in a few hours. There is no use experimenting—just set machines as I ordered."

Edison was unable to get away from Glenmont for any length of time because, he told Mallory, "Wife expected to be ill soon." He was, indeed, responsible for the expected illness, and on August 3, 1890, a son, named Charles, was born.

When, a few days later, Edison arrived at Ogden, he found the operation in a terrible state. The material he was dealing with now was far different from the clear, soft sand of the beaches, or the crumble-dry ore he had tested in the machine at West Orange. The heavy clay and roots and brush ripped from the earth clogged the screens. The material that sifted through was pasty-fine and damp. To do anything with it, the current had to be increased greatly, but when it was, the powder stuck to the magnets like lumps of dough. The instruments overheated and became worthless. The ore in the concentrate averaged only 37 to 53 percent.

The dust was horrendous. Men staggered blinded and choking about the machinery. Edison's "dust proof bearings" jammed and burned out. The huge conveyor belts cracked and split. The faces of the rollers were ground down and had to be replaced every few days. By the latter part of August, operations were at a standstill. Edison ordered a whole new set of equipment installed. From December to February, however, the weather was so bad that nothing could be done, so the mill could not be started up again until the following spring. In Michigan, Mallory, dealing with much higher quality ore, was having better success. But early in December the mill burned, and operations there were shut down also.

Edison was far from depressed. On maps he searched for more mines from Maine to North Carolina. He went to Sudbury, Ontario, a new copper- and nickel-mining area, to investigate properties, for he was using large amounts of both metals. He consulted Andrew Carnegie, who replied: "My Dear Wizard: nickel properties are getting plentiful as blackberries, we shall have plenty of that material, a little goes a long way."

For nearly a year the Ogden mill was quiescent, yet kept expanding, a giant, dark monster gnawing away at the resources of its creator. Expenditures were $20,000 to $30,000 a month, two thirds borne by Edison. Italian and Hungarian laborers were recruited on the docks as they landed from immigrant ships, and loaded onto the train for Ogden. Paid a dollar a day, they were worked in two shifts of eleven hours each. Toward the end of April, 1891, the mill began running, and in the third week of May John Fritz of the Bethlehem Steel Company headed a party to inspect the operation. Fritz agreed to take 100 tons of ore daily, and another steel company in Danville, Pennsylvania, later ordered 50 tons a day. But Ogden was badly located in relation to steel mills, so high railroad freight charges added significantly to the cost of the ore. The delivered price at Danville was $6.35 per ton, which was economically marginal even at a time of historically high prices. Yet to make a profit Edison had to sell 4000 tons of concentrate a month, and to obtain that he had to crush 30,000 tons of rock. (The rock averaged 20 percent iron, but about a third of that was not extracted in the crushing process.)

Buoyed by the Bethlehem order, Edison decided to buy the Ogden mine outright, and paid $75,000 to its Philadelphia owners. The man in charge at the mine was Harry Livor, a capable, outspoken engineer who had won Edison's gratitude by helping him get credit during the Pearl Street days, and had later headed the Edison Tube Works.

Livor would take no guff from anyone, including Edison. Edison, to whom any kind of confrontation was physically painful since it set his stomach churning, had therefore gone to Ogden only infrequently and left his devices to Livor. Livor was obtaining concentrate of 66 to 67 percent iron, and if he could raise that by another two or three percentage points the mill could be said to be a success technically, if not economically. The phosphorus content, however, rose proportionately with the iron. On June 10 Edison warned Livor: "Bethlehem complains iron running down, phosphorus running up. Be careful or we will be ordered to stop shipping."

Under the pounding of the huge boulders and sharp rocks, the equipment wore out at an alarming rate. Everything was going haywire. Magnets burned out, rollers and belts collapsed. The machinery performed capriciously—the concentrate from one load would be good, from the next terrible. Heat, humidity, swarms of insects, and the awful dust made working conditions abominable. Skilled workers refused to stay. White powder covered everything, killed the trees, and even damaged the crops of farmers several miles away. The stockholders were unhappy.

"Everybody is hounding me about expenses," Livor complained to Edison on June 18. He was unable to understand the bugs that kept popping up: "Things baffling that I want to confer with you about."

P. F. Gildea, a mining engineer retained by Edison, suggested: "Mr. Edison will soon have to give up his ideas of quarrying and resume mining if he wishes to be successful in concentrating." It was costing about $1.83 a ton to produce the Ogden ore. Yet, Gildea said, "There are few if any mines that can't produce ore 40, 50, and 60 per cent on the surface for less than $1.00. The wearing and tearing of machinery, the extra power as well as the extra help, throw lean ore, less than 30 per cent, in the shade. We must return to the seams or veins which are worked as mines, not quarries."

"This man is a damned fool!" Edison exploded, and promptly fired Gildea. He quarreled with Livor, who quit. Deciding he would take hold of the operation personally, Edison told Insull on June 29, 1891: "I am going to Ogden tonight to stay."

Edison's arrival boosted everyone's morale. "Your pocket will reap great benefit by his presence here," one of the managers wrote to Robert Cutting. "I am not bold enough to doubt any proposition that he would submit."

It was the kind of challenge Edison loved to come to grips with. He could ramble about in a grimy, working environment in which he felt more comfortable than at Glenmont, where Mina emphasized propriety and culture. Here he could chew tobacco to his jaws' content, and smoke cigars, which he ordered in batches of more than $700, like a man trying to kill time, or himself. For most of the next sixteen months Edison was at Ogden from Tuesday morning until Saturday evening. A post office was established at the site on his request. The name of the place was changed officially from Ogden to Edison.

Batchelor, now entirely toothless, the old fire in him reduced to a flicker, had been vacationing in the United States and Europe for several months, but came to Ogden on his return in early November. Edison had appointed a new superintendent, Owen J. Conley. He had shut down and entirely revamped the plant, thrown out a huge amount of material, and installed a cable system hauling four-ton skips. "I see no reason for any alarm," Batchelor commented.

Yet when the mill started up again, the proportion of concentrate in the iron ranged only from 31 to 49 percent. The concentrate was so fine that on the open railroad gondola cars a considerable amount blew away, and in the furnace much of it went up the chimney before it reached the melting point.

The difficulties were aggravated by the isolated location and the foreign origin of much of the working force. Many of the men did not speak English. There were living accommodations at Edison for only a few, so most walked from Ogdensburg, three miles away. There, crowding together eight and nine to a room in boardinghouses, they gambled, drank, fought among themselves, and were simultaneously exploited and abhorred by the local populace. At Edison it seemed, sometimes, as if a Tower of Babel were under construction. The buildings, impressive on the outside, were generally foundationless and jerry-built. A swarm of men were working on the skeleton of an A-frame stockhouse when an order issued in one language was misunderstood in another, and the men holding a guy rope let go. The framework crumbled as if made of matchsticks, and several men at the top plummeted sixty feet to the ground. Two were killed, and two seriously injured.

Edison himself and Mallory were badly shaken up a year or two later when they crawled through a manhole into the bottom of an eighty-foot-high dryer to determine why the ore was not falling to the bottom. As soon as they were inside, several tons of crushed rock slithered down around them. By the time they were dug out they had half suffocated. It was a hazardous operation, and injuries were almost routine.

As the winter winds howled in, Edison shut the mill down again. Ruefully, he once more sent men out to make a survey of the Long Island shoreline for black sand. To deal with the problem of dampness, he decided he would dry the crushed ore before dropping it past the magnets. To cope with the fineness of the ore, he invented a briquetting machine to press the material into pellets of forty-five to fifty pounds each.

It was as if Edison had gone to work on the electric light by installing a full-blown system of platinum lamps, then added one device after another to them, not under laboratory conditions but in attempted commercial operations. It was, of course, fiercely expensive and wasteful. By midsummer, 1892, Edison had spent $850,000 ($100,000 of which could not be accounted for at all), although the entire facility could have been duplicated for $250,000. Naturally, it was more and more difficult to extract money from investors, so Edison was selling blocks of his General Electric stock. With every piece of apparatus added, the chance of something going wrong increased. Through most of 1892 things proceeded so badly that all visitors were barred from the site. Not until October was a test run made.

After all the revisions, the amount of iron in the concentrate was 42 percent.

Obsessed with the ore mine, Edison let everything in electric lighting slide. He regarded the development of an invention as a contest, Edison against the elements, and until he achieved what he considered success he was relentless in his attack. Earlier in his career, when he depended on others for financing, he often satisfied himself with a device that worked in the laboratory. Whether it met the far more rigorous test of commercial practicality was not his concern. With the magnetic ore process he was, for the first time, essentially on his own. He not only had to make his invention work, but make it work profitably.

Yet there was more to it than that. A successful ore-extracting operation would be justification for his failure to produce an alternating-current system; and his failure to produce an alternating-current system was an indication that developments in electricity were passing beyond Edison's grasp. More and more, electricity was the domain of mathematicians and physicists, whose complex formulas and analytical abstractions were more alien than a foreign language to Edison. He either had to admit that this was a task for which his limited education had left him unprepared and give someone, such as Dr. Kennelly, his head; or he could procrastinate and fight a rearguard action, contending direct current was superior to alternating and not lacking for the performance of any operation.

He continued, consistent with his character, in the second course. In July, 1890, Insull reminded Edison that at the Association of Edison Illuminating Companies meeting the previous summer Edison General Electric had promised to come out with an AC system within six months. A year had passed without progress. Insull thought it of the utmost importance that the promise to the illuminating companies be kept, and that Edison move ahead swiftly with the development of an AC system. "The matter of Alternating apparatus is a very serious one indeed," he advised Edison.

"Give this to John Ott!" Edison waved off the message.

Edison had invested $54,000 in the Cataract Construction Company, which intended to develop hydraulic power at Niagara Falls and transmit it to Buffalo. He and Thomson-Houston were involved in an AC-DC competition. When Johns Hopkins University professor Henry Rowland, who was retained to do a study, reported in favor of AC, Edison barked: "I should like to see about a dozen professors set down to a banquet of boiled crow."

With Edison's DC system, thirty-six power stations would have to be constructed to service New York City. Yet, more and more, citizens objected bitterly to the constant traffic of wagons bringing coal and hauling ash away, to vibration and noise, and to the tremendous outpouring of smoke and noxious fumes. Threats of lawsuits were innumerable. Edison maintained, "There is plenty of places in the slummy streets for central stations," but slummy streets were not always conveniently located.

In Chicago, the earthshaking power of the Edison station was so great that the adjacent Home Insurance Building vibrated continuously, and the workers were in a permanent state of agitation. "Smoke belching forth from the chimney constitutes such a nuisance as to be at times almost unbearable—without exception the worst chimney in the city," one of the occupants complained.

The many poses of Edison: industrialist, 1892 (left); "Bowery bum" thinker in the laboratory library, 1903 (above); humorist and storyteller (below).

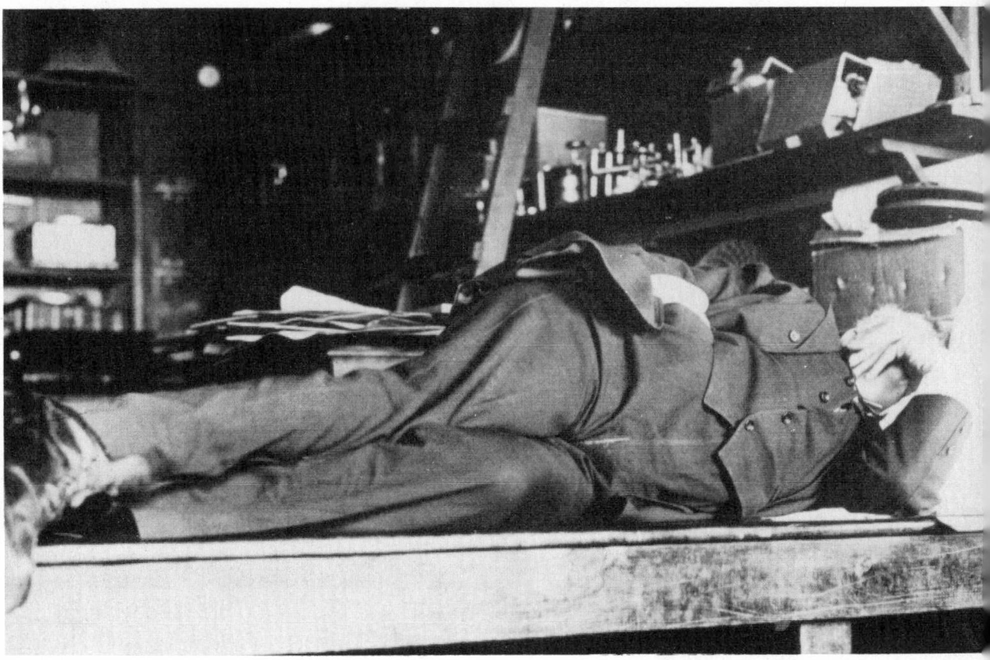

Edison, who could go to sleep any time on anything, takes a snooze on a laboratory table (above). He tries to find out what's wrong with a bus powered by his alkaline storage battery (below), and, on the facing page, inspects a submarine (top). A devastating fire swept through the Phonograph Works in 1914 (bottom).

12-11-14

William C. Anderson, manufacturer of the Detroit Electric (above), reintroduced Edison to Henry Ford. Below, Ford (left), Edison (center), and Luther Burbank (right) were surrounded by throngs at the 1915 Pan-Pacific Exposition. Facing page: Ford became a strange combination of godson-godfather to Edison (top). During one of their camping trips Ford snapped a picture of Edison and John Burroughs (bottom), and on another occasion he operated a logging railroad locomotive while Edison stood atop the cowcatcher (middle).

Miller Reese Hutchinson (above, standing to the left of Edison) had ambitions to become Edison's heir to the laboratory, but Mina (right) thwarted his plans. Standing behind Mina is Tom Edison, Jr. Mina made sure that it was Charles, seated at Edison's desk in the laboratory (below), who took over the enterprise.

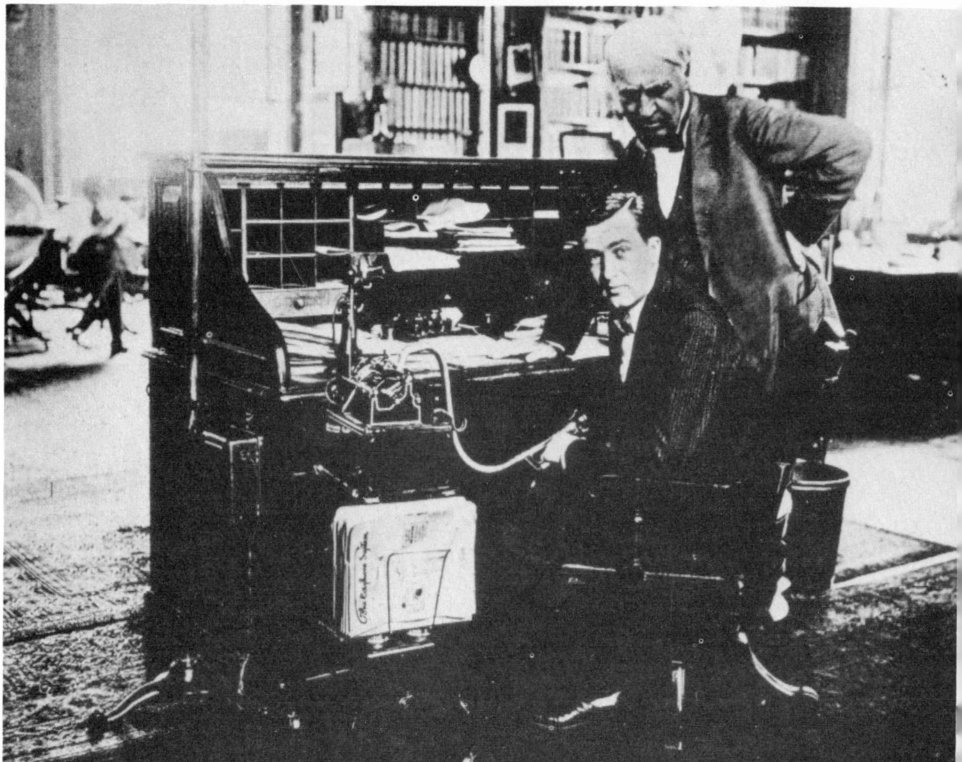

Edison greeted President Hoover at the Fort Myers laboratory (above). Edison, out hunting goldenrods with Harvey Firestone in the spring of 1931 (below), stuck to his milk diet.

Edison and Mina were photographed at Fort Myers on his last birthday, February 11, 1931 (above left). The last picture taken of him was when he emerged from his doctor's office in July, 1931 (above right). Tens of thousands of people waited hours to pay final tribute to him on October 19–20, 1931 (below).

Edison tried to develop a lamp using a 200-volt current so that a direct-current station would be able to double its coverage to a maximum of six square miles, but the immediate effect of his work was to stall the erection of additional stations.*

The cost of constructing six miles and sixty cars of his low-voltage, direct-current railroad came to an estimated $300,000—far more than an alternating-current road—and, of course, the final reckoning could be expected to be higher. But when Villard kept pressing Edison on an AC system, Edison gave what appeared to be his final word:

"The use of alternating current instead of direct current is unworthy of practical men."

Villard could not afford nor had he the inclination to humor Edison further. In his attempt to save the North American Company, he was ready to write off his foray in electric trust formation as an unsuccessful adventure. On December 1, 1891, he dropped a note to Charles Coffin.

During the next few weeks Villard held a number of secret meetings with the Thomson-Houston president. Edison was told nothing. On February 3, 1892, the executive committee of the North American Company authorized Villard to liquidate all of the company's holdings in Edison General Electric, and five days later Villard asked Sprague, who had become a consultant with Thomson-Houston after his company had been merged into Edison General Electric, to work on the electrification of the Northern Pacific terminal in Chicago.

The bankers were brought into consultation. They were delighted not only at the proposed consolidation, with its prospects of increased profits, but looked favorably upon the replacement of Villard by a more stable and businesslike president. Twombly, with his Boston background, contacted the firm of Lee Higginson & Co. in his native city.

Villard was anxious to sell out, Twombly to buy, Coffin to merge, and J. P. Morgan to end "ruinous competition." Only Edison was dead set against a combination. But he and Mina together now owned no more than 5 or 6 percent of the outstanding stock, and everyone was working together to present him with a fait accompli.

During the first week in February, rumors began abounding, and on the eighth of the month Sherbourne Eaton worriedly called Edison's attention to the fact he had not been devoting his efforts to electric lighting, as called for by his contract. "If we fall into the hands of the enemy they may raise the point sometime that you do not give one half of your time," Eaton warned.

Two years before, Edison could have had Thomson-Houston on any

* In Chicago, for example, the Edison Company needed to add a second station in time for the World's Fair of 1893, but Edison delayed the project from May, 1890, to April, 1892. In response to repeated inquiries about a starting date from the frantic Chicago executives, he wrote that he was not to be hurried: "When I am done you will see that my forethought is so long that it sags in the middle."

reasonable terms. Now he was powerless to affect the course of events. So, closing his mind to them, he continued to pour his energy into the mine as if nothing else were going on. When a committee of three Edison General Electric directors—Morgan, Twombly, and D. O. Mills —met with three Bostonians to work out the terms of the merger, Edison was off in the Jersey highlands.

The market value of Edison stock was $14 million, compared to $10 million for Thomson-Houston. Coffin, however, by a policy of easy credit and acceptance of the securities of local companies at inflated values, had expanded Thomson-Houston's business to the point it exceeded Edison's. By dealing in alternating current, Thomson-Houston had installed more than twice as many central stations as Edison General Electric, and, even with Edison's acquisition of Sprague, was ahead in electric-traction construction. Although the Insull-Kruesi-run Schenectady plant was the jewel of both companies, and Upton's lampworks and Bergmann's factory were not far behind, Coffin could make a case that Thomson-Houston was the stronger company and deserved better-than-equal treatment. Villard, Twombly, and Morgan, with the desire to merge uppermost in their minds, were not inclined to argue.

Under the terms of the consolidation, Thomson-Houston stockholders were to receive $18 million and Edison General Electric stockholders $15 million of the $50 million in shares of the overcapitalized new corporation. (Seventeen million dollars' worth of stock remained in the company treasury.)

Since Villard was out, the logical man to assume the presidency of the new company was Coffin. Insull and Upton appeared too valuable to be replaced. But, to dilute Edison's influence, other top management positions went to Thomson-Houston men. The trickiest problem was to decide on a name for the new company. Thomson didn't get along with Houston (who was inactive anyway), and Edison and Thomson disliked each other as much as competitors with hearty egos could. It seemed better to follow the German example and suppress personalities. The new company would be called simply: General Electric.

That, to Edison, was a stunning blow, the hardest part of the events to reconcile himself to. He was named a director of the new concern, and was retained on the board for nearly a decade. But after attending one meeting in August, 1892, he never went to another. "I may not be held responsible for the acts of an organization in which my voice is but one amongst a great many," he had said some time before. "I will not go on the board of a company that I don't control."

CHAPTER 27

The Cyclops

In four years Edison had gained a laboratory and an isolated iron-concentrating mill at the cost of an industrial complex. He had been toppled from his dominant position in the electrical world. After an expenditure of nearly a million dollars, his magnetic ore-separating process appeared to be a flop. To an ordinary person such a combination of events would have been utterly, even incapacitatingly, disheartening, and he might well have retired with his still-considerable fortune.

Edison, however, was anything but ordinary. "Spilled milk dont interest me," he remarked. "I have spilled lots of it, and while I have always felt it for a few days, it is quickly forgotten." No matter what the setback, he never lost his supreme confidence in himself, nor the belief that if he pursued his goal persistently enough success would come in the end. He had a remarkable psychological resiliency that, by always placing failure on some dastardly extraneous force, never on his own limitations, enabled him to bounce back.

"Tate," he said vehemently after he returned from the iron mine in October, 1892, "I'm going to do something now so different and so much bigger than anything I've ever done before people will forget that my name ever was connected with anything electrical!"

He would remodel Ogden/Edison on a massive scale. The capacity of the mill would be increased from 1000 to 5000 tons a day. He intended to control every magnetic mining property in the East, and ordered the search renewed with a fury: "Keep right at it steady and cover the whole country from Easton to Vera Cruz, and further if any attractions." Retiring to the third-floor lecture room in the laboratory, he converted it into a drafting office, and had his desk moved up from the library. Surrounded by spittoons and a half-dozen assistants, he set out to design huge new crushing rolls, dryers, bricking machinery and ovens, conveyor belts, elevators, and screens. "You never will find anyone to assist you in engineering knowledge," he maintained, since all the engineers he had consulted had warned that the machinery would destroy itself in crushing the rocks. In March, 1893, as soon as the weather improved, remodeling started at the site.

The Panic of 1893 overtook the work. Ore prices fell to half their previous highs, and the steadily mounting production of the Lake Superior mines presaged a glut for the indefinite future. No one was interested in buying the briquettes. Iron makers were not enthusiastic about them because of their fineness, because they could mix them with other ore only in proportion of about 1 in 20, and because "The universal opinion of the ablest furnacemen [is] that the briquettes will not reach the melting zone without disintegration."

Edison paid no attention, but churned ahead like a Cyclops with his eye riveted dead ahead. "This talk about present ruinous prices of pig iron is simply mediocrity at the end of its rope," he expectorated. All iron makers needed to do was to modernize their methods.

After more than a year of reconstruction, the mill was ready for a test run in April, 1894. The 237,000-pound crushing rollers commenced rumbling. Tons of rock were poured into the hopper. As the boulders tumbled between the rollers, they cracked with earsplitting, cannon-shot booms. The machinery shook and vibrated with jackhammer force. As the pounding continued hour after hour, cracks appeared in the foundation. The cement crumbled. The machinery settled . . . leaned . . . tilted.

Edison, his lack of engineering knowledge costing him heavily, had miscalculated the strength of the foundations required. All the machinery would have to be ripped up, and the work begun over from the bottom up.

Edison never hesitated. Liquidating his entire holding of General Electric stock at $33 to $43 per share—rock-bottom prices—he rebuilt the crushing plant. Another year passed. Ogden/Edison became an immense facility, some of its frame buildings rising as high as seven stories out of the rank forest carpeted with ferns and moss. Spreading over the countryside were thirty-one conveyors, six elevators, four stock houses, innumerable shafts, pulleys, and hoppers. Miniature locomotives pushed cars of ore along narrow-gauge spurs. The shrieks of whistles and the clanking of machinery were punctuated by blasts of dynamite. Lumbermen acted as outriders, clearing the land of trees. Behind them snaked a pipe from which the water, under tremendous pressure, was used to flush the underbrush and debris off the earth. Day and night the world's biggest steam shovel gnawed away, creating a giant cut. Edison was fascinated by it and sat hour by hour peering down into the great trench, from which extraordinary shrieks and grunts emanated.

"Wouldn't you think he was alive?" he shouted to a visitor. "Always seems to me like one of those old-time monsters or dragons we read about in children's books." In truth, it seemed like a lumbering dinosaur, its great neck bending down to bite at the earth, its three steel prongs protruding fanglike, its body trembling and steam shooting from its nostrils when it met resistance.

"We are making a Yosemite of our own," Edison said proudly. "We will soon have one of the biggest artificial canyons in the world."

From the freight cars the rock was lifted to the hopper by electric cranes. One set of crushers passed it on to another, until it was pulverized to fine powder. The powder was carried by an elevator to the top of an immense dryer. When it emerged at the bottom, a conveyor belt propelled it to the building housing the separator. There it was sifted down past 480 magnets arranged in three sets. The separated iron, looking like black sand, was transported by another conveyor to the briquetting machines, where it was mixed with oil and baked. The white, gritlike by-product meanwhile was carried off to an immense craneway whose pivoting neck spewed it out like flour. Artificial dunes of the material marched across the land and were blown off by the wind. To breathe the powder-laden air, the workers had to wear pig-snouted masks.

Everything was mechanized. It was an impressive technological accomplishment, and those who saw it were amazed. "I do not believe there is much doubt but that he will be an enormously rich man," one visitor, no friend of Edison, remarked.

During the spring and summer of 1895 the mill rumbled on. The cost of operation was $1200 a day. The briquettes ranged from 65 to 67 percent iron, but contained ever-changing percentages of other minerals. The binder worked badly, and the oven proved defective. There was virtually no market for the briquettes. Storms howled about the mountain, poured water into the machinery, and made the material too wet to handle. Giant roots torn out of the ground jammed the works. Edison had invested $1.25 million in the place, Robert Cutting —who had died in the meanwhile—nearly a quarter million, and others a half million dollars more.

Income from the mill was a pittance. Edison's resources were exhausted. In August he had difficulty meeting the payroll. On the twenty-second the men, wandering about coated with alabaster, refused to continue. Abruptly, Edison shut down the mill and dismissed the work force of 300.

Going to his friends, associates, anyone who would listen, he tried to raise $250,000 to revise the briquetting process and build new ovens. There were no takers. On October 11 he advised Frederick P. Fish of General Electric:

"I am pretty close in to the end of my mill biz and will soon be able to come back to work."

Fish, in his thirties, was a Boston lawyer who had been general counsel for Thomson-Houston and had then assumed the same position at General Electric. Both he and Coffin were well disposed toward Edison. With a low-key, kid-gloves approach they tried to reconcile him to the new state of affairs. The newspapers, doing a great deal of speculating—almost entirely false—reported Edison had been frozen out in a Wall Street coup to form an "electrical trust." A surge of public sympathy was generated for Edison. After the shutdown of the ore mill in October, 1892, Edison was not favorably disposed toward anyone. Wanting, momentarily at least, to have nothing more to do with electric lighting, he decided to withdraw from the joint Edison–General Electric exhibit planned for the 1893 World's Fair. General Electric, however, needed Edison's good will.

"I can't help feeling that you would not willingly do us the great harm that your withdrawal from the Chicago exhibit will inflict upon us," Fish appealed to him, "unless you felt that our relations were not going to be of the close and intimate character which we both expected when at work upon the contract. Now my dear Mr. Edison, we must have and maintain just those close relations. You know we cannot do without your cordial and hearty cooperation, and we know, and hope you do, that our profit and advantage will be your profit and advantage.

"Now about this World's Fair. We will do anything and everything we can to make your part of the exhibit the bright particular star of the constellation, for the brighter the star is the more we shall shine of reflected light. Tell me just how to meet your views, and we will meet them in every way."

Edison reflected and relented. Under an agreement negotiated after the merger, he was supposed to remain a consultant to General Electric. He and Upton together still held several major lighting patents, and General Electric was to pay them 3 percent of gross profits and 12½ percent of net profits from all lamp sales—obviously a very large sum, which Edison and Upton divided between themselves according to their own formula. The company, furthermore, committed itself to pay $600 a week (later reduced to $400 a week at Edison's request) for support of the laboratory. Even though Edison was doing virtually nothing in electric lighting and telling people, "I am out of that line of business," his career and his emotions were too closely linked with General Electric for him to divorce himself entirely. Edison became good friends with Coffin, and took an especial liking to the diplomatic Fish. Eight years later, when the lawyer was named president of the American Telegraph and Telephone Company, Edison told Coffin: "I have always had a higher opinion of Fish than I think he entertains himself."

Coffin could well be generous, for the Edison General Electric properties represented about three fourths of the assets of the new company. In the aftermath of the Panic of 1893, more than half of Thomson-Houston's book value of $10 million proved worthless and was written off. The Lynn, Massachusetts, facilities were shut down. Lamp production was shifted to Edison's plant at Harrison (formerly East Newark), and the remaining operations were consolidated at Schenectady.

Unshackled from Edison's idiosyncrasies, production at the lamp factory improved dramatically. A few weeks after the merger, one of Edison's old experimenters, John Howell, applied the flashing process to the bamboo filament and succeeded in greatly increasing the quality of the lamps. In 1895 General Electric, by combining the Sprague and Thomson-Houston patents, placed three electric locomotives in limited service on the Baltimore & Ohio Railroad. Two years later the company constructed an eighty-one-mile power line carrying 33,000 volts of alternating current from Santa Ana Canyon to Los Angeles.

At Schenectady, Insull had done an estimable job with the Machine Works. In contrast to Coffin, he had operated on a cash and short-term-credit basis even while continually expanding the plant, and the property was by far the most valuable in the entire organization. He had never gotten along with Villard, and had disapproved the whole thrust of Edison's operations after 1889. Edison had continued to depend upon him for financial management, and Insull was the president of the Phonograph Works and the vice-president of all the other Edison companies. Simultaneously, he was more and more independent, and did not hide his annoyance at Edison's methods of conducting business.

In early April, 1892, Edison ordered Insull to reduce expenses at the Phonograph Works. Insull told Edison it could not be done. Thereupon Edison, in high dudgeon, not only fired the whole work force and ordered the factory closed, but dismissed all except a few key men at the laboratory. Insull, worn out in his difficult position vis-à-vis Edison and not looking forward to having his independent management of Schenectady curtailed under Coffin, jumped at the offer to take charge of the Chicago Edison Company. Resigning his various presidencies and vice-presidencies at Schenectady and Orange, he departed, the most capable albeit non-too-personable manager Edison ever had.

CHAPTER 28

An Indefensible Transaction

Throughout his career, Edison was not only an inventor but a manufacturer and a businessman, and there was constant interaction between his various roles. The full cycle carried him from an invention to the establishment of a business based on that invention, then on to further inventions required for the viable conduct of the enterprise. The cyclical pattern was clearly established in the development of the electric light and power business; and after 1892 Edison seldom deviated from it.

In the wake of the formation of General Electric, Edison told Tate: "A man never makes any money until he has passed forty. He don't get sense till then." In the future, Edison decided, he would keep sole control and market all of his inventions himself. That meant, of course, that his activities as an industrialist would have to expand, and he would have less time to function as an inventor. It would also require more follow-up, pedestrian creativity, and decrease the opportunities for pioneering spectaculars. Yet, even though his two greatest outpourings of inventive energy, from 1878 to 1884 and from 1888 to 1892,* were behind him, it is clear that, at the age of forty-five, he suffered no diminution of creative capacity. By the end of 1892 he had received 682 patents, and during the remainder of his life he would add 410 more.

After withdrawing from the electric light and power industry, Edison concentrated his efforts primarily on two endeavors: ore milling and recording. The ore mine was a disaster. And during the first half of the 1890s, the phonograph seemed almost its equal in misfortune. For a half-dozen years the whole recording industry was in a chaotic condition

* He was issued 282 patents during the first period and 234 in the second.

in which the various enterprises had one thing in common—general insolvency.

Edison's 1889–90 model combined the recorder and reproducer in one instrument. But to make it practical, Jonas Aylsworth, the chemist, had to devise a new stearite or "hard soap" compound for the cylinder, and Edison was required to greatly refine the sapphire stylus. (Since there were not enough natural sapphires to go around, he developed an artificial jewel.) Nearly two years passed before the model was ready for the market. Meanwhile, even Upton was thrown for a loss by the existing machine. He didn't know how to make corrections, he didn't know he could shift the diaphragm or hear what he had recorded. He complained to Edison he couldn't get enough on the cylinder, and Edison discovered he was operating the machine at nearly twice the prescribed speed.

"This is a type of complaint everywhere," Edison said, putting the onus on Jesse Lippincott. "It's about time the North American got up a simple book of instructions."

The phonographs could be bought with a battery-powered electric motor, an electric motor worked from the house current, a water-powered motor, or a foot treadle—but none of the motive powers was satisfactory. The treadle, while simple and least given to malfunction, was awkward and made it hard to keep the cylinder turning at a constant speed. The house current, or "electric light," motor was a disaster. "The electric light machine gives great dissatisfaction," Insull told Edison. "Even at Schenectady where you could expect to find some expert talent they have been unable to use the electric light machine satisfactorily, and have returned them."

The water motor, which gulped four gallons a minute, was generally impractical since it required the phonograph to be placed in the kitchen or the bathroom. For a concert in the parlor forty or fifty feet of hose, run from a faucet and back to a sink, was necessary, with the attendant danger of a connection breaking and the guests having to wade their way out. The water, furthermore, gurgled through noisily. German piano prodigy Josef Hofmann, to whom Edison sent a machine, remarked that using the water motor was like recording on a rock atop Niagara Falls.

In fact, except with the elaborate apparatus available at the laboratory, the phonograph, no matter what the model, was useless for recording music. Even at the laboratory, the attempts to record produced such abominable scratching and nasal sounds that one prominent pianist turned ash white upon hearing the results, and artists almost unanimously divorced themselves from the machine because of its infidelity.

Irritated by the flood of criticism, Edison, who was turning out 100 to 200 recorded cylinders a day, declared he had permitted the recordings to be made only to accommodate the North American Phonograph

Company. His staff, he said, "had selected only those which were absolutely perfect and discarded others." But the business was taking up too much of his time and too much space, and he was tired of complaints. With typical rationalization, he told Tate: "I don't want the phonograph sold for amusement purposes. It is not a toy. I want it sold for business purposes only."

On January 25, 1891,* he announced: "I have today closed my Music Room and discharged the staff."

During the next year a number of the local phonograph companies started recording their own cylinders. By emphasizing novelties and popular items, they built up a small but profitable business. (A lively trade developed also in pornographic and obscene material, as for example the purportedly secret recording of a husband's dalliances with the maid.)

In early 1892 Edison decided there was enough money to be made to compensate for the criticism, and prepared to renew the production of records. One of the oddities he intended to record was the activity of sunspots. At Edison he put up forty telephone poles rigged with wires, a crude forerunner of the great parabolic antennas developed to receive signals from outer space. But a whirling storm blew the whole installation down.

More mundane, but successful, was a line of "darky" cylinders. On the ferry from New York Edison met a black man who could whistle astonishingly, and Edison marketed him as "The Whistling Coon." "Row at a Negro Ball" was another title. The action started with the playing of a fiddle and banjo, progressed to whiskey drinking and an altercation over a girl, and ended with the drawing of razors, the sound of pistol shots, and the arrival of police.

"The business will evidently take years to build up, the public being rather backward in taking it up," Edison ruminated. As a promotional measure, he sent phonographs to President Harrison and other prominent men. Impulsively but injudiciously, he had promised a phonograph to the German emperor during his audience in September, 1889. When, as he completely revamped the machine and the cylinder, the instrument still had not arrived a year later, the German press gibed at Edison as "A Man of Promises."

"I know you are not particularly sensitive to this class of criticism," Insull cautioned, "but it might do our Phonograph business a great deal of harm in Germany."

Although only about 5000 phonographs were sold or leased in America and Europe yearly, Edison's invention was doing well in comparison to Tainter's. The graphophone was as marketable as a lead boat, and even distress sales produced few takers. By far the biggest

* The announcement was dated January 25, 1890, but the year evidently was 1891.

loser was Lippincott. He had spent more than $1.3 million to obtain rights to the two recording instruments and put them on the market, but received back only $725,000 from franchisers. His entire fortune was dissipated. With his spirit broken, he went to California in November, 1890, to recuperate. A few months later he suffered a stroke, and died on the West Coast in the summer of 1891. The North American Phonograph Company was near bankruptcy.*

Lippincott's policy had been not to sell machines but, like telephones, lease them for forty dollars a year. (The battery cost an additional twenty dollars.) After Lippincott left for California, however, the North American Company started selling phonographs outright. The new policy brought the company into conflict with the Automatic Exhibition Phonograph Company, which had obtained rights from Edison to exploit the phonograph in public places by means of a nickel-in-the-slot device invented by Jim Gilliland, Ezra's younger brother. (Edison, despite his break with Ezra, continued to have business relations with members of the Gilliland family, and was one of the largest stockholders in the Automatic Exhibition Company.) Since anyone could now buy a phonograph and set himself up in business, the Automatic Exhibition Company claimed its rights were being infringed, and sued the North American Company.

This was but the opening battle in more than fifteen years of internecine legal combat, as the entire structure of the recording-machine industry came apart. Edison was the pivotal figure. His perception of the business world had grown ever more jaundiced, and he had always had a cynical conception of humanity—"Everything is beautiful but man alone is vile," he had once remarked to Insull. There was no question that a good many of the men he was dealing with in the phonograph field were unprincipled, and that, as the ore venture ate up his money, he felt himself engaged in a struggle for survival. Yet by adopting the piratical viewpoint that strength and ruthlessness were the overriding criteria for success in business, he alienated so many people that he severely damaged himself both personally and commercially. As the

* A company headed by two Bostonians who had tried to develop a line of phonographic toys was already out of business. The only item ever put on the market was a large talking doll designed by Batchelor (who had two daughters). But producing a miniature phonograph for the innards of a doll subjected to normal handling by children was far beyond Edison's technology. Dolls were returned to stores like waifs.

Edison promised to develop a new movement, but demanded more money. When this was not forthcoming, he attached all the tools, machinery, parts, and dolls of the Toy Company at the Phonograph Works. The officers of the company were bewildered, and the wounded stockholders, numbering some 150, who had invested $67,000, couldn't understand what was going on. "The doings of this company from first to now is damnable and a scandal . . . rotten with hidden corruption," one stockholder wrote. But the doll was deceased.

multitude of suits matured, the acerbic and none-too-upstanding executives of the Edison United Phonograph Company, which held the rights to Edison's invention in Europe, coined the telegraphic code name "Dungyard" for him.

In the late spring of 1892, Edison decided he would revamp the entire structure of the phonograph business, and establish himself as its autocrat. Having already assumed the presidency of the North American Phonograph Company, he appointed his secretary, Tate, vice-president, and worked out with him and Thomas Lombard, general manager of the company, a plan to dismantle the franchises and form a national sales organization.

The franchisers were to become the agents of the national company and, in return for the money they had paid Lippincott, receive 10 percent royalty on the sales of all phonographs in their districts. Tate and Lombard traveled around the country for several months selling the plan. It was not met with enthusiasm. But since the business seemed moribund and the franchisers were unable to protect their territories anyway, it was a choice of Edison's deal or none, so nearly all accepted.

The principal holdout was the Columbia Phonograph Company of Washington, D.C. Its president was Edward D. Easton, whom Insull called "the most intelligent man in the phonograph business." Easton had obtained control of the American Graphophone Company, and both companies were now operated by the same management. Columbia was selling phonographs where it pleased, both in the United States and abroad, and was thus a competitor to the North American Phonograph Company, as well as to Edison United.

A new rival, meanwhile, entered the field in the form of the gramophone. Emile Berliner, a young German immigrant residing in Washington, D.C., had already developed his own versions of the microphone and of the telephone transmitter, though his claims were inferior to those of Bell, Edison, Hughes, and a number of other inventors. When both Edison and Tainter neglected to put a disk machine on the market, Berliner filled the breach, and brought out a crude instrument that could be cranked by hand or worked by a spring motor. Berliner's machine was a "hear gramophone"—it could only reproduce, not record. Technically, it was far inferior to the phonograph. But from a sales standpoint it was simpler, far cheaper, and much less cumbersome than its competitors.

Edison recognized the potential of such an instrument. "A very large number of machines go into private homes for amusement purposes," he remarked in June, 1893, two and a half years after he had told Tate he wanted phonographs sold for business use only. "Such persons do not attempt to record. They simply want to reproduce. It has always been my idea that one of the greatest fields for the phonograph was in the household for reproducing all that is best in oratory and music but I have never got anyone to believe it until lately."

Edison set out to develop a less expensive machine for the home market. Basically, it was designed for reproducing only, although an attachment could be bought that would permit recording also. Playing time was increased by doubling the number of threads per inch from 100 to 200. Edison developed a new motor greatly reduced in weight. He designed, furthermore, a "long-playing phonograph" that would handle five cylinders. (This instrument was not a commercial success.)

By the spring of 1894 he was ready to introduce his new model. He did not, however, want the instruments to get into the hands of the Columbia Phonograph Company. Even less was he willing that the North American Phonograph Company should continue paying to the American Graphophone people a $10 royalty on every phonograph sold, as called for by Lippincott's contract. The American Graphophone Company had manufactured no machines since the beginning of 1891, and in two and a half years had had gross receipts of only $2500. It seemed to Edison that it was his money which was supporting an otherwise defunct competitor, and he was not inclined to continue the subsidy.

Edison not only was the president of the North American Phonograph Company, but held life-and-death power over it. The company had no reserves, but was existing from receipts to payments. Edison was still owed the final $69,000 installment (including interest) of the original $500,000 purchase price of the Edison Phonograph Company, he held $300,000 in bonds of the North American Company, which were delinquent as of May 1, 1892, and he claimed $83,000 from North American for laboratory expenses. He had only to demand payment of any of the items and the company would be forced into receivership.

His relations with Lombard and Tate, furthermore, were increasingly strained. Edison had complete control of the Phonograph Works, which manufactured the machines that the other companies marketed, and his markup was so high that he had everyone in a rebellious mood. Tate complained that Edison was adding 100 percent to labor, 20 percent to material, and 50 percent for profit: "This bill is out of all reason, and the North American Phonograph Company simply cannot live if the Works are going to make charges against them on any such basis. You simply cannot afford, as President of the North American Phonograph Company, to permit this. It is not just that the N.A.P. Co. should be made to pay for the incompetency which is displayed at the Works."

This was an indirect rebuke to Edison, since he would wander into the Phonograph Works, decide this operation or that could be done better in some other way, switch workers around and order them to change their methods, then depart without telling anyone in authority what he had done, and so in his inimitable fashion leave chaos in his wake. Edison always had a difficult time with men who had been close to him and were familiar with his quirks. Knowing his aversion to disputes, they would stand up to him, and Edison more often than not backed down.

Edison, however, had a Machiavellian way of striking back. In response to the complaints about the Phonograph Works, Edison in April, 1894, hired thirty-one-year-old William Gilmore, who had been Insull's hard-driving assistant at Schenectady, to take charge of the operation. The phonograph business, Edison agreed, was "in very bad shape indeed." Then, however, he pointed his finger at his accusers. "Lombard and Tate have not, to my mind, proved successful, and have not exercised that economy made necessary during the past year due to the general depression. They have not pushed the business as should have been done."

Edison concluded that the time had come to rid himself of the entire structure put together by Lippincott, and thus end the necessity of making payments to the Graphophone Company and the bother of fielding complaints. As soon as Gilmore was on the job, Edison wrote Tate:

"I want to put the North American Company into bankruptcy and sell phonographs direct from the factory regardless of the local companies. I enclose my resignation as President." He informed the banker, Theodore Seligman, who was a member of both the boards of the Edison United Phonograph Company and the Phonograph Works, that the upheaval was caused by his discovery of irregularities in the running of the Works. When Seligman inquired what these were, Edison refused to respond.

Seligman fired off a frigid letter: "Your positive refusal to give a company director . . . specific knowledge of facts of maladministration of the affairs of the company . . . is to say the least strange."

"There is some infernal work going on in the U.S. and it looks as if the Phonograph business which might have been one of the finest in the world is going to be utterly ruined," an officer of Edison United in England complained.

Tate tried to remonstrate with Edison that by placing the North American Phonograph Company in receivership and stripping the local entrepreneurs of their entire investment—after negotiating new contracts with them less than two years before!—he was destroying his credibility. When Edison was not to be swayed, there was nothing for Tate to do but follow Edison's example and resign. He refused an offer to take up once more the thankless, nerve-racking job of Edison's secretary—a decision that may have saved his life. Edison, at the last moment, vacillated, and for three months left the company dangling from the edge of insolvency.

Finally, on August 6, he demanded payment on the bonds. The North American Phonograph Company went into receivership, and the phonograph business came to a virtual standstill.

Edison was the principal creditor, he had been the president of the company, and he intended to purchase the assets. Conflict of interest was rather obvious. Edison's attorney, Richard Dyer, advised him success in the bankruptcy action was not possible "without the assistance of first class New Jersey counsel."

The case, in other words, would have to be "fixed." Dyer suggested "Judge" Howard W. Hayes, a Newark lawyer and Yale graduate, who had an intimate grasp of New Jersey politics and knew all the judges. Hayes cooked up a scheme that, for the moment, served Edison well. But when the day of reckoning arrived, more than a decade later, Frank Dyer, who succeeded his older brother as Edison's attorney, characterized it as "a perfectly indefensible transaction. The advice of Judge Hayes was almost criminally insane."

Under Hayes's scheme, Edison used John Ott's younger brother Fred —who had been Edison's stand-in when Mary had gone to dances in Newark and had become such a close sidekick he was nicknamed "Santcho Pantcho"—to form a dummy corporation, the Fred Ott Manufacturing Company. At the sale of the North American Phonograph Company assets in February, 1896, Ott, offering $50,000, was the only bidder.

Ten days later, Edison, paying Ott, bondholders, and creditors a total of $135,000, got back the phonograph rights he had sold to Lippincott for $500,000. (Of this, he had actually received $435,000, but the entire deal was worth more than $600,000, since Lippincott had paid off the stockholders of the old Edison Speaking Phonograph Company.) In the process, the local franchisers appeared to lose their entire investment.*

As a successor to the North American Phonograph Company, Edison organized the National Phonograph Company, and set up a nationwide organization of dealers and jobbers under his control. Except for the Columbia, New York, and New England Phonograph companies, which were strong enough to survive without official ties to Edison, the other local companies became defunct.

The beneficiaries of the vacuum Edison created in the industry for nearly two years were his struggling competitors. The American Graphophone Company and Columbia Phonograph Company introduced cheap spring-motor machines selling for ten to twenty dollars. Although as late as 1894 Edison declaimed against the spring motor, "It may be OK for fakers but the future phonograph will be run with a battery,"

* Lippincott's wife and children were destitute. When Edison was appealed to on their behalf, he replied: "The Works of course can do nothing but should I succeed in the biz of the phono I would be disposed to do something for Mrs. Lip." After a time, he gave her a $250 monthly allowance.

he was forced by the continuing mechanical problems of his battery-powered machine and the success of the competition to introduce a spring-motor phonograph, priced at forty dollars, in 1896. At the same time, since neither he nor the Graphophone Company could put out a machine without violating the other's patents, they came to a cross-licensing agreement.

The introduction of the spring-motor phonograph resulted in a dramatic increase in sales. In 1895 Edison grossed only $20,000 in the United States. The next year sales more than tripled. In 1899 they topped $500,000, and in 1903 they surpassed $1 million for the first time.

The sales of the amusement phonograph brought a rapidly increased demand for records. Through the latter 1890s, duplicates continued to be made mechanically, an expensive and laborious process. Artists were recorded in a big red barn a short distance from the laboratory—paid one dollar for a master, they could earn fifteen to twenty dollars a day. Then, in 1900, Edison finally perfected his process of gold-plating masters in a vacuum by substituting an induction coil for the electric arc, and in February, 1902, the mass production of records began. The new records did not have the Punch-and-Judy twang of the mechanical reproductions. An entirely new dimension was added to the phonograph business.

Edison concentrated heavily on popular selections, including an ever-growing number of the well-selling "darky" cylinders. But while these selections went over great in small towns and in rural areas, they were not accepted by the millions of Americans of foreign-born extraction in the cities. Edison, therefore, dispatched a team to Brussels to open a European record-making department. In 1901 the Edison United Phonograph Company went into bankruptcy, enabling Edison to take control of Continental business as he had of American. Large shipments of records and phonographs were sent to Antwerp. The records, however, were unsuitable for the European market, and the Edison phonographs were too expensive, so they piled up in a warehouse. Eventually, the warehouse was destroyed by a fire, and since the stock had been heavily insured, the company was able to show a profit.

Other offices were set up in London, Paris, and Berlin. In general, however, Edison's phonograph fared badly in Europe against Berliner's gramophone and the disk recordings, which were about a third cheaper.

By the latter 1890s, disks were making rapid headway against the cylinders everywhere. Headquarters of the Gramophone Company was moved to Philadelphia, and a Camden, New Jersey, machine-shop operator, Eldridge R. Johnson, was retained to manufacture the motors. In 1901 Johnson, obtaining control of the gramophone, organized the Victor Talking Machine Company.

Since Edison manufactured only cylinders and Victor only disks, the two companies were complementary, and were drawn together by their competition with the graphophone. Johnson became Edison's friend. Victor marketed its line through Edison dealers. Yet Victor siphoned off more of Edison's business than the graphophone. Victor's "tone arm," introduced in 1902, was a great technical advancement. The disks were considerably louder than the cylinders, and far less subject to breakage. Fifty disks could be stored or shipped in the same space taken up by a handful of cylinders. Victor gained enormous prestige and publicity, if not profits, by signing opera stars like Caruso and Melba for fancy prices. (Caruso received $45,000 in 1908.) Victor captured the sophisticated, middle-class consumer, and sales of disks and gramophones accounted for two thirds of business in the cities. Seven years after its formation, Victor had eight large buildings and nearly 800 employees at its Camden plant.

Edison tried to recoup ground by developing a long-playing and unbreakable cylinder. He succeeded, bringing out the Blue Amberol record, which played four minutes instead of two and a half, in 1908.* Victor, however, immediately countered by recording on both sides of the disk instead of one, thus providing five minutes of playing time to Edison's four.

The problems of Edison's phonograph business were compounded not only by his high prices but by Gilmore's insistence that the goods be fair-traded. A private detective, Joseph McCoy, was employed to keep tabs on the dealers.† Wherever McCoy found a dealer selling below the established price, the man was hounded and taken to court. Dealers could not even reduce the prices of records that were out of date and were simply cluttering up the shelves.

"The fact that there are cheaper goods on the market has never been considered in fixing prices for Edison's goods," Gilmore pontificated. The policy, however, was self-defeating. Dealers had their money tied up in stock they could not sell, and the great mass of ordinary people—precisely the population Edison was trying to reach—drifted away to competitors.

Edison had the worst image in the industry, one that was not improved when the remnants of the North American Phonograph Company franchisers gathered their forces to strike back. In September, 1900, a special committee of the local phonograph companies was formed under the leadership of Columbia Phonograph president Edward Easton to reassert the rights they had bought from Lippincott. The Columbia, New York, and New England Phonograph companies, claiming they

* The compound for the record was developed by Jonas Aylsworth, and Edison paid him $35,000. The long-playing, 200-thread record was the work of the Phonograph Company's Walter Miller, who was known as "the Emperor of Recording," and received $10,000.

† As a youth, McCoy had gone to work at the lamp factory in Menlo Park. In 1882 he had been sent to work as a spy in the United States Electric Light Company plant. Nicknamed Gumshoe, he was Edison's private eye for nearly half a century.

had been arbitrarily and illegally deprived of their franchises, charged Edison with fraud. Edison was subpoenaed as a witness in the suit of the New York Phonograph Company. For months, using Keystone Cops maneuvers, he avoided service, until one day his car broke down and the process server managed to catch up with him.

When he appeared as a witness, he repeated his evasive performance of the electric-light trials. Quite deaf, he smiled a lot, said little, and pleaded a great lack of memory. "Business matters are distasteful to me," he declared. "When the lawyers say sign, I sign."

"Do you not own 11,000 shares of the Edison Phonograph Company?" the opposing counsel asked.

Edison, laughing, scratched his head. "You'll have to ask my bookkeeper. I don't remember."

When he was requested to examine several patents, he refused. "I do not know they are true copies. Certified copies should be obtained."

When the judge ordered him to produce several subpoenaed documents, Edison asserted they could not be found. The judge, increasingly irritated, threatened Edison with a contempt citation if he didn't come up with the documents.

"I'd rather do almost any kind of a day's work than be a witness," Edison pleaded. "A lawsuit is the suicide of time," he maintained; and, in an effort at suicide prevention and the termination of the suit, he tried but failed to obtain control of the New York Phonograph Company.

Early in 1905 the court ruled that, despite the bankruptcy of the North American Phonograph Company, all the contracts with the local companies were still in force. The New York Phonograph Company was entitled to half the profits on all phonographs sold in the city since 1894.

Edison appealed the decision, and in the meantime ignored it. The New York company hired Edison's old attorney and confidant, John Tomlinson. In March, 1906, Tomlinson obtained an injunction against Edison's New York jobbers and dealers, prohibiting them from doing business unless they paid the New York Phonograph Company. Edison ignored that court directive, too. Finally, in the first week of February, 1908, the New York court held the officers of Edison's phonograph company to be in contempt, and imposed a massive fine unless Edison complied immediately with its order. Since the decision was based on the local companies' rights to Edison's patents, Edison was also enjoined from selling any of his gold-molded records in New York until October, 1909 (when Lippincott's contract with the local companies ran out), and was forced to eliminate a dozen features from phonographs sold in the city until that time.

Edison's phonograph business in the nation's largest market was thus crippled for eighteen months at a time when the company was staggering from the inroads of disks and the effect of the 1907–1908 recession. In the United States, Edison's business declined 50 percent. Dealers complained that Edison was stagnating, that the company assembled machines in a slipshod fashion and made no attempt to appeal to women, and that "It was one of the squarest houses in the country." Frank Dyer, who came more and more into the spotlight as Edison wrestled with the legal problems of the phonograph, felt: "Something has got to be done or the National Phonograph Company will be out of business before three years."

Edison decided to close up the European recording studios and manufacturing plants, and centralize production once more at West Orange. Gilmore, whose uncompromising, hard-nosed management characterized the 1894–1908 period at the plant, collapsed from exhaustion, and was replaced by Dyer.

Dyer's initial problem was to settle the case of the New York Phonograph Company. In fourteen years Edison had grossed $12.5 million from phonographs and records in the affected territories. His profits in New York had been $1,136,000, so that the New York Phonograph Company was entitled to $568,000. During the fall and winter of 1908–1909, the case was before the Court of Appeals. Dyer was not sanguine of the outcome. Neither, however, was Tomlinson. Tomlinson offered to settle for $200,000.

On February 23, 1909, Dyer urged Edison, who was wintering in Florida, to jump at the offer. Legal fees to continue the case, he advised Edison, would run a minimum of $100,000, so the actual cost of settlement would only be $100,000.

"The thing we have most to fear," Dyer said, "is that the United States Court of Appeals may switch over and decide that the rights of the New York Phonograph Company are based on contracts and not on patents," in which case the company would be entitled to an even higher settlement.

Dyer urged Edison to make an immediate decision. But Edison, basking in the Florida sunshine, fudged, fidgeted, and threw in several "ifs" that required further negotiations. The talks were still in progress when, on March 16, the Appeals Court rendered precisely the decision, based on contracts, that Dyer feared. "Judge Noyes's decision was a knockout blow under the belt," he informed Edison. Tomlinson immediately raised his demand to $500,000.

Dyer managed to bring Tomlinson down to $425,000, then once more faced the necessity of convincing Edison "that distasteful as the settlement is, it is the best thing to do. If they had an inkling in any way of the extent of the business, the money demanded would be greater." Furthermore, Dyer pointed out, an even more adverse decision had been avoided only because the intimate relationship between Edison and Fred Ott had not been definitely established. But this would, undoubtedly, come out in further litigation. Rather than wait for Edison to once more make up his mind, Dyer wired him: "Unless you telegraph to the contrary, I will accept Tomlinson's offer."

CHAPTER 29

The Kinetoscope

Yet, during the very years that the phonograph business was in utter disarray, Edison introduced a goggle-eyed public to a mutation of the phonograph that was to change the character of public entertainment throughout the world.

Photography had been in its adolescence during the years Edison was growing up. The pictures taken by Civil War photographers were received with greater amazement than a Michelangelo painting. While living in Newark during the early 1870s, Edison had acquired a stereopticon, a device that projected three-dimensional pictures. Other owners of stereopticons were men like Zenas Wilber and Uriah Painter; and Edison, whose collection was extensive, frequently swapped pictures with them. Photographic equipment and darkrooms were incorporated into both the Menlo Park and West Orange laboratories.

A few days before Christmas, 1887, W. P. Garrison, a member of the culture-promoting New England Society, wrote from Boston to ask whether Edison could entertain the members of the society in West Orange that winter. Edison replied he would be happy to. He was fitting out a large room in the new laboratory for projection lectures, and would be glad to furnish the men and apparatus.

The lecturer who came at the society's invitation on February 25, 1888, was Eadweard Muybridge. An Englishman, Muybridge had been retained by the Federal government to take photos of Alaska after the territory's acquisition from Russia in 1867, and upon the completion of the assignment had settled in San Francisco. In 1874 he had shot his wife's lover, but in a sensational trial had been acquitted of murder. Four years later Governor Leland Stanford had hired him to settle a $25,000 bet that a horse in full gallop had all four feet off the ground.

Muybridge rigged up a series of cameras with trip wires at Stanford's farm in Palo Alto (today Stanford University) and proved that the governor was right. Three years later the photographs were published in a book, *The Horse in Motion.*

Muybridge thereby was launched on a career of sequential photography. Fifty years earlier, Joseph Plateau, a Belgian, had invented the "Wheel of Life," which gave the illusion of motion to sequential pictures mounted on it. In 1877 Charles Reynaud, a Frenchman, improved upon this by constructing a zoopraxiscope. Pictures were affixed to the inside of a horizontal wheel and reflected by a mirror in the center, so that a number of people could watch at the same time. Muybridge's photos were applied to both the Wheel of Life and the zoopraxiscope, as well as to a third gadget of a like nature, the zootrope.

When Muybridge, working at the University of Pennsylvania on a $35,000 grant, appeared at West Orange, Edison was striving to produce a commercial phonograph. In spirit, he and Muybridge were both showmen. The demonstration caused Edison to recall a comment of the *Scientific American* when it had introduced the phonograph to the world ten years before: "It is already possible to throw stereoscopic photographs of people on screens in full view of an audience. Add the talking phonograph to counterfeit their voices and it would be difficult to carry the illusion of real presence much further." Why not, Edison asked himself, combine and synchronize a zootropic device with a phonograph, so that people could see a demonstration of whatever they were listening to?

Edison bought a set of ninety Muybridge plates, each of which cost one dollar and contained a series of about three dozen photographs depicting motion. Muybridge titled them "Animal Locomotion," and there were remarkable and Disneyesque pictures of trotting and jumping horses, deer, kangaroo, camels, jaguars, and eagles in flight. But, very much aware that art was a great deal more marketable when spiced, Muybridge concentrated on animals that were human, and as naked as their counterparts in nature. Of the 733 plates he offered, 219 were of animals and birds; 211 of men, of whom 205 were mostly or entirely unclothed; and 303 of women, of whom 243 were totally or transparently nude. They included a naked baseball player and a naked cricketer taking batting practice; naked boxers and jiujitsu combatants; naked fencers; a naked tumbler; a diaphanous-gowned fan dancer; innumerable bare, buxom women performing a variety of tasks; two nudes frolicking in their Saturday-night bath; and a mostly unclad mother receiving a bouquet of flowers from her naked daughter. These pictures were, of course, not displayable in mixed company, but they made great conversation pieces for men of wealth in their smoking rooms, and subscribers included Cornelius Vanderbilt, J. P. Morgan, Augustus Belmont, and Anthony Drexel.

Edison stashed his collection in the laboratory library and during the next few months evolved the concept of reducing Muybridge's pictures to "pinhead" size, 1/32 inch, and photographing the strips in a continuous spiral on a cylinder or plate, which would be run in synchronization with the phonograph. (In considering the alternative of a plate, Edison was thinking of the disk phonograph.) That his scheme was impractical was not as important as it might seem—more frequently than not, his inventions started out technically in the wilderness.

The employee who became his associate on the project was twenty-eight-year-old William Kennedy Laurie Dickson. Dickson, raised in France by Scottish-English parents, had written Edison in 1879 in an attempt to get a job at Menlo Park, but his application, like hundreds of others, had been ignored. Shortly thereafter he had gone with his mother and stepsister to visit relatives in Virginia, and there had married a woman, twenty years older than himself, of an upper-middle-class family. Coming to New York in the early 1880s, he had made repeated, unsuccessful attempts to see Edison. Not until the fall of 1885 had he struck the right chord.

He had then been working for several months with a group of associates on a new type of insulation developed from a South American gum much cheaper than gutta percha. Dickson, however, wrote Edison that, for Edison's sake, he would be willing to undercut his associates. He suggested Edison make an offer.

This was a proposition Edison, who was having problem after problem with insulation, took to heart. Dickson, bringing his gum—which might have been cheaper, but was no more effective—went to work for the Edison Tube Works,* where he was made assistant to the manager, Harry Livor. In 1887 he was shifted, along with Livor, to the ore-milling project. He became Edison's leading expert and experimenter on the magnetic separator. A considerable portion of the work was as much his as Edison's. Of artistic temperament (his stepfather had been a painter), somewhat effeminate, a dandy with a thin, upward-curled moustache, he was an excellent photographer who took not only hundreds of pictures at the laboratory but innumerable shots of Edison and his family.

He was, therefore, the logical choice for the phonograph-cum-photograph project. By October, 1888, Edison was ready to file the first caveat, and was hunting for a name for the new device. He wanted a word to represent "a moving object," and Edison's attorney, Sherbourne Eaton, involved his friends in the search. Edison insisted on including the suffix "graph," to which Eaton's partner, Eugene Lewis, objected on the grounds it referred exclusively to written language. The Greek word for motion was "kinesis," so Lewis suggested *kinesiphan,* "motion

* Since Dickson's means of entry into Edison's service was none too commendable, Dickson in his own account obscured it and predated it by four years.

appearing," or *kinesikad,* "moving image," but he was not optimistic about the results.

"Suppose I try Sanskrit or Russian?" he suggested to Edison.

Former South Carolina Governor Daniel H. Chamberlain's entry was *kinesigraph,* and, with Edison making minor adjustment, it became the kinetograph. This was to be the apparatus for recording the photographs. To the device for watching them, Edison applied the name *kinetoscope,* or moving view. On October 17 he filed his caveat:

"I am experimenting upon an instrument which does for the Eye what the phonograph does for the Ear, which is the recording and reproduction of things in motion, and in such a form as to be both Cheap practical and convenient." He anticipated getting 180 microscopic photos per revolution, a total of 42,000 on one cylinder. This would provide twenty-eight minutes of viewing time.

Edison's caveat had as much relation to technical practicality as a trip to the moon shot out of a Jules Verne cannon, but through the winter and spring of 1888–89 the kinetograph remained an intriguing secondary project to the phonograph and ore milling. Edison and Dickson discovered that photos mounted on a curved surface were thrown out of focus, so that the "cylinder" would have to be fashioned of numerous planes. Thus the work continued sporadically until Edison departed for Europe.

While Edison was gone, the direction of the entire art of photography shifted dramatically. During the previous months, John Corbutt of Philadelphia and George Eastman of Rochester, New York, had been competing to develop a celluloid-based photographic film. Corbutt's film was the first on the market, and Edison received a shipment in late June. But Eastman's, when it came out two months later, was three fifths the thickness of Corbutt's, and more flexible. By the time Edison returned from Europe, on October 6, 1889, Dickson had had a chance to familiarize himself with it. Because the vibrations from the machine shop interfered with the photographic work, Dickson also acquired a separate studio, attached to the ore building.*

In Paris, at the fiftieth-anniversary dinner honoring Daguerre, Edison had had opportunity to talk with the most distinguished European photographers. He met Étienne Marey, who had for some time been photographing people and animals in motion. Marey used a paper-based film band, which was stopped and started twenty times per second by means of an electromagnet. Edison went to Marey's shop and saw his camera. He also quite likely came to know Reynaud, who was now animating drawings by placing them on perforated celluloid moved by a sprocket wheel. In Germany, Edison had the chance to familiarize himself with Ottomar Anschütz's electric tachyscope. Anschütz mounted his photos

* The ore building was a fifth ancillary structure that had been put up at the laboratory.

on a wheel and lighted them with the intermittent flashes from a Geissler tube to obtain a highly realistic effect. There were, in fact, as there had been in incandescent lighting, numerous experimenters in the field.

Ore milling was still Edison's priority project. The phonograph was second; and the kinetograph continued as an adjunct to it. As Edison combined the information he had gathered in Europe with Dickson's report on Eastman's film, he recalled the experiments that had led to the birth of the phonograph. In one of the first versions he had tried to record sound on a band, and had sketched a machine with spools to move the band along. It was evident that the cylinder concept of the kinetograph was faced with almost insurmountable obstacles, but that Eastman's new film would be ideally adaptable to strip photography. Returning to the early phonographic design as a basis for propelling the film, Edison perforated the celluloid strip along one edge and passed it over sprocket wheels. The camera shutter and the movement of the strip were to be controlled by an electric motor and a polarized relay that would enable ten pictures a second to be taken—the film was to be stationary the instant the shutter opened, then would be jerked forward to the next frame in $\frac{1}{100}$ second. Copying Anschütz, Edison employed a high-intensity light to flash in conjunction with the opening of the shutter. Finally, the whole apparatus was to be synchronized with a phonograph so that sight and sound were recorded simultaneously.

In November, Edison filed his fourth kinetograph caveat, and Dickson wrote to Eastman for six 54-foot sheets of celluloid, the largest available. Dickson worked on the kinetograph through the winter ore-milling hiatus. When Edison returned from his health-restorative and gold-prospecting trip to North Carolina on March 18, 1890, Dickson was able to greet him with an animated, speaking image of himself that said: "Good morning, Mr. Edison, glad to have you back. I hope you are satisfied with the kinetophonograph."

Three weeks later, Edison, at the request of Mrs. Villard, opened a display of some of his Paris exhibits at a benefit staged by the New York Exchange for Women's Work at the Lenox Lyceum. At a single, private showing he unveiled an apparatus that projected moving images of people onto the ceiling.

The viewing machinery, however, was clearly unsatisfactory. The camera, if it existed at all, was even more primitive—Dickson conducted a good part of his early experiments with copies of Muybridge's photos. The film was grainy and the emulsion had a frustrating habit of peeling from the celluloid. With the onset of spring, Dickson was constantly in demand at the ore mill, and Edison finally became irritated that no work on the magnetic separator could go on without him. Thus it was that further progress on the kinetograph was delayed until the winter of 1890–91. Then Dickson succeeded in photographing a Maltese youth,

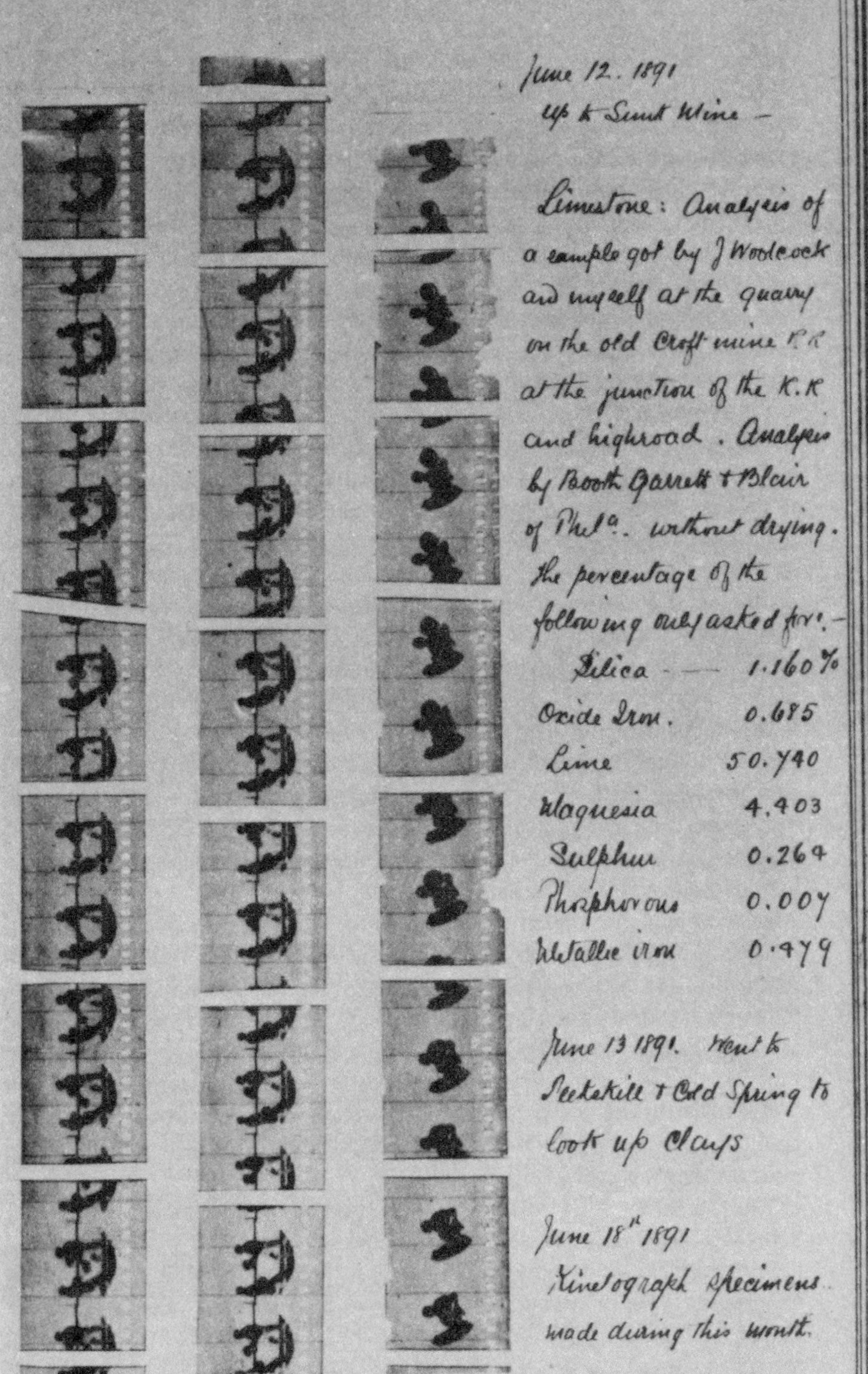

153

June 12. 1891
Up to Sunk Mine —

Limestone: Analysis of a sample got by J Woodcock and myself at the quarry on the old Croft mine R.R at the junction of the K.K and highroad. Analysis by Booth Garrett & Blair of Phl^a. without drying. the percentage of the following only asked for:—

Silica —	1.160%
Oxide Iron.	0.685
Lime	50.740
Magnesia	4.403
Sulphur	0.264
Phosphorous	0.007
Metallic iron	0.479

June 13 1891. Went to Peekskill & Cold Spring to look up Clays

June 18th 1891
Kinetograph Specimens made during this month.

William K. L. Dickson's notebook for June, 1891, demonstrates how work on the ore mill and the kinetoscope proceeded side by side. The first two film strips at left depict a boxing match, the third shows a juggler.

Joseph Sacco-Albanese. Dressed in a ballooning white blouse, he pretended to be a juggler and a clown in a performance of a few seconds that acquired the name "monkeyshines."

On the afternoon of May 20, Edison opened the laboratory to 147 members of the National Federation of Women's Clubs whom Mina entertained for lunch at Glenmont. Among the sights they were treated to was one Edison called "My latest novelty." On the floor was a pine crate with a silver-dollar-sized peephole in the top. Through this the women looked and saw the magnified figure of a man who bowed, smiled, and took off his hat, while a nearby phonograph repeated his voice. Dickson and Edison had increased the images to forty-six per second—about three times the number actually required for the illusion of motion—and their longest strip lasted only a few seconds. The strips were only a half-inch wide, and the size of the pictures was too small for clear reproduction. But Edison, whose ore mill was momentarily in good operation, was jaunty:

"When you get your basic principle right, then it's only a question of time and a matter of details. The details can all be worked out after you get the germs."

By the end of June, however, the mill was in disrepair again, and for the rest of 1891 and the first nine months of 1892 Edison and Dickson spent the major portion of their time at Ogden/Edison. Dickson had to squeeze experiments on the kinetoscope and kinetograph in between work on the ore equipment. He enlarged the size of the pictures to 1 square inch, and the width of the film to 1 $\frac{9}{16}$ inches. John Ott designed a nickel-in-the-slot device for the kinetoscope, and Edison intended to introduce the machine at the World's Fair.

In October and November, 1892, however, the patent office rejected several of Edison's claims, including one for taking the pictures stereoscopically. An appeal would cost sixty-five dollars. Edison, who was planning to spend hundreds of thousands of dollars to expand the ore mill to gigantic size, was reluctant to put out the money.

"Can't I soak this thing for a while to see if there is any money in it first?" he asked his patent attorney, Richard Dyer. Dyer replied the application could be allowed to remain dormant until May, 1894, without endangering the patent.

On December 31, Edison received word that the patent office had approved the application for the apparatus to exhibit photographs of moving objects. His impulse was to "soak" that, too. "Let it lay as long as possible," he instructed Dyer.

Tate and Dickson, who were good friends, appealed to Edison to take the patent out. Tate and Lombard, who still headed the North American Phonograph Company, intended to exhibit the kinetoscope along with the phonograph at the World's Fair, whose opening was only a few months off. Edison relented, and Dickson went on a crash program to prepare a kinetograph camera and kinetoscope viewing instruments. He experimented, simultaneously, with showing the photographs on a screen, and achieved a projection about ten inches square. The projection room was draped in black, and there was a distinct aura of the supernatural as the lilliputian figures mysteriously appeared and disappeared.

For the fair, Dickson focused his energy on the peephole apparatus, in which the pictures were magnified to two and a half inches. Night after night he came to the lab and tried to achieve "the harrowing task [of establishing] harmonious relations between kinetoscope and phonograph." At the end of January his health broke down and he suffered what Edison called "an alarming sickness." Edison shipped him off to Fort Myers for two months.

Dickson left behind "a perfect model ready for manufacture—it only remains for Mr. Edison to decide on a lamp and motor, the machine to be run with gears as of old or with a chain—both tests are at his command."

But Edison, who all his life had charged ahead, hesitated. He was losing money at the Phonograph Works. He had lost a fortune on the ore mill, and was closeted upstairs in the laboratory germinating new plans. "Nothing can be done as it's impossible for me to give it attention just now," he remarked. A great deal more work was required on the kinetograph (camera). On the kinetoscope, the fifty-foot endless strips provided a viewing time of only fifteen to sixteen seconds. Synchronization of the film with the phonograph was proving an impossibility. Thus Edison vacillated month after month, until it was too late to manufacture the machines for the fair.

When Dickson returned, Edison once more tied him up on the ore mill. It was not until the late fall of 1893 that Dickson was able to turn his attention full time to the odd-shaped building that had taken form on the laboratory grounds the previous December and January. Roughly oblong and all but windowless, its central portion humped steeply to two stories. Entirely encased in tar paper, it resembled a brontosaurus's sarcophagus. The sharply pitched roof over the central portion could be raised like a skylight, and the entire structure was mounted on a circular track so that it could be rotated to catch the direct rays of the sun. Having cost $638 to construct, it was the first motion-picture studio in the world. Because of its resemblance to a police paddy wagon, it was nicknamed "the Black Maria."

The kinetograph camera, a large, cumbersome, immobile piece of equipment, was mounted in the studio. Though the camera was still plagued with bugs, Dickson began painstakingly photographing short strips of film. At first only one or two actors could be filmed at a time, and often a week was required to obtain fifteen seconds of usable footage. Much of this consisted of retakes of Muybridge's subjects. Fred Ott, who was assisting Dickson on the ore-milling project, gained fame by merely standing in front of the camera and sneezing.

Though the kinetoscope failed to make it to the World's Fair, Tate and Lombard talked at length with Frank Gammon, secretary of the awards committee, and Gammon enthusiastically ordered twenty-five machines at a cost of $1000. Gammon's young and wealthy brother-in-law, Norman C. Raff, joined the group. They were impatient to get the kinetoscope before the public, but Edison could think only of ore milling. Finally, on February 13, 1894, Tate and his friends could restrain themselves no longer, and Tate made Edison a proposition.

The group would provide all the funds to complete the work on the camera and pay the wages of Dickson and other personnel. They would take over Edison's commitment to pay Dickson a percentage of receipts.* They offered Edison a $10,000 bonus, and guaranteed him royalties of $10,000 a year. They would purchase kinetoscopes from him for $60, a price at which he could realize a profit of 200 percent, and give him a similar high markup on kinetographs, once these were ready for manufacture. Altogether, Tate was ready to turn over to Edison $16,000 within a few weeks.

Seemingly, it was an offer that Edison, strapped for cash, would find difficult to refuse. Only a week after he received it, he wrote Muybridge: "I have constructed a little instrument I call a kinetograph [i.e., kinetoscope] with a nickel and slot attachment and some 25 have been made but I am very doubtful if there is any commercial feature in it and fear that they will not earn their cost—these zootropic devices are of too sentimental a character to get the public to invest in."

Edison, of course, was not being entirely frank. He himself had not been too excited about the kinetoscope, but if others were fired up, he could sniff money in the air. He had had nothing but trouble with selling and franchising the rights to his inventions, and had concluded that in the future he would retain complete control. He was soon to throw the North American Phonograph Company into bankruptcy, and part ways with Lombard and Tate. He consequently did not even dignify the proposition from Tate and his associates with the formality of a reply.

Edison delivered the first twenty-five kinetoscopes that Gammon and Tate had bought in early April, but without phonographic linkage—so the silent movies were born. Tate set up ten machines in a former shoe store at 1155 Broadway. (Of the other fifteen, ten went to Chicago and five to Atlantic City.) Their presence was advertised by an illuminated dragon with electric eyes. The place was scheduled to open Monday, April 16, 1894. But late Saturday afternoon such a throng gathered in front that Tate said to his brother: "Look here, why shouldn't we make that crowd out there pay for our dinner tonight?"

Five machines were lined up against one wall, five against the opposite, and the charge to see each row was a quarter—no mean sum when 25

* Dickson received substantial payments from Edison on the insulation deal and for his work on the separator. After he left Edison, he was paid thirty dollars a week for nearly forty years.

cents was a skilled worker's hourly wage. The length of the film on each machine was still limited to sixteen seconds. Yet by the time Tate closed up at 1:00 A.M. the kinetoscope parlor had taken in $120.

Edison raised the price of the kinetoscopes to $250, but this did not discourage interested parties. Frank Z. Maguire, a former associate of Tainter who had switched allegiance from the graphophone to the phonograph, and a friend of his, Joseph Baucus, acquired European rights. By November numerous kinetoscope parlors were operating in London, Liverpool, Copenhagen, Paris, and elsewhere. Crowds were immense. The bulk of the clientele was made up of working-class men, and the most popular strip was one showing two prizefighters. Prizefighting, because of its bare-knuckles, conflict-to-the-death heritage, had the same status as cockfighting, and was outlawed almost everywhere in the United States. It had, naturally, a forbidden-punch allure. A group composed of Virginians Otway and Grey Latham, Enoch Rector, and Samuel Tilden, Jr. (the nephew of the 1884 Democratic Presidential nominee), approached Edison with the proposition to turn out prizefighting pictures.

Edison, who was an aficionado, responded favorably. The Kinetoscope Exhibition Company was formed. Rector went to work with Dickson at the laboratory to increase the running time of the film to one minute, and Edison devised a new, large "prize fighting kinetoscope," for which he charged $300 to $500 (depending on the purchaser, and what the traffic would bear).

In July, Michael Leonard fought six one-minute rounds against Jack Cushing in an eight-by-eight-foot ring set up in the Black Maria. The next month the Kinetoscope Exhibition Company began showing the film, one round per machine, in six kinetoscopes placed in a parlor near New York's City Hall.

Public reaction was so profitable that the Lathams ordered seventy-two more machines, and Edison sent a message to William Brady, the manager of "Gentleman Jim" Corbett, a bank teller who had won the world heavyweight championship from John L. Sullivan two years before. Boxing, because of its underground nature, was not a money-making livelihood, and Corbett was acting in a Broadway play, *Naval Cadet*. Brady proposed to Corbett that he make the film for the publicity: "We'll have to get a man you can knock out with one blow as soon as they give the signal."

A candidate by the name of Pete Courtney, who agreed to the match for twenty-five dollars, was found in Newark. "He's a fighter, and he's as big as you are, but he's an awful boob," Brady told Corbett.

The match was scheduled for September 7, 1894, in the Black Maria. The first round was shot at 10:00 A.M., but more than an hour was required between rounds to change the film. The day was warm. The sun was absorbed by the tar paper, and blazed through the open roof.

The interior of the Black Maria steamed like the Black Hole of Calcutta. Finally, at 4:00 P.M., Dickson had the camera ready for the sixth round. Corbett, worrying that he would not be able to knock out Courtney—who was kept unaware of his fate—when the signal was given, induced his opponent to switch from eight-ounce to five-ounce gloves. Dickson chalked an X on the floor, and told Corbett:

"Be sure when you hit him to stand on this chalk mark. Otherwise you won't be in focus."

With Edison looking on, chuckling, Corbett got the nod after forty seconds of the round. Courtney swung away wildly. Corbett feinted with his left, then unleashed a powerhouse right. He caught Courtney high on the head, but merely stunned him. Courtney staggered back toward the ropes.

"You're out of focus! You're out of focus!" everyone screamed at him.

Courtney, having no idea what they were yelling about, swiveled his head drunkenly from one side to the other. Someone pointed him in the right direction and gave him a shove, and he plummeted back into focus. Another Corbett right impacted with him where X marked the spot, and he went down as if poleaxed.

"Say, Corbett, you're pretty good. But I don't think you could do it again," Courtney ventured after he was revived.

Four days later a New Jersey judge charged the Essex County Grand Jury with investigating the fight and indicting all connected with it—but Edison managed to have the inquiry quashed.

Nevertheless, Dickson, the world's first cinematographer, disliked shooting boxing matches. He wanted to concentrate on more artistic and worthwhile subjects. He thought himself the master of the Black Maria, and resented the intrusion of the Phonograph Works' new manager, William Gilmore, whom Edison placed in charge of the kinetoscope business when all phonographic work was halted in 1894.

"We are compelled to take boxing matches," Edison told Dickson in mid-October. "When Gilmore requests you to take films of this character please do so."

Dickson wanted to get on with the development of the projector, which remained half completed. But Edison, with his incipient, highly profitable business of building kinetoscopes, was not the least interested. For Otway and Grey Latham, on the other hand, the peephole machines were costly and of limited utility—projection on a screen promised greatly increased revenues. In December, 1894, the Lathams, Dickson, and Dickson's friend Eugene Lauste (who had worked on the phonograph) quietly formed the LAMBDA Co. to construct a projector.

Working in a small, fourth-floor room in the shadow of the Brooklyn Bridge, the group succeeded during the winter in making both a camera and a projector, which cast a picture about the size of a window frame. To cope with the tearing of the film, which was proving a problem as the strips became longer, the Lathams looped it, a solution that was soon adopted by others. Rumor of what was going on reached Gilmore, and on April 2, 1895, he, Dickson, and Edison had a showdown. The upshot was that Dickson left the laboratory.

Although the projector had a number of faults, the Lathams exhibited it to the press, and by late spring were selling territorial rights and staging their own fights on the roof of Madison Square Garden. Headlines like "Kinetoscope Outclassed" appeared in the newspapers. But Edison pooh-poohed the projector as old hat and said he had a better machine.

With the kinetoscope, Edison had, for the first time, come onto the market with a fully developed device rather than an invention still in or barely past the birth pangs. He sold more than 900 of the peephole instruments. Performers clamored to be filmed, and the trains to West Orange were filled with showpeople who did their routines for no remuneration except a sumptuous dinner. Buffalo Bill, Annie Oakley, three dancers from "The Gayety Girls," a lion tamer, and a tightrope walker went before the camera. The laboratory grounds—now mostly deserted—were filled with the laughter of girls.

After a year the initial curiosity in the kinetoscope began wearing off—people could, after all, see the real performers cheaper in the theater. Raff and Gammon searched for a new element, something sensational that could not be done live.

As early as 1893 an Englishman had written to Edison suggesting the film could be reversed so as to make sequences appear to go backward, and in late August, 1895, Alfred Clark of the Kinetoscope Company had the notion of stopping the camera in mid-action. Going out to West Orange, he staged the beheading of Mary, Queen of Scots. The principal part was played by a laboratory employee, Robert L. Thomas, whose head was saved when the camera was halted just before the crucial moment and a dummy was inserted in his place.

This film was followed by a lynching scene in the laboratory yard, then by a scalping. The combination of violence and sensationalism revived public interest. Momentarily, the kinetoscope business picked up.

It was clear, however, that screen projection was a much more viable medium. And with Edison fiddling, Raff and Gammon were burning. In 1893 a Washington, D.C., realtor, Thomas Armat, had gone to the World's Fair, where he had seen an Anschütz tachyscope. Inspired, he and an associate, Charles F. Jenkins, initiated work on a projector, which they were able to exhibit at the Cotton States Exposition in Atlanta in September, 1895. The underlying principle on which it operated was the same as that of the Edison-Dickson camera—a relatively long pause during which a frame was shown, followed by a quick forward movement of the film. On December 8, Armat gave Raff a private demonstration in his workshop basement.

Armat's "Vitascope" was by far the best projector yet developed. Raff went to Gilmore and demanded that the Kinetoscope Company be allowed to introduce it for showing the Edison films. Raff was faced, of course, with Edison's aversion to being connected with anything not developed in his own domain. During a discussion of several weeks, pressure was put on Armat:

"In order to secure the largest profit in the shortest time it is necessary that we attach Mr. Edison's great name to this new machine," Raff wrote Armat. "Mr. Edison has no desire to pose as the inventor of this machine, yet we think we can arrange with him for the use of his name and manufactory."

So Armat was induced to take the cash and let the credit go.

On April 3, 1896, both Edison and the newspaper reporters saw the Vitascope for the first time when it was previewed by Armat at the lab. Sixteen years had passed since Edison had been incensed when David Hughes had appropriated credit for the microphone under circumstances far less clear-cut, and in another twenty-five years Edison would write Armat: "I had a projection machine, but when you came on the scene I saw you had a much better one than mine, I dropped my experiments and built yours, which was the first practical projection machine." But on April 3 the reporters were led to believe the machine was Edison's invention, and when it debuted publicly at Koster and Bial's Music Hall on the twenty-third it was advertised as: "Thomas A. Edison's latest marvel, the Vitascope." Edison was in a box and Armat manned the projector as the audience cheered the short scenes used as "chasers" between acts.

Armat's brother remained at the Phonograph Works to supervise production of the Vitascopes. But Edison now went a step further in the appropriation of the invention. By early July he had devised a different movement for the machine. "Put the Vitascope model already made somewhere where no one will see it, perhaps you better take the new movement off and put it in the cupboard," he instructed John Ott. On July 29 he took an order from Maguire for fifty "Projecting Kinetoscopes" or "Edison Projectoscopes." (Hence the term "projector.") On November 15, after having built eighty Vitascopes, Edison introduced his own model at Koster and Bial's, and stopped making payments to Armat.

Armat sued, and for the next five years the industry was like the Barbary Coast, with everyone stealing, and no inventor able to get along without some pilfered portion of machinery.

Edison, in the throes of saving money and letting his patent applications "soak," had neglected to take out patents in Europe—where, in any case, they would have been challenged by Marey, Anschütz, Reynaud, and others. Within a year or two of the kinetoscope's appearance, Europeans began offering competition. Louis Lumière introduced a portable camera that was a more versatile instrument than Edison's, and adopted a film speed of sixteen frames per second, which was to become the industry standard.

Dickson, after leaving Edison, combined with two other men to form the KMCD Company. This evolved into the American Mutoscope Company. KMCD developed a camera taking pictures with an area eight times the size of Edison's. Dickson filmed the *Empire State Express;* and when the American Mutoscope program opened at Hammerstein's theater on October 12, 1896, the effect of the great train seemingly thundering straight into the audience had people falling off their seats and stampeding into the aisles. But Mutoscope, using fourteen times the amount of film as Edison's system, could compete only in the largest theaters. Dickson went as the company's representative to England a year later, and came back only infrequently to the United States.*

The transfer of moving pictures from the peephole to the screen created a sensation greater than their original appearance, and firmly established film as a new form of popular entertainment. Shock effect, sensationalism, sex—and outraged reaction to them—all manifested themselves in the formative years. When Raff and Gammon put a long-playing kiss from a Broadway hit, *The Widow Jones,* on film, a Chicago publisher roared:

"Such things call for police interference. The immorality of living pictures and bronze statues is nothing to this. I want to smash the Vitascope."

* In 1908, after Gilmore left Edison's employ, Dickson and Edison were reconciled. For a time Dickson even discussed the possibility of returning to work at the lab. He appeared as a witness in two motion-picture patent suits, and his testimony was highly favorable for Edison.

CHAPTER 30

X-rays, Wireless, and the Conquest of Mars

Edison gave only peripheral attention to the nascent motion pictures. He had a great liking for vaudeville, and enjoyed the short, vaudeville-like sketches that, for many years, were all the screen had to offer. But for two decades there didn't seem to be much money in movies, certainly not as much as in ore milling. Over the span of his career, Edison received only nine motion-picture patents, compared to sixty-two for magnetic mining. During 1896, when films were first commercially projected, Edison was energetically pursuing an entirely different endeavor—the investigation of a new form of light.

In the fall of 1895, at the very time Edison was telling General Electric he was ready to go back to work, Wilhelm Roentgen, a professor at the University of Würzburg, discovered that when a high-voltage current was passed through a Crookes tube some photographic plates that happened to be nearby became fogged. Roentgen enclosed the tube in black paper, so that no visible light escaped. Yet a nearby fluorescent substance was made to glow. The intensity of the current was increasing the energy of the electrons (though the electron, in fact, was still unknown) and shortening the wavelength of the emissions, which passed through some substances but not others. Roentgen found that when he placed his hand over a photographic plate he could see his bones. When he announced his discovery at the end of the year he attributed the phenomenon to unknown, or *X*, rays.

News of the finding generated worldwide excitement. Edison forgot his disappointment with the ore mill, his disillusionment with electricity, and his vow to have nothing more to do with electric lighting. Once again full of enthusiasm, he wired Kennelly (who had left the laboratory two years before to form a consulting partnership with Edwin J. Houston in Philadelphia): "How would you like to come over and experiment on Rotgons new radiations. I have glass blower and pumps running and all photograph apparatus. We could do a lot before others get their second wind."

Kennelly declined, but Edison, without bothering to learn how to pronounce or spell Roentgen's name, was off with an outpouring of energy reminiscent of twenty years before. Everyone was searching for a material that would register the rays after they had passed through a body or an object, and some crude and shadowy fluoroscopic screens had been constructed. Edison was as good at chemistry as he was bad at engineering, and he had an excellent idea of what substances were promising. On March 13, only seven weeks after initiating his investigation, he sent a message to England: "Please inform Lord Kelvin that [I] have just found calcium tungstate properly crystallized gives splendid fluorescence." Within two weeks he sent a fluoroscopic screen to Michael Pupin, a Serbian-born Columbia University professor who was subsequently to receive the first Nobel Prize in physics.

"It is a beautiful instrument," Pupin wrote Edison. "Your success will be greeted with great delight by all scientific men. Please accept my sincere thanks for the princely gift."

Popular imagination, already stimulated, was heightened by the news of Edison's work. One man wrote requesting fluoroscopic goggles so he could see through the backs of playing cards. Another, who since early childhood had eaten broken glass, tacks, and knife blades, so becoming known as the "boy ostrich," wanted Edison to take a picture of his stomach to find out what had happened to all the junk. A third related his brother had swallowed his false teeth while sleeping six months before, and hoped Edison would be able to locate them. The *St. Louis Post-Dispatch* reported Edison was developing a nickel-in-the-slot X-ray, and some peephole fluoroscopes made their appearance at the 1896 New York Electrical Exposition. William Randolph Hearst indicated great personal interest, and asked Edison as a special favor "to make a cathodeograph of the human brain."

Medical interest was intense, centering especially on locating bullets and other foreign objects in people's bodies. Edison received scores of requests for help. He experimented with creating a fluorescent paint. He resuscitated the electromotograph, and hoped that by finding the right chemicals he might employ it as a cathode-ray lamp or an X-ray tube. On May 19 he filed a patent application for a fluorescent lamp activated by bombarding calcium tungstate with X-rays, and he continued the experiments for General Electric until July, 1910, when he wrote: "The project with the Fluorescent lamp is hopeless."

He was regularly at the laboratory until after midnight, and back early in the morning. (But when he went home for dinner at noon, he often remained for a siesta that lasted until late afternoon.) Day after day, with usually the same four assistants, he sat behind a battery of cathode-ray lamps and fluoroscoped people who stood before one of his screens. He suggested the apparatus might be used to find cavities in teeth, and physicians experimented with X-rays as a cure for a variety of ailments. The emanations were said to be a dermatological boon. "The X-ray is a skin beautifier. Better than any cosmetic, and lasting in effect," the *New York World* reported. Dr. William Norton, a well-known New York doctor and electrotherapist, came to the laboratory to employ the rays in an effort to cure John Ott, who was growing lamer and lamer as the consequence of his fall down Bergmann's elevator shaft ten years before.

During the autumn of 1896 there was an explosion of attempts to restore sight to the blind. A St. Louis doctor played host to a continuous stream of sightless pilgrims as though he were conducting a miraculous shrine. The press published sensational reports portending the end of blindness—some persons were able to detect the brilliant light and occasionally even see the shadow of an object. Though Edison deplored the false hope offered, he jumped in and conducted the tests along with other scientific men and doctors. The blind person was given a fluoroscope to hold before his eyes. The roar of the high-voltage current was followed by a sharp crackle as electricity filled the Crookes tube. Sparks hissed as they jumped between the platinum poles. A violet glow emanated from the tube and fell onto the screen. The person might say excitedly: "I see the light," or dumbly shake his head.

The first indication that danger lurked in the invisible rays came in November. An army captain, J. Mc. A. Webster, had been accidentally shot, and the bullet had lodged near his spine. In an effort to find it, doctors examined him two to three hours daily for several days with X-rays. His hands and abdomen were fearfully blistered. His beard dropped out, and the nail of his thumb fell off. A New York doctor suggested: "It is probably eczema." U.S. Surgeon General Sternberg declared: "We have as you know abundant evidence of the good, practical and entirely safe results from the application of the rays." Other medical men were similarly sanguine.

By the summer of the following year, however, more and more reports were coming in of giddiness, headaches, vomiting, diarrhea, and general disability suffered by people who had been exposed. The skin on the hands of Clarence and Charles Dally, brothers who had regularly participated in the experiments with Edison, was wrinkling and charring. Edison, though he usually sat behind the lamps, was experiencing difficulty with his eyesight. His skin was peeling, and gobs of his hair, now almost entirely white, fell out. Thoroughly scared, he halted the experiments.

For Clarence Dally it was too late. During the next five years cancer developed on his fingers, spread to his left hand, and then up his arm. The arm was amputated. A few months later, cancer appeared on the fingers of his other hand. The fingers were amputated; then the hand; then the arm. In 1904 cancer broke out on his face. A few months later he died.

"There is no remedy, but if not burned too much it will heal in time, but a very long time," Edison responded to an inquiry. He himself was suffering from strange lumps in his abdomen that doctors could not diagnose, and he was afraid they were cancerous. After a time, however, they disappeared, without anyone's knowing their nature or their cause.

The X-ray experiments caused C. H. Woodbury, a Bell Telephone Company engineer, to remind Edison that at the Philadelphia Electrical Exposition in 1884 Edison had demonstrated one of his incandescent lamps containing a platinum plate between the arms of the filament. The plate had interrupted and drawn off part of the current, and Woodbury thought there might be some connection between the invisible flow in that lamp and the invisible X-ray.

"Am working in that direction now, there is something missing in the X-ray business," Edison responded in late February, 1896.

Ever since the development of the incandescent lamp, Edison had been plagued by the rapid blackening of the interior. Despite thousands of experiments, he still did not know the cause. Upton, in October, 1882, had been the first to note: "The blue on the lamp always appears on the positive pole. The blackening of the wire always occurs at the negative pole."

During the next year Edison had established, by interposing a platinum plate, that part of the current was being thrown off by the wire and was flowing from the negative to the positive pole. He had patented the tripolar lamp, though he had no idea what its usefulness might be.

"I never have time to go into the esthetic part of my work, hence have done very little with it," he responded to an inquiry in 1884. "But it has I am told a very important bearing on some law now being formulated by the Bulged headed fraternity of the Savanic world = I will send you a half doz Lamps if you want to have a little amusement."

Earlier, at Menlo Park, Edison had given William Hammer the task of constructing a lamp with a long, cylindrical globe placed between the poles of a magnet to determine what effect magnetism had upon the direction in which the carbon particles were drawn. When Hammer went to England in 1882 to help set up the Crystal Palace exhibit, he showed his notes and sketches on the "Edison effect," as it was known, to Professor J. Ambrose Fleming, who had been retained as a consultant by Edward H. Johnson. Fleming performed some tests on his own, and on

May 26, 1883, read a paper, "Phenomenon of Molecular Radiation in Incandescent Lamps," before the Physical Society of London.

During the next decade, experiments on the curious reaction occurring in incandescent lamps continued, and interest intensified with the discovery of X-rays. Between 1895 and 1897, Joseph J. Thomson, a British physicist, constructed a vacuum tube with two charged plates and a fluorescent screen. The electric discharge, or rays, given off by the flow of the current registered as dots on the screen. By measuring the deflection of the dots, Thomson was able to hypothesize the existence of a previously unknown particle of electricity: the electron.

Concurrently, a young Bolognan, the son of an Irish mother and a wealthy Italian father, assumed the pursuit of another neglected Edison experiment. Although Edison had conducted his Grasshopper and space-telegraphy tests in 1885, he had not received his patent on "Means for Transmitting Signals Electrically" until December, 1891. Less than three years later, twenty-year-old Guglielmo Marconi picked up where Edison and German scientist Heinrich Hertz had left off. Marconi, ignoring expert opinion—as Edison had done so often—that wireless telegraphy was impractical, set up his apparatus on his father's estate. In his early attempts he could produce a spark over a distance of only thirty feet. Little by little, incorporating knowledge that had come into existence during the last few years, he increased the distance to three hundred yards, then to two miles.

When he was unable to interest the Italian government in forming a wireless-telegraph company, Marconi went to England. William H. Preece was still the electrician at the Post Office, and Marconi solicited his help. In 1897 he succeeded in transmitting eight miles across the Bristol Channel. The following year the first commercial link was established over a distance of thirty miles, and Edison, unaware of how closely Marconi was following his path, remarked: "Wireless telegraphy is in the experimental stage and I think it will not be long." In 1899 Marconi bridged the English Channel. The next year he constructed a 20,000-volt transmitter of fifty wires supported by 200-foot masts in Cornwall, and in 1901 he went to St. John's, Newfoundland, where he set up a receiving station with kites and balloons. On December 12 he was able to pick up a faint three dots—the Morse-code "S" that was the prearranged signal.

Edison, when informed, was incredulous—he was still thinking in terms of his own experiments sixteen years before, and was now cast in the role others had played when he had brought the incandescent bulb into existence. (Meanwhile, the Anglo-American Telegraph Company, acting like any enlightened monopoly, prepared to sue Marconi for infringing on its territory.) When Marconi heard of Edison's skepticism, he sent him a cheerful telegram: "Hope soon to show you wireless working between United States and Europe. Wish you happy Christmas."

The relationship between Edison and Marconi might have gone no further, but in three articles in the *Saturday Review*, starting April 5, Professor Silvanus P. Thompson held that Edison, with his 1885 experiments, was the originator of wireless telegraphy. The article piqued the interest of a number of people; but not Edison.

"I do not remember anything in these patents Marconi would desire," he responded to an inquiry about whether the Grasshopper patents could be purchased.

Edison's memory, however, was—as was not unusual with him—slightly foggy. William Hammer showed Edison's patent to Marconi's American representative, who, when he saw how much it had anticipated Marconi's technology, was startled. Hammer hinted that other parties were interested in obtaining the patent. In Roanoke, Virginia, Reginald Fessenden, who had been a young chemist in the West Orange laboratory before he had defected to Westinghouse, was experimenting on wireless transmission for the U.S. Weather Bureau. In 1902 he joined General Electric, where he supervised the building of a high-frequency transmitter that was installed at Brant Rock, Massachusetts. On Christmas Eve, 1906, Fessenden made the first voice and music broadcast, and shortly thereafter he was able to reach Scotland.

When differences of opinion arose in the negotiations between Edison and Marconi, Edison asked Columbia University Professor Francis B. Crocker to evaluate his patent. Crocker replied:

"The whole arrangement of the apparatus is very similar to that now employed by Marconi. The distance to which you could transmit messages was less than that reached by the Marconi system. The fact, however, that you were able to transmit two miles was sufficient to prove the operativeness of your invention; and Marconi's ability to signal for two thousand miles is simply the result of further improvement accomplished during the last ten years. In conclusion, I consider that your patent covers important features in wireless telegraphy."

The opinion made it clear that Edison had occupied the ground on which Marconi was building. The Marconi Wireless Telegraph Company of America agreed to pay Edison $60,000, half in cash and half in stock, over a period of five years for the patent. Edison was appointed a member of the technical board of the company. If Edison, for once, was not greedy, it was because some of his other enterprises were prospering, he was negotiating only for a patent, not an invention, and neither he nor anyone else perceived the enormous future of radio.

"I think Marconi will make a success commercially," Edison remarked, "but like all new things it requires time." Not, in fact, until 1910 did the company make its first, small profit.

Marconi's success with wireless telegraphy stimulated Ambrose Fleming to return to his studies of the "Edison effect." Wireless signals consisted of alternating current, and as such had no effect on a telephone receiver. Fleming, retracing Edison's footsteps, inserted a positively charged plate into what was essentially a light bulb. Electricity entered the bulb as alternating current, but when the electrons discharged from the filament were attracted to the positive plate the current emerged unidirectional. The tube was therefore a *rectifier*, and facilitated the detection of the wireless signals on a telephone receiver. Two years later an American, Lee De Forest, placed a third element, called the grid, in the vacuum tube. This resulted in the *amplification* of the weak current induced by the signals, and completed the development of the chief elements necessary for radio reception.

Thus, curiously, in the history of radio Edison was one of the men who played a pioneering role, but left the development of practical application to others, while in incandescent lighting and motion pictures it was Edison who benefited from the groundwork of earlier inventors. Since invention does not, by and large, consist of a sudden flash of genius, but rather of a viable idea carried through to ultimate realization, inventors wrestling day after day and year after year with difficult and frequently disheartening technical problems have a tendency to suffer a form of intellectual battle fatigue. Once they have come to too many dead ends they assume that there is no exit, and that the problem is unsolvable. It is at this point, sometimes a few years later when progress has been made in ancillary technology, that someone else takes up the work where it was previously abandoned. The ultimate invention is the result of a creative relay, and is not, as popular acclaim would have it, the ingenious output of one man.

Edison had accepted the limits he had encountered in wireless telegraphy as an insurmountable barrier, and had lost interest in further experiments. Certainly, it was not lack of imagination that caused him to abandon the subject—one of his ideas in the early 1890s was to devise a submarine telegraph operated by "etheric force" or light waves shorter than ultraviolet, the messages to be recorded as dots and dashes on photographic plates.

Edison's vivid visions sparked the enthusiasm of a writer, George P. Lathrop, the husband of Nathaniel Hawthorne's daughter, Rose, with whom Mina had become friends during her stay in Boston. Lathrop suggested in 1888 that he write a series of articles based on Edison's experiences, to be titled "The Memoirs of an Inventor." Edison, however, had no desire to have anything written about his past, so the project turned into a futuristic one. In May of 1890, Edison gave his assent to a book for which he would provide the science and Lathrop the fiction, though the division of labor wasn't all that clear.

While some of Edison's concepts were far out, others were remarkably prescient. He proposed the manufacture of artificial silk in great quantities from cellulose, and predicted the appearance of electric tricycles, electricity converted from sunlight, color photography, space travel, telegraphic communication with Mars, a hypnotizing machine, and huge telescopic lenses. He suggested the existence of color music and a seventh sense. The plot of the book, he thought, could revolve about the formation of a Saharan sea that causes the earth to shift on its axis, or a gigantic volcano that creates a lush valley populated by highly civilized people in the Antarctic, or a humanoid Amazonian society descended from a lost colony of men who had interbred with apes. Some of his ideas, especially in chemistry, were detailed, accompanied by copious notes, and highly plausible.

With Edison's agreement, Lathrop negotiated a contract for a series and a book with *McClure's Magazine,* and obtained an advance from the publisher. To his chagrin, however, he was progressively made aware that access to Edison was only slightly more difficult than to the Dalai Lama, and it was not until October, 1890, that he was able to talk to the inventor.

Lathrop was delighted with the results, and after going through his notes, told Edison the material was enough for the book. But Edison demurred and said he needed to provide a great deal more. Throughout the winter Lathrop made repeated attempts to see Edison. But Tate always responded that Edison was away, or that he didn't know where Edison was—sometimes Lathrop couldn't even track down Tate.

More and more frustrated, Lathrop went ahead and wrote five chapters. In May, 1891, he sent them to Edison for his approval. Edison misplaced them, and it was six weeks before Mina hunted them down. At this point Edison went off to the ore mill "to stay," and Lathrop was left with an October deadline from his publisher.

As letter after letter from Lathrop to Edison went unanswered, the writer became disillusioned. "The book has been promised to the public a long while, and my reputation is involved," he appealed to Edison on August 10. "McClure has made me certain payments on the work, which I am not in a position to refund; and I ought not to be left liable to refund them through delay on your part. I stayed a whole month in Orange this summer, and the net result was about 15 minutes conversation with you regarding the book, which did not help. For it is impossible to write with enthusiasm or success when you cannot look at what I have written, or supply details, or even talk about the story. I will ask you to realize what it is to me to be forced to hang out like a dog waiting for a bone. I have never been placed in such a position before and shall take exceedingly good care never to be led into a similar predicament again."

Momentarily penitent, Edison said he was sorry but, dogged by problems at the mine, he just couldn't get to the book. He would give Lathrop the money to repay McClure. Lathrop declined it. There were further promises and further delays. One of Edison's futuristic devices, the kinetoscope, materialized. By the end of 1891 Lathrop and McClure gave up. As a sop, Lathrop was paid by Edison and General Electric to write the brochure for their exhibit at the World's Fair.

Six years passed before the literary trail was picked up by William Randolph Hearst. In 1896, after Hearst became interested in the X-ray, Edison prevailed upon him to try out the autographic telegraph, and for a brief period Edison had the same kind of special relationship with the *New York Journal* as he had had twenty years before with the *New York Herald*. Arthur Brisbane, the *Journal*'s managing editor, asked Edison to cooperate in an outer-space series. When Edison declined, Brisbane went ahead without him, and in the winter of 1897–98 produced "Edison's Conquest of Mars." When Martians attacked the earth, Edison invented a disintegrating ray (à la the X-ray), and it was "Edison to the Rescue of the Universe."

But while Hearst had the inventor bounding about the heavens as the savior of civilization, Edison was enthusiastically turning his attention back to his own, terrestrial, science-fiction monster: the magnetic ore mine.

CHAPTER 31

Invitation to Disaster

Edison's return to the ore mill was catalyzed by the biggest boost his ore process had yet received. On January 23, 1897, the president of the Crane Iron Works at Catasauqua, in eastern Pennsylvania, gave Edison the result of tests made with the briquettes:

"The yield gradually increased in proportion to the quantity of briquettes used, until the yield reached 138½ tons per day, an increase of fully 33 per cent. The quality of the iron steadily improved. The iron was, in fact, the strongest and toughest foundry iron we have ever made. Another advantage is that with the same amount of fuel we can smelt a larger quantity of ore. We are satisfied that with a continuous run of briquettes in the furnace, we would effect a saving in smelting cost of not less than 75 cents per ton. We trust that you will soon be able to give us a large and regular supply. I feel quite sure that you will have no trouble in marketing all you can make."

With such a stimulant, Edison set about to get the mill running again.* He ordered the construction of eight new brickers and ovens. During the long hiatus, the equipment had deteriorated, and he needed the equivalent of a small bank's assets to resume operations. He himself had very little cash. He borrowed a total of $53,000 from Batchelor, Bergmann, Upton, Reiff, and Richard Dyer. Frank Z. Maguire, a friend of the Coates family (the founders of the Coates Thread firm and of Coatesville, Pennsylvania), induced them to invest $80,000. Edison mortgaged the Phonograph Works for $300,000. He induced Villard to invest in a special $75,000 syndicate.

The facility was so immense that a minimum of 5000 tons of ore

* On his fiftieth birthday, which he spent at Ogden/Edison, Mina sent him a cake shaped like the mill and adorned with tiny electric lights and figures of miners.

had to be produced daily to break even, and 10,000 to show a profit on the capital investment. That meant more than 70,000 tons of rock had to be run through the machinery every twenty-four hours—a gargantuan undertaking. The margin for error was so small and the potential for breakdown so great that the operation seemed like an invitation to disaster.

Yet during the spring and summer of 1898 it appeared as if luck had once more settled on Edison's shoulder. Walter Mallory, bearded, congenial, and professorial, was Edison's right-hand man on the project. Mallory lacked administrative ability, but he performed the role—as Edward H. Johnson had once played it—of Edison's court jester, and kept Edison's morale at a high pitch. Upton, who had left the Lamp Works in 1894, came aboard, and started applying his efficiency engineering. The Portland Cement industry was expanding rapidly, and Upton found he could readily sell sand, the operation's by-product, to cement manufacturers and builders. The sand soon attained such a reputation for purity and excellence that Upton could not keep up with orders—it seemed another example of the Edison serendipity. Edison decided to go into the cement business himself, and designed a small crushing machine for cement clunkers.

"I feel sanguine that this machine will be an epoch in the Portland Cement industry," he said. Following a successful test at Ogden/Edison, he was jubilant: "Our troubles are over!"

Euphorically, he returned also to his dream of separating gold, and in January, 1898, leased 54,000 acres of low-grade, gold-bearing land near the famous Ortiz Mine in Santa Fe County, New Mexico.

"This is certainly the biggest thing I ever invented—this electrical process for extracting gold from sand," he told a Chicago reporter, and explained he intended to take out $800 million worth of gold by first magnetizing the ore. He devised a briquetting process for anthracite coal. He proposed to build a plant to manufacture his rock-crushing machinery. *McClure's Magazine* announced: "Edison's Revolution in Iron Mining."

Reality impinged again quickly. Not until midsummer of 1898, after refurbishing expenditures of a half million dollars, was the mill ready to run. Everything went wrong. Water and sand ran down the conveyor belts into the machinery. Gears broke. Oil lines got clogged, and bearings burned up. An elevator leaned like the Tower of Pisa. Conveyor belts broke and wrapped themselves around drums, jamming the works. In late November a blizzard swept over the highlands and made the mill all but inaccessible. The pipes and machinery froze, and men failed to come to work. Edison added layer after layer of underwear to keep himself warm, dropped his outer clothes in the middle of the room like a moth shedding its cocoon when he went to bed at night, and on one occasion was so exhausted that he fell asleep before he reached his bunk and crashed to the floor.

"I am nearly discouraged up here and I will have to shut down for the winter," Edison, once more out of money, wrote Villard's private secretary. "I am sure Mr. Villard does not understand yet the full meaning of this enterprise, it is the pioneer of a far-reaching and important industry now at the critical point where a few thousand dollars is all that is necessary to show results that will be surprising."

The pitch remained the same. Edison seemed constitutionally incapable of acknowledging failure. Tantalizingly, during the next year the price of iron kept rising until it returned to its pre-depression level. Edison built seventy-five small houses for workers, installed heating in some of the mill buildings, and again revamped the plant.

In the spring of 1900, the monstrous wheels turned once more. But the mass-produced briquettes were just a shade too low in iron content and too high in phosphorus. Since they had to be porous to admit air in the blast furnace, they became waterlogged in the open railroad cars. During cold weather they froze, and were rendered useless. In the harsh climate, the old frame buildings were disintegrating. The machinery was shaking itself off its foundations. Much of it had become obsolete.

In eleven months Edison spent $380,000 to earn $51,000. When bad weather came in October, Edison dismantled the plant. He would, he said, install all new equipment, and rebuild with steel instead of wood.

This time, however, there was no reprise. In eleven years Edison had spent $3.2 million, two thirds of it his own money. In dismantling the plant he raised $175,000, enough to pay off the company's debts. But since he did it without consulting or informing the board of directors, he left a bad taste, and brought another lawsuit on himself.

The gold-mining venture had also fizzled out. Two ill-starred offspring of Ogden were, nevertheless, just in their formative years.

In 1897 Edison, Bergmann, and Herman Dick (the son of A. B. Dick, who marketed the mimeograph) formed a concern for licensing the ore-milling process in Europe. The next year the Edison Ore Milling Syndicate was organized in England. Headed by Sir John Lawrence (the chairman of the Linotype Company and the Sheriff of London), it included some of the nation's leading men in the iron trade. Edison was paid $70,000 in cash and $330,000 in stock.

The syndicate's original intent was to exploit magnetite deposits in Scotland. But their interest was diverted to a huge mass of medium-grade ore in Norway's Dunderland Valley, just south of the Arctic Circle. Most of the deposit consisted of hematite, not magnetite iron, so Edison invented a process for magnetizing the hematite prior to separation. On January 13, 1900, Edison wired Lawrence (whose telegraphic code name was, appropriately, *Obsession*): "Property has immense value, better form company with own people, don't license."

Edison designed the Dunderland plant, and sent an engineer, William

Simpkins, to build it. Edison was supposed to have final approval of all features. But his disregard of economics and habit of procrastinating put a severe strain on the transatlantic supervision. When it became evident that it was impossible to receive prompt consideration and decisions from Edison, Sir John made a secret agreement with Simpkins and put him in full charge. When Edison realized that he was being cut out and that the Dunderland plant was taking shape in the image of Ogden/Edison, he began to have premonitions of disaster.

"Things are not going right with the mill designs, etc. of Dunderland," Edison told Dick. "There is I believe going to be some terrible errors made. Simpkins has no conception of the problems or the difficulties. I will not be responsible for a single thing the drawing of which has not been approved by myself. I am not to be bluffed into approving what from actual bitter experience I know will not work. The company will learn too late what a sad mistake has been made."

Simpkins pleaded he was faithfully trying to carry out Edison's ideas within the constraints of time and money. But the Dunderland company was doomed to repeat, almost step by step, the experience of Ogden/Edison. By 1907 $2.5 million had been expended. The magnetite separator produced ore averaging less than 55 percent iron and too high in phosphorus for use. In nearly six years of operation only 29,000 tons were shipped. Simpkins was fired. In the recession of 1907–1908 the company went into receivership, and Simpkins committed suicide.

"He at last realized he had made a gigantic failure of the concern and Lawrence is entirely to blame," Edison remarked to Dick. "Every influence was used to keep me from having anything to do with the mill. I gave the Dunderland people plenty of warning of what the outcome would be, but no attention was paid to it. You need not use any of this material unless I am attacked or insinuations are thrown out that I have any responsibility."

As had been the case with Ogden/Edison, efforts were made to revive the mill. Over the course of several years the company was reorganized, and more and more money pumped in, Edison contributing his share. Edison's separator was removed, and an entirely new magnetic device, built by a Krupp engineer, was substituted. By July, 1914, Dunderland had become a combined British-Krupp operation, and the mill appeared ready to run again. Then an assassin fired a shot in Sarajevo, and in so doing put the coup de grace to Dunderland as well.

The other offshoot of the ore mill fared little better.

After testing the cement-making machinery at the mill in 1898, Edison built a large mock-up of an immense cement plant at the lab. The Edison Portland Cement Company was organized with a number of the same stockholders as the ore-concentrating works. Edison bought a large tract of limestone-bearing land near Stewartsville in the Delaware River Valley of western New Jersey, the center of the American cement industry. The all-steel plant was highly mechanized, and contained a number of innovations, including 150-foot-long rotating kilns, the largest ever built. The twenty-seven buildings occupied an area a half mile long and a quarter mile wide, and were connected by a 25-foot-deep tunnel through which an electric railroad carried the materials.

Three years and $1.5 million, far more time and money than originally contemplated, were required to construct the facility. In July, 1902, the plant started up. Five months later the grinding machinery broke down and operations were halted. Two months after that, while repairs were under way, a horrendous explosion flashed through one of the buildings. Six men, including the manager, were killed, and nine others seriously injured. Work was not resumed until fall.

Once again, Edison and the other investors had to keep shoveling money in. By 1905 their total investment was $3 million. The facility was both a high-cost and high-volume operation, an impossible combination in that it could be run profitably only when both demand and prices were high. When auditors were brought in from Price-Waterhouse early in 1906, they recommended that the plant be shut down and the company dissolved.

Edison roared defiance. If his plant could only be run successfully at high capacity, he would create his own demand.

"Mr. Edison intends to do wonderful things with cement in the way of cheap buildings," William Gilmore told Johnny Randolph, Edison's secretary. Edison went to work designing molds for mass-producing concrete houses. He estimated a three-story, six-room house would cost $1200—provided the houses were built in lots of a hundred or more on locations where sand and gravel could be taken from the site itself: qualifications that at once cast a pall on the project.

Two years later the number of molds required for each house had multiplied to 2300. A builder would have to invest a minimum of $175,000 in molds and equipment—an impossible demand in an era when most single-family residences were put up by small contractors one or two at a time.

The project dragged on for seven years before Edison, unable to stimulate a scintilla of interest from builders or real estate men, abandoned it without pouring the cement for a single concrete house. Before the end came, however, Edison developed the idea that a concrete house should have concrete furniture. He proposed making concrete refrigerators and concrete pianos, and did, in fact, cast several concrete phonograph cabinets. To mark the final resting place of the inhabitants of the concrete world, he devised a concrete tombstone.

The Recession of 1907–1908 generated cutthroat competition in the cement industry, and forced the closure of Edison's plant. Twenty years before, Edison had derided the concept of an electric-industry monopoly. But now he took the lead in the formation of a cement trust. In January, 1909, an agreement was signed by all but three cement companies in the Lehigh Valley of Pennsylvania and New Jersey to divide production and fix prices. Two years later Edison devised a project to combine the entire American cement industry into a cartel, and personally presented the plan to President Taft.

Between 1914 and 1916, the plant had to be shut down once more. Scores of new cement plants were being built, and the orientation of the industry was shifting westward. A new, "wet cement" process was devised, placing older plants like Edison's, which used a "dry" process, at a disadvantage. In 1920 the company went bankrupt, and in the reorganization all the stockholders lost their investment.

Edison would not give up. Only in exceptional boom years did the plant operate at a profit. Nineteen hundred and thirty brought another bankruptcy, and another reorganization. The company never paid a dividend, and had cumulative losses (aside from the initial investment) of approximately $5.5 million. A bemused auditor wrote:

"The history of the Edison Portland Cement Company, even during the war inflation period, was a succession of operating losses. In most instances this company would have been dissolved after a short period of financial reverses. However, due to Mr. Edison's well known tenacity of purpose it was continued indefinitely."*

* The company was finally liquidated in the mid-1930s, and the machinery shipped to a nitrate mine in Chile.

CHAPTER 32

The Inevitable Hour

The ore mill took its toll of Edison in a variety of ways—the final years of the 1890s, when he made his last attempt to get the mine operating profitably, were the least creative of his career until the closing decade of his life. In his fifties he no longer had as much resilience, either physically or psychologically. His illnesses grew more frequent, and he took longer to recover from them.

Everywhere among his numerous relatives and acquaintances the effects of age were manifest. His brother, Pitt, was worn out. For twenty years he had relied on Edison to rescue him from financial difficulties. When Pitt was in his fifties, Edison helped him buy a farm; but Pitt neglected the farm to go hunting and set traps. Eventually, when the mortgage holder threatened to foreclose, Edison purchased the farm outright for his brother. In June of 1890, Pitt, a heavy drinker, was felled by heart disease, and soon afterward he was discovered to be suffering from cancer of the abdomen. Edison, after going to see his brother one last time, sent him a phonograph and a case of Imperial Tokay wine. In January, 1891, Pitt, not yet sixty years old, died.

Sam three years earlier had come down with pneumonia, and was given up for dead. "Have best doctors and do everything for him possible," Edison wired to Port Huron.

Within two months Sam overcame the doctors, and was once more making a sprightly appearance around town. In the winter of 1891, Edison shipped him to Fort Myers to take his mind off Pitt's fatal illness, and paid Jim Symington ten dollars a week to be his companion. In Florida Sam and Symington found Ezra Gilliland's half of the property in perfect order, but Edison's caretaker was letting his side go to ruin. Sam set to work several hours a day to straighten up the place, and by the time he returned to Port Huron, five months later, he was ready to renew the pursuit of his fortune.

He bought a half interest in a cockleshell tug for $400, and proposed, like a geriatric Huckleberry Finn, to run the boat through the Great Lakes, down the Mississippi, and into the Gulf of Mexico, where he would go tramp-steaming from port to port.

It was, as Symington said, "a project wildly insane." But Sam had enormous vitality. Although he was growing quite absentminded, he never lost his sense of humor, and drank rotgut whiskey of a quality and quantity to kill a Hooting Hollow moonshiner.

For three years Edison sent Sam and Symington to Fort Myers for the winter. Sam had "a contempt of dudes," and would show up at Glenmont with one ragged suit and three shirts in the last stage of dilapidation. Mina suffered through his and Symington's presence. Edison, though genuinely fond of his father, ignored him as he did everyone else, and Sam seldom caught a glimpse of him.

Then, in his ninetieth year, Sam declined rapidly. He was given "to fits of extravagance, and then of absurd economy," Symington reported, indicating that this was perhaps an inherited trait in the Edison family. Milk and whiskey kept him going for more than a year. On February 26, 1896, when he was ninety-one and a half years old, he died. Edison, in the midst of his X-ray experiments, did not plan to attend the funeral. But his sister, Marion, wired: "Alva, I think you had better come."

Edison, heeding her admonition, discovered Marion herself was not well. He prescribed quinine with whiskey and water, but this gave her so much gas that after a few doses she stopped. Four years later she, too, was dead of cancer. Of his immediate family, only Edison and his three unacknowledged young half sisters survived.

More and more of his contemporaries were succumbing: Jay Gould (aged fifty-six) in 1892; Jim Mac Kenzie (fifty-seven) and Frank Pope (fifty-five) in 1895; John Kruesi (fifty-six) in 1899.* When Sam died, Edison searched out Gray's "Elegy Written in a Country Churchyard," and copied the verse:

The boast of heraldry of pomp and power
All that beauty, all that wealth ere gave
Alike await the inevitable hour
The path of glory leads but to the grave.

For the Miller family, also, the 1890s were a time of tribulation. In April, 1893, Lewis Miller suffered a heart attack, though he attributed his illness to "winter diarrhea," and returned to work within a few days. He was managing the affairs of not only his own Buckeye Mower and Reaper Company in Akron but his brother Jacob's firm, Aultman

* Kruesi left an estate of less than $60,000, substantial for the time but not large considering his thirty-year association with Edison and General Electric. One of his great-grandchildren is former Tennessee Senator William Brock, now national chairman of the Republican party.

and Miller in Canton. Following the Panic of 1893, the latter company failed, and Miller was forced to put up virtually his entire fortune, $320,000 in stock, as security for creditors. Mina lent her father $100,000 in bonds that Edison had given her. But even with Mina's loan, Lewis could not pull the business out, and the creditors ultimately foreclosed. The Miller family was reduced from wealth to very modest circumstances.*

Mina's two younger brothers, John and Theodore, were little older than Edison's two sons by Mary, Tom and Will. The spread between the oldest, John, and youngest, Will, was only five years, and the four boys were treated, in effect, as brothers. They were all enrolled in St. Paul's, a fashionable New Hampshire prep school, and came to Glenmont on holidays.

In 1898, Theodore, at twenty-two the youngest Miller, graduated from Yale and enrolled in New York University Law School. He was an amateur actor, an accomplished violinist, and the all-around star of the Miller family.

When, on April 21, 1898, the United States declared war on Spain, Theodore, in the spirit of adventure, joined Theodore Roosevelt's Rough Riders. The first week in June, Lewis came to see Theodore off, and there was a big family gathering at Glenmont. A couple of weeks later, John joined the navy as an engineer and was assigned to the U.S.S. *Marblehead,* which sailed almost immediately.

Pitted against the Spanish navy, it was safer to be on a U.S. warship than in a rowboat in Central Park,† but the army was a mismanaged, misdirected, misgeneraled mélange of amateurs and stiff-necked regulars. On July 1, Theodore was among the troops thrown against Santiago. Seriously wounded, he died seven days later.

Theodore's death was an intense shock to Mina and the Miller family —even the fact that Mina gave birth to another child (christened Theodore) the day after the telegram was received was little consolation. Mina's oldest sister, Jenny, suffering from a rheumatic heart, died a few months later. Lewis Miller, dispirited, came down with severe abdominal pains in February, 1899. He was placed on a train from Akron to New York, where, on arrival, he underwent a major operation. On February 17 he died.

Mina ached for her husband's support. But he had bottled up his feelings long ago; triumph and tragedy, success and failure were all met

* In the midst of the crisis, Mina's older brother Robert involved himself in shady financial dealings and narrowly escaped being sent to jail. When William McKinley, with whom the Millers were close friends, was elected President, Robert went to see him, but was unable to obtain the high position in the administration that he hoped for. (He was ultimately appointed postmaster in Ponce, Puerto Rico.)

† In destroying the Spanish fleet in Manila Bay and off Cuba, the navy suffered not a single serious casualty.

with the same overt lack of emotion. He immersed himself in his work, and remained ever a curmudgeon, ready to twist any straw so as not to have to display sympathy. When once, late in life, he received a message, "You will be grieved to learn that Professor Wilson and his wife were almost instantly killed in an automobile accident," he snorted: "It dont say he was killed, does it. Says *almost*."

By the time their second child had been born, Mina had accepted the fact that she was not going to have a conventional marriage; and though she might periodically wish otherwise, in general she made the best of it. At breakfast Edison immersed himself in the *New York Times*, which had to be laid out in pristine condition for him; so she took up the *Tribune*. He noticed no one and nothing that went on about him, and on occasion rose from the table without having any idea whether he had eaten or not. If they were invited out to dinner, Mina accepted conditionally, on behalf of herself and "our uncertain quantity, Mr. Edison." He had made her financially independent, and was quite content to have her independent in other ways, too. Had it been left to her, she would gladly have served him hand and foot day in and day out. But what he desired most of all was not to be bothered with anything except his interest of the moment. So she traveled without him, went shopping alone to New York—where she frequently stayed the night at the Hotel Normandie—and involved herself in cultural and religious affairs.

Starting in 1901, they regularly spent February and March at Fort Myers, which had developed from an isolated frontier community into a boom town. Edison was unable to sleep on a train, so often interrupted his journeys to pass the night in a hotel. He acquired the portion of the property belonging to Gilliland—who died in 1903—and year by year invested more money. He purchased a thirty-six-foot electric launch and a large, gasoline-powered boat. He built an extensive dock. He added an irrigation system, and a swimming pool. The bamboo and fruit trees he had planted in the 1880s had grown huge, and provided the seclusion of a tropical retreat.

During the early years of the twentieth century Edison, for the first time, occasionally seemed disheartened. The iron-ore venture had failed. The situation at the cement plant was not promising. Things were not progressing well at the laboratory. He was embroiled in phonograph and kinetoscope suits. Clarence Dally was dying a horrible death from cancer as a consequence of the X-ray experiments; and Edison feared he would follow his assistant before long.

The time Edison had previously spent at the ore mill in Ogden was now occupied with the cement plant at Stewartsville. His residence was a solitary, rickety house that came to be known as "the Monastery," deep in the hills. Although there was nothing to do at night, he stayed up until two or three o'clock every morning. During the daytime he remained for hours in a chair, head slumped on his chest, daydreaming and nodding off. Periodically his fingers searched out his abdomen, and felt for the strange lumps.

One night in early October, 1904, he sat as usual in the rocker on the porch of the Monastery. Behind him, inside the house, the light of a flickering oil lamp was accompanied by the wan flames of a log burning in the fireplace. From the huge tree overhead caterpillars dropped onto Edison's grease- and concrete-splattered linen duster. Edison, his thumbs twisting convulsively, his thoughts far distant, paid them no heed. Handsful of them, green against the dirty white, crawled over him as if he were a corpse. Suddenly, intuitively, he became aware someone was watching him. Jerking himself to a standing position, he shuddered.

Edison's strong constitution had been battered by high blood sugar, exposure to chemicals and X-rays, deteriorating teeth, and lack of exercise. His resistance, perhaps even to a greater degree than most people's, diminished when he was depressed or suffered a psychological shock—undoubtedly that was one reason why he tried to suppress his emotions and developed few close relationships. In the winter of 1905 he became critically ill. All through the first part of January his eyes hurt him, he suffered terrible headaches and earaches, and spent most of the time in bed. On January 23 the doctor diagnosed a virulently inflamed abscess on his mastoid bone. He was rushed to the Manhattan Eye, Ear, and Throat Hospital, where he was operated on the same night. Mina went with him, and stayed at the hospital to nurse him. "I feel like an imposter here," she lamented to fourteen-year-old Charles, "but I guess I will brazen it out. What are we coming to with trained nurses and hospitals for every ailment? Mothers and wives seem to be unnecessary articles nowadays."

Edison, however, was like a puppy in her hands. He dreaded the painful changing of the dressing, and would endure it only if she were there to hold his hand. When he was sick, or in Florida, or sought to escape from the frustrations of a problem he was unable to solve, he plunged into popular magazines and read all of the short novels in them. He was especially fascinated by detective stories. One of his favorite writers was Arthur Reeves, creator of Craig Kennedy, because Reeves used a variety of scientific gimmicks in his tales.

"If you want points for this kind of story drop over some time," Edison wrote him.

Edison had gained considerable experience with detectives since he had first hired Jay Gould's former sleuth to infiltrate the United States Electric Light Company in 1880. A few years later, William J. Burns had tried to track down Edison's bamboo hunter, Frank McGowan, after he disappeared from the streets of New York. Edison retained Pinkerton detectives when there were labor troubles at the laboratory or in his plants. In May, 1901, he brought in the Pinkertons following his receipt of a kidnap threat printed with a rubber stamp on brown wrapping paper:

"If you dont put $25,000 gold for next Thursday night at 12,30 at foot of sign Hahne & Co. in Central & Essex Avenues Orange N.J. We will kidnap your child if you notify the Police We shall do you same."

The detectives suspected an Italian liquor-store owner (though no one was ever prosecuted), and Edison later hired a bodyguard for himself and his family.

Edison's personal private eye for nearly half a century was Joseph "Gumshoe" McCoy; and for a period in the 1900s McCoy's principal task was to keep track of the aberrant, alcoholic, and sometimes illegal activities of Thomas Alva Edison, Jr.

CHAPTER 33

Children of the Man of the Century

Edison's attitude toward young people was as full of contradictions as his personality in general. Remembering his own adolescence, he always retained a warm spot for boys earning their way in the world. He wanted his own children to do well so that he could be proud of them. He had let Mary splurge money on them—indeed, they were spoiled and received an exaggerated sense of their importance as the progeny of the great inventor.

But Edison never interacted with his children. He was not disposed to sacrifice his time to play with them, except on special occasions such as the Fourth of July, when he set off the biggest firecrackers he could buy or devise. Edison's lack of involvement with his children was, in truth, the norm in an era when the work week was sixty hours, and fathers were to be left alone when they returned, frequently disgruntled, from their jobs.

Mary's illness had placed an emotional burden on the children. In the last two or three years of her life, when her health deteriorated markedly, Dot was at Bordentown Academy, and to a large extent insulated from the affairs at home. Will was young and fussed over by Alice, Grammach, and the servants. Not yet six at his mother's death, he emerged without notable psychological bruises. Tom, however, was between the age of six and eight—highly impressionable years. He knew his mother was very sick, that he must not make noise because she had terrible headaches, or that he could not see her because she was in a state of nervous prostration. But he did not understand the reason. Himself a sickly child, with the Edison family's tendency for bronchial problems, he whined, demanded attention, and gradually took his mother's symptoms for his own.

When Mina became the mistress of Glenmont in the spring of 1886, she had a certain naiveté about her role as stepmother. In a large family such as the Millers, the older children helped out with the younger as a matter of course, and she assumed that being mother to Tom and Will would be little different than being older sister to John and Theodore. Although the boys had not had the intense religious instruction or the discipline of the Miller household, Mina's father and Edison agreed on the upbringing they were to have. Edison, like Lewis Miller, was very conscious of the importance of education—he himself was periodically frustrated by the lack of his own. When he railed at education, it was because he thought that it was ineffective and not practical enough. Mina worked with the children on their reading. And Tom, who discovered that at last he had a real and loving mother, utterly adored her.

Dot, on the other hand, detested Mina, and Mina was at a loss how to cope with her. It was more than the normal tension between a girl and her stepmother. Mina was but seven and a half years older than Dot, and in many ways Dot was the more sophisticated. In the home of Grosvenor P. Lowrey, where she had sometimes stayed with her mother, she had met some of New York's elite families. She had regularly gone to the theater and the opera, and knew far more of nightlife than Mina. She was a more accomplished dancer and musician than her stepmother, and could talk back to her in French, which Mina did not understand. But she was also developing into a rather homely girl, who appeared all the plainer next to the lovely Mina. Dot was jealous enough to spit venom. The best Mina could do was keep her away from Glenmont as much as possible. For three years Dot was enrolled in an academy in Connecticut; and as soon as she turned sixteen, Mina had her shipped off to Europe with the two youngest Miller girls.

Mina was delighted when Dot remained in Europe after the Paris Exposition. But, traveling with her companion, Elizabeth Earl, she spent money at an alarming rate. Edison had intended to use Mary's $50,000 estate for the education of his children. Dot, however, went through $35,000 in two and a half years. Edison, deciding he had better call a halt, ordered the travelers home. Mina delayed the inevitable by detouring the two women to Fort Myers. Dot, nineteen years old, did not settle into Glenmont until March, 1892.

Mrs. Earl, who had gained the impression that companioning Dot was going to be a lifelong sinecure, continually fueled her jealousy. Mina, Mrs. Earl charged, was depriving Dot not only of her father's affection but of the patrimony that rightfully belonged to her and her brothers. Dot, pursuing her extravagant ways, ran up bills everywhere. She and her brothers shared two traits: they were completely undisciplined and had not the vaguest idea how to handle money. Edison gave her an allowance of $125 a month, and told her she would have to take care of her own expenses. Additionally, he deeded the Menlo Park property (which he had obtained back from Batchelor) to her, so that she could have the income from the farmland, and would have a house of her own in which to live.

(Menlo Park had been occupied by the Holzers until 1890, when William Holzer was appointed manager of a new Edison General Electric lamp factory in Canada, and for two years after that by Dot's grandmother, Margaret Stilwell.)

Edison, away at the mine much of the time, felt satisfied he had found a solution to the rivalry, and avoided seeing Dot, who was reduced to writing him letters, to which he did not respond. She was so unhappy that after she reached her twenty-first birthday in the spring of 1894, she returned to Dresden, where she had been ill with smallpox and had many friends.

The most important of those friends was Karl Oscar Oeser. In his early thirties, he was a lieutenant in the Kaiser's artillery. Since a lieutenant's pay was only $50 a month, a German officer—even at a time when European workingmen's wages were 50 cents a day—could not support a wife unless either he or she was financially independent. Oeser's parents, though well educated and of the middle class, were not wealthy. Dot, bearing a famous name and supported by an income of $200 a month from Menlo Park and her father, made an attractive alternative to bachelorhood. Oeser was authoritarian and correctly romantic. He bullied her a good deal—and she loved it.

The imperial government required that any foreign woman marrying one of its officers post a perpetual $10,000 bond, so that she would not return home and lure her husband off from the fatherland. On July 24, 1894, Dot wrote her father, "I at last love someone better than myself." Asking his permission to marry Oeser, she pleaded for him to attend the wedding, so people would believe she was his daughter. Edison, who was fond of Dot, replied he wanted her to be happy, but not rush into a marriage she might regret.

"I did not believe you cared enough for me and my affairs to write as you did," Dot answered.

Edison commissioned his stepuncle Sim, no longer needed for scouring up iron mines, to investigate Oeser and determine if he was a fortune hunter—though truthfully Edison's fortune at this point was scarcely worth hunting.

(At the end of 1894, when J. P. Morgan dissolved his partnership with Drexel and established a firm under his own name, Edison's new secretary, Johnny Randolph, asked whether he should close out the account. "Better transfer the account to the new firm, have always had an account there and don't want to break it," Edison instructed him. Edison's balance at the time: $27.16.)

From Dresden, Sim reported to Edison: "He is a nice gentlemanly man. I have seen a good deal of him and like him very much, and think it is a true love match."

The marriage could not take place until a year after the banns were posted, and was set for October 1, 1895. Edison sent nineteen-year-old Tom as a surrogate to give the bride away. Edison had a good deal of difficulty raising the requisite $10,000, but added $1500 for Dot's trousseau. She spent that plus another $800, then forwarded the bills to her father.

With candlelight dinners, and the captain proffering her his carriage, and the regimental band coming to serenade, her life could not have been more romantic. Romance, however, was not enough. Money was the stuff that made Dot's world go round. The $10,000 posted with the German government gave her an income of $50 a month, equal to her husband's. But she had expected her father to continue her $200 monthly allowance as well; and, when he did not, she requested that he give her $600 a month, the amount he had provided while she was traveling in Europe—a sum with which she practically could have bought the regiment. When Edison was not forthcoming, she wrote him she would very much regret having to sell her mother's diamonds and the house in Menlo Park, but . . .

Edison agreed to let her have an extra sixty-seven dollars a month, news of which Dot managed to keep from Oscar and her in-laws, so that Oscar's mother hand-wringingly wrote Edison:

"Your daughter she is grown up as a americain lady and well educated, now she try's to get accustom to german live and german manners, but she must feel herself very unhappy for want of money. 5000 mark yearly ($1200) is a good deal of money, and it might be sufficient for a german officer and his wife, but it is not enough for a americain lady, who is too much accustom to what we call a high life. I beg you to give your daughter at least 50 dollars a month pin money for her very own else she would live in great trouble and sorrow."

Edison didn't bother to answer, as he hadn't answered any of the previous letters from the Oesers. He was quite content to let Oscar and his family remain strangers. He didn't see Dot again for sixteen years, but his relations with her were good compared to those with Tom and Will.

The two boys, ignored by their father, had been as out of place in the classical, disciplinarian atmosphere of St. Paul's as Edison himself would have been. Edison was anxious that they learn, and sometimes even dropped them off at the school himself. As the sons of a famous father, they were expected to be brilliant. When they turned out to be less than ordinary, they were taunted and isolated by their schoolmates. Will, who resembled his father physically, and had picked up many of the mannerisms of his uncle Pitt, developed a hide like a porcupine and bluffed his way through. But Tom, pathetically sensitive, wanting to please but failing, was hopeless.

"I feel *very badly indeed* papa so unhappy I dont know what to do. I have failed in every attempt," he wrote his father. When letters like that brought only admonitions that he must buckle down and do better, he fantasized how well he was doing and wrote Mina he had won six events in track, setting four records.

Still terribly immature at the age of sixteen, he puffed himself up: "I am somewhat like Caesar never satisfied till I am at the head." Afterward, he had to explain why his report card showed him at the bottom of the class. "It is not my fault. I do all right in my studies but it is my *absence* that counts. I always feel sick either in my head or on my person somewhere. Something will have to be done with me, mamma. I will have to give up. I cannot stand it."

Edison, who at the same age had been starting on his telegraphic career, and all his life tried to cover up his illnesses because he regarded them as a sign of weakness, could not abide Tom's whining and his hypochondria.

By January, 1893, when Tom was seventeen, the futility of his continuing at St. Paul's was obvious, and he went to work in the machine shop at Ogden/Edison. He stayed at "the White House," Edison's cottage at the mill, where he amused himself by playing the piano. He painted, but was shy about his canvases. "I am but a youth with youthful desires and ambitions to become a landscape artist—the only one desire and aim I have in this cruel and unappreciative world," he told Mina when he was eighteen.*

He desperately wanted to please his father, but felt entirely overawed by him, and was unable to speak to him at all. In conversing with Edison but receiving no answer, a person never knew whether he had not been heard, or had been heard but was being ignored. One might find oneself speaking louder and louder in the direction of Edison's good ear, only to be withered by a glance into a whisper.

"I dont believe I will ever be able to talk to you the way I would like to—because you are so far my superior in every way that when I am in your presence I am perfectly helpless," Tom wrote his father.

Like his dead cousin Charley twenty years before, Tom tried to pattern himself after his father, and their handwritings scarcely could be told apart.

* Occasionally Tom's paintings, which show some talent, turn up; and since he signed his name "Thomas A. Edison," people mistake them for his father's, and are amazed the inventor could have been an artist, too.

"I often wonder am I his son?" Tom confided to Mina. "What have I done? to be deserving of such a great and honorable Father—truely he's a father—but entirely different from others—and it is his genius—I have to blame—and if God will let me—I hope to some day—mother—to be able to please him—in the way he deserves—I have no genius—no talent and no accomplishment. I will try my very best—to make him think that the Tom Edison Jr. of to-day—is only for to-day—to-morrow—he will be different."

Except on holidays, Tom stayed at Ogden/Edison even after the mill shut down in August, 1895. Tall and gawky, with the Edison family's Jimmy Durante nose, he felt self-conscious and out of place among people. His loneliness, moroseness, and sense of inadequacy led him to alcohol—he started nourishing his fantasies from a bottle. More and more he looked upon the mine as his own. "I am getting along very nicely with my separator— or rather father's—and fully expect to have it finished in about two weeks," he wrote Mina in his fourth year of exile.

Finally, as 1897 dawned, he decided, "I ought to be somewhere—where there is a little more life."

Mina urged him to come back to Glenmont, but he did not like the life of dinners, dances, and receptions she expected him to engage in. His adoration of Mina, furthermore, seemed increasingly unfilial, and would have made it unwise for him to live in the same house with her.

"I love but I am not loved," he lamented to her. "There is only one that I feel loves me and she is only a girl and a girl does not and cannot realize what love is. Write me of the dearest mother for I love you so. One who looks upon and loves you as I do can well consider your words precious. Even if you wrote me every hour of the day your letters would be far between—yes—ages between each one."

He asked his father to help him find a position. But Edison was indifferent. Tom, in his misery, was drawn more and more back to his mother's side of the family. The Stilwells, and especially Mary's brother Charles, shared Dot's resentment of Mina. They felt Edison had abandoned them, even though he had paid for the education of Jenny, Mary's younger sister, and given Grammach an allowance of $100 a month. Thomas Edison, Jr., was a potent name, and through Charles Stilwell Tom met several New York sharpies. They blew Tom up as much as his father had deflated him. In October, 1897, Tom, deluded with grandeur, informed Mina:

"What I am undertaking is to establish and control the market in incandescent lamps. I am going to make a grand success of it or take the fate that I deserve." He would cover the United States with 10,000 agents for the "Thomas Edison Jr. Improved Incandescent Lamp."

"I have several very valuable things which are and will be a great thing for the people of the world. I wonder what father will think when he hears about this? He very probably wont believe it as is perfectly natural for him. It will occur to him why didnt he think of this himself. He has a wrong impression of me. You may doubt my word—but of course you can do as you choose. I shall either make you feel proud of me or die in the attempt."

Liberated from his father's domination, Tom's personality changed. Even his handwriting took on an entirely different characteristic, resembling Mary's. His antagonism toward his father began to emerge through the cover of respect and affection. Edison had used most of the money that remained from Mary's estate to purchase a piece of property for the Edison Industrial Works, and had given Tom a mortgage on it. In April, 1898, when the principal and interest were several months overdue, Tom wrote Edison's secretary, Johnny Randolph, that he did not want to foreclose on his father. "I state this not for the purpose of coercion or threatening but simply to show you that I would be the last person in the world to take steps to injure my own father."

Thomas C. Martin, the editor of *Electrical World* and *Electrical Engineer,* had in his early twenties worked in the Menlo Park laboratory. He had known Tom when he was a baby, and was aware of the stresses involved in having Edison as an employer or a father. He himself had left Edison in 1879 to become the editor of a paper in Kingston, Jamaica. Since his return to New York as editor of the electrical journals his relations with Edison had been cordial. He was sympathetic to Tom, and thought it would help Edison if Tom had someone to turn to for counsel. He told Edison he had asked Tom to serve on the Electrical Exhibition Commission, and intended to take an interest in him and help him all he could.

"You are not helping me," Edison replied with displeasure, "by putting my son on the Com, he is being spoiled by the newspapers. His head is now so swelled that I can do nothing with him, he is being used by some sharp people for their own ends. I never could get him to go to school or work in the Laboratory, he is therefore absolutely illiterate scientifically and otherwise."

Edison scarcely realized he had just shut off the last possibility of a productive life for Tom. Tom, cashing in the mortgage on the Bloomfield property, made up for his previous years' monastic existence with a wild plunge into New York's nightlife. His hangout was The Casino, a musical-comedy house, where there were plenty of girls. He impressed a cute Irish bit actress, appearing in *La Belle Helene,* with his name and his stories of his work with his father. In February, 1899, Tom and Mary Touhey were married by a Catholic priest.

It soon turned out that their principal common interest was alcohol. She frequented the Tenderloin, boozed around with prostitutes and pickpockets, and boasted of her close relations with Thomas Alva Edison, the inventor. About a year after Tom's marriage to her, he discovered she was having an affair with his best friend. Leaving her, he went completely to pieces. Sleeping in a Fourth Avenue saloon, he joined the nameless, faceless bums staggering about the streets of the city.

William Holzer, who had lost his position as manager of the Canadian lamp factory in 1896, enlisted Tom in the formation of the Edison Steel and Iron Process Company. Claiming they had a secret process to treat iron and steel, they sold hundreds of shares of stock at twenty-five dollars per share—which, before long, turned out to be worthless. In December, 1899, they announced a new scheme to inaugurate telephone service across the Atlantic. This was followed in the spring of 1900 by the formation of the International Bureau of Science and Invention, which promoted Tom as "The Inventor's Confidential Friend." He lent his name to other enterprises: the Edison Jr. Electric Light and Power Company and the Thomas A. Edison Jr. Chemical Company, which marketed items like "Wizard Ink Tablets" and "Edison's Magno-Electric Vitalizer."*

Tom, picked up off the streets, was bundled around the country to publicize the Vitalizer as a cure for rheumatism, consumption, paralysis, and locomotor ataxia. Charles Stilwell, believing from what he heard of Edison's health that the inventor would soon die, bought the Edison name from Tom and planned, after Edison's demise, to prohibit all of the Edison companies from using the Edison name without a license from him.

Edison put Gumshoe McCoy on Tom's tail, and obtained injunctions against the use of the Edison name in association with any of Tom's companies. Tom, who had passed a number of bad checks and was being investigated for mail fraud, pleaded with his father to be allowed to go back to work at the laboratory, but Edison refused to have anything to do with him. Tom drank so heavily that by December, 1902, he was seriously ill. Charles Stilwell took him in, and from Stilwell's house Tom wrote his father on December 29:

"I am convinced you are about as well informed as to my personal affairs and actions as I am myself so I will talk to you man to man realizing that our hatred toward each other is very intense. I was forced to enter into agreements to save myself from absolute poverty. In each case the parties to these agreements took advantage of me for the sole purpose of revenging themselves upon you—knowing the injury that would be done you.

"I never dared to ask you for advice for from the very first you gave me sufficient cause to consider you my worst enemy and I still consider you today as such. I have been condemned for things that never existed. I have been wrongfully accused. I gave you an opportunity to stop all these crooked enterprises bearing my name—but evidently I

* Other concerns capitalizing on Edison's name included a corset manufacturer and the old Polyform company, still going strong peddling the patent medicine Edison had concocted at Menlo Park.

took it that they were not injuring you sufficiently to grant my request and give me a job.

"I will sign any reasonable agreement with you in which case you can dictate your own terms—an agreement which will deprive me of all future rights to the name Edison for the purpose of obtaining money."

The letter was first seen, of course, by Johnny Randolph, who thought that the breach would become irreparable if Tom's letter reached his father. Tom's offer to come to an agreement seemed a positive step if it could be couched in less antagonistic terms, and Randolph offered to meet Tom alone to discuss it with him. Tom, who was both looking and acting slightly mad, mistook his intention and replied:

"Your desire to see me alone—causes me to believe that such a meeting is for no good purpose—and that your intentions are to do me bodily injury—you will find me no coward—I am quite sure you would receive better satisfaction by dealing with a stronger man—but I will do as you wish—even though I may be carried there—"

The duel Tom envisioned did not come off. When Randolph deemed that Edison was in the proper mood, he suggested the possibility of an agreement with Tom. The result, in June, 1903, was a contract negotiated by Judge Hayes, Edison's Phonograph Company attorney. Edison consented to pay Tom thirty-five dollars and Charles Stilwell (who was going blind from glaucoma) twenty-five dollars a week, and they, in turn, pledged themselves to enter no business without Edison's express approval.

Tom attempted a reconciliation: "I wanted to stick to you from the first—I hope my willingness to act perfectly fair in our recent agreement convinces you that I have not been as antagonistic towards you as I have been painted to be—I am willing to work hard and faithful. I can do no more than ask you if there is something you could place me at in some of your many enterprises."

"You must know," Edison replied, "that with your record of passing bad checks and use of liquor it would be impossible to connect you with any of the business prospects of mine."

For two years, Tom, who was to drift in and out of alcoholism all his life, alternated between drinking himself nearly to death and being dried out. He had picked up most of his father's phraseology, and could speak as convincingly of futuristic inventions as Edison himself. He had a keen sense of the ludicrous, and of course knew intimately all the ridiculous, mythological things that had been written about Edison, his methods, and his inventions.

"His present diaphragm," Tom told a reporter, "was discovered in an odd manner. One day he sat down on a man's derby hat and as he did so there issued from beneath a peculiar noise. He immediately tore out a piece of the felt and used it as a diaphragm, with successful results." The *Paterson Press* printed the story in all seriousness, as if Tom were the oracle of Orange.

In time, Tom adopted the name Burton Willard, so as not to be identified with his father. In November, 1905, he wrote Edison: "My entire system has forever and eternally rid itself of the poison that was hastily eating my life away—and a new form of manhood has enveloped and transformed me. I have long been an enthusiast in the interest of agriculture and I find my greatest ambition is to possess a farm and start a mushroom business. Now my idea is first to have you purchase me a farm—taking a six per cent mortgage."

Edison was still a soft touch when appealed to right, and when he was not in danger of imminent attachment himself. Tom got his farm—a place near Burlington, New Jersey—though a toadstool or two was the nearest he ever came to producing mushrooms. Early in 1906 his wife died, and in July he married Beatrice Heyzer, a big-boned, awkward, and mawkish girl, who had nursed him back to health.

Now that Tom no longer was a thorn or had ambitions of his own, Edison was complaisance itself. What Tom asked, Tom could have. "Let Dyer draw money when requested to carry out Tom's wishes," Edison ordered. He increased Tom's weekly allowance to fifty dollars, subscribed to a dozen popular magazines for him, supplied him with a car, furnished the money to keep the old farmhouse in repair, and paid his doctor bills when Tom periodically drank himself sick, came down with his terrible headaches, or was involved in auto accidents. Tom puttered about, but seemed incapable of earning a living or even making rudimentary repairs at the farm. Beatrice petted him, indulged him, and agonized: "It is too sad for words to watch him fight and struggle against nature." Periodically McCoy checked up on him just to make sure he did not resume his wanton ways.

Mina went to see him when he was in the hospital in Philadelphia. Early in 1909 he and Beatrice were invited to Glenmont, and the reconciliation was complete. A year and a half later Tom resumed the Edison name, talked to his father, and promised to help straighten out Will. "I feel it is my duty, and anything I can do for my father at any time is always a pleasure. I want him to be happy and have as little care as possible."

Will had the same incapacity for business as most of the members of the Edison family, and made a botch of everything he tried his hand at. Like Tom, he sponged off his father all of his life, but his personality was the opposite of his brother's. He was gregarious, had a devil-take-the-hindmost temperament, and face to face could intimidate his father the way Charley had once done. His character resembled Pitt's. In July, 1898, after Theodore Miller was fatally wounded, Will joined the army engineers, and arrived in the Caribbean after the action was over but in time to come down with tropical fever. For a time he attended Sheffield Scientific School. In the fall of 1899, he married Blanche Travers, the spoiled, complaining daughter of a Washington, D.C., physician.

For three years, Will and Blanche led the life of high-class vagabonds. In the fall of 1901 they acquired an old motor-driven scow and, together with Tom, another man, and two girls, headed down the Atlantic Coast. In mid-December they were shipwrecked off Beaufort, South Carolina. On the return trip north they stopped at Manteo, North Carolina, and for a few months Tom and Will helped Reginald Fessenden with his wireless experiments.

Will's main interest was automobiles. In August, 1903, he got $2000 from Edison to open an automobile agency in Washington, D.C., for Ford, St. Louis, and Royal cars. Less than a month later he complained: "This is the hardest town in the country to start a business in," and requested more money. Edison sent $250. Two weeks later Will reported: Business is tremendous. Please send another $150.

Aside from those sums, Edison provided Will with a regular weekly income of forty dollars, twice the salary he paid chemists and engineers at the lab. Will, however, was a good-time boy who entered all the doors the name Edison opened for him. He and Blanche competed to see who could spend the most money to keep up appearances. In December, 1903, Blanche, angling for an increase in the allowance, wrote Edison:

"Do you realize that we are the children of *'The Greatest Man of the Century'*—and for them to live, as they should, upon an income of forty dollars per week, takes very much more ability than I can display—you might just as well ask the children of the King of England or the Czar of Russia—you would not have us live as Tom in a cheap boarding house with plain people—I won't live in a detestable boarding house."

Edison replied angrily: "I see no reason whatever why I should support my son, he has done me no honor and has brought the blush of shame to my cheeks many times. In fact he has at times hurt my feelings beyond measure. For more than fifteen years I supported my family on less than two thousand per year & we lived well. Let your husband earn his money like I did."

The auto agency in Washington folded within a few months, and for the next several years Will bounced back and forth between automobile ventures, dog breeding, and bird raising. He was happiest as a mechanic, but Blanche was chronically dissatisfied and nagging. Month after month either she or Will appealed to Edison for money. In 1907 Will announced he had perfected a new spark plug "that is going to create one of the real sensations. I have shown the plug to many of the racing men, to Ford, Scarritt, Splitdorf, and others. Nothing but praise. Ford alone will take 16,000." While Blanche leased an elaborate house in Yonkers and stuck Edison with the rent, Will asked his father to put up the money for production of the plug.

"You seem to be a hopeless case—you better get back on a farm—I have reached the limit in your case," Edison retorted.

"It takes you a year or more to come to a conclusion concerning an invention and about one minute where I am concerned," Will snapped. "I will not return to any farm as I have a fine position as well as a meritorious device in my spark plug. You never hear of the good things about me but there is always some damn bastard always ready to knock me."

Edison capitulated and offered to help. Will formed the Edison Auto Accessories Company, and rented a barn to start production and raise pheasants. Piling up bills, he directed them to his father in the same manner—but without the productivity or positive results—that Edison had once leaned on Harrington. "Could you persuade father to give me an order on his tailor for two suits of clothes," he requested Frank Dyer, who headed the Edison organization. "If so I will step out of the bread line. My present wardrobe would be turned down by a Mexican hairless dog."

Bills piled up. Creditors obtained attachments. "I am going to sit on the porch with a shotgun and repel boarders," Will told Dyer. "It's a mighty proud thing to be the son of so great a man but not a happy proposition. It does not take a wise man to know that my father has absolutely no use for me and he can be assured I will not bother him in the future."

"You are a willful and headstrong boy, full of pride, conceit, and vanity," Dyer replied. "I have no doubt that if we turned $100,000 over to you it would be spent in idle foolishness within two months. Suppose he should take you at your word and let you go adrift. You know perfectly well that you could not support yourself by honest means. Come down to the earth and realize your own capabilities and limitations."

In the spring of 1910, Will's property was attached and auctioned off at a sheriff's sale. Edison financed Will's retreat to Salisbury, Maryland, where Blanche had spent her girlhood. Within a few weeks, Will was reporting, "Governor Jackson's son and Mr. Miller, one of the town's richest men, are my great friends and both have fine yachts." So of course Will went about acquiring an impressive boat of his own.

"Blanche has not had a set of furs in four years. Daughter-in-law of the great wizard," Will waxed sarcastic to Dyer in January, 1912. "Cheer up. The worst is yet to come, as I am nursing a very sore spot and its getting larger and larger every day. Dont think I am going to try and make ends meet on a stinking forty dollars a week much longer, do you?"

Edison had raised Dot's allowance back to $200 a month and, all in all, was spending about $15,000 a year on support of various members of the family and doles to friends and former associates. He purchased the house in Milan, Ohio, where he had been born, and pensioned Sim

Edison off to it. He was providing Pitt's daughter, Nellie Poyer, who was widowed and having as much trouble with her children as Edison was with his, $150 a month. (One of Nellie's sons was a psychopath who, after repeated criminal acts, was committed to a mental institution in Ann Arbor, Michigan.) He gave his cousin, Lizzie Wadsworth, who had won his heart by paying back the $200 she had once borrowed, $25 a month, and Lizzie suggested: "You must try Christian Science so if I send you a mince pie you could eat it with comfort." He periodically sent $50 to his grandfather's brother, John, who took to his deathbed suffering from a purported aneurism of the aorta every time he needed money. The Edison family had, in fact, proliferated so greatly, and was so representative of the American people with their unmet medical needs, psychological and financial problems, and old-age poverty, that Edison could have incorporated a charitable institution for none other than his aunts, stepuncles, cousins, nieces, grandnephews, et al.

From time to time he received requests for contributions to other inventors or their descendants, and was reminded of how lucky he had been. He gave $250 to Georg Ohm, when the inventor of Ohm's Law was aged and penniless. He sent $100 to the widow of W. E. Staite, who had experimented with carbon and developed an iridium lamp. He became a member of the committee to provide funds for Moses Farmer's daughter, Sarah, who at the age of sixty-five was crippled and without support.

He was generous toward the men who had been close to him in the 1870s and 1880s. He lent Josiah Reiff tens of thousands of dollars. He told Edward H. Johnson, who had dissipated his money in stock market speculations and was ill with heart disease, "You can always have a little slice now and then for personal use." He gave Johnson $3000 in 1904, $1000 in 1907, $750 in 1908, $1000 in 1910, and lesser amounts at other times.

The requests, naturally, were always heaviest when the economy was in depression. In the first part of 1907, Frank McLaughlin, who had hunted platinum for Edison in 1879 and been connected off and on with Edison's ore-minining ventures and phonograph business while himself engaging in gold mining and real estate in California, came east to raise money. He asked Edison for $400. After he returned to the West Coast, he ran out of cash again. Early in 1908 he shot and killed his daughter and himself in a San Francisco hotel room.

"If I should loan money to all that ask me it would burst the Bank of England," Edison said. In an era when families were large and men were laid off for long periods through no fault of their own, many of the appeals were heartrending.*

* In the early 1890s, for example, Lillie Clifton, whose husband had done all the painting for Edison when he moved to Menlo Park, appealed to Mina. When Clifton had suffered lead poisoning and had had to give up painting, he had

The responsibility for sifting through the letters, picking the ones worthy of consideration, and approaching Edison at the right psychological moment was Johnny Randolph's. Randolph had started as an office boy with Edison in the early 1880s. When Insull had gone to Schenectady and Tate had taken over as Edison's private secretary, Randolph had been put in charge of Edison's business office in New York. In 1893 Edison had brought Randolph to West Orange to replace Tate, and had given him power of attorney to transact all business.

Like all of Edison's private secretaries after Stockton L. Griffin, Randolph was short, slight, myopic, and balding. The position was a powerful one. Randolph, who was kind, sympathetic, and liked by everyone, was not really suited for it. He had to deal with Edison's immensely complex and often deficit financing, his voluminous correspondence, his appointments, a good deal of his business, and a large part of his family relations—handling Edison's affairs would have turned Solomon into a lunatic and produced an ulcer in a robot.

Insull and Tate had used the job as a stepping-stone to executive positions in various Edison enterprises. Randolph, however, took over at a time when Edison decided to keep control over all his companies himself. Figurehead boards of directors were established consisting of William Gilmore (the head of the Phonograph Company), Walter Mallory (Edison's aide-de-camp for ore milling and the cement works), and Randolph. In actuality, the directors never met. Gilmore ran the operations. Any critical policy decision was Edison's alone.

In the first part of February, 1908, the New York court ruled against Edison in the phonograph suit, and Randolph was one of the directors threatened with a massive fine unless Edison immediately complied with the court's order. Edison was enjoined from conducting much of his phonograph business in New York for a year. The court's decision, following on the heels of the depression, sharply curtailed production at the Phonograph Works.

A third of the workers were dismissed. The appeals for assistance turned into a flood. Randolph wrote letter after letter for Edison: "I have been compelled to lay off a lot of my old men and cannot do any-

---

taken a six-dollar-a-week job at the lamp factory. On this he had supported his wife and six children. Eventually, several months after the factory moved to East Newark, his pay had been raised to twelve dollars a week. When the oldest boy was twelve he joined his father in the factory at two dollars a week. After Clifton had worked in the plant for eight years, Upton replaced him with someone willing to do the same work for six dollars. Now suffering from mercury poisoning contracted in the lampworks as well as lead poisoning, Clifton was unable to obtain another job. When Mrs. Clifton, a genteel and well-educated woman, wrote Mina, Clifton had been unemployed for two years and his family were beggars. "Dear Madam," Mrs. Clifton said, "I trust you will excuse me for troubling you. But surely there must be something to do. He will be glad of anything to earn an honest living."

thing." "I am sorry that I have no work for you. I enclose $10 as a loan, pay if you can; if not I will forgive you."

Randolph lent some of the men money himself, though—as Insull had remarked long before—the job of Edison's secretary was not the road to riches.

Although outwardly he seemed all right, Randolph was being overwhelmed by the relentless pressure. He was more and more distracted, and did odd little things, like paying one bill three times.

After finishing work at the office on Saturday, February 15, he went home and killed himself.

He left behind a wife and several children, as well as a shocked group of friends. "Randolph always seemed to be steadiness and faithfulness itself," Reiff mourned. Nellie Poyer felt as if she had lost a member of the family.

Edison, no matter how he might rationalize, could not escape at least partial responsibility. A week after Randolph's suicide, Edison took sick and underwent his second ear operation.

CHAPTER 34

A Most Extraordinary Course

Between the ages of twenty-five and seventy, Edison was embroiled in one lawsuit after another. From Gould and the other robber barons he had learned to take the cash, the stock, and whatever else was in his interest, and not worry about the terms of the deal. On one occasion he lamented: "When trouble arises all my contracts had clauses in I never remember having noticed at the time I signed."

His method of dealing with legal actions, also patterned after Gould's, was to drag them out year after year, and sometimes decade after decade. He considered that delay was the next best thing to a dismissal, and might eventually lead to one. He settled only after every alternative had been exhausted.

Not until 1892 was the dispute over the half interest he had sold in his duplex to Boston businessman E. Baker Welch in 1868 resolved. In 1878, Welch, claiming that the quadruplex was an offshoot of the duplex, demanded $20,000. When Edison rejected the demand, Welch, in 1884, filed suit for a quarter million dollars. The case was not pressed, however, until Welch hired Benjamin Butler, who had been Edison's attorney in the quadruplex action.* Eventually, in April, 1892, Edison agreed to pay Welch $13,000.

The most unusual of the litigations involved Edison's sale of the quadruplex rights to Gould, after first having disposed of them to Western Union, and the subsequent flimflam Gould had pulled upon purchasing the Automatic Telegraph Company.

Gould's conjunction with Western Union had placed the case in limbo, suspended, it seemed, for all time. Since none of the parties was entirely

*** Butler had been alienated from Edison by Edison's treatment of Butler's friends, Hubbard and Painter, over the phonograph.**

blameless, and the gains to be derived from prosecution of the case were questionable, it appeared doubtful that further action by anyone would be worth the effort.

For more than a decade Josiah Reiff subscribed to the common view. He was an intimate of J. P. Morgan and of other financiers. He controlled two or three nearly worked-out copper mines, which provided him with status but little money. During good times he lived well enough; but with every recession he scraped bottom. Finally, in 1895, when he was broke, he changed his mind and reinstated the suit against Gould.

The action had been dormant since 1883, and Gould had died in 1892; but Reiff hoped to obtain recompense from Gould's estate. Since no attorney was willing to take the case on contingency, and Reiff had no money himself, he once more dropped the litigation when his finances improved. His interest revived when, in 1904, a lawyer, John Notman, agreed to prosecute without charging a large set fee. Edison was willing to join Reiff in the suit, so long as no money was required:

"I am already loaded up with notes to the limit. Would be afraid to put more in the bank. I can lend you $300 for *personal* use but do not ask me for money for biz purposes just now." Twenty-nine years after the suit was first filed, it was reactivated.

The case was argued before Judge John Hazel of the United States District Court in the summer of 1905. Judge Hazel issued an interlocutory decree ruling that Gould and the Atlantic & Pacific Telegraph Company had fraudulently appropriated and infringed upon Edison's patents. Reiff and Edison were entitled to collect damages, and the case was referred to a master to determine the amount.

The issuance of the decree was held in abeyance while Notman negotiated with the attorneys for Gould's heirs. In June, 1906, they offered to settle for $100,000. Notman thought he might try for $150,000. Reiff, however, refused to consider such a sum, and Edison was even more adamant. He would take no less than $750,000.

"The acts of Gould and the Telegraph Company impoverished both Reiff and myself. I had to begin practically at the bottom of the ladder," he told Notman. More than money, he was looking for vindication. "I will take all the chances to disprove the ridiculous intimation that the A & P Telegraph Company abandoned the use of the Automatic because of its inefficiency."

When Reiff and Edison refused to take his advice, Notman resigned from the case. Reiff suggested Edison countersign $30,000 worth of notes in order to continue the action, but Edison retorted: "I wouldn't endorse notes for anybody. You will have to find some other scheme." J. P. Morgan made his own counsel's advice available to Reiff. In December, 1906, Edison lent Reiff $1700, the case went before the master, and Edison's and Reiff's new attorney, W. B. Burnet, asked for $4.5 million in damages—$1,623,000, plus interest.

Every month starting in late 1907 Edison advanced Reiff sums varying from $300 to $900 to pay the premiums on his life insurance. On December 28, 1908, the master issued his report. The Automatic system had, in fact, a theoretical value of $1,623,000. But so little use had been made of it that Edison and Reiff had not been injured, and were not entitled to any damages.

Edison and Reiff appealed. On February 4, 1911, the United States Court of Appeals threw the case out. The justices, remarking, "The course of the case has been most extraordinary," ruled that the United States Circuit Court for New York lacked jurisdiction, and that the suit should have been filed not for patent infringement but for equity.

"Dyer, what does this mean when translated?" Edison asked.

It meant, Frank Dyer replied, that because George Harrington was legally a resident of the District of Columbia, the case should have been originally filed in that jurisdiction. The cause on which it should have been prosecuted was that of Gould's failure to carry out the contract, by which action Edison and Reiff might have been injured, not of infringement of the patent, by which they had not.

On the surface, Reiff, seventy-three years old, was still full of fire. He planned to appeal to the United States Supreme Court. In actuality, he was disheartened. Three weeks after the ruling he came down with a cold. This suddenly turned into pneumonia, and on February 28 he died at the home of Mrs. J. Hood Wright, his nearest relative.

Edison was named an honorary pallbearer, but since he had an aversion to funerals, he declined: "I am feeling ill and do not believe I should attend the funeral."

On the very day of Reiff's death, Edison had sent another $482 check for a life insurance premium. Since October, 1888, Edison had advanced Reiff more than $43,000, and he expected repayment from the estate. He was shocked when informed that claims against the estate totaled $340,000. The only hope of getting his money back was to carry the case to the Supreme Court. On December 9, 1912, however, the court refused to review it.

CHAPTER 35

Wet Electricity

One by one Edison's companions fell by the way. Batchelor had retired in the mid-1890s. He devoted himself to his investments, his family, travel, reading, and solitaire. He jotted down a wide variety of questions about nature, subjects for investigation, and ideas for inventions, but did nothing about them. By the latter 1900s he had, like Edward H. Johnson, a classic case of progressive heart disease. His doctor, unable to diagnose the cause of the pain in the left side of his chest and throat when he exerted himself, treated him with electromagnetism, a vibrator, antacid, and liver elixir. On January 1, 1910, Batchelor suffered a heart attack and died.

It was forty years before, when Edison and Batchelor had come together, that Edison's meteoric rise had started. Batchelor was the closest friend Edison ever had. They had struggled side by side with the same problems, they had experienced the failures and shared the successes; theirs was an intimacy that required no expression. With Batchelor's death, part of Edison's own life died, and he took the news hard.

Upton was in retirement, and spending much of his time in California. Johnson meandered from one unsuccessful endeavor to another. Only Sigmund Bergmann was, like Edison, still fully active, and the two men continued to have close ties.

Following the reorganization of Edison General Electric into General Electric, Bergmann had formed a new concern, the General Incandescent Arc Light Company. Although Edison had once told Bergmann, "You are too fierce to make money," he was always glad to be able to borrow from him.* Throughout the middle 1890s, Bergmann lent Edison money. Sometimes, in turn, Edison lent Bergmann money—that is, they wrote notes for each other, which they discounted at banks, so that in reality they borrowed money from the banks on each other's credit.

In the autumn of 1897 Bergmann fell seriously ill. Leaving his junior partner in charge of the company, he went to Germany for an extended rest. In Berlin, he was the proprietor of the Bergmann Conduit Company and the Bergmann Motor and Dynamo Company, formed in the early 1890s. After he recovered, he expanded the businesses, and began installing isolated lighting plants. Edison sold him the license for central Europe on the ore-mining process. And when the inventor, at the close of the century, turned his attention back to electricity, Bergmann became his partner.

Edison detested horses. He considered them irresponsible animals whose minds always worked counter to those of their riders or drivers. Once his carriage, driven by his coachman, was involved in a serious accident with another on the way from Glenmont to the laboratory, and Edison was laid up for several days. On nice days Edison would occasionally walk down the hill from the house, stopping to browse with the cows munching on the lower lawn, just so that he would not have to trust himself to the caprice of a horse.

He had, consequently, a personal interest when the *Scientific American* reported in 1889 that Karl Benz had powered a tricycle with a one-cylinder gasoline engine. "Horses are a necessary evil," Edison declared. "I don't consider horses safe. I like a motor with a governor on it. In 25 years from now electricity will have superseded horsepower in New York. The horse will have become a luxury, a toy, and a pet."

Edison ordered Arthur E. Kennelly to start work on a 7½-volt, ½-horsepower electric motor for a trike. Although there were too many other things going on at the lab and the storage batteries were too inefficient for the project to get rolling, no idea ever left Edison's mind, and he kept it in reserve while powered vehicles developed gradually during the 1890s. In 1891 William Morrison of Des Moines built the first electric carriage. Two years later the Duryea brothers pop-popped along the streets of Springfield, Massachusetts, in a gasoline-powered buggy. Colonel Albert Pope,† the largest manufacturer of bicycles in the United States, initiated commercial production of automobiles during the latter half of the decade. In two years his plant produced 500 electric- and 40 gasoline-powered cars.

* Bergmann was also a canny poker player, who had cleaned up in the occasional games Edison and his companions played on Saturday nights. Once, when the group entrained together for an electric association meeting, the game continued nonstop from New York to Chicago.

† No relation to Frank Pope.

By 1899 the industry had progressed far enough for the New York Edison Company to sponsor the first electric-automobile parade in the world. Perceiving that there was money to be made, Edison could sit back no longer. In February he ordered a tricycle from the Pope Company for experimental purposes. He next bought a car with an internal combustion engine, but told reporters that his own vehicle, on which he had been experimenting for six months, would be much better:

"The French naphtha machines will not be in the same class with mine and the inventors will hang their heads with disgust. The mechanism of my machine is far more practical."

He predicted the upkeep of his electric runabout would be one fifth of the expense of a horse, though in 1893 he had remarked: "Don't think storage batteries commercial. I stick to my own business and never go outside for fear of being busted."

He often complained he had never had much luck with "wet electricity." He continued to hold American rights to the Lelande battery, the best copper-oxide battery so far developed, but this was not powerful enough for running a vehicle. In 1896 a Frenchman, Louis Krieger, had patented a nickel-oxide battery, and Edison intended to improve on this and construct an ordinary primary battery (a battery that cannot be recharged) so that it would act as a storage battery. Many inventors had tried to make such a battery. All had failed.

The basic difficulty with a storage battery was that it returned only 80 percent of the power put into it, so that it was on its face inefficient. Existing batteries were all alike in that they had lead electrodes and an acid electrolyte. Edison proposed to use an alkaline electrolyte with nickel and iron or nickel and cobalt electrodes. He and Jonas W. Aylsworth, who directed a staff of several university-trained chemists at the lab, hypothesized that such a battery would be more durable and provide more power per unit of weight.

In developing a battery, Edison was on familiar territory. He did not have to contend with the abstractions and arithmetic of alternating current, but returned to the comforts of direct current and his favorite science, chemistry. (Somehow the math involved in chemistry never bothered Edison.) Far from the cut-and-try methods he had used in experimenting with the phonograph and the incandescent bulb, or the build-and-run endeavors of the ore-milling plant and the cement works, his approach to the battery was a methodical advance from the foundation of existing knowledge. Within little more than a year the experiments seemed successfully completed. Edison made his first batteries with nickel-oxide (positive) and iron (negative) electrodes, and a potash (20 percent potassium hydroxide) electrolyte encased in shiny steel. He outfitted Hamilton McK. Twombly, Jr.'s, runabout, and the Vanderbilt heir was highly pleased as he sped at seventeen miles per hour over a range of thirty-five miles. Kennelly read a paper before the American Institute of Electrical Engineers, claiming tests of the new battery showed it to be three times as efficient as lead batteries. In February, 1901, Edison turned down an offer of $3 million cash to market the battery. Instead, his new Edison Storage Battery Company issued $500,000 worth of bonds to erect a plant, and he set about expanding his Silver Lake facilities in nearby Bloomfield to manufacture the chemicals.

(Edison had purchased the forty-seven Silver Lake acres in 1889 with the view, as he told Tate, of establishing an Edison industrial complex. Declaring, "My men and myself are abnormally fortunate in managing factories," he had proposed to erect a large chemical works, copper refinery, machine works, and typewriter and telephone factory. The vision had dissolved in the problems of the ore mill, and in 1894 the New Jersey Board of Assessors had considered it "strange that Mr. Edison has lost all recollection of the formation of this company." During the 1890s Edison tried but failed to sell the property several times, and until 1900 the small plant was used principally to manufacture the Lelande battery.)

Several railroads switched to the battery, and orders poured in from the owners of electric vehicles. "In consideration for past and future aid in helping me to perfect the battery," Edison assigned Aylsworth 7 percent royalty. He signed contracts with Bergmann for production of the battery in Germany, and with Morgan's European partners for exploitation in England and France. To obtain an independent supply of nickel, he formed the New Jersey Mining and Exploration Company, and initiated a nickel hunt similar to his iron survey. The leading investor in the company was Charles M. Schwab, the first president of the United States Steel Corporation. John Miller, Mina's younger brother, headed the exploring party that devoted most of 1901, 1902, and 1903 to surveying promising areas for magnetic nickel ore in Pennsylvania, Connecticut, and the region of Sudbury, Ontario. (The men walked thirty yards apart over the terrain, and took dip-needle readings every ten yards.) Edison designed a small nickel-concentrating works on the pattern of the just-abandoned ore mill. He loved nothing better than to go prospecting, and began spending much of his summers camping out in the wilds of Sudbury, the world's largest nickel-mining region. He applied for seventeen leases, but was blocked by timber owners who controlled the granting of mineral rights.

The early successes with the battery, meanwhile, were followed by disappointments. A German journal pointed out that the alkaline battery produced only 1.1 volts per cell, compared to 1.95 volts for a lead-acid battery, so that it required twice the space of a lead battery. "The value of this invention is rather doubtful," the *Technische Rundschau* remarked.

More bothersome were a host of bugs that popped up as Edison tried to initiate commercial manufacture. The potash ate the solder away, and the cans leaked. The machinery to fill the battery's pockets didn't work.

The rubber molding cracked. There was a great buildup of gas, and the batteries frothed over. Inexplicable short circuits occurred, and the battery lost power after three or four months.

By November, 1903, Bergmann, who had formed a stock company and built a plant, was exceedingly anxious, and offered to lend Edison up to $50,000. Edison turned him down: "The phono works is doing very well and keeps me supplied. You know I am no slow coach and there are very good reasons for delay."

A year later Edison still hadn't found the bugs and was in a quandary. Battery production was shut down entirely while six men and twelve boys ran round-the-clock tests. About 500 electric cars in the United States were equipped with Edison batteries, and Edison had inspectors out servicing them and checking them to try to trace down the difficulty. Frequently the batteries were returned to the factory to have cells rebuilt.* Finally, at the end of 1905, after counting up 10,296 experiments, Edison discovered that the short circuits were caused by elements in the battery separating, seeping out of the pockets, and setting up secondary circuits. The spaces between the plates were too small, and in many cases the sides came in contact. "No one would ever have suspected the cause," he noted. "It will be difficult to fix but I have no fear."

Deciding to revamp the entire battery and substitute cobalt for the iron electrode, he sent John Miller back to Canada to search for cobalt. Although he was told there was no shortage of cobalt—a new source had recently been discovered in Oregon—he set out in May of 1906 with Miller, Fred Ott, and several other assistants on an auto prospecting trip through the Blue Ridge Mountains of North Carolina and Tennessee.

During the past five years he had become a confirmed automobile buff who went on outings every Sunday. The greater the speed and the wilder the ride, the better he liked it. He depended on someone else to do the driving—he himself was as helpless a driver as he had been a locomotive engineer. The one time that he took the tiller (this was before the advent of the steering wheel) he promptly directed the car into a tree.

He had changed his mind about French cars, which he thought superior to the American "ginger bread" models.

"I have solved the automobile problem," he announced in 1902. "I can make an automobile that will go so fast a man cannot sit in it. The speed of storage battery machines is unlimited." He planned to sell his car for a mere $350.

"Edison Will Make Automobiles the Poor Man's Vehicle," Hearst's *New York Journal* headlined.

Edison scoured the electric and automobile industries in his effort to build a practical electric car. He got two motors and an engineer, Alex

* The story that this service was performed free is not true. Typically, the cost to the owner of one of the many-celled batteries was about $200.

Churchward, from General Electric. He borrowed a carriage from Studebaker. He financed the building of two electric vehicles by the Birmingham Electric and Manufacturing Company, headed by John M. Lansden, Jr. He induced the Pope Motor Company to send an engineer to help draft a design for a car. The drawings were three fourths completed when Colonel Pope, suspecting Edison wanted to use the plans to build a competing electric car, recalled the man.

Edison protested that he only intended to help the business along. It was a touchy point. His primary goal was to become the supplier of batteries for the whole electric-vehicle industry. (There were, at the time, five manufacturers.) For that reason, he did not want his name associated with any auto-making venture, and when Will organized the Edison Automobile Company in Washington in the fall of 1903, Edison blew up: "You have now done just what I told you not to do, by organizing the company you are doing me a vast injury because every Auto company will now believe that I am interested in manufacturing autos and will fight shy of batteries."

Yet Edison, convinced "There isn't a single rig on the market that is any good," kept planning to produce his own vehicle if he could not find one to his satisfaction. Financed by Edison, Lansden moved to Newark and formed the Lansden Electric Vehicle Company. Edison eventually purchased the company outright for $35,000, but kept the acquisition quiet.

Edison's Blue Ridge prospecting party traveled in a Grout touring car and two White steamers, which he had acquired through William Hammer.* (Edison found the steamers impractical, and got rid of them after the trip.) On completing the journey, Edison told reporters: "My discovery [substituting cobalt for iron] means a revolution in the electrical world. I can reduce the cost of city traffic 55 per cent and cut the weight of storage batteries in half."

Another year passed before Edison was issued his patent for a cobalt-nickel alkaline battery. Simultaneously, his testers and chemists discovered that the iron-nickel battery, which had been all but abandoned three years before, was very much better than everyone had thought. All that was necessary was to change the potash solution once a year, give the battery a long, slow charge, and 90 percent of its original capacity was restored. Two hundred and fifty vehicles were still running on the batteries.

Edison decided, nevertheless, to concentrate on the cobalt battery. On June 12, 1908, he wrote: "At last the battery is finished and all tests made except the jarring test which is being run now."† He tried substi-

* Hammer had married the daughter of Thomas White, the firm's president. Before turning out cars, the company had manufactured sewing machines.

† The jarring test consisted of running a car sixty miles a day over cobblestoned streets and of dropping and raising a cell a half inch 1,776,000 times on a machine.

tuting lithium hydroxide, a better conductor than potash. Lithium, however, was much more volatile and gas-generating, and far too expensive. The entire cobalt battery failed to come up to expectations. After six years Edison decided to go back to the nickel-iron potash battery. On March 15, 1909, Edison noted it had taken him "seven years and $1,750,000 to find out how to make a battery that will last four years."*

While all the hoopla was concentrated on the new battery, Edison's Battery Supply Company continued to turn out and show a nice profit on the Lelande lead-acid batteries. The alkaline battery was superior to the lead-acid in that it could be charged in half the time, did not deteriorate with quick charges, had up to a fourth more useful capacity, lasted three to ten times longer, and weighed about a third less. Its drawbacks—some of which were not to become apparent until after extended use—were that it was virtually useless in cold weather, could not be exposed to hard use, and produced only four fifths of the power of a comparable lead-acid battery when new. Although it weighed less, it required more space than a lead battery, and needed more maintenance. Its biggest handicap was its price—it cost three and a half times as much as a lead battery. In a $1400 vehicle, the battery accounted for $600 of the price.

"An investment, not a running expense," Edison advertised, attempting to neutralize the price differential. Bergmann, who for years had been hard put to placate the stockholders of the German Edison Accumulator (Battery) Company, complained he was unable to sell the alkaline battery in competition with lead. Electric-vehicle makers were annoyed by customers who, dissatisfied with the performance of lead batteries, kept inquiring about Edison's revolutionary new battery.

One manufacturer responded: "We cannot furnish Edison's latest battery. He has never developed a battery that can be used in an automobile. This Edison cobalt storage battery is nothing more than newspaper talk."

Because of the much greater size of the alkaline battery, vehicles had to have their battery compartments rebuilt before it could be installed. Salesmen didn't even want to mention the alternative of an alkaline battery for fear of spotlighting the problem of electrics and negating the sale altogether.

Edison's Lansden Company, on which he was losing $50,000 a year, was handmaking only about three dozen trucks and taxis annually. They

---

One day a visitor came to the laboratory, and as he was sitting in the library with Edison, he heard a crash outside. A workman entered and reported: "Second floor OK, Mr. Edison." Edison, acknowledging, said: "Now try the third floor." Noticing his guest's puzzled expression, Edison straight-facedly explained that he was testing the storage batteries by having them pitched from the windows.

* The sum was a slight exaggeration. Edison's actual experimental investment was $1,212,000.

were much higher priced than competitive makes, and generated complaint after complaint about poor workmanship.

"Pound Lansden on his price. Pound him right along!" Edison exhorted. Friction increased to the point that, in June, 1911, Lansden quit and joined the Electric Truck Division of General Motors. Shortly thereafter Edison liquidated the company.

Edison drafted his own copy to promote the battery: "The electric vehicle for trucking and pleasure has been kept back for years for the want of a reliable storage battery. Edison has filled this gap by his new battery. Nothing now prevents the rapid progress of Electrics. Gasolene cost $32 per ton. 3 ton of coal turned into electricity and stored in an Electric will give as much power as a ton of gasolene 9 dollars instead of $32. Gasolene driven cars have no legitimate place in city traffic. As the electric street car has displaced horses so will the self propelled Electric displace them in every other kind of city traffic."

As a promotional stunt Edison sent a Bailey and a Detroit Electric from New York to Boston, Mount Washington, and back. The cars failed to make the top of the mountain, stopping one mile short, and had to take numerous breaks on the journey to cool their commutators. So the results of the tour were mixed.

Despite the lack of a definitive performance, William C. Anderson, manufacturer of the Detroit Electric, was impressed with the output of the battery. Anderson, a few years younger than Edison, felt a kinship with the inventor, for he had also grown up in Port Huron. It was there that he had organized his carriage company, later moving it to Detroit. He became the only electric-vehicle producer to give preference to Edison's battery. Edison and Anderson joined together in a promotional campaign. Edison promised: "There will be a hot time in the Electric Vehicle biz in advertising."

Starting early in 1909, Edison's battery plant operated around the clock. Yet the backlog of orders kept growing. By September, 1911, Edison was selling ten times as many batteries to Anderson, who was the single largest electric-car producer, as to all other auto manufacturers combined.

One knotty problem—an echo of the AC-DC controversy twenty years before—was that most electric power stations were now producing alternating current, and Edison needed to develop a rectifier before the batteries could be charged.

"I can see that I will have a tremendous job in pioneering the battery among our central station people on account of the bad reputation the Electrics have attained because of the lead battery," Edison told Insull, who, as president of the Chicago Edison Company, was now one of the leading figures in the power industry. "It seems too bad that we must go through the long dreary educational process like we did with the Light and other things. But they will in time understand that the greatest market for electricity will be the electric vehicle. The Central Stations in the large cities could have an all night load inside of 10 years greater than the present peak for Lights and Motors." Speaking allegorically, Edison envisioned the nuzzling of "many electric Pigs to your big Electric Sow. Now Sammy you are the one man in the electric field who is followed by the trade and you should do your best to stear the Centrals into going for this business. It may be slow but it will come as certain as the tides."

Edison had, in bringing the alkaline battery to fruition, given the battery industry almost as much of a push forward as he had the lighting industry with the development of the incandescent bulb a generation before. Left behind were the disasters of ore milling and concrete; in his sixties he was once more full of pep and enthusiasm.

He sent a "confidential" note to William S. Andrews at General Electric: "I am going back into the Electric biz again to show some people that I resent the treatment received." Caution evanesced in hyperbole. Attending the meeting of the National Electric Light Association for the first time in years, he told reporters he was developing a battery that could run a vehicle, be charged in five to ten minutes, and would fit into a suitcase.

Anderson blew up: "It is impossible for you to know the conditions and handicaps we are up against. People are anxious that something different be invented than even your present Battery and any intimation that there is something going to be perfected immediately holds up the sale. This report will cost the sale of a good many batteries." Anderson wondered if Edison could not talk less and produce more. "The day has long since passed when you promised to do something for us in the rectifier line."

Despite such irritations, Edison and Anderson continued to have close ties. Anderson attended the same Episcopal church as Henry Ford in Detroit, and knew the auto manufacturer well. The Model T, introduced in the fall of 1908, was on its way to immortality. It carried no battery, but derived its electrical power from an ingenious magneto. Like all internal-combustion-powered cars before 1912, it had to be hand-cranked to start. This required a powerful arm as well as quick reflexes, should the crank happen to kick back. The mulish tendency of gas-powered cars led to numerous and sometimes grisly accidents, and was considered one of their chief drawbacks. In 1911 the entire industry was searching for an automatic starting mechanism.

Ford had had a transitory encounter with Edison at the meeting of the Association of Edison Illuminating Companies on August 15, 1896. Ford, at the time chief engineer of the Edison Company in Detroit, had just test-driven a gasoline-powered "quadricycle," his first car, and Alexander Dow, superintendent of the company, amusedly introduced him to Edison as a renegade. Edison, who at this period looked upon internal-combustion-powered cars more favorably than he did later, issued a few words of encouragement. Ford, sixteen years younger than Edison, departed glowing.

Edison soon forgot the meeting had ever taken place; but from that time on Ford worshiped Edison. Ford's next encounter with a member of the Edison family was in early 1907, when Will tried to interest Ford in his spark plug. Properly encouraged, Ford, thinking Edison had been as impressed with their meeting as he himself, wrote on February 18:

"My Dear Mr. Edison: —I am fitting up a den for my own private use at the factory and I thought I would like to have photographs of about three of the greatest inventors of this age to feast my eyes on in idle moments. Needless to say Mr. Edison is the first of the three and I would esteem it a great personal favor if you would send me a photograph of yourself."

"No Ans," Edison scrawled across, though this was a time when he was sending photos to most people who asked for them. In 1907 Ford was but one of more than 150 manufacturers of motorcars, and the only association the name conjured for Edison was that of a rival in the automobile field.

By 1911, however, Ford produced one fourth of the cars manufactured in the United States, and when William G. Bee, the sales manager of the Battery Company, learned from Anderson of Ford's adulation, Bee immediately sent Ford a signed picture of Edison and invited him and Anderson to visit West Orange.

For several months Bee and Anderson worked together to arrange the meeting. Finally, in late December, 1911, Edison grudgingly assented: "Guess I will be here on the 9th [of January]." Not to bruise Ford's ego, Anderson made him think that Edison had issued a personal invitation.

After Anderson brought Ford into the laboratory library on the cold January day in 1912, Ford monopolized Edison so completely that Anderson departed chagrined, having been unable to transact any business at all himself. Ford asked Edison to design a battery, starting motor, and generator for the Model T, and immediately set about modifying the engine compartment of the car to accommodate them.

Edison convinced Alex Churchward, who had left General Electric, to design a starting and lighting system. The first week in May, Charles Kettering, whose adaptation of a cash-register motor made the 1912 Cadillac the first car with a self-starter, came to the lab to talk to Edison. Other manufacturers, like Hudson, made inquiries. Ford promised Edison $4 million in battery orders yearly for the Model T.

Edison developed a starter unit with a chain drive connected to the crankshaft in front of the radiator. But the battery did not provide enough starting power, and in cold weather it died entirely.

Edison was trying to solve the battery problem, bring out a new disk phonograph, and introduce sound movies all at the same time. It was more than his income could handle. In December, 1912, Ford lent Edison $100,000, with interest, against delivery of the batteries, and during the next two and a half years he advanced a total of $1.2 million.

During the winter of 1914–15, however, the battery was still unable to budge the Model T engine. Shortly afterward, the experiments were turned aside by the world war. They resumed in 1919, and continued, without positive results, through the 1920s. Ford obtained his battery and ignition system from another source.

The self-starter negated the principal advantage the electrics had had over gas-powered cars—no longer were ladies dependent on muscular chauffeurs to crank their motors. The electrics could not compete with Model T's in price, in ease of maintenance, in convenience, or in economy. Eight hundred pounds of fully charged Edison batteries would not take a driver as far or as fast as a Model T's gas tank. On even moderate grades the electrics came to a standstill. Eventually, Edison asked Anderson:

"Did it ever occur to an automobile builder that the motor should fit the battery and that hilly towns and flat towns should be treated a little different. Of course I could make and will in time make a battery to fit a flat town motor so it could be worked in a hilly town but the easiest way is to change the motor."

Electric-car owners developed exhaust-fume humor. A New Jersey doctor wrote: "I took great comfort from the fact that even if a $5000 limousine did whisk past me at a 40 mile gait when I was laboriously doing 4 miles per hour, still my ride was costing me so much more than the other fellow's that I could look down with disdain upon him."

By the end of the war electrics were becoming curiosities, and steamers were as rare as puffing dragons. Only in urban trucking, where many stops and starts were required, did electric vehicles have some vogue. In powering these, Edison came close to holding his own, and in 1928 there were 6645 electric trucks operating with lead batteries and 5795 with Edison alkaline batteries.

Even less successful was the effort of an enterpriser, R. H. Beach, to apply the Edison battery to the driving of self-propelled streetcars, railroad cars, and locomotives. During 1911–12 Edison worked in conjunction with the Baldwin Locomotive Works and General Electric. He built oversized cells, and envisioned locomotives equipped with battery tenders that would be exchanged every ten miles. He proposed to construct a subway of battery-powered cars in Chicago. Beach built a sizable number of self-propelled cars and discovered "that cars can be equipped with the Edison Battery providing the service required of them in miles per day would not exceed about sixty. With this mileage the battery will be as represented and durable, but if the service requirements are much greater than 60 miles per day, then the battery is almost certain to be overheated and consequently destroyed." Beach found this out the hard way, and within five years his company went bankrupt.

The battery's most promising use was in areas where dependability and long life were of primary importance. Edison designed a miner's lamp that significantly increased safety and was a remarkable success. The batteries were employed for operating telephone and wireless equipment. Railroads adopted them for their signals. The navy used them for floating cranes and for powering gun turrets. Edison designed an electric canoe. A Norwegian proposed to build ships that were one huge storage battery, charge them at a Scandinavian waterfall, then sail them to England and sell the electricity. Edison rigged up a 3-million-candlepower searchlight with batteries at Glenmont; and terrified people who had never seen anything like it swamped police switchboards with reports that the earth was being attacked by rays from outer space and that the second coming of Christ was at hand.

In 1912 Edison made a net profit of $300,000 on the batteries, which ultimately became the most successful product that he marketed.

The most enthusiastic booster of the battery was Miller Reese Hutchinson, whose entry into Edison's life generated a storm at West Orange. Hutchinson, the scion of a landed Alabama family, had been born on Mobile Bay in 1876. He had, like Edison, an early inventive flair, and was granted his first patent at the age of eighteen. Experimenting on a deaf friend, Lyman Gould, Hutchinson in 1895 held the earpiece of a telephonelike device connected with a battery to Gould's ear and spoke the words "Hello," "Mama," and "Jesus." Gould heard. Four years later, Hutchinson gave his first public demonstration of the hearing aid, named the Acousticon. In May, 1902, he was asked to go to England to exhibit the device to the deaf Queen Alexandra. He visited the queen at Buckingham Palace on several occasions to adjust the Acousticon for her. Handsome and straightlaced, Hutchinson cut a swath through afflicted European royalty. After his return to the United States, he settled in New York.

In 1904 Mina bought an Acousticon for Edison. "It's all right when it keeps in adjustment, but it does not always do so," Edison remarked. With the instrument his hearing was, indeed, quite good. "Last night I listened to a concert and heard it all," he told a friend.

He was not bothered at all that Hutchinson had succeeded with the Acousticon, while he had not developed the aurophone he had worked on at Menlo Park. He had, in reality, not wanted to succeed: "Deafness has been of great advantage to me as my business is thinking. I don't think it would help my nerves or give me much pleasure to have my hearing restored. Now everything is quiet and my nerves are perfect and I am satisfied."

Edison used the Acousticon only on special occasions; but Hutchinson was a periodic visitor at the lab. He had come for the first time in the summer of 1897, been given a peek at Edison eating dinner in the library, and had gone away inspired. Like most inventors, he lived an uncertain existence—the Acousticon company went into receivership in 1903. But he was talented and had prominent friends—especially among U.S. Navy officers. Alfred I. Du Pont knew him well, and told Edison he had always been impressed with his genius. During the early 1900s Hutchinson was, like Edison, trying to develop a storage battery, and in 1907 he discussed purchasing Edison's Silver Lake plant for $50,000 —pure bluff, since he scarcely had $5 to his name. In 1908 he devised an electric tachometer that could be used to measure the speed of ships. He had two or three meetings with President Theodore Roosevelt, as well as with Admiral George Dewey and other high-ranking officers. He invented the electric Klaxon horn, came to the lab to present Edison with one of the devices, and talked to him about using the alkaline battery to power it. Hutchinson found Christmas, 1909, "lean," but in 1910 royalties from the Klaxon came to $42,000. He was working on a half-dozen other inventions, and managed to get in to talk to Edison about each one. In July of 1910 he told Edison he was developing an automobile self-starter, and asked for exclusive rights to an ignition battery. Edison, refusing, advised him not to worry—if he perfected the starter, Edison would provide the battery for it. Five months later Hutchinson impressed Edison by bringing several navy officers to the lab and securing an order for Edison's battery from the navy. Edison agreed to give him a 20 percent commission on sales of batteries.

The relationship grew rapidly closer. Hutchinson brought all the important men he knew in to see Edison—including a commander of the Russian navy—thus impressing both Edison and the visitors. He coined the slogan "rugged as a battleship" for the battery, and began promoting it for use on both American and foreign submarines. On the first of April, 1911, Edison asked Hutchinson to move his office into the laboratory, and offered him all the space and experimental facilities he needed. By summer Hutchinson was Edison's constant companion and driver. They went together in 105-degree weather to see J. P. Morgan, Jr. They motored about in the sultry July evenings, and dropped in at baseball games, to which Edison took a sudden liking. They stopped at roadside inns to drink beer, then returned to work in the laboratory after midnight. They discussed ideas and held long talks on a variety of subjects. Sometimes Hutchinson would take Edison home at 2:00 A.M., sometimes at 6. Hutchinson filled the role that Batchelor had once played. With Batchelor dead, Hutchinson became the junior partner, whose presence stimulated Edison.

In November they went to Washington to confer with Admiral Dewey and President Taft—Hutchinson convinced Edison to spit out his wad of tobacco when they reached the porticoes of the White House, but Edison put another plug in as soon as they stepped back outside.

Hutchinson impressed Edison with his stamina and willingness to work long hours. Occasionally Edison let him take a nap in his bed, tucked away in one of the library alcoves. On New Year's Eve, 1911, Hutchinson sat daydreaming in the laboratory: "Am ensconced here, right next to the greatest living inventor and prepared to step into his shoes when he passes away. Brilliant future ahead of me."

When General Motors bought rights to the Klaxon for $143,000, Hutchinson lent Edison—who was spending $200,000 a year running the laboratory—$50,000. In August, 1912, Edison appointed Hutchinson his chief engineer, and gave him responsibility for operation of the lab.

For a quarter century the laboratory had existed as an unstructured and disorderly rats' nest. Edison had stocked every conceivable substance, but after a few years no one knew where anything was, and often it was cheaper and quicker to order something than to go hunting for it through the storerooms. The rodents thrived on the hops, rice, animal skins, walrus hides, and moth-eaten cat hair, so that as soon as the lights went out and quiet settled upon the building it was transformed into a bustling community of rats.

The machine shop itself was seldom entirely vacant. Edison frequently kept some men working until midnight; and a few, either as a matter of convenience or because they didn't get along with their wives, never went home but rolled up in blankets amid the machinery. Paperboys were given the run of the lab, and gamins, hitching rides on wagons, darted in and out of the yard and around the buildings. When a new employee asked what the rules were, Edison spat tobacco juice onto the floor and retorted:

"Hell, there ain't no rules around here!"

Hutchinson thought it was time there were some. He rearranged the entire third floor of the laboratory. He inspected the machinery, and found most of it obsolete or too old for accurate work. Edison said he did not want to buy any new: "We are going to be a laboratory and not a factory."

Hutchinson installed a time clock. He made out card number 1 for Edison, 2 for John Ott, 3 for Fred Ott, and 200, the last one, for himself. He informed Edison: "I have put up a Klaxon horn running at high power and shrill note for calling anyone from the yards etc. when wanted in a hurry."

"So they told me," Edison responded dryly.

Hutchinson decided he would clear out all the "bum actors," and, at the same time, give merit raises long overdue to some men—in the past the only people who had received raises were those who had both the opportunity and temerity to collar Edison. He suggested to Edison that a whole bunch of inexperienced men in the drafting room be discharged and replaced with two or three good designers. Edison, surprisingly, agreed with him: "Three good men who can sketch and are older men with experience is worth two dozen young fellows with no experience."

A female secretary had infiltrated the male community of the laboratory, which created a problem since there was no rest room available for her. "The toilet for the female stenographer would cost about $100," Hutchinson advised Edison. "Cheaper to fire the stenographer and hire a man."

Edison, however, would not countenance her dismissal: "Where could we put the ladies' toilet?" he asked.

Hutchinson suggested that an epileptic be let go because he was a danger to himself and others. But Edison, who knew the man was his mother's sole support, ordered him kept on.

If a man was loyal and had been with Edison for any length of time, Edison would retain him even if he was unproductive. John Ott, unable to walk and often ill, was assured of a lifetime job. One man made a habit of only appearing three days out of every six, but Edison directed: "Pay Hooper even if he gets sick now and then." Philosophically, he was a liberal. When the Edison Illuminating Company of Brooklyn initiated a profit-sharing and pension plan, he complimented the head of the company:

"Good for you. Why shouldn't business men use their brains a little to help out the hard working man with a family who is doing the best he can with his limited intelligence. Just as much money will be made and it will be able to look any man in the face."

When it came to his own operations, however, Edison was not so progressive. He was unalterably opposed to labor unions. He continued to pay only two dollars for a 10-hour day to machinists and a dollar and a half to laborers. When, in September, 1903, the workers walked out, he broke the strike with Pinkerton detectives. (Twelve boys in the packing department who quit at 8:00 A.M. had been replaced by noon.) His penny-pinching was indiscriminate: ten years after coming to work for Edison, John Miller still received a salary of seventy-five dollars per month. Edison refused to pay more than twelve dollars to twenty dollars a week to chemists, even though wages elsewhere were considerably higher.

If the men were dissatisfied, they could leave—and they did. His principal source of academically trained men was European immigrants, who were conditioned to lower wages. But they found it hard to adjust to the undisciplined atmosphere of the laboratory, to Edison's arbitrary ways, and most of all to his habit of badmouthing men and needling them to the breaking point. Edison was already talked about like a legendary demigod of invention, and the refined college men expected to come into the presence of a leonine figure. Instead, they were met by a rumpled, stained, unorganized man with unrefined Western pronunciation and a tendency to get words mixed up with his tobacco juice. Sometimes wearing a bowler, he had a distinct resemblance to a Bowery bum. Since he was irritated by punctiliousness and theory, he took glee in seeing just how outlandish a story he could make these wide-eyed college men believe—a practical man would at once realize that he was being put on.

One day he challenged Martin A. Rosanoff, a chemist he had hired on the spur of the moment, to say what material had made the first filament for the incandescent lamp. "You couldn't guess it in a hundred years. Limburger cheese!" Edison rasped. "Now, can you show me a book of theoretical chemistry that explains why Limburger cheese must be good for the incandescent lamp?"

In hiring men he still sometimes made expansive, if indefinite, promises of rewards as a way of getting them to work for low wages. But he was now so secretive he would not even tell experimenters what was in the product they were supposed to improve. "Everybody steals in commerce and industry," he maintained. "I've stolen a lot myself." And he did not want somebody coming in and walking out with one of his secrets. He told Rosanoff, who was supposed to work on the phonograph wax, "The records are all lost, and I have clean forgotten what's been done, so you'll have to start all over again."

When Rosanoff expressed regret to one of the old hands about the loss of the records, the man snorted: "Are you really innocent enough to believe that? Do you know where you'll land if you believe everything the Old Man says? You'll land with both feet in the green cheese of the moon, and there you'll stick. The Old Man is all for taking, and for giving nothing. You just catch that bird giving away his hard-earned trade secrets to an innocent phenomenon like you who happened to breeze in here."

Edison took particular delight in using Rosanoff for a pin cushion. As Rosanoff worked month after month without coming up with anything, Edison gibed: "How is *theo*retical chemistry this morning?"

Rosanoff finally threw up his hands in frustration. "My problem is nothing but a wild goose chase—a problem without a solution, and the sooner we abandon it, the less of your money and my life will be utterly wasted!"

"Maybe you know best, but that's not my opinion," Edison retorted.

"My experience is that for every problem that the Lord has made me he has also made a solution. If you and I can't find the solution, then let's honestly admit that you and I are damn fools, but why blame it on the Lord?"

"Whatever you have in your head must arise there by spontaneous generation," Rosanoff ground out silently. "To get something in from outside one would have to drill a little hole in your thick skull and pump it in under pressure."

Edison himself was reassuring: "Negative results are just what I want. They are just as valuable to me as positive results. I can never find the thing that does the work best until I know everything that *don't* do it." In actuality, Edison had decided to get rid of Rosanoff, and one morning sent a message to him that he was dismissed.

Rosanoff, who had, like so many of Edison's acquaintances, developed a love-hate relationship, was very much hurt by the peremptory discharge. When one executive took Rosanoff's side, Edison declared: "I consider him dangerous and of the same type as K. L. Dickson only even more impracticable—he has worked over a year and practically got nothing. I have made up my mind that he is the last foreigner in the chemical line."

In reality, Edison kept right on hiring foreign-born chemists.

With his own nose glued to the test tube, he had little sympathy with anyone who did not display equal dedication; and he could be a veritable Scrooge. When Hutchinson asked him: "Could we shut down at 12 noon Christmas Eve to give men time to buy presents?" he answered: "This is a mere excuse. We could shut down at 4 p.m."

Hutchinson attempted to close the lab to all unauthorized personnel as a security measure, for he considered submarines to be one of the largest potential users of the alkaline battery, and had the blueprints for American, Russian, and Argentine subs in his office. All were being built by the Electric Boat Company of Connecticut, the world's largest manufacturer of submarines. Initially, the company tried to obtain an exclusive tie-up with Edison for the alkaline batteries; but when Edison refused, relations cooled. Hutchinson was nonplused: "If we only had time to get out a submarine, it would be an easy job to clear the Electric Boat Company up in a short while."

A submarine could go fifty-five miles underwater on an alkaline battery and only thirty-three miles on a lead battery. Edison's battery, however, was three times as expensive, and at 40° its output dropped to 15 percent of capacity. In Germany, Bergmann was so discouraged he indicated to Edison he wanted to end the battery tie-up, even though he had had great hopes for submarine business.

Hutchinson faced an uphill sales job, made even more difficult by

Edison's practice of demonstrating the battery with one electrolyte, potassium hydroxide, but switching to a cheaper and less efficient one, sodium hydroxide, when shipping the batteries to unsuspecting customers. Hutchinson, pulling all the strings he could and having a long talk with Assistant Secretary of the Navy Franklin D. Roosevelt, succeeded in convincing the navy to install the alkaline battery in the E-2, the first submarine the navy had ever constructed itself, not purchased. The battery was being discharged in January, 1916, when a spark set off a dangerous accumulation of hydrogen gas. Five men were blown apart and seven others injured.

Testimony at the navy court of inquiry indicated that elementary safety precautions had been ignored, that there had been no ventilation, and that the battery had been gassing violently without anyone's paying attention to it. The navy, however, issued a report designed to protect the reputation of the service. The court forwarded a memo to the secretary of the navy recommending that the battery be subjected to exhaustive tests before other submarines were equipped with it.*

The disaster gave the battery a black eye, and brought to an end its installation on underwater craft.

* Edison settled for $66,000 a half-million-dollar damage suit filed by the relatives of the men who were killed.

CHAPTER 36

A Trust Bust

Whenever Edison had developed a major invention, a company—sometimes more than one—had grown up out of it. Usually there was one concern, such as the National Phonograph Company, that was the marketing organization and another, like the Edison Phonograph Works, which did the manufacturing.

(An exception was the motion-picture division, which, since it was an offshoot of the phonograph, remained under the control of the Phonograph Company.)

After paying off the New York Phonograph Company and bringing the long litigation to an end, Edison's company attorney, Frank Dyer, suggested that the time had come to pull the various companies together into one enterprise. The settlement, Dyer advised Edison in 1909, "removes a black cloud that has hung over the business for years and makes it possible to reorganize the business under one corporate name with you as the real and acknowledged head."

The phonograph, kinetoscope, cement, and battery-manufacturing companies, as well as various lesser concerns, were tied together as Thomas A. Edison, Inc., with Dyer as president.

At the same time, Louis Hicks, the attorney who had handled the phonograph case for Edison, pointed out that since Berliner's gramophone patent expired in February, 1911, the opportunity had arrived to put a disk machine on the market. If Edison could produce a machine without a horn and with a needle that did not have to be replaced with every change of record, he could steal a march on the Victor Talking Machine Company.

Ten days before Christmas, 1909, Edison decided to go into the disk business. Thirty years earlier, at the time he had abandoned the phonograph for the electric light, he and Batchelor had sketched a design for an "improved phonograph" that was, in fact, a disk machine. Now Edison himself took on the work of producing diamond points for the needles, and Aylsworth was charged with concocting a new compound for the disks.

Edison's plans became general knowledge during the summer of 1911, although the disks and machines were not ready for introduction until a year later. Eldridge Johnson of Victor expressed his regrets that after the long harmonious relations between the companies they were now to become competitors. At Columbia, Edward Easton did not savor the thought of being squeezed further between Edison and Victor. Early in the fall he approached Dyer, and offered Edison control of the Columbia and American Graphophone companies if Edison would guarantee the companies' investors $400,000 a year.

Dyer favored the takeover. "Our phonograph business is in a most unhealthy and hazardous condition," he pointed out to Edison. The Edison sales organization was huge and unwieldy, and out of all proportion to the business conducted. The acquisition would bring the amount of business in line with the number of people. It would encourage jobbers and dealers, who had deserted Edison in droves, to return. It would provide Edison with a viable operation in Europe, where Edison's sales were vanishing. It would give him a monopoly in the cylinder field, and fill a yawning gap in his projected disk business, since Edison intended to turn out only high-priced, deluxe machines and disks. The greatest demand, of course, was in medium-priced items, which Columbia produced. Victor was preparing to put up a bitter battle against Edison; and without the Columbia line, Dyer warned, "Our position in such a fight is not strong."

Edison, however, was suspicious. If the companies were doing well, why should Easton offer to turn them over; and if they weren't, wouldn't they just be an additional drag? So Edison let the matter drop.

His new diamond disks were introduced with great fanfare. But the records retailed for $1 each, and the machines for $60 to $450. The cheapest model was no good, and Dyer wrote another executive in Europe: "Our men here are disappointed in this machine and predict very little business in it."

As with most new products, there were a good many bugs—the amount of surface noise was especially annoying. Ever the maverick, Edison produced a record entirely set apart from the disks of other companies. It was, as he promised, "indestructible"—but in making it so he created an article more a discus than a disk. A quarter of an inch thick, it weighed, depending on the size of the record, from ten ounces to a pound and a half. It could not readily be played on other makes of machines—and this lack of interchangeability was precisely Edison's intent.

Mina found it "strange that father never likes the fine artists, always some mediocre musician. I always feel he will never get any music until he gets some decent musical man."

Edison, considering himself to be his own best talent scout, decided he himself would select the artists and the music to be recorded. "The general public have been for years calling for louder and still louder records," he remarked, "but I have always maintained that anyone who really has a musical ear wanted soft records." He hired a female pianist to play through the world's entire repertoire of waltzes. Every night for weeks and months on end the woman banged them out as loudly as possible on the piano in the huge den (more properly a salon) at Glenmont. Edison seldom wore his Acousticon, but sometimes bit into the piano to hear more clearly. The rest of the family was driven almost batty, and Edison wondered why the house suddenly was deserted every night.

While Victor had acquired the top talent, the Edison Phonograph Company had gained a reputation for searching out the bottom. Edison, trying to improve the repertoire, sent teams to Europe to record all performers of any reputation. Listening to the records at West Orange, he selected the opera singers and the pieces he wanted them to sing, but failed to take into account the language differences. When the artists appeared in London for the recording session, not only did some not sing the same operas, but they did not sing the arias that they had in common in the same languages. The result was that the quartet from *Rigoletto* was performed by a Russian, a Frenchman, a German, and an Italian, each in his own language, and the sextet from *Lucia di Lammermoor* was inscribed in a babel of six different tongues.

Edison was, simultaneously, trying to develop a fourteen-inch long-playing record. In conjunction with this, he returned to his idea of synchronizing the phonograph with the kinetoscope.

Through the latter 1890s and early 1900s the movies had remained little more than vaudeville fillers, and there was not much money to be made. Boxing matches continued to be staples. When Jeffries fought Sharkey in November, 1899, Biograph outbid Edison for the right to film the fight. James White, the head of Edison's kinetoscope division, decided to take bootleg pictures of the match. Smuggling his camera crew into the twentieth row of the packed, steaming Coney Island Arena, White managed to film all twenty-five rounds. But when he and his crew returned to a nearby roadhouse after the match, Jeffries and his retinue suddenly came pounding up the stairs. The man with the negative swung out the window on a rope and then, with Jeffries in pursuit, dodged through the dark alleys and byways of Brooklyn. In the morning the film came to rest at West Orange.

Another common practice was to re-create, or fake, newsreels of sensational events, such as the execution of spies, and Edison staged his own Boer War in the woods of New Jersey. The first attempt to use film to tell a story of any length was made at the Edison studio in 1903 with the production of a narrative, *The Life of an American Fireman.* This was followed by Edwin S. Porter's *The Great Train Robbery,* a turning point in motion-picture development. Ten minutes in running time, the film was one of the longest that had ever been shot. A Jewish vaudeville actor, who adopted the name Bronco Billy Anderson and had had limited riding experience on a wooden stage horse, was hired to play the leading role, but had difficulty maintaining an upright position on his mount. Edison borrowed a train from the Erie & Lackawanna Railroad, and when the train crossed a high bridge over the Passaic River the director had a dummy thrown off. The "corpse" landed directly in front of a streetcar, which screeched to a halt, its passengers yelling, "Murder!"

The Great Bank Robbery, Camel Caravan to Peking, How a French Nobleman Got a Wife, Uncle Tom's Cabin, and *Honeymoon at Niagara Falls* were some of the Edison releases that followed *The Great Train Robbery. Parsifal,* with a running time of approximately twenty-five minutes, was the longest of the early films. But it was common practice to chop the films up and sell them in pieces, or to splice parts of one film into another. Piracy was widespread, and portions of a film shot by one producer sometimes appeared in the middle of another's.

In 1905 Edison bought a 100-by-100-foot lot in the Bronx and erected a studio, which he deeded to Mina. (Studios in the early days resembled outsized greenhouses.) The structure was at the periphery of the Bronx Park, which was used for outdoor scenes. The location was more convenient for actors—who were paid five dollars a day—than West Orange. Mack Sennett made a transient appearance there, and D. W. Griffith was hired to play a mountaineer who rescues a baby from an eagle. In 1909 the Edison studio filmed Mark Twain's *The Prince and the Pauper,* the first work by an internationally recognized author to go before the camera. Shortly thereafter, Edison acquired the rights to Robert Louis Stevenson's stories. Orville and Wilbur Wright granted Edison exclusive rights to film their flight trials for the Signal Corps. Commander George Dyer, the brother of Frank Dyer, wrote scenarios like *Tale of the Torpedo Boat* and *Sea Hounds,* for which the navy provided ships and personnel.

In 1901 the courts confirmed Edison's basic kinetoscope patents, and he won his suit against the Lathrops and the American Mutoscope Company. Thomas Armat, however, was adjudged the inventor of the camera, and Edison and Mutoscope were ordered to stop infringing on Armat's patent. So far as operations in the wild and flea-flicking industry were concerned, the rulings had no effect. In November of that year, Armat suggested to Edison that they should combine for their mutual interest:

"Years have gone by and your proper profits in the business have not yet I believe materialized. The same is true of me. The situation can be made to bear the *full* success of our hopes in but one way—by the formation of a trust. This combined action would establish a real monopoly. The way things are now the woods are full of small infringers who are reaping that which belongs to yourself and ourselves. There is big money in *all ends* of this business if properly conducted and little otherwise. As things are now business runs by spurts. If there happens to be a yacht race or the assassination of a president there is a good run on films for a few months. Then it drops down to a demand that keeps the large force busy about one-fourth the time while much money is wasted in experimenting with costly subjects that the public will not buy. In the case of a trust a closely organized department under a competent director would at once order the films desired and then consume his own order."

So long as the hard-nosed Gilmore pursued his steamroller tactics, Edison turned such appeals aside. But with Gilmore's departure, Edison changed his mind. Frank Dyer believed, like J. P. Morgan, that competition was unprofitable. On February 1, 1908, the principal parties in the motion-picture industry concluded a licensing agreement. Edison was to receive 50 percent of all fees, and Armat and Mutoscope-Biograph were to split the other 50 percent. Eastman Kodak was to have exclusive rights to manufacture film. In December, the agreement was formalized with the organization of the Motion Picture Patents Company, embracing all the recognized producers.

In the eyes of its participants the trust was so successful that on the night of December 18, 1909, the first anniversary of its formation, the members threw a love feast for Edison at the Hotel Plaza in New York. The banquet room was filled with roses, rose-colored lights, champagne, and caviar. Only one element was missing: the guest of honor. As the hour grew late and he did not appear, a committee was dispatched to West Orange. They found him in the library, where he was fast asleep. Shaking him awake, they bundled him off to New York.

The Motion Picture Patents Company formed the General Film Company to handle distribution, and attempted to take over all the film exchanges, but was thwarted by the refusal of William Fox, one of the biggest, to go along. The MPPC tried to drive all the independent producers and nickelodeon operators out of business. Gumshoe McCoy hired goons, who signed up as extras in unlicensed productions, then broke up the sets and smashed the cameras. In 1910 the MPPC controlled 5300 theaters, while 4200 remained independent. Two years later, however, the total number of theaters had grown to nearly 13,000, and most of those added were independent. The MPPC was collecting three quarters of a million dollars in fees annually, but was such a blatant attempt to establish a monopoly that in January, 1913, the United States Justice Department filed suit.

The greatest damage suffered by the MPPC was of its own making. Immediately upon its formation, the organization established a censorship committee to pass upon all films produced by members. The committee reflected upper-middle-class Protestant mores. *Variety,* on the other hand, identified movies as "the poor man's amusement." The preponderance of the film audience was made up of young, working-class people who preferred ribaldry and belly laughs to sermons.

While independent producers were driven out of New York by the goon squads, creative men like Edwin S. Porter quit because of the heavy-handed censorship. They fled to the location in the United States most distant from New York, and found there not only some degree of security but year-round sunshine ideal for movie production. *The Count of Monte Cristo,* made in 1908, was the first film shot in Southern California.

The people who gathered around Los Angeles were showmen and artists, they were fiercely competitive, and their only commitment was to make pictures that were popular. While Edison and the other members of the trust counted their royalties, the West Coast independents leaped ahead in the art of filmmaking with innovative techniques, better stories, and the development of "features."

Edison established film and record committees to critically review the product before it was released. But the performance of the record committee was so abysmal that he finally appointed himself as its head, and the actions of the film committee were even worse. Dyer placed one of his friends, Horace Plimpton, in charge of movie production. Plimpton, a former rug merchant, considered Hammerstein an impresario second only to himself, and had an unpleasant allergy to criticism.

"We sit in the Film Committee week after week," one of the executives complained, "and pass pictures we know will get us nothing but unfavorable comments and cancellations. We haven't the power to throw out the distinctly bad pictures, nor the courage, because as poor as they are they represent a certain sum of money invested in negative production. Four times out of five I leave the meetings feeling I have had pictures jammed down my throat."

The bluebloods, however, were satisfied. A member of the Ohio Censorship Board wrote: "It is always a relief to me when our operator puts on an Edison comedy. I know that it is going to be a good, clean, moral picture. I have never had to order one scene cut from an Edison picture."*

* Some communities nevertheless banned motion-picture theaters altogether. Among them was the town of Montclair, adjacent to West Orange, which was determined "to preserve the purity of its intellectual and moral atmosphere."

An Edison executive, displaying the insight of a dodo bird, thought: "It is things like this that make it so hard for us to determine where we are at and who is right."

In fact, the minutes of a special meeting of the film committee in January, 1915, summed up the situation succinctly: "Our comedies have a bad reputation and our dramas do not have distinctive popularity."

Edison did not involve himself in the dramatic side of production, but continued to devote himself to technical improvements. From 1900 on he tried to develop color film. (Some scenes, in those days, were hand-colored.) Starting in 1909, he worked to bring out a home kinetoscope (amateur movie projector), as well as sound movies. The home kinetoscope was first demonstrated in New York in April, 1912, but Edison was unable to develop a market for it. A year and a half later he turned it over to Hutchinson. But Hutchinson never pushed it, and it fell by the way.

In August, 1909, Edison noted to himself: "There are more than 40 patents on combining phonographs with motion picture film. None are successful. There is a wide gulf between an idea and its reduction to practice." Two years later his other, optimistic self reasserted itself. "The speaking picture is already perfected," he declared.

The device, which he named the kinetophone, was in actuality scarcely off the drawing boards. It was little more than a loud-speaking, long-playing phonograph with a lever that permitted the machine to be slowed down or speeded up. It sounded, all too often, as if the performers were speaking with a mouthful of mush.

"A little more ginger in the tent would help," Hutchinson suggested.

"All right, put some in," Edison responded, and in the summer of 1912 came to an agreement with the newly formed American Talking Picture Company to produce fifty kinetophones.

When delivery of the machines began, in January, 1913, Hutchinson told Edison: "This entire apparatus is the most unsatisfactory product we have ever turned out." The machinery was hard to operate and easily damaged. It could not be worked by the projectionist, but required the hiring of an additional man. The kinetophone was placed next to the screen, where it was controlled from the rear of the theater by means of a fish line. Unfortunately, Hutchinson reported to Edison, "It develops that *rats* have a very pronounced appetite for the string we use on the kinetophone."

Edison's talking pictures were introduced into eleven theaters in New York, Chicago, St. Louis, and Milwaukee on February 17. But any of a dozen elements could throw the sound out of synchronization with the film. The effect of an operator trying to speed up the kinetophone so as to catch up with the actors' lip movements was as hilarious as watching a woman's voice come out of the mouth of a man. The whole melodramatic technique of silent movies was unsuited to speaking actors. Audiences alternately hooted and roared with laughter.*

The kinetophone was a disaster. On July 17, 1913, Edison and the American Talking Picture Company came to an agreement: Edison released the company from its contract, and the concern's officials promised to keep their mouths shut and not make derogatory comments.

Stewing over the failure, Edison told Hutchinson a few months later: "I am going to erect a studio here to perfect the kinetophone to the limit. We are the only ones who can do it. I want your cooperation to show the theatrical people that scientific people can beat them at their own game and produce things that will open their eyes." The new studio was completed in May, 1914. But the productions that emanated from it were no more satisfactory.

* Warner Brothers used essentially the same system as Edison in *The Jazz Singer,* a 1927 musical without speaking parts (except for the final line), and the 1928 drama *Tenderloin,* which contained three speaking scenes. In one of these the heroine, Dolores Costello, was trapped by the fiend in her bedroom. He advanced with a menacing leer: "I will give you your choice. You can tell me where you have hidden the money or . . . or . . ." The leer broadened.

Miss Costello, biting her knuckles and gasping, shrank back against the wall: "Not that! Not that!"

The reaction of the audience, which had accepted such theatrics straight-facedly in silent movies, was hysterical laughter. Not until 1930 was the modern system of sound synchronously recorded on the film inaugurated, and a new style of acting born.

CHAPTER 37

The Faustian Soul

In general, the implicit accommodation that Edison and Mina reached early in their marriage worked satisfactorily. Edison was the master of the laboratory. Mina, after her early intrusions, ventured into that male sanctuary only sporadically, though she persuaded her brother, John Miller, to stay on there as her unofficial representative despite his low salary.

Edison's views on women were progressive. Electricity, he thought, would help "develop woman to the point where she can think straight. For ages woman was man's chattel and in such a condition progress for her was impossible. Now she is emerging into real sex independence and the result is a dazzling one. Matrimony will become a perfect partnership. We shall stop the cry for more births and raise instead the cry for better births."

But Edison never learned to talk to women. He much preferred the company of men, with whom he could exchange ribald stories and tell dirty jokes, which he copied and stashed away in his library desk like a squirrel hoarding nuts. He felt most comfortable in a jumbled, junk-filled atmosphere where he knew where everything was—if someone moved a bottle or a bolt he had left weeks before in some odd corner, he would stir the entire laboratory into an uproar. At Glenmont, Mina curbed some of his more unsanitary habits, but at the lab he could be followed by the tobacco juice he squirted onto the floor. When Mina asked him why he did not use a spittoon, he answered that a spittoon was hard to hit but the floor was difficult to miss.

Since the laboratory workweek continued to be a full six days, Edison was never really "home" at Glenmont except on Saturday nights and Sundays. There he spent much of his time in the book-lined upstairs sitting room adjacent to his and Mina's bedroom. Edison had a desk in one corner, Mina in the other, and the children occupied the far end. He read a great deal, but continued to like activity, which did not disturb him in the least, going on around him—the sitting room was the center of family life. Parcheesi was the only game he could be coaxed into playing. The rules were variable. Edison changed them from moment to moment to give himself the advantage—it mattered less to him how he played the game than whether he won or lost. The others did not protest—he was, indeed, treated like an absolute monarch. Mina had an ironclad rule that whatever Edison wanted, Edison got. She even dressed herself and Madeleine according to his tastes. (He hated red dresses, which Mary had been partial to, and disliked elaborate pieces of jewelry.)

The less he needed to concern himself with the problems of daily living, the happier he was. He let Mina order his suits for him, and gave her an allowance of $12,000 a year for his room, board, and other requirements. She ran the household like an executive, from behind her desk, which had an array of concealed push buttons. Her fiscal year started the first of May—the date she had originally taken over at Glenmont in 1886—and her annual budget varied from $20,000 to as high as $63,000.* She met the outlays with Edison's contribution, the interest from the bonds he had given her early in their marriage, the rental income from the Bronx studio and a New York office building (10 Fifth Avenue), and dividends from her large holdings of stocks and bonds in Edison's enterprises.

She devoted her time to the Red Cross, the Daughters of the American Revolution, the Missionary Services, the Sunday-school organization, the hospital guild, and to literary, art, and music circles. She subscribed to the New York, Philadelphia, and Boston orchestras, and entertained their conductors at Glenmont. "It drives one nearly wild with all the engagements," she complained; but in reality she loved them. A fair representation of the world's famous people passed through the house—Helmholtz, Marconi, Stokowski, the Wright brothers, the King of Siam, and Eleanor Roosevelt (who regularly attended Chautauqua). Woodrow Wilson came the year before his election to the Presidency, but caught hardly a glimpse of either Edison or Mina.

Aside from the famous, the house was always harboring relatives and friends—Mina sheltered people the way some women collect stray cats. "I seem to be able to surround myself with dependents," she noted. The visitors took up Edison's habits, and the first floor of the residence was

* The $63,000 budget was for 1905–1906, and contained $24,000 for repairs and renovation. Included were personal expenditures of $3100 for herself, $900 for Edison, $2200 for Madeleine, $800 each for Charles and Theodore, $3900 for food and incidentals, $1000 for fuel, $950 for travel, $400 for entertainment, $600 for doctor bills, $3300 for salaries, $1655 for taxes, $3700 for charity, $3100 for gifts, and $2500 for Christmas presents.

sometimes filled with snoozing people. "Papa is asleep in the den. Miss McWilliams lies in the living room on the lounge. Miss Hess does the sensible thing and goes to bed and does not lie about on lounges with her mouth open," Mina wrote.

Later Mina was to regret she had not spent more time with the children, but had left much of their upbringing to governesses and preparatory schools. "I feel I have made such a failure of being a mother. Every day I grieve over it for the opportunity will never come again. I was blind," she told Charles.

She was an eternal pessimist, who would have counted mosquitoes in Paradise. She was extraordinarily, if schmaltzily, affectionate—she called Edison "Dearie" and the children were "Deariettes." She thought culture and religion—to which Edison was oblivious—the most important facets of life. While the children were growing up, she read to them every evening before supper. Sunday mornings she bundled them off to the Orange Methodist Church, and Edison was left alone to work his way through three or four newspapers. Sunday suppers were a special time. Edison liked young people, and the children would often invite friends. Left alone, Edison would take no part in what was going on, but sit, quite contented, isolated in his thoughts. If, however, he were made a fuss over, he would thaw out and start to spin yarns, and he could be a delightful storyteller.

Madeleine, the oldest of Edison's and Mina's children, resembled her father more than any of his other offspring. Though he was not so radical as to consider that Madeleine might participate in the affairs of the laboratory, he encouraged her to take a physics course at Bryn Mawr. (That turned out to be a miscalculation.) Peppery, intelligent, and articulate, she played tennis, exercised a delightful wit, and sprinkled her conversation with "hells" and "damns"—a daring display of emancipation, especially when manifested by the granddaughter of Lewis and Mary Valinda Miller. Having been cared for by "Madamsell," she spoke fluent French. Her nose was atypical of the Edison and Miller families, and her father teased her about it.

*"C'est mieux d'avoir un nez qui regarde la lune, qu 'un qui vous tombe la bouche,"** she responded.

In the summer of 1910, when Madeleine was twenty, she vacationed on Martha's Vineyard. "It's hotter here than the front porch of the residence of Satan, and I'm bored beyond endurance," Madeleine wrote her mother. "Tennis is like being boiled in oil. The only really amusing thing we've done to-day is to decide what kind of husbands each of us is to have. Barbara is to have a poet. His name is Rudolph. Their eldest son is to bear the name of Rhubarb after both parents. Mine is a minister—that seems to be preordained."

Mina worried that in her boredom Madeleine might forget prudence.

* Better to have a nose looking at the moon than one falling in the mouth.

Madeleine replied: "I don't think you need bother about familiarity with men. I think you can trust me now—but you can put in a warning every once in a while if it makes you feel better."

The only male around was John Eyre Sloane, the son of a prominent Catholic doctor, who was acquainted with Edison. John was so quiet and so much a member of the party that the girls thought of him not as a man but as one of themselves. Madeleine beat him at tennis. She asserted she had never found anyone whom she could beat before, and it depressed her.

John's reaction was somewhat different. He began to entertain thoughts of playing doubles. Within a few months he and Madeleine were unofficially engaged, much to the distress of Mina.

"I am hoping Madeleine will see her mistake and understand herself better later on," Mina told her mother in June, 1911. "I am sick over the whole affair but if it is to be I am praying for grace." Madeleine was studying Catholicism, though she felt herself too much of a free spirit to convert.

"I am sure it will make poor Thomas very unhappy," Mina shook her head, though her own feelings were stronger than those of her husband, who was more against religion in general than any in particular. "I am sure that it will make a difference all around. If only some very agreeable protestant American could rush her enough as I think Madeleine is first infatuated with the idea of having a little babe. A dear little mother she will make but I would not want the little thing to look like the Sloanes."

Mina gave ground grudgingly, hoping all the while she could kill the romance by procrastination. Not until June, 1914, were Madeleine and John Sloane married by a monsignor at Glenmont. Mina, of course, staged a splendid wedding. The drawing room was transformed into a chapel filled with hundreds of orchids from her greenhouse, and an orchestra played at the lawn dance afterward.

Madeleine would have liked to have her husband associated with her father's business. But Edison wanted no more relatives working for him; so John Sloane concentrated on the Sloane Aeroplane Company and School of Aviation.

Theodore, the youngest of the children, was a bright, affectionate, introverted boy. Once, when in knickers, he created panic at a party by unleashing his pet alligators among the guests. His nickname was Baba, and when he was small he loved to have Mina climb in and "cuddle" with him when he went to bed. At the age of fourteen he emulated his father by acquiring a printing press and starting a newspaper, *The Edison Works Weekly*. He editorialized: "When you get real mad at anyone, dictate a letter to him on a talking machine; tell him exactly what you think of him—and then smash the record." Like most teenagers he was fascinated by cars, and got into periodic scrapes

—which Miller Reese Hutchinson helped get him out of by taking care of his tickets.

Charles, who had been born in 1890, was Mina's favorite. Called Toughy, he had inherited the Edisonian propensity for ear ailments and inaptitude for formal education, but had the looks of a movie star. Like his father, he began growing deaf at an early age. When he was eighteen he underwent an operation for mastoiditis. His grades at Hotchkiss, the exclusive Connecticut prep school he attended, were abominable. Sociable and good-natured, he disliked "grinds" and cut classes. A few days before he was to graduate, he was in danger of not completing his courses. "Why Charlsie how embarrassed we should be to get up there and not have you get your diploma," Mina pleaded with him. "Papa is coming up too. Oh, Charlsie, don't disappoint us. That would be terrible."

In an age when grades counted less than a father's checkbook, Charles enrolled at MIT. Girls competed for his attention, and he regarded them as "succulent morsels." He played the piano well, and composed extemporaneously. He liked drama and poetry, and wrote good verse himself. A member of the polo set, he went to hops, movies, and ball games more often than to classes. He had an allowance of $125 a month, but periodically ran short of money, and would write his father for "a little check," or occasionally "a big check." Edison kept a supply of plaid-patterned cloth in his desk, and when he tired of the requests responded by sending Charles a square of small-patterned plaid for "a little check" and large-patterned plaid for "a big check."

Though Edison hated cigarettes and Mary Valinda Miller was a charter member of the WCTU, Charles smoked constantly and, like his half brother Tom, acquired an early taste for alcohol. Mina pleaded with him: "Please let the drinks alone. Oh, dearie, if you should let that monster get the better of you it would break both your father's and mother's heart." But Charles was hooked. He fought a lifelong, often losing, battle against alcohol.

Charles was the only child who really got along with his father, and Mina smoothed the way for him. From the age of fifteen, Charles was one of Edison's chauffeurs. Since Charles had neither scientific nor inventive bent, he was not a threat to Edison, who, once he made up his mind, would tolerate no contradiction, and coiled with rage at the thought of an argument with someone who might know more than he. When Charles dropped out of MIT after less than three years of none-too-serious attendance, Edison was unperturbed.

The summer following Henry Ford's appearance at the laboratory, the Edisons drove through Canada to visit the Fords in Dearborn. Charles and Edsel, Henry's son, liked each other. Both felt more comfortable away from their fathers' domination, and they began to have visions of going into business together. Then, in late 1913, Charles went to visit the

Uptons, who had moved to the West Coast. Coming down with California fever, he involved himself in the Ocean Pier Amusement Company in San Francisco, and elsewhere took an option on 1560 acres of land. He asked Hutchinson to join him in putting together a syndicate to buy the land for $66,000. But Hutchinson, playing the role of older brother, helped Mina and Edison persuade Charles to abandon the speculation.

Mina, who grew plump and matronly early in life, hated the sticky, hot New Jersey weather, to which Edison seemed more and more immune as he grew older. Even the long Sunday drives that Edison loved were primarily journeys of obligation for her.

Sometimes she invited friends to the house on Sundays as a means of diverting him from the outings, but it made no difference to him.

"Thomas says he is going out in the car just the same," Mina told her mother. "He makes out he is so abused if we ask him not to go I have not the courage to keep him."

He did things in streaks—during one period he insisted the family accompany him to vaudeville shows in Newark two or three times a week. Mina, who thought the entertainment vulgar and whose conscience reminded her she should be at a prayer meeting, averted her eyes when they passed the church.

She acquired a place on Monhegan Island, off the coast of Maine, and spent much of the summer there or in other northern climes, like Banff, leaving Edison behind in the laboratory.

During 1912 Mina was away more than at any previous time in her married life—her mother was terminally ill, and Mina remained with her in Akron. She liked to feel that she was missed at Glenmont, but suspected her husband hardly knew she was gone. Edison, addressing her as "Darling Partner" and "Darling Billy," signed himself "Your Lover forever" and tried to reassure her:

"Now Billy darling dont be a poet, too sentimental—you must know that I love you to pieces & be satisfied & do not always complain that I am not loving you Enuf—"

Mina, in her fantasies, still waited for the day Edison would be transformed into as passionate and single-minded a lover as he was an experimenter.

"I wish I had a big love," she declared wistfully. "Everybody is so good to me but my heart is so pinched."

Even Charles felt compelled to reassure her: "You've got a beau that loves every inch of you, even though you try to persuade yourself he doesn't—In your heart of hearts you will have a mighty hard time to ever doubt it. Give your heart of hearts a chance. It knows more than all the minds in the world put together."

Mina's mother died in October. Except for Mina, none of the surviving children was well off, and they fell to squabbling among themselves over the $70,000 estate. (Ira had been borrowing money from Mina for some time, and owed her several thousand dollars. Robert, the postmaster in Ponce, Puerto Rico, had once more come under suspicion over his financial dealings, and had died shortly after being dismissed from office in 1911.) The only family member everyone trusted was Mina, so she was appointed executor of the estate.

In the months that Mina was away, Edison was at the laboratory nearly all of the time. He was attempting to alter the battery so that it could be used with an ignition system in the Model T, he was bringing out the disk phonograph, he was trying to mate the phonograph with a movie projector, and he was still plodding along on the project to cast concrete furniture and houses. Hutchinson competed with him to see who could stay up the latest. Edison headed what came to be known as "the Insomnia Brigade."

When Edison needed something done, he wanted it done immediately, be it two o'clock in the afternoon or two o'clock in the morning, and he thought nothing of calling men out of bed to come down to the lab. Five or six hours of sleep, his own nightly ration, was enough for any man, he contended, though he took no account of the frequent naps he engaged in at the laboratory. It was his habit, while waiting for a model to be made or a man to carry out an assignment, to lean back in his chair, cross his feet on his desk, spit, think, spit, think deeper, spit again, nod off, jerk his head up with a start, then gradually drop into a sound sleep. One time when he was driving Fred Ott ragged, Ott, sticking his finger under Edison's stentorious nose, wailed: "Look at this son of of a bitch. He tells the world he never sleeps, but he is fast asleep like this pretty near all the time. He just don't believe in nobody else sleeping!"

William H. Meadowcroft, who had replaced Randolph as Edison's secretary, complained: "Edison can find work for you in three minutes that will take three months to accomplish." After less than two years on the job, he collapsed and had to take a long vacation.

Frank Dyer found that occupying the company presidency was like sitting on a wild carnival ride with Edison throwing the switches. Dyer was an Ivy League country clubber, and suffered from a psychosomatic appendix that periodically sent him to sanitariums for long cures.

Edison was deriving $88,000 in salary and $300,000 in profits from Thomas A. Edison, Inc. (The profits did not include the cost of running the lab.) But the expenses of introducing the disk phonograph and the kinetophone, combined with the slump in the phonograph business and Edison's huge outlays for experimentation, presaged a shortfall of $800,000 in income versus expenditures during the coming year—a deficit that was providentially made up by Henry Ford's advances. Edison refused to take Dyer's advice on such matters as the takeover of the Columbia Phonograph and American Graphophone companies. When Edison returned from his mother-in-law's funeral, he and Dyer had a bitter quarrel. Dyer's health gave out on him again, and he resigned. Edison took over the presidency of the company himself.

New Year's Eve, 1912, Edison and Hutchinson were working on the third floor of the laboratory. At midnight, when the whistles started blowing, Edison grabbed the horn of a kinetophone, put it to his ear, and went to the window to listen. It was a beautiful night, with the thin coating of snow on the ground shimmering like ice. Edison seemed very lonely, and Hutchinson suggested they go to the Fireman's Ball.

Edison shook his head. Returning to his experiments on electroplating disks, he wiped potassium cyanide over a record. About quarter past twelve he became dizzy and nauseated from the fumes, and decided to go down into the library to take a nap. He had, in reality, come close to being fatally poisoned, and remained two days in bed at Glenmont. When Edison resumed his vigils with Hutchinson, Mina could bear it no longer. She did not, of course, know of Hutchinson's ambition to be heir to the lab. But Edison was sixty-six years old, and it seemed clear to her that he could not sustain the pace Hutchinson was driving him to. Hutchinson's influence was growing, and threatening to complicate her plans to have Charles succeed as head of the enterprise. On the evening of Friday, January 24, she called Hutchinson up at the laboratory and asked him to come to the house for a talk. He refused. This was a miscalculation, and before the night was out Mina had him standing on one of her Oriental carpets at Glenmont.

"We understand each other now," Hutchinson mused afterward like a chastened Satan bested in a match for the Faustian soul of Edison.

While Mina could make an impression on Hutchinson, her attempt to have Edison curtail his activities failed. Involved in business affairs as well as experimentation, he continued his midnight hours. Mina, seeing no hope of getting him home, went down to the lab and followed him around with a pillow and quilt, which she spread out for him wherever he happened to lie down. By October, 1913, he was exhausted. Suffering a nervous collapse, he did not recuperate until after his Florida vacation the next spring. Mina convinced him to let Charles handle his correspondence, and Hutchinson was given strict instructions that while Edison was at Fort Myers he was not to be upset by reports from West Orange, no matter how bad the situation was.

It was, indeed, so bad that Hutchinson informed Charles: "I am afraid it is going to make the Old Man sick." Before leaving, Edison had insisted on having his own plans rather than an architect's used to construct an addition to the Silver Lake plant, and the result was, just as it had been at the ore mill, that the whole structure had to be torn down and new foundations put in.

Furthermore, Hutchinson reported, "Things have gone rotten since his departure in the disk. What on earth am I going to do when he writes me wanting to know why in Hades he doesn't hear from such and such a thing? He would certainly jump on the train if he knew the full extent of the trouble in Silver Lake."

Then there was worse news. The laboratory, the chemical and phonograph works, and the motion-picture division all worked with highly volatile materials. Fires broke out repeatedly, but had always been caught at an early stage. On March 28, however, a rheostat overheated in the Bronx studio, and before the flames could be checked the building was gutted.

One director, exploiting the debacle, filmed the fire and set about writing a scenario. But, as it developed, it was only a preview.

"Cigarettes, they're deadly," Edison maintained. "Tobacco is not harmful after a man is fully formed, but the use of paper and tobacco together in the form of cigarettes is extremely harmful and causes degeneration. It seems to affect not only energy but the moral sense." Once, when he found a pack of cigarettes in the lab, he pinned it on the bulletin board with the note: "A degenerate, who is retrogressed toward the lower animal life, has lost his pack." (When Edison returned some time later, the cigarettes were gone and a cigar had been substituted.) Consequently, many of the men sneaked smokes in out-of-the-way places, and these were often full of combustible materials.

At quarter past five on the afternoon of December 9, 1914, fire was discovered in a film-inspection booth. The man present tried to put it out himself, and by the time he called for help flames were sprouting from all sides of the old wooden building. Workers carried flammable materials from nearby buildings to structures on the periphery, which were considered safe. The fire was in the factory portion of the complex, not near the laboratory, and was centered on the old wood-and-brick phonograph buildings Edison had put up in 1888. These had no fire walls. The fireproof vaults turned out not to be fireproof at all. Sparks set the roofs of nearby buildings afire, and the flames leaped from upper story to upper story. Chemicals, oils, celluloid, and wax created a spectacular spectrum of flaming color, punctuated by sporadic explosive puffs. The Orange fire department appeared, but it was undermanned and the mains in the area were entirely inadequate for the huge industrial compound Edison had created. Edison's own water tanks were three quarters empty. The plant's fire chief and the town's fire chief didn't talk to each other, and there was no cooperation between the men.

By seven o'clock, more than a half-dozen buildings were crackling. The light thrown against the sky could be seen as far away as the Passaic River. Fire departments from Newark and surrounding communities rushed to the scene—only to stand helpless for lack of water as the whole system of mains collapsed. The materials carried out of the central area fueled the flames as the peripheral buildings became involved.

The fire, on the other hand, inexplicably passed over 2000 gallons of high-proof alcohol. Two hundred thousand dollars' worth of phonographs were saved. The diamond disk master molds and presses were not injured. Edison watched the flames from the storage-battery building, which was undamaged. Thousands of spectators lined the streets. Not until 11:00 P.M. was the fire controlled.

Edison stayed to receive a preliminary damage report, then went up to Glenmont at 1:00 A.M. The laboratory was intact, but thirteen factory buildings had been destroyed or gutted. Initial estimate of the loss was a million dollars.

Edison did not go to the laboratory the next day, but after a long sleep applied himself to pondering the task of reconstruction. Henry Ford, who was in New York, came out to see him. Firemen were still stalking through the ruins, dousing embers, but the total damage was not nearly as bad as first feared. In the more modern steel-and-concrete structures, flammable materials had been consumed, but 88 percent of the structures and 85 percent of the machinery in them had been saved.

Someone brought Edison a photograph of himself that, although it had been in one of the destroyed buildings, had escaped unscathed. "Never touched me," he scrawled across it.

Focusing searchlights on the area, Edison set 2000 of his 6000 employees working around the clock to clear the rubble. Insurance companies paid $220,000 for the stock that was lost. Ford lent Edison an additional $100,000. One week after the fire, the debris was gone. On January 22, 1915, the Phonograph Works went back into production.

CHAPTER 38

Prometheus

Edison still responded to a crisis such as that of the fire with verve—he had been crisis-oriented all of his life. But he was more and more easily fatigued. By October he looked haggard, and Mina was glad when the time came for them to board the private railroad car that was to take them to the West Coast, where an "Edison Day" was scheduled at the Pan-Pacific Exposition. Edison continued to be pepped up by the sight of locomotives and the feel of wheels clicking across the rails. "Papa looks better already," Mina wrote to Charles as the train lanced across the prairies.

It was Edison's first major sortie outside his usual haunts since he had reluctantly returned to Europe in August, 1911. Mina had been badgering him to take her again since the middle 1890s, but he had kept putting her off.

Finally, he sighed: "I am now cornered and must carry out my promises."

They had returned to Paris, where Edison caused a stir by standing before a modern painting, ejaculating, "Punk!" and shuffling off. He hired a Daimler and a chauffeur, and off they went on a grand motor trip of central Europe. They traveled, often on dusty roads, through the west and south of France, then on through the Swiss Alps. In Switzerland they met Dot, and Edison was introduced to her husband. Before they parted, Edison promised her $2000 to buy a car. Looping through northern Italy, they went to Vienna, then to Budapest, where they were greeted by Francis Jehl, who had worked on the electric light at Menlo Park.* Edi-

* Jehl had gone to install the electric lighting system in the theater in Brünn, Moravia, in 1882. He married there, and had remained in Europe.

son, appalled by much of what he saw, created a verbal dust storm:

"Of all the follies of man that great palace in Budapest is the worst," he declared. "Millions are wasted there in artistic things and baubles. Yet 50 miles from it I saw a woman in harness with an ox pulling a plow." After visiting a monastery in Prague, he remarked: "No one in the world knows where he came from, why he's here, or where he's going. Religion is a hopeless piece of insanity."

Mina scrunched down lower in the back seat, and off they rattled to Germany. Edison admired German industry but deplored the architecture. He praised the Kaiser but called the King of Greece a stock gambler. He liked French wines and German beers, and told a writer for the *Literary Digest*: "Natural wines and beers do not lead to drunkenness; on the contrary, used exclusively, they take the place of alcoholic poison and produce a temperance people." He affirmed the statement to the *American Brewers' Review*. But when a temperance organization called him to task, he declared it "a perverted quotation. I stated that while the wines of France and beer of Germany had only 3 or 4 per cent alcohol, they drank so much that they managed to get enough alcohol to stupefy them all right."

In Berlin, Edison was welcomed by Sigmund Bergmann, who had risen to the position of confidential counselor to the Kaiser. Bergmann's immense concern now outstripped Edison's works. Edison, in fact, could claim a relationship with all three of Germany's electrical giants—he had long had close ties with Emil Rathenau of German General Electric, and Johann Schuckert, the head of Siemens-Schuckert, had worked in his Newark shop in 1870–71. (Edison, however, had not recognized him when they bumped into each other in a London fog in 1873.)

Since Edison's triumphal tour in 1889, honors had poured in on him. He had received the Albert Medal of the British Society of Arts in 1892, the Rumford Medal of the American Academy of Arts and Sciences in 1896, the John Fritz Gold Medal in 1908, the Rathenau Gold Medal in 1913, the Civic Forum and Franklin Institute gold medals in 1915, and numerous lesser decorations. Princeton made him a Doctor of Science and New York University a Doctor of Laws. The great men of the world were honored by his presence. Columbia University president Nicholas Murray Butler, who presented the Civic Forum Medal to him, had been one of the investors in the original Edison Speaking Phonograph Company, and was to be a Democratic candidate for President in 1920. When Insull staged a dinner for Edison in Chicago in 1912, the guests included the governor of Illinois, the mayor of Chicago, future Vice-President of the United States Charles Dawes, inventor Cyrus H. McCormick, General Electric president Charles Coffin, and thirty-eight publishers, chairmen, and presidents of corporations.

Edison's fame was greater perhaps than any other person's in the world. A North Carolina man made a bet that a letter with nothing but

Edison's picture on the envelope would be delivered to him. It arrived at the laboratory nineteen days after it was mailed.

Only in the environs of West Orange, where his achievements were screened by the smoke pouring from the stacks of the plant, and the luster of the light was tarnished by the continual struggles over money, was Edison not especially admired. His factories were excruciating generators of noise and pollution—one petition protesting the poisonous fumes emitted by the Silver Lake plant carried 1500 signatures and was sixty-five feet long. One man threatened to kill Edison if the plant were not cleaned up. The Civics Club of the Oranges filed a complaint. Bloomfield residents formed an association to take action, but could get nowhere with the Board of Health—Edison, through Gumshoe McCoy, was paying off every public official of importance. Assessors were particular targets. "We can keep our assessments down by having an understanding with the Assessors," McCoy had reported in 1889, regretting there was no way Edison could obtain an exemption from taxation altogether. "The Assessor for Bloomfield Township called to see me the other day and assured me that he would keep the assessments down to the lowest possible point. We do not want to be at the mercy of these Assessors, we ought to have a proper understanding with them, and make them our friends."

Edison was only adhering to the pattern of any shrewd industrialist of the period, but that did not endear him to the residents of Silver Lake and West Orange, who every year saw the towns' principal industry assessed at a small fraction of its value while they struggled to maintain schools and public facilities. When a proposal was made to the West Orange Board of Education in 1911 to name a new school after Edison, the resolution received only one affirmative vote.

That was but one more example of the dichotomy between Edison the industrialist and Edison the inventor. The kudos were always for the inventor; the brickbats for the manufacturer. No one could improve on Henry Ford's declaration that Edison was the world's greatest inventor and worst businessman.

Since their meeting three years before, Ford had been behaving like a planet that has adopted Edison for its sun. He visited Edison at Fort Myers, sent him a Model T for Christmas every year, and became his escort when Edison arrived in San Francisco for the exposition. They went together to Luther Burbank's farm in Santa Rosa, where Edison wrote in the guest book: "Thomas A. Edison. Business: Inventor. Interested in Everything." They arranged to meet their wives at the Japanese Tea Garden at noon, but did not show up until 3:00 P.M.

"So you see, he is like all other husbands," Mina generalized. "He forgets, but he expects me to understand."

The San Francisco telegraphers staged a huge dinner for him. All the men had telegraph keys by their plates, and the only communication permitted was by clicks. Edison lowered his head close to his instrument as the messages came in: "Mr. Edison, we forgive you for inventing the quadruplex"; "Mr. Edison, your Morse is copper-plated"—the highest praise one telegrapher could offer another. Halfway through the dinner a waiter served Edison a mammoth apple pie and a bottle of milk. All the while 50,000 people waited in the fog on Market Street to catch a glimpse of the inventor.

One day 60,000 schoolchildren paraded before Edison in the Court of the Universe. During the past twenty-three years, while Edison had been almost totally dissociated from electrical development, the cities of the world had been transformed into tiaras of light. The beaconed hills of San Francisco shimmered in the mirror of the bay—no setting could have dramatized more clearly the evolution of darkness to beauty. Edison's remoteness from the electrical industry gave him a stature contrasting sharply with his image in the phonograph business—he had become the Prometheus of light. From the crest of Nob Hill Edison looked down on a city ablaze in his honor. For an hour rockets shot up like pillars of flame from the Hearst Building, disappeared into the fog, then burst into myriad shooting stars. Suddenly Edison saw his own face outlined against the black arch of night—high in his hand he held a glowing incandescent lamp. In the constant struggle at West Orange he had had little opportunity to reflect on how the world regarded him. In 1889 he had been honored; in 1915 he received a public ovation.

"This is the greatest thing that ever happened to me," he declared.

Six years before, the commissioner of public health of Queensland, Australia, had suggested to Edison the production of educational films.

"The introduction of kinetoscope pictures in schools would be an epoch in the common school," Edison had responded. "You couldn't keep the children away even by the police. Knowledge and the school would then be an attraction."

Following up, Edison had organized an educational film division, and even started up experimental classes in which motion pictures were the only learning material used. "I am spending more than my income getting up a set of 6,000 films to teach the 19 million children in the schools of the United States to do away entirely with books," he said; and from San Francisco he traveled down the West Coast, visiting schools everywhere he went.

He was greeted by thousands of children throwing flower petals in his path. He spent one night at the Lick Observatory on Mount Hamilton, and visited Stanford University, whose president, Dr. John Branner, reminded him that he was one of the men who had gone out hunting bamboo.

"Why, bless you!" Edison, who had quite forgotten, exclaimed.

At Universal City, in the San Fernando Valley, Edison laid the cornerstone for a new, electric motion-picture studio. It was, in reality, as if he were placing the tombstone on the freshly dug grave of his own film enterprise. It was the year D. W. Griffith produced *The Birth of a Nation,* which was to become Edison's favorite film. The product whose commercial value Edison had questioned had developed into such a huge attraction that its stars had greater incomes than Edison—Mary Pickford was receiving $800,000 and Charles Chaplin $670,000 a year. (Edison himself paid Marguerite Clark $250,000 a year, or $25,000 a picture, though the total cost of each film was only $50,000.)

The activities of the Motion Picture Patents Company had driven the vast majority of talented people to Southern California. On October 15, 1915, after Edison started on his trip west, the New York Circuit Court ruled that the MPPC was a violation of the antitrust law. The General Film Company was already heading toward bankruptcy—in 1916 it lost more than a half million dollars, and had an indebtedness of $2.5 million. Edison and the other members of the trust were completely out of step with the trend toward features of five or more reels, which were replacing the one- to three-reelers that had been the staple.

Edison's executives moved desperately to disentangle themselves from the alliance. They negotiated an agreement to release one film a month through Paramount Pictures, but the deal fell through when there was a corporate shake-up at Paramount. They talked with producers in California about shifting filming to the West Coast. Charles Edison, especially, was anxious to keep the division going—backed by his mother, he started the Thimble Theater at 10 Fifth Avenue to hobnob with actors and singers and scout for talent.

But by January, 1918, the kinetoscope division was spending two dollars for every one it was earning. On Wednesday, February 6, filming concluded on a six-reeler, *The Unbeliever,* and the decision was made to shut down production permanently on the following Saturday.

Edison did not take the loss too hard. During the months that the kinetoscope division was in its final decline, he was spending little time at the laboratory.

The outbreak of war in Europe in August, 1914, had had an immediate effect on Edison, dependent as he was on a variety of German chemicals. No one, however, had a better background to cope with the crisis. He himself had been experimenting with chemicals for forty years, he had a staff of chemists at the lab, and all the experiments on materials for records had already resulted in some by-products.* When the supplies from Germany were cut off, Edison initiated production of

* In 1912 he, Aylsworth, and Dyer had formed the Halogen Products Company to market the flameproof and moistureproof halowax discovered by Aylsworth.

his own carbolic acid and phenol, used in pressing records. By the end of the first week in September, he was ready to manufacture in quantity. To replace German benzol, he built the first benzol plant in the United States at Johnstown, Pennsylvania, then added another in Bessemer, Alabama. He supplied carbolic acid to the Monsanto Chemical Company, shellac to General Electric, and pyrax to Goodyear. He synthesized the aniline dyes on which Germany had held a monopoly, and was swamped by the worldwide demand. His entire production for 1915 was sold before the year was half over. His plants at Silver Lake, Johnstown, and Bessemer furnished not only commercial customers but the American army and navy with critical supplies of exotic chemicals ranging from acetyl paraphenylenediamin to xylol.

When Edison had returned from Germany in 1911, he had been full of praise for German business and German government. "The best government in the world is that of a benevolent despot of great mental capacity, of which Emperor Wilhelm of Germany is a type," he had said. He preferred German cartelization to American competition. "We have been compelling people to compete. Competition results in the destruction of weaker concerns and the control of the trade of the country by the stronger. The time will come when a few individuals of great concerns will control the country. Competition of this kind is war. It means death to the weaker. Cooperation means life. Germany has legislated to produce prosperity. Our legislation has resulted in depression. We waste where they economize."

After the outbreak of the war, he could see no advantage accruing to either side: "If they are not crazy over in Europe, I don't know what crazy men are. Nothing short of the Almighty can stop this war. It is necessary to settle for all time that Dynastic Militarism shall disappear from the earth."

His ties, however, were primarily with Germany. (Dot's husband was an officer on the Western Front.) His expressions had a pro-German slant until the sinking of the *Lusitania,* which he called a horror. A few submarines on both sides were equipped with alkaline batteries, but Edison said he would not apply himself to the production of any implement of war except in the service of the United States. The navy, he suggested, was badly in need of developing antisubmarine weapons, and the armed forces in general would benefit from greater inventiveness. On July 7, 1915, Secretary of the Navy Josephus Daniels wrote him confidentially:

"I think your ideas and mine coincide. There is a very great service that you can render the Navy. One of the imperative needs at the Navy is machinery and facilities for utilizing the natural inventive genius of Americans to meet the new conditions of warfare, and it is my intention to establish a department of invention and development. You are recognized by all of us as the one man above all who can turn dreams into realities. What I want to ask is if you would be willing to act as an adviser to this board."

Hutchinson, actually, had done considerable maneuvering to prompt the offer, seeing in it an opportunity to cement the ties between Edison and the navy, and thus increase battery sales. But Edison's own relations with the navy went back thirty years. In 1886, he had joined with Charles E. Munroe, chief of chemistry at the U.S. Naval Institute in Newport, Rhode Island, to develop a torpedo.* In conjunction with Gardiner Sims, who was building steam engines for his power stations, Edison had formed the Edison-Sims Electric Torpedo Company to construct weapons, torpedo boats, submarines, dynamos, and other naval machinery. He had designed what was probably history's first remote-controlled weapon—a torpedo with a maximum range of one mile and a guidance system worked by cable. Five of the devices were tested between 1886 and 1890, but their cost—$5000 apiece—and limited utility had brought the project to a close.

In the fall of 1915 the Naval Consulting Board was formed, with Edison as chairman. Hutchinson and Frank Sprague were among the scientists and inventors included as members.

The appointment caused Edison to momentarily reconsider his lifelong Republicanism. He had had some acquaintance with most of the Presidents and Presidential contenders since 1878. William Jennings Bryan had made several records for the Phonograph Company, and was anxious to give his golden tonsils further exercise, but Edison demurred: "We don't want any records of his just now if ever—they never sold." In December, 1911, Edison suggested to President Taft that the President could use the kinetophone to film and record his speeches in the forthcoming election, thus being able to campaign in fifty cities simultaneously.

A few months later, however, Edison and Mina decided to support Theodore Roosevelt and the Bull Moose Progressives. A suffragette offered to speak at the laboratory; but Edison, although he supported women's right to vote, was dubious about the effect of women's freedom to speak.†

Mina, generally conservative, didn't approve of the militancy displayed by women. "The American man will see that the American woman has what she asks for in a reasonable way. In fact, she gets what she asks for whether she is reasonable or not."

Edison, though engaging in monopolies and deploring competition,

* Edison was introduced to Munroe by his staunch advocate, Professor George Barker of the University of Pennsylvania, whose daughter Munroe married.

† A straw poll of the men at the laboratory found thirty-three supporting Wilson, twenty-four Socialist Eugene Debs, seven Roosevelt, and six Taft.

tapped the philosophical liberal in himself to speak out against the businessman, and suggested that the trust-busting proposals of the Progressives should be made even stronger.

In May, 1916, Daniels enlisted Edison to march in the administration-sponsored Preparedness Day parade up Fifth Avenue, and a few weeks later arranged for him and Henry Ford, a Democrat, to have lunch with the chairman of the Democratic party. When a friend remonstrated with Edison for being drawn into the Democratic orbit, and pointed out several Wilson blunders, Edison retorted: "Perhaps, but I notice he always blunders forward."

Edison still had a good portion of the happy hooligan in him, and when he and Ford got into a conversation about their respective physical prowess, Edison pointed to a chandelier with numerous globes in the center of the room and said: "Henry, I'll bet you anything you want to bet that I can kick that globe off that chandelier." Whereupon he wound up like Nijinsky and with an over-the-head kick shattered the globe to pieces.

Ford, trying to duplicate Edison's acrobatics, missed.

"You are a younger man than I am," Edison gleefully jabbed at Ford during the meal, "and yet I can outkick you."

If Daniels expected Edison to offer the sage advice of an elderly inventor or preside at circumlocutious board meetings, he was, of course, misled. During 1916 the Naval Consulting Board accomplished nothing, but after Wilson's reelection relations with Germany worsened. Edison decided to plunge in. Announcing he would devote the entire facilities of the laboratory to the armed services, he gathered a staff of young engineers and volunteers from colleges, and charged at the navy like a torpedo.

One idea after another emanated from his head: sound ranging to find hidden guns, oleum cloud shells to blind and suffocate the enemy, torpedo nets to be let out by ships at the sound of a torpedo, a water-penetrating projectile, an underwater searchlight, and a chain of submarine-detection buoys stretching down the Atlantic Coast. Each buoy, occupied by three men, would contain sonar apparatus and wireless, and be able, like a submarine, to surface and submerge.

The inventor was concerned about the rapid deterioration of the rifling in the barrels of heavy guns, and devised a smooth-barreled weapon with turbine-headed projectiles, and shells that put out fins after they were shot. The navy lent him an old one-pounder, and Edison, who early in life had several times come close to blowing himself up, had a great time with it.

After the American declaration of war, the White House assigned four Secret Servicemen to Edison, though whether this was to protect Edison from the Germans or the navy brass from Edison was open to debate. In the late summer of 1917, the navy placed a 200-foot converted yacht at his disposal, and Edison went to New London, Connecticut. He asked Mina to come with him, and they lived in cramped quarters aboard the ship.

"I detest it on the boat and long to be home," Mina informed Charles. "I wish I knew just how much and what Papa wants me out here for. To father it does not seem to matter too much but home does mean a tremendous lot to me. The more cluttered the place the better contented father seems to be. I could kill Hutchinson for ever getting him into this mess."

On the waters of Long Island Sound, Edison tried out a sea anchor which he and his team had invented to enable ships to turn quickly at sight of a torpedo. The anchor, it developed, had no effect on a ship's turning radius, but worked marvelously to bring the vessel to a dead stop—which did not help matters at all!

Through the fall, Edison, suppressing his queasy stomach, plied the Sound. "It looks like a winter's job as far as father is concerned, as you know father," Mina noted to Charles. "He constantly gets new ideas and that leads to more experimenting and halfway never counts with him. It looks hopeless to me and I get more and more discouraged with that Navy proposition. Water being one of the things he never cares for. Is it not the perverseness of life to always get the thing you think you would never stand for."

In early November Edison asked for information on the location of ships that had been sunk, and was told that no such statistics had been compiled. The British always used the same shipping lanes, and Edison theorized that Germans were capitalizing on the British habits. Taking three men, he went to Washington and spent two months examining the records. From these he compiled detailed strategic shipping plans, and proposed they be tested in comprehensive war games of ships versus submarines.

Edison's confrontation with the navy pitted an irresistible inventor against an immovable bureaucracy—Edison could think only of why things should work, the admirals solely of why they wouldn't.

"He was suffering keenly from disappointment," Edison's secretary, Meadowcroft, remarked to Mina. "Only those who know him well realize how greatly he was discouraged. I am looking forward to the day of Mr. Edison's triumph over bureaucracy whose reports have ice water instead of red blood in their veins."

In February, 1918, Edison, dropping Mina off at Fort Myers, moved his operations to the naval station at Key West. He took with him as an assistant Theodore, who wanted to join the army, but whom Mina was determined to keep out. Edison found the atmosphere in Key West more congenial than in New London or Washington, but in practice he accomplished little.

The West Orange laboratory developed a bombsight, and Edison, who long had maintained "The helicopter is the only correct principle," organized a company to manufacture helicopters. But the inventor, with his futuristic notions,* was as out of place as Buck Rogers in the naval establishment. He submitted forty-two projects. None was accepted. A year after the end of the war, he wrote Secretary Daniels: "I do not believe there is one creative mind produced at Annapolis in three years. When you are no longer Secretary I want to tell you a lot of things about the Navy that you are unaware of."

* Thirty years before, at the time of the AC-DC controversy, Edison had suggested that an enemy force could be decimated by spraying it with a jet of water laced with alternating current.

CHAPTER 39

The Freethinker

By August of 1918 Edison had lost all illusion that his work would have an impact on the war, and looked forward with relish to his annual two-week auto camping trip. He had gone on the first of these in 1916, after Ford had introduced him to naturalist John Burroughs, a childhood pal of Jay Gould. Burroughs had related his experiences as a camping companion of Theodore Roosevelt, and Edison had enlisted him for a tour through the Adirondacks and New England. They had gone out again in 1917, but the 1918 tour was to be more ambitious. Edison planned to retrace his 1906 route through the Blue Ridge Mountains, and invited Ford, Harvey Firestone, and Edward Hurley, a high official in the Wilson Administration, along as well.

"The invitation to go gipsying with Mr. Edison and Mr. Ford appeals to me," the eighty-one-year-old Burroughs wrote Hurley. "I have a pure strain of the vagabond in me. If you went along to tell fortunes we might pass as a real gipsy troop. Mr. Edison could play the magician and Mr. Ford the watch and clock tinkerer. I do not know about Mr. Firestone, maybe he could pass as an umbrella mender."

Meeting in Pittsburgh, they departed like a caravan of sheikhs in three large touring cars followed by three trucks and a retinue of servants. (Edison was the owner of nine cars, seven gasoline-powered and two electric.) A Japanese chef headed the kitchen staff. Even lunches under the trees were sumptuous repasts served on white linen with solid silver ware. The tents were equipped with electric lights. Only flush toilets were lacking.

Edison insisted it was his prerogative to pick the campsite nightly. Between Horse Shoe Run and Lead Mine, West Virginia, Edison chose what appeared to be an isolated spot amid the woods. The group had just sat down to dinner when a bell clanged through the darkness and a train steamed up only a couple of tree trunks away. A horde of lumberjacks materialized like hungry locusts from a logging railroad. Gathering in the penumbra of the lights, they stared silently at the elegant diners.

The next day the campers took possession of the locomotive. Edison assumed his spot atop the cowcatcher. Ford—who had started his career as a steam-engine mechanic—took the throttle, and Firestone became the fireman.

Word spread around the countryside about the procession of potentates, and when they reached Parsons, ten miles distant, half the people in the vicinity were lining the street. Edison bought all the chocolate in town to fortify himself against that evening's spartan dinner at the Elkins Country Club. Newspapers caught hold of the story, and proclaimed them to be "Five Wizards in the Wilds." At Johnson City, Tennessee, a huge crowd greeted them as if they were the wise men magically appearing out of the East.

Edison chipped rocks with a hammer, and inspected the fragments for gold and cobalt. He and Ford investigated every brook in the interest of harnessing water power for industry. At Weaverville, North Carolina, Firestone gave a speech, as if they were on the campaign trail. (Ford, in fact, ran for the U.S. Senate on the Democratic ticket in Michigan that fall.) Twenty-three hundred miles after starting out, they ended the journey in Hagerstown, Maryland.

The next August, Ford was missing, but Edison, Burroughs, and Firestone motored through the Adirondacks. They were camping by an abandoned farmhouse near Saratoga Springs when three women suddenly popped up and began berating them. Firestone tried to calm them.

"We're responsible people," he said. "My name is Harvey Firestone. The man over there, reading the paper, is Thomas A. Edison. And—"

"I suppose next you are going to say," one of the women snapped and pointed at Burroughs with his long, white beard, "that man is Rip Van Winkle."

Burroughs, who noted a slight disparity between Edison's pronouncements and his behavior, chronicled in his journal:

"Oh Consistency, thy name is not Edison.

"10 a.m. Edison not up yet—the man of little sleep. He inveighs against cane sugar, yet puts two heaping teaspoonfuls into each cup of coffee, and he takes three or four cups a day. He eats more than I do, yet calls me a gourmand. He eats pie by the yard and bolts his food."

Burroughs died in 1921, and that summer the party consisted of Edison, Ford, and Firestone. At a campground outside Hagerstown, they were joined by another well-known sportsman, President Harding. By then, even the women in the families were permitted to share the hardships of the wilds.

In 1923 Edison and Mina went camping in the north woods of Michigan. Edison's health, however, was deteriorating, and he no longer felt well enough to take long auto trips. During the summer of 1924 the Edisons, the Fords, and the Firestones combined for a less adventurous journey through New England. At Plymouth Notch they stopped off to see President Coolidge. Ford had been mentioned as the Democratic nominee to oppose the President, but had relinquished his none-too-plausible candidacy, and camaraderie flourished as Coolidge and his guests drank whey and nibbled cheese during the warm afternoon. From Plymouth Notch, Edison and Mina went alone to retrace their path through the White Mountains, then rejoined Ford on his yacht off Portsmouth, New Hampshire.

Through the war years and immediately thereafter, Edison took little active part in the operation of Thomas A. Edison, Inc. Although Charles H. Wilson was the general manager, Stephen B. Mambert, an efficiency engineer who had joined the company as financial executive in 1914, exercised the principal power. The disk-record division was the weakest in the company, and Mambert raked the executives: "We have men who are drunkards—absolutely soused the greater part of the time. We haven't the slightest idea whether we have made money or lost money."

Edison's phonograph business was in a parlous state. "The discontent of users and dealers with the record situation is quite serious and something needs to be done," Edison was told in May, 1917.

To give good results, the reproducer of the disk phonograph had to be kept in perfect adjustment. Purchasers were given a guarantee that these adjustments would be made free of charge, so each jobber had to employ at least one repairman. By the end of 1917, Edison had sold a cumulative total of 250,000 disk phonographs, and the annual expenditure for making the repairs was a half million dollars. "The cost of repairing reproducers might some day reach staggering proportions," an executive pointed out, and the company decided to reduce its lifetime guarantee to one year.

During the postwar boom, Edison's phonograph sales and profits were the highest ever—in 1919–20 the division had net earnings of $2.8 million on sales of nearly $21 million. Even so, disk records steadily lost money.

"Well, what's the use of saying anything about them?" Wilson rhetorically asked Charles.

Charles, supported by Mina, was steadily accumulating power. By 1918 he was "chairman of the board."

"I *want* them to know that you are head," Mina declared. "I do want your hand at the helm without any doubt about it."

Mambert had accomplished what Mina had been unable to—he had ousted Hutchinson. Hutchinson, Mambert told Edison, "was entirely insincere in his oft-repeated show of affection for the company." Hutchinson promoted only his own interests, and had kept the factory tied up with special orders on which the company made no profit. His privileged status caused dissatisfaction among the other employees. The Federal government did not want to deal with middlemen, and balked at paying Hutchinson's 20 percent commission. One Federal official accused the company of behaving like "unpatriotic profiteers with a flag in one hand and a gun in the other." Another hinted that the government would commandeer the plant if Edison refused to do business except through Hutchinson. Charles perceived Hutchinson as a potential stumbling block, and as Edison grew indifferent, Hutchinson concluded he would not survive to occupy Edison's chair.

With Hutchinson disposed of, Charles turned his attention to Mambert. Mambert disapproved of Charles's bon-vivant, cocktail-hour approach to business. Charles retaliated by making Mambert feel as unappreciated as possible. Edison told Charles: "Mambert is the most valuable man I ever had, for more than 40 years neither myself or any of the men around me have ever been able to get banks to give us credit. Mambert's methods and showings have now enabled us to borrow very large sums, give us good standing, and the banks have confidence." Mambert was working sixty-five to seventy-five hours a week, and receiving $10,000 a year. In 1919 Edison told Charles to raise him to $12,500; but Charles did nothing. Four months later, Edison chided Charles: "I was quite surprised when I found out that Mambert was not getting the $12,500 per year I told you to put him at. Fix it up and see he gets the back pay."

Both Edison and the company had done very well during the war, largely from the sale of chemicals and batteries. Edison himself had accumulated one million dollars in Liberty Bonds, and had no financial anxieties for the first time in thirty years. In 1919 Thomas A. Edison, Inc., paid dividends of nearly $3 million, and Charles was all for expanding the business. Edison, however, had seen too much money evanesce during his life. In his seventh decade he was courting caution. In July of 1919 he laid down the policy: no expansion requiring capital; the business to be perfected and cheapened; all bills to be discounted; debts to be liquidated; surplus to be accumulated.

"I think as far as financing goes, that Mambert should give his consent to any extra investments, otherwise I cannot hold him responsible," Edison told Charles.

Charles had cut a lively figure in Greenwich Village. (One night Edna St. Vincent Millay sat down next to him and started writing on the back of an envelope: "My candle burns at both ends.") Although, in the spring of 1918, he married his Boston girl friend in a surprise ceremony at Fort Myers, his marriage did not make him less hyperactive, and at times he seemed as if he were trying to find a candle with three ends. For lack of other excitement he rode the rods, and set the surface of the hockey pond at Hotchkiss afire with chunks of carbide.

"I have the odor of gasoline in my nostrils and speed! speed! speed! is the password," he exulted, as he smoked, drank, and drag-raced other cars in the streets. Not behaving precisely like the image of a thirty-year-old chairman of the board, he maneuvered against his father and Mambert, and tried to impose his own ideas on production and planning.

The crunch came in the latter part of 1920, when the nation sank into a severe recession. In December, the battery plant and phonograph factory shut down. The chemical works were hard hit by the resurgence of German exports, and employment fell by three fourths.

During the summer of 1921, Edison was forced to put up his million dollars in Liberty Bonds as security to borrow $800,000 to tide the company over. More crusty than ever, he jumped on both Mambert and Charles. When Charles attempted to defend his own ideas, or question the validity of one of his father's, Edison gave Charles to understand that the business had survived for a long time without him, and could be run without him again. Charles retreated to a resort, there to contemplate momentary submission and the arrival of another day, or the likelihood of an exile paralleling Tom's and Will's.

"It's been a pretty strenuous 18 months for both of us and yours truly has experienced a kaleidoscope of rude awakenings," Charles wrote his father on September 12. "There were times when I felt you had stuck the spurs in so deep that I'd surely bleed to death.

"Perhaps the very furiousness of your attacks and the all inclusive scope of your damnations makes a fellow rush this way and that without any particular plan except to save the women and children. This course of action probably gives the appearance of opposition to your ideas, whereas it's only a case of being rattled.

"Since about last January I have not opposed in principle one solitary thing you have wanted to do, but I have actually been heartily in sympathy. Naturally little differences of opinion on details are bound to exist if I do any thinking for myself at all. But just because I argue on this or that point doesn't mean that I am opposed to the whole idea. Lots of times when we are talking together and you tell me I am arguing and disputing I'm not at all but merely bringing up certain cases to get your thoughts.

"There was a time when I thought that I knew something about organizing and you didn't but that seems a long time ago now. What I want you to believe is that for sometime past any false pride in the air-castle organization I helped to construct during the past few years is gone—completely, absolutely, unequivocally gone. Also that I look to you and only to you for leadership. That when I know you have the correct facts and *all* the facts before you I'd take your judgment and tell the world to go to the devil."

Mina was always there to help calm the waters for Charles. Even when, a few months later, Charles got himself into trouble again, she was reassuring: "Don't worry, put it down as another lesson. You are all right Charlsie dear and Papa is as proud of you as a peacock. He realizes that everybody makes mistakes and says that you are learning."

In his seventy-fifth year Edison could still conceive original ideas, but his lifelong introversion had been exacerbated to the point that he was a veritable mental anchorite. If Mina invited guests for dinner, he would say scarcely a word, and afterward might go to sleep on the couch. People pretended not to notice; but Mina was so embarrassed she stopped issuing invitations.

Any attempt to influence him was hopeless. After 1920 the phonograph business collapsed, and the orientation of the industry turned in an entirely new direction. Edison, however, plodded straight on, heedless of all evidence he was heading into a mercantile desert.

In 1916, David Sarnoff, contracts manager for the American Marconi Company, had recommended that transmitting stations be built for broadcasting speech and music, and that "radio music boxes," capable of receiving several different frequencies, be manufactured. In November, 1920, KDKA in Pittsburgh inaugurated regularly scheduled broadcasts, and stations quickly sprang up in other cities.

By 1921 there were about a quarter million radios in operation in the United States. In November, Aylsworth, reminding Edison that he held some very strong patents on vacuum tubes,* suggested that the company begin producing radio equipment in competition with General Electric. "The plant practically exists today in the idle equipment in the phonograph works," Aylsworth noted.

During the first part of 1922, sales of radios accelerated as rapidly as sales of phonographs declined. A Detroit dealer wrote Charles: "Salesmen report the loss of many sales to people who have bought radio outfits and they seem to feel that if we had some kind of radio outfit that would work in with the phonograph that it would mean the sale of many instruments."

But Edison was adamant in his contention that "the radio craze" was only a temporary phenomenon. "It will die out in time so far as music is concerned. But it may continue for business purposes," he predicted.

He tried listening to radio with a set of headphones, but had to turn the volume up so high that static erupted like thunderclaps in his ears. The limited programming provided nothing that interested him, and he believed people would always prefer to listen to a record of their own choosing rather than be forced to accept someone else's selection. All of his life he had strained to deal with sound; and he was not prepared, at an advanced age, to cope with a new source.

* An "Edison effect" lamp patterned after one exhibited at the Philadelphia Exposition was successfully substituted for a vacuum tube in a radio by NBC on November 21, 1932.

Yet neither would he admit that he was tired; and he most definitely would not trust a major project to anyone else. He had plunged fortunes into inventions and their development, but now he only wished to preserve what he had. He shut his mind to all the signs of radio's growing popularity, to its economic advantage for consumers over the phonograph, and to the rapid diversification of programming. Charles could not dispute with him. Letters poured in from dealers all over the United States; and all Charles could say to executives in the phonograph division was: "Get out our stock of replies and see what fits the case."

By 1925 four times as many radios as phonographs were being sold. A Midwestern jobber wrote: "Nearly every Edison distributor you have, Mr. Edison, has been obliged to add radios to his line to keep his head above water and his books out of the red."

It made no difference. "Our sales will drop this and next year very much and only increase when the Radio Craze is over," Edison rasped. "Don't make any connection with the radio gang of crooks."

In early 1925 Victor decided to go into the radio business, and Edison could not have been more pleased: "Charles—we will be unique. The only hill and dale and only straight phono."

The records of Victor, Columbia, and Brunswick all employed a lateral cut and were interchangeable, but Edison's records, made with a vertical cut, could not be played on other machines without special attachments. Edison was selling no more than 2 percent of the records in the United States, and had over a two-year backlog piled up in warehouses. The cylinder business, which had depended on rural and small-town sales, was skidding into extinction as masses of people moved into the cities.

Mambert, a proponent of drastic reorganization, was being pushed to the outer edge of the executive circle. "Mr. Edison has men around him, Mr. Charles has, who are yes-men," one of Mambert's friends told him. "They are worse enemies to you personally and to the business as a whole than men who take graft or steal cash." In the spring of 1924, Edison, prodded by Mina, asked Mambert to resign so that Charles could assume his job, and production was cut back severely.

Edison blamed the layoffs on the political situation in Washington. "We are forced to curtail production due to business conditions over which we have no control. There seems to be little relief in sight unless business men capable of dealing with business problems are elevated to public officials," he declared. But if Edison had complete confidence in businessmen, he had none in financiers: "Everybody wants to know what is the matter. If one reads the daily letters sent out by the brokers in Wall Street he will soon learn they certainly don't know."

In the fifty-plus years since his own appearance on Wall Street, Edison had acquired a broad understanding of finance, but his views were still those of a Midwestern Greenbacker. In 1923 he published a booklet, "A Proposed Amendment to the Federal Reserve Banking System," in which he suggested formation of a Farm Commodity Credit Corporation that would guarantee farmers interest-free loans on security of their crops, the amount to be set at 50 percent of the average selling price during the last twenty-five years. Despite his suspicion of financiers, he discussed the scheme with Bernard Baruch, whom he had met in Washington during the war, and Baruch commented:

"With the proper check upon it, I cannot see any weakness in your plan. Let me congratulate you upon the directness with which you went at this subject."

"Friend Baruch," Edison thanked him, "you are the first man who has had imagination enough to throw off the trammels of the money religion and analyze the proposed scheme like an engineer."

Baruch was one of the nation's leading Jews, and a particular subject for attack in the sleazy anti-Semitic "International Jew" series financed by Henry Ford and produced by Ford's private secretary, Ernest Liebold. Edison, because of his association with Ford and a misinterpreted remark about the importance of the role Jews had played in Germany's conduct of the war, was splattered by the fallout. In fact, Edison had been associated with several Jews (mostly notably, Bergmann and Rathenau), and his views were, if not enlightened, at least moderate:

"The Jews are certainly a remarkable people, as strange to me in their isolation from all the rest of mankind, as those mysterious people called gypsies—while there are some 'terrible examples' in mercantile pursuits, the moment they get into art, music, science and literature the Jew is fine—he has been persecuted for centuries by ignorant malignant bigots and forced into his present characteristics and he has acquired a sixth sense which gives him an almost unerring judgment in trade affairs and got himself disliked by many. I believe that in America where he is free that in time he will cease to be so clannish, and not carry to such extremes his natural advantages."

Having all his life had difficulty managing money, Edison had come to think there must be some extrasensory perception involved.

Such anti-Semitism as he displayed was less a matter of bigotry than of antisectarianism in general. He reacted negatively to all organized religion. He followed the freethinker's path he had been led to by Tom Paine, and could grow lyrical speaking about him:

"His Bible was the open face of nature, the broad skies, the green hills. He disbelieved the ancient myths and miracles taught by established creeds. What you call God, I call nature, the supreme intelligence that rules matter."

Through the 1920s Edison supported the Freethinkers Society, welcomed its president to the laboratory library, and praised him for "dispelling the clouds of superstition and breaking our bondage to a mythical religion." Mina, meanwhile, an unreconstructed fundamentalist who worked successfully for the passage of a blue law in New Jersey and thought evolution a plot of Satan, entertained Methodist bishops at Glenmont.

Through much of his life Edison was attracted by mysticism. After the phonograph came into existence, he could almost sense a mystic force moving about in the universe. In 1878 he had come under the influence of Madame Elena Blavatsky, the founder of the Aryan Theosophic Society, who was then residing in Manhattan. Madame Blavatsky amalgamated spiritualism, Hinduism, and Buddhism, and tried to assimilate science into theology. Edison, after attending several meetings, received a diploma of fellowship, and attempted to answer Madame Blavatsky's question, "Is gravitation a law?" by thinking a pendulum into movement.

Later in life Edison denied he had ever had anything to do with the Theosophists, but he modified the Theosophic concept of reincarnation into an occult idea of his own. Memory, he theorized, consists of subparticles of matter, somewhat like electrons. These particles travel through space, and perhaps even come from outer space, bringing to earth the wisdom of other worlds. Coalescing into swarms, they lodge in people's brains, thus creating intelligence. Edison could see no other explanation for instinct or heredity than that these "little people," as he referred to them, pass from one human being to another. Genius, he postulated, is the result of some particularly fortunate grouping of these "little intelligences."

Sometimes, he believed, the billions of these intelligences within a single person cannot agree, and generate conflict among themselves. "They fight out their differences," he said, "and then the stronger group takes charge. If the minority is willing to be disciplined and to conform there is harmony. But minorities sometimes say: 'To hell with this place; let's get out of it.' They refuse to do their appointed work in the man's body, he sickens and dies, and the minority gets out, as does too, of course, the majority. They are all set free to seek new experience somewhere else."

Edison was uncertain whether these infinitesimal intelligences, once they leave a body, separate or stay together. But he thought it quite possible that, like a swarm of bees, they travel united through space until they find another host in whom to settle, in which case a human being's personality never dies, but is merely transferred.

"Now the great question arises," Edison mused in 1921, "how can we get into communication with a group of highly educated units traveling in groups and holding the memory or personality of a man—might not photographic plates produce figures—a host of combinations could be tried—can we find apparatus that the whole mass of the group following the law of gravitation be able singly or jointly to move and give signals which our ear or eye can appreciate?"

Edison conducted some experiments along these lines, but failed to obtain any positive results—which of course did not discourage him. Henry Ford introduced him to a parapsychologist, Bert Reese, who, in numerous sessions in the laboratory library, convinced Edison of his psychic powers. Edison even vouched for Reese to a skeptical Houdini, and tried, without success, to duplicate Reese's telepathic results by winding electric coils about his own and three other persons' heads.

During the first years of the 1920s, Edison's friendship with men like Ford, Presidents Harding and Coolidge, Secretary of Commerce Hoover, and Secretary of the Treasury Andrew Mellon made him as visible a figure as he had been a generation earlier. Newspaper reporters were constantly on his trail; and since he suppressed no opinion—even if five seconds before he had had none—he was often in the headlines. Such ideas as his concept of memory swarms set editorial writers feverishly inserting new ribbons into their typewriters. His most controversial issuance was his "intelligence questionnaire."

Even before the organization of the educational film division, Edison, the recollection of his own school days rankling him, had had a keen interest in education. During the 1880s and 1890s he was (in name) the dean of the National School of Electricity, a correspondence institute, in Chicago. In 1886 he donated a dynamo to help establish a school of electrical engineering at Cornell University, and eight years later the dean wrote him:

"It has become the largest and I think the best-equipped, and you ought to get an occasional glimpse of it in order to appreciate your own work. I am inclined to think that you will feel that you have taken part in no other enterprise that exhibits so much promise of real and permanent usefulness."

In 1888 Edison presented an electric lighting plant to MIT, and in 1909 he contributed to a scholarship drive to enable Boston newsboys to attend the institute.

The business collapse of 1920–21 led him to conclude that his own employees, from cleanup men to executives, were grossly deficient. "It costs too much to find out if a man is a good executive by trying him out on the job," he said. "I made up my mind that we should have to have a formal test. This brought up the question of what we should look for. What is the most important qualification for an executive? When I call upon one of my men for a decision, I want it right away. It isn't convenient for me to wait, and certainly it isn't convenient for a whole department to hang in the air for an indeterminate period waiting for an executive to find something out that he might have had right in his head. The very first thing an executive must have is a fine memory. Of course it does not follow that a man with a fine memory is necessarily a fine executive—he might be an awful chump in the bargain. But if he has the memory he has the first qualification, and if he has not the memory nothing else matters."

Edison had in his own head a prodigious collection of miscellaneous facts—one day Rosanoff, searching for a lengthy price list of waxes, had stumbled upon him asleep in a closet.

"What do you want?" Edison had asked as he woke with a start.

Rosanoff explained.

"You don't need any price list, I'll give you the prices," Edison had grumped, and run down the whole list of waxes, naming the price of each. By the time Rosanoff gathered himself together and stumbled off, Edison was snoring again on the decrepit mattress.

But the information in Edison's head was as disorganized as his business affairs. Though he had an immense knowledge of things in general, what he knew was often deficient in detail and filled with error.

(He asserted the Germans were not a musical people, explaining that Beethoven was a Dutchman and Wagner was Jewish. He considered Gutenberg to have had the greatest influence on the history of the world, but in describing Gutenberg's work he said: "All he did was saw a bar of wood with contiguous letters on it into blocks." In reality, Gutenberg had constructed brass matrices, enabling letters to be cast individually, and so had become the inceptor of mass production.)

To test the knowledge of every job applicant, Edison had a questionnaire, consisting of 150 items, prepared for each different position. A college graduate was expected to know: "What city in the United States is noted for its laundry-machine making?" "Who was Leonidas?" "Who invented logarithms?" "Where is Magdalena Bay?" "What is the first line in the Aeneid?" "What is the weight of air in a room 20 x 30 x 10?" A mason was supposed to answer: "How many cubic yards of concrete in a wall 12 x 20 x 2?" "Who assassinated President Lincoln?" A cabinet maker was asked: "Which countries supply the most mahogany?" "Who was the Roman emperor when Christ was born?" A carpenter was posed the questions: "Name 20 different carpentry and joining joints." "What are the ingredients of a Martini cocktail?"

Edison professed he did not expect an applicant to answer all of the questions; he would be satisfied with a score of 90 percent—a result that would have required an IQ of no more than 180. When his expectations were not fulfilled, he reiterated his lifelong conviction that schools, and especially colleges, spent too much time on abstract and irrelevant material and not enough on "facts."

"Why teach them Latin? Latin is a dead language," he said rhetorically. "The professor himself don't know how to order in Latin a sirloin steak with potatoes. Who the hell uses Latin outside the Catholic

Church? And there nobody understands it except the pope, so even he can only use Latin when he is talking to himself."

Actually, Edison's views on education were as diverse as his opinions on everything else. He despised "parrotlike repetitions," and liked "the Montessori method. It teaches through play. It makes learning a pleasure. It follows the natural instincts of the human being." He wished schools would stiffen their standards, so that "The system of learning today and forgetting permanently tomorrow would go out of fashion." Despite his mocking of abstract thinking, he declared: "Einstein has shown the world the sort of thought it needs. The more Einsteins we can get, the better. I wish we had an Einstein in every branch of science."

When educators and editorial writers tore his intelligence questionnaire apart, and the *New York Herald Tribune* called him illogical for attacking college education while sending Theodore (who did none too well on one of his father's questionnaires) to MIT, Edison turned right around and praised colleges with faint damnation:

"Colleges do not deserve all the condemnation I have been credited with heaping on them, yet they measurably fail. I am in favor of the college. That is where I get some of my best men. I have 60 of them now, but that is 60 culled out of 2000. There is something wrong with the college system. The present system does not train men to think."

During the second half of the 1920s, Thomas A. Edison, Inc., thanks primarily to the battery division, had some of its best years. While the net profits of $2 to $3 million a year were small compared to the standards of post–World War II industry, they were magnified by the fact that all the money went to Edison and Mina, who continued to live far more modestly than most multimillionaires. Edison decided he would do something to improve the quality of college graduates, and instituted the Edison scholarship competition. Forty-nine applicants, one from each state and the District of Columbia, were to be chosen.

As judges he appointed himself, Ford, George Eastman, Charles Lindbergh (fresh from his honeymoon), Dr. Perry, the headmaster of Phillips Exeter Academy, and Dr. Samuel Stratton, the president of MIT. Working from nine thirty one morning until two thirty the next at Glenmont, they graded every answer. Lindbergh, the only true conservative among the group, became so enraged by one contestant's mild suggestion that a different distribution of wealth was inevitable in the United States that he wanted to throw the boy out of the competition, even though his score was second highest overall. The millionaire Edison and the billionaire Ford laughed off such "radicalism."

Solicited on how to cope with the communist menace, Edison replied: "If they talk from soap boxes, let the business men raise a fund and go into the soap box business."

CHAPTER 40

Conspiracy

At the age of sixty, Edison had revealed his schedule for the next thirty years: "From now until I am 75 I expect to keep more or less busy with my regular work. At 75 I expect to wear loud waistcoats with fancy buttons; at 80 I expect to learn how to play bridge whist and talk foolish to the ladies. At 85 I expect to wear a full dress suit every evening at dinner. And at 90—well, I never plan more than 30 years ahead."

In the winter of 1923, however, when he turned seventy-six, he was seriously ill. It was not just his perennial pulmonary problems; his entire system seemed out of kilter. After coming down with a cold in January, he remained in Florida until May, but was still not feeling well when he and Mina returned to West Orange. His handwriting was shaky, and he lost interest in things he previously would have snapped at like a hungry fish.

"Charles, you go into this, I am too sick," he noted when a question regarding the phonograph came up in March. In addition to continuing his daytime napping, he regularly slept until 10:00 A.M. Only after a late August visit to Ford in Dearborn, where he once more discussed a starter system for the Model T as well as other projects, did he recover his zest.

"Papa seems like another man now since he has matters in his own way," Mina informed Charles. "We will have to toe the mark now I can tell you. Charles, you will have to visit the Ford plant. It is a great organization but too appalling."

But by winter, Edison felt chronically unwell again. Sometimes he slept all afternoon. He continually worried about his health, and was less anxious only when periodically relieved by a doctor's report. His tonsils kept getting infected, his teeth were aching and abscessed, he suffered terribly from gas and constipation, and his stomach seemed occupied by an internal ore-milling machine.

His interest in amateur medicine had never flagged—he compounded tauraethyl ammonium to dissolve uric acid and relieve gout, and tetraethyl ammonium hydroxide as a cure for rheumatism. He consumed buttermilk germs and lactic acid tablets to settle his stomach, and told one of his acquaintances who had heart disease: "Americans dig their grave with their teeth. Your heart will be OK if you let up on eating and spend one third of your time collecting funny stories." He theorized that warts were the work of vagabond cells: "Say, for example, a lot of cells get away from a toe and land way up on the nose. They don't know where they are or how in hell to act. So they go crazy and start building a toe, because that's all they know how to do—see? We call it a wart, but it's nothing but a piece of toe in the wrong place."

Along with such books as *Criminal Psychology,* H. G. Wells's *Outline of History,* and *Rocks in the Road to Fortune* (a tale of disastrous mining ventures), he perused dozens of medical volumes. Shortly after he was sixty he read Elie Mitchnikoff's *The Prolongation of Life* and C. G. Minot's *The Problem of Age, Growth, and Death,* and acquired the notion that drinking curdled milk was good for the kidneys and that starving the bowels was good for everything else.

"If people would diminish the amount of food they eat to a point where the nutritional value of the contents of the large intestine is very low," he asserted, "the bacilli of that tract would starve to death and not make toxins to poison the whole system."

From time to time throughout his life he went on fad diets—at one point he ate mostly dried herring, sardines, hard dry toast, and stewed prunes—though he always managed to get his ration of pie and sweets. In 1924 he launched a determined attack on "the traveling poison factory" of his lower bowel, cut back his food intake, and ate practically no roughage.

He improved little, but lost so much weight that he turned into a shrunken, wizened old man. Finally, in 1926, he consulted a specialist, Dr. Hubert Howe. Since Edison did not like to be probed, Dr. Howe was forced to make much of his diagnosis from the voluminous notes Edison provided him.

"We can reach tentative conclusions now, but others would require investigation," the doctor said.

"I feel sure for years you have had a chronic ulcer of the duodenum, the history is perfectly typical.

"It may be that there has been some contraction at the pylorus that prevents emptying. The X-ray study would clear up these points. I do not feel that a reduction in diet accounts entirely for the loss of weight. We cannot be too sure the recent year of constipation is due entirely to low diet and this should be studied.

"The question of pulling the teeth in my opinion depends on whether there are undrained pockets. This can be shown by X-ray study of the teeth.

"If you will follow a careful regimen you can be properly relieved of pain in the stomach but it must be rigidly adhered to for months."

Edison, however, burned once, would have no part of X-rays. Although thirty years before he had suggested the X-raying of teeth, he decided now to simply have all of his teeth pulled. The new dentures were uncomfortable, and he took them out at every opportunity. Often he forgot where he had put them, and then most of the laboratory force was assigned to hunt them down. On other occasions he popped them back into his mouth absentmindedly and then, gagging, spit them out again. "They get all mixed up with chemicals and everything, and sometimes they taste something awful," he complained.

To cope with his ulcer, he went on an exclusive milk diet. Dr. Howe tried to convince him he must eat a reasonable balance of foods, but the only thing he would take additionally was an occasional orange.

He lacked the energy to work more than a few hours a day, and in September he and Charles swapped positions—Charles became president and Edison chairman of the board. His stomach was no better. He had increasing difficulty walking, and signaled his approach by the loud scuffing of his shoes on the floor. Though he still could project flashes of his previous self, as when the Crown Prince of Sweden visited the lab, he had less patience than ever and saw fewer and fewer people. "It's one damn thing after another—get me out," he told Charles when someone asked for an appointment after his return from Florida in May of 1927. "I am not feeling well enough."

The next month Charles gathered the courage to inform his father: "I've been taking the bull by the horns and making decisions without bothering you—and unless you tell me you want something different, I'll continue to do so."

The laboratory remained Edison's exclusive domain. Very little was being done, and with operating costs nearly $100,000 a year, Charles hoped to get his father's permission to cut back. "The Laboratory is sort of like a ship with all sails set waiting for the captain to give it orders," Charles pointed out. "If the orders aren't going to come for a while we can roll down a few of the sails. If it is possible for you to give some idea of the *kind* of work you may expect of the Lab. in the near future we can go ahead and adjust things accordingly."

"I want to perfect super phono, Electric Recording, Ford starter, finish up blank machine, and use Fred Ott and Joe as experimenters on Rubber," Edison replied. "That is about all."

The production of rubber was Edison's latest project. In 1922 the British and Dutch had set off an immense brouhaha by combining to restrict rubber exports from Ceylon, Malaya, and the East Indies as a means of propping up world prices. Ford bought 3.7 million acres in the Amazon to plant rubber trees, and later added 8000 acres in Florida for a plantation. Secretary of Commerce Hoover obtained $500,000 from Congress to initiate experiments on producing rubber in the United States as a means of breaking the monopoly. United States Agriculture Department scientists in California planted and began testing a variety of cacti, trees, and vines. In 1923 Edison started experiments with milkweed, and the following year he built a factory for making the hard rubber that was used for battery casings. But it was not until 1926 that he seriously entered the rubber race, and planted 300 Madagascar rubber vines at Fort Myers.

There were hundreds of plants with sap of rubbery consistency. The problem was finding one that not only would produce rubber in amounts of commercial quality, but could have the sap extracted mechanically—for the cheapness of labor in the Far East made the employment of American hand labor uneconomic.* Edison, therefore, while compiling an exhaustive catalogue of thousands of plants, worked simultaneously on a rubber-extracting machine.

Mina hoped that the project could be just a hobby. But articles soon appeared in the press. Edison felt once more compelled to perform. Even though he was seriously ill, he went at it with as much dogged determination as he had left.

"Everything has turned to rubber in our family," Mina complained. "We talk rubber, think rubber, dream rubber." Edison regarded goldenrods as particularly promising, and liked to go on drives, stop, and pick the flowers from the side of the road. Eventually, one of his workers discovered a species of goldenrod growing ten to twelve feet tall in Florida. In 1929 Edison and Ford formed the Edison Botanic Research Company, and Ford put $500,000 at Edison's disposal.

By then, however, the British-Dutch attempt at cartelization had collapsed, and rubber prices plummeted. The half million dollars Ford allocated was more for the purpose of keeping a very old and sick man occupied than for any realistic endeavor of producing natural rubber in the United States.

In the West Orange laboratory the area where the rubber experiments took place was known as "the hay fever room." Despite Charles's intention to curtail activities at the lab, they continued much as always, with what was increasingly a superannuated crew. Edison's secretary, Meadowcroft, was only six years Edison's junior. His health was almost as bad as Edison's, and he was gradually losing his sight. The library had been

* The chemistry of synthetic rubber was already understood, but the process was as yet too expensive to compete with the production of natural rubber.

a playhouse for New Jersey's flies and mosquitoes during the first quarter century of its existence, and in 1916 Meadowcroft had fought a losing battle to have screens installed.

"Nothing doing!" Edison had snapped. "Why in God's name do we want so many screens when for ventilation not more than three windows need be opened. I will stand for three only."

Even during the purgatorial summers, Edison insisted the windows should remain closed and the shades drawn "to shut out the heat." Ivy, in truth, had grown over the windows so completely that they could not have been opened without the employment of a pair of shears.

The 7500 volumes contained in the library filled perhaps a fourth of the shelves, and the remainder of the space resembled the storeroom of a pack rat—there were phonograph records, crystals and rare minerals from the X-ray experiments, boxes of light globes, cases and cases of various types of ore, samples of plants, models of machinery, trunks and crates of unsorted documents and letters, and an almost indescribable variety of paraphernalia.

Edison's rolltop desk was a litter of material and pigeonholes crammed with papers—all sacred. When, at the age of eighty-two, he returned after an absence of some months, he cried out:

"Meadowcroft! Where the hell are those moving pictures I told you to look for, why don't somebody find 'em when I keep asking for 'em. Who the hell now has been cleaning house around here and losing every God damned thing I want!"

Every two hours he swiveled from his desk, opened a Thermos on the conference table in back of him, tucked a napkin into the front of his shirt, and slowly sipped two cups of milk, which he demanded must come from the brown cow from Parsippany, and from none other. His own Napoleonlike visage looked down upon him, and the hands of the huge inbuilt clock silently advanced. Oblivious of the people moving around him, he might sit for an hour or more enveloped in the cocoon of his thoughts. Sometimes a ray of light found its way through a high window and rested on his thin white hair. "I am long on ideas but short on time. I only expect to live to be about 100," he remarked.

There were days when he took three or four naps of an hour or two each. Meadowcroft, the only person authorized "to kick Edison in the slats," would grope his way over to the corner alcove and call, shake, and poke him—sometimes for several minutes—before being able to awaken him. The men walked around with their hats on, spit on the floor, and drank water from a communal cup, as if the twentieth century had not yet intruded here. Sometimes the life-sized statue of a bare-breasted, laurel-crowned Orpheus gripping an Edison phonograph record—nicknamed "Charlie Chaplin and his custard pie," but resembling a seated Grecian discus thrower—seemed more real and vigorous than they.

Edison's personal experimenters still came to talk to him, and shouted into his ear. But increasingly operations went on without his knowledge. When Theodore had graduated from MIT in 1923, Edison had been proud of him: "I have a lot of problems I can't figure out myself. He can do the work with mathematics. He's a pretty fair experimenter, too, in physics." But as soon as Theodore went to work, the inevitable happened. Edison's views differed so much from Theodore's that, no matter how hard Theodore tried to please him, Edison became irritated. In 1927 Edison refused to believe the results of a series of experiments, and twice yelled at Theodore before a group of people that Theodore was deliberately trying to mislead him. The affectionate, sensitive Ted was hurt to the quick.

"I got to feeling so badly that I couldn't sleep, couldn't eat, felt sick and pepless and like Hell in general. I *couldn't* do my best work in that condition," he later wrote his father. "I thought it would save your nerves as well as mine if I went ahead and did the best I could on my own account and then let you judge by the *results*."

Theodore was attempting to produce a viable model of the long-playing phonograph, and to adapt it for "Cine-music," a system intended to replace orchestras in small movie houses. But the technical difficulties of developing a canoe-shaped diamond stylus that would not jump the infinitesimally fine grooves (there were 450 threads to the inch) were enormous. Theodore was a deliberate, meticulous worker who, Charles declared, "steadfastly refused to be kidded, cajoled, razzooed or buffaloed into allowing some halfbaked gimcrack to be offered to the trade." But with the development of talking pictures, the whole project became moot.

"No dying moans of dead animals goes into our jazz," Edison chortled, but the quality of the records was more and more irrelevant as the quantity sold kept dropping. The company had a huge phonograph factory, with far too little work to justify it. The obvious solution was to manufacture radios, a move that Edison refused to countenance. If radios were to be produced at the factory, the groundwork would have to be laid without Edison's knowledge. So Charles enlisted Mina's help in the conspiracy.

In 1919 General Electric had purchased the American Marconi Company, and renamed it the Radio Corporation of America. General Electric had then negotiated agreements with AT & T, Western Electric, Westinghouse, and the United Fruit Company, the holders of most of the other radio patents, and concentrated all of the patents in RCA.* Through the early 1920s, RCA licensed about twenty-five manufacturers to produce radios and radio parts, but then stopped issuing new licenses with the explanation that the business was becoming too fragmented.

* Following an antitrust action, RCA was divorced from General Electric in 1933.

Charles approached General Electric and RCA, and argued that Edison was deserving of an exception. In February, 1928, however, RCA turned Charles down.

The only alternative was to acquire a company that already had a license. No concern that was doing well was available for purchase, so Charles entered negotiations with a Newark firm, the Splitdorf Corporation. Splitdorf was a crazy-quilt company that produced a variety of items like motors and spark plugs in addition to radio parts. In 1927, its losses were $1.7 million on sales of about $4 million. Its stockholders were delighted to discover someone willing to assume the mess.

Edison was at Fort Myers all spring, and Mina wanted to break the news to him gently. One day in late April when Mina knew Charles had sent his father a telegram announcing the acquisition, she and Edison stopped in the telegraph office on the way to the movies. (Edison went to the movies several times a week, and would tolerate none but happy endings.)

"I was in dread lest father dear would go all to pieces before everyone," Mina recounted to Charles. Edison read the message, and for a long time stood staring blankly. The long period of suspense almost caused Mina to break down. "I just dreaded the silence. Oh the relief of that moment when he gave me the telegram to read!"

Finally he remarked: "It don't mean much bugs anyway."

Mina told him Charles wanted to name the radio the "Edison."

"Why yes, Charles is running it, isn't he?" he said absentmindedly, and went on to the theater.

CHAPTER 41

The Path of Glory

Edison was eighty-one years old. He had outlived not only most of his contemporaries, but to a large extent his time. He had become well known in his twenties, and was world famous by the age of thirty. In 1928 he was already as much a figure of history as George Washington—there were parents who had read about him in school books that their children were now reading. With the road to Miami opening in the spring, thousands of tourists flocked into Fort Myers and clustered along the fence to catch a glimpse of him.

Of Edison's Newark and early Menlo Park days, only John and Fred Ott were still alive—Johnson had died of progressive heart failure in 1917, and Upton of abdominal cancer in 1921. But there were quite a few survivors of the 1880s. William Hammer, who had arrived at Menlo Park in December, 1879, was not only the outstanding showman of the Edison organization, but its unofficial historian. He compiled the most complete collection of incandescent lamps and other electric-light memorabilia in America. In 1918 Hammer was the force behind the organization of the Edison Pioneers. Each February 11, Edison's birthday, the group gathered at the laboratory or in Newark.

In 1924 the Pioneers started talking about erecting a monument at Menlo Park. Without money, the project was confined to conversation, and had not progressed at all when the Pioneers met with Edison on his eightieth birthday at the Robert Treat Hotel in Newark. Then, suddenly, the proposal was given an inadvertent push by the developers of a scheme in Milan, Ohio, Edison's birthplace.

Spearheaded by an Ohio Appellate Court judge, Roy H. Williams, and the publisher of the Milan *Ledger,* F. A. Day, town boosters formed a Thomas A. Edison Memorial Tribute Association to acquire several hundred acres of land surrounding the Edison home site. On this they planned to establish a ten-million-dollar College of Technology and Research. Henry Ford, Harvey Firestone, and other prominent friends and associates of Edison were invited to participate. Edison sent Gumshoe McCoy to investigate the project in the summer of 1927, and in October informed Ford: "The proposed tribute is utterly repugnant and distasteful to me and contrary to my ideas of propriety."

Over the years, Ford had become more and more of a strange combination of dutiful godson/benevolent godfather to Edison, his family, and Thomas A. Edison, Inc. Edison had gradually though none too energetically been paying back $1.2 million he had borrowed from Ford between 1912 and 1915. In 1925 Ford decided simply to write off the $750,000 still outstanding on the debt.* (Three years later he began advancing more money for the E-mark, lead-acid starting battery which Charles successfully brought out.) There were continuous exchanges of information between Edison executives and Ernest Liebold, Ford's private secretary and all-around factotum. Ford regularly sent cars—Model T's to the children, and Lincolns to Mina and Edison. He even took an interest, through Liebold, in Tom, who invented a "period timer" for Ford cars and trucks. This, Tom claimed, greatly improved performance and mileage. At the end of the world war, Edison allowed Tom to come to work in the laboratory, and financed his Ecometer Company. Ford engineers tested the device at Dearborn in 1921, but the cars constantly stalled. Edison's battery division helped line up dealers, but the first 200 timers sold to the public were all returned as worthless. In 1924 the company went into bankruptcy.

During the fall and winter of 1927–28 Ford was busy designing the Model A that was to replace the Model T, but meanwhile he started thinking about means to memorialize Edison. He had already decided his next project would be the creation of a museum of Americana adjacent to his Dearborn factory and estate. He intended to assemble all the products of American ingenuity, and house them in authentic surroundings by acquiring and relocating historic buildings. Those he could not buy, he would duplicate.

When William Meadowcroft, Edison's secretary, learned of Ford's project, he wrote Liebold that there had never been an exhibit of all of Edison's inventions together in one place, and suggested that Ford might like to undertake it: "Mr. Ford seems so desirous of making his museum one of world renown that I could not forbear mentioning the above idea."

Ford, of course, was delighted by the suggestion—the question was how to get Edison to agree to it and the Pioneers to relinquish their col-

* The loan was secured by stock in Thomas A. Edison, Inc., but the arrangements were, to say the least, informal. When Edison sent Ford a check, he often asked the auto magnate not to cash it until informed that it was covered; and Ford, who had a habit of stuffing large sums of money into his pockets and forgetting about them, genially complied. Sometimes Edison did not release a check until a year and a half after he issued it.

lection. Two weeks later, on May 17, 1928, Ford was in Philadelphia to check out Independence Hall, and there met William Hammer. Ford told Hammer he intended to reconstruct Menlo Park, duplicate the Fort Myers laboratory, and install a replica of the Jumbo dynamo at Dearborn. By early June he had bought the Menlo Park property from Edison and Dot—through all the years, Edison had retained twenty-two acres surrounding the old laboratory site. The laboratory and office buildings had been used at intervals for various purposes—a Sunday school, the Menlo Park firehouse, and finally a dynamite storehouse—but they had gradually been vandalized and were now tottering. The apparatus, models, and other material that Edison had not considered worth transferring to West Orange had disappeared over the countryside. Unlike Ford, Edison had no interest in memorabilia. "My mind runs to the future, not to the past," he said.

Ford appealed to Edison's view of the future by proposing to spend $10 million—or $50 million if necessary—to create an Edison Institute at Dearborn. This institute would not be some promotional scheme, but would occupy the principal place in the museum. It would provide instruction for boys in the history and art of American manufacture. On September 24, 1928, Edison traveled to Dearborn to lay the cornerstone for the institute and the Greenfield Village complex.

In Washington, D.C., meanwhile, a group of Emile Berliner's partisans had unwittingly focused attention on Edison by claiming that it was Berliner, not Bell or Edison, who had invented the telephone and the microphone. When they had announced they planned to honor Berliner with a fiftieth-anniversary celebration in 1927, Hammer and Frank Dyer had launched a campaign in behalf of Edison. Hammer pointed out that although the Wright brothers and Charles Lindbergh had had special Congressional gold medals struck for them, Edison had not been honored. New Jersey Congressman Randolph Perkins introduced a resolution, which passed easily. Secretary of the Treasury Andrew Mellon scheduled the presentation of the medal to Edison for the evening of October 20, 1928, in the laboratory library.

Over the years, October 21 had gradually fixed itself in Edison's mind as the date when the incandescent lamp had been fashioned, though Upton in his letters had made clear that it had not been until mid-November, 1879, that Edison had "succeeded in obtaining the first lamp that answers the purpose we have wished it for."* As late as November 25, Edison had signed and sent off to England his application for a system using a platinum lamp. Not until the twenty-eighth had a stable horseshoe lamp been made.

* Since the light had been brought to fruition step by step and there had not been one day on which to focus, the men who had commemorated the twenty-fifth anniversary had held the celebration on February 11.

Nevertheless, when Edison had first formulated his account, three decades after the event, he said:

"The problem was solved—let me see—October 21, 1879. We sat and looked, and the lamp continued to burn, and the longer it burned the more fascinated we were. None of us could go to bed, and there was no sleep for any of us for 40 hours. We just sat and watched with anxiety growing into elation. It lasted about 45 hours."

There had, of course, been no lamp that lasted forty-five hours during this period. Edison always had a tendency to telescope events, and he was seldom able to sort out times and dates. When he testified in the electric light trial in 1889, he had been unable even to pin down the month in which the first successful carbon filament had been devised. The forty-five hours to which he referred had been spoken of by others who were there as a "death watch," and was, in actuality, the period when everyone awaited word of Charley's fate in Paris. Charley's death had occurred at a time when the outcome of the incandescent light had hung in the balance; and for Edison the events had become inextricably interwoven. When work had resumed on Sunday evening, October 19, Upton had measured the light emitted by a carbon stick at 40 candlepower, and it had become evident to Edison that carbon could be maintained in a vacuum. The discovery made on the twenty-first was that even a carbonized piece of thread would emit light (though of only ½ candlepower), and that it would provide the relatively high resistance of 113 ohms!

Following the experiments of October 19 to 21, Edison had not had an incandescent lamp; but *he had felt confident that he could construct one!*

Upton had summarized the situation succinctly: "He is always sanguine, and his valuations are on his hopes more than on the realities."

Yet it was perhaps fitting, considering the record of Edison's career, that the accepted anniversary of the incandescent light came to be not a date when Edison in fact had a bulb, but a day on which he was still reaching for the future.

Thus, on October 21, 1928, 135 guests crowded into the library, and another 500 assembled in the laboratory. The British ambassador returned the original phonograph, which had made its way, via Colonel Gouraud, to a British museum. Fifty radio stations broadcast the 9:00 P.M. ceremony. Secretary Mellon cited Edison as "Illuminating the path of progress through the development and application of inventions that have revolutionized civilization."

On October 22, everyone went to look at the relics of the old buildings at Menlo Park. Ford told Mellon that as soon as he had moved them he would donate the land for a national park. Gerard Swope, president of General Electric, began laying plans for the fiftieth-anniversary celebration of the incandescent light the following year. Edison approved—he had never lost his paternal feelings toward General Electric. In 1922 he had visited Schenectady and been given a founder's reception. Charles Steinmetz, who in 1901 had established the world's first corporate laboratory after General Electric stopped supporting Edison's research, had showed Edison around. Steinmetz characterized the inventor as "the best informed in all fields of human knowledge. His knowledge of the electric field and electrical application embraces a wider range than any other man."*

Henry Ford, on the other hand, regarded the General Electric plans as placing a damper on his own designs for Dearborn. He still had not succeeded in getting the Pioneers to turn over their collection to him. On January 5, 1929, Ford, trying to persuade Edward Adams, the president of the organization, wrote: "We shall not only have a museum filled with objects of education and historical value, but it will be an *operative* museum, and in addition a school stimulating initiative for serious creative work."

A few days later, Edison gave Ford his support, and the Pioneers agreed to the transfer. Ford, jubilant, strode up and down Edison's library. He was determined to upstage General Electric, and hold the celebration and the dedication of the Edison Institute simultaneously at Dearborn. "I'll show 'em! I'll kidnap their whole party!" he exclaimed.

Reluctantly, General Electric chairman Owen D. Young and president Swope agreed to Ford's plans, providing General Electric was to be the co-host of the jubilee. Construction of Greenfield Village and the museum was speeded up. Ford hired Francis Jehl, the lone survivor of the 1879 experiments aside from Edison, to supervise the re-creation of the lab's interior. Jehl had worked from 1882 to 1922 in Brünn and Budapest, the last twenty-five years as an engineer for General Electric. His European wife had died in 1915, and the inflation following the war had wiped out his savings. Edison had sent him $300 to enable him to return to America.

General Electric retained Edward L. Bernays, one of the nation's leading public relations experts, to orchestrate "Light's Golden Jubilee." Bernays, a nephew of Sigmund Freud, later recounted the experience: "You recognize that he can be made a myth, so you start myth building."

Bernays promoted the organization of a committee of forty-one prominent Americans, headed by President Hoover. He established a speakers' bureau supported by the National Electric Light Association to provide

* Steinmetz, deformed at birth, was a hunchbacked mathematical genius who had emigrated from Germany and first met Edison at the Chicago World's Fair in 1893. Edison offered him a plug of chewing tobacco, which Steinmetz, quietly gagging, slipped into his pocket and preserved as a memento.

speakers for all organizations that asked for them. He arranged for editors of prominent journals to interview Edison. He distributed 10,000 copies of a song, "Thomas A. Edison—Miracle Man," composed by George M. Cohan. He propelled one of the greatest bandwagons in American history into movement, and soon had people in a large part of the world clamoring to get on.*

Yet while everything proceeded toward the great climax, the star of the celebration grew feeble. With his milk diet he was starving not only his bowels but himself. He exhibited unmistakable signs of malnutrition. As he lost weight, his skin developed the folds of a limp flag. In late August, 1929, he contracted pneumonia. This developed into nephritis. The prognosis for a run-down, eighty-two-and-a-half-year-old man felled by two critical illnesses for which there were as yet no drugs was bleak.

But, like his father before him, Edison defied the prognosis. He would never be well again, and he had longer and longer periods of reverie when he noticed no one, but by the end of September he was walking.

Edison and Mina arrived on Ford's private railroad car two days ahead of time, and spent Sunday morning touring the village. Mina played mother to Edison, buttoning his coat and looking disapproving whenever he said or did something she did not like. He was not supposed to chew tobacco, but he had secreted a plug in his pocket; and, like a small boy with a bar of candy, he took a nibble whenever her attention was diverted.

Mina's sentiments about the celebration were mixed. That evening, sitting in her and Edison's room in Ford's palatial home, she jotted down some of her thoughts. "I do not approve of disturbing historical spots. Feel sad to think Seminole Lodge [Fort Myers] was disturbed by taking out the little laboratory.† Feel Menlo Park would have been much more a tribute to Dearie if it had been left in Menlo—here it will in time be forgotten. Feel that Menlo and Electric Light Idea is made so prominent that all the other giant inventions and work is almost forgotten. It is simply a plaything for Mr. Ford and no one will have the same sentiment. I think it a most magnificent thing of Mr. Ford but I regret the method. We must collect all interesting things for Orange Museum. Orange laboratory must be kept the real shrine."

The next morning Edison, Mina, Ford, and many of the other important guests boarded the four-car train pulled by the replica of an 1860 locomotive machined in the shops of Ford's own railroad, the

* When Bernays visited the reconstructed laboratory at Greenfield Village, Ford proudly showed him around. Reaching the outhouse, Ford pointed: "That's Tom Edison's straight-shot privy from Menlo Park."

† Ford built Edison a much more elaborate laboratory at Fort Myers in exchange. "Dear me," Mina remarked when she was disturbed by Ford's workers dismantling the place, "I do wish he would keep out of our back yard."

Detroit, Toledo & Ironton. The train chugged out to the junction to greet President Hoover and his party. It was a cold, rainy day, but the gleaming engine with its immense cowcatcher and huge, V-shaped stack casting whirls of smoke into the gray air had a magnificence that the little local on which Edison had worked had never possessed. A news butcher hawked a basket of fruit. Edison took it from him and went down the short length of the car. "I'll take a peach," President Hoover said, and proffered a coin.

Bell clanging, the train pulled into the station at Greenfield Village. The dignitaries gingerly descended the slick, steep steps of the cars. Crowds of reporters and onlookers clustered about. Charles Schwab selected a green apple out of a barrel and munched on it. John D. Rockefeller, Jr., stopped to chat outside the cobbler's shop. Edison, Hoover, and Ford posed for fifteen minutes together in the rain—Mina tried to hold an umbrella over Edison's head, but he waved her away.

Four hundred guests were present that evening in the outsized replica of Independence Hall. Among them were Marie Curie, Orville Wright, George Eastman, Lee De Forest, Cyrus Eaton, Harvey Firestone, Dr. Charles Mayo, Hiram Walker, Adolph Ochs, Fielding H. Yost, Henry Morgenthau, Walter Chrysler, and most of the greats of the auto industry. Edison, grinning and enjoying himself, waved his hand to people he recognized. Ford, near the end of the table on the dais, kept covering his face with his hands, and when called upon to acknowledge the plaudits rose blushing and quickly sat down again.

The official ceremonies began at 7:30 P.M. and were carried by 144 stations around the world, the largest network up to that time in radio history. Edison and the President were taken to the spic-and-span laboratory. ("We never kept it that clean," Edison remarked, in a distinct understatement.) Jehl, awaiting them there with a replica of the 1879 pump, told Edison the lamp had a good vacuum. Edison ordered him to seal it. A radio announcer, Graham McNamee, gasped melodramatically:

"It is now ready for the critical test. Will it light? Will it burn?" A breathless pause: "Ladies and gentlemen—*it lights!* Light's golden jubilee has come to a triumphant climax!"

High in the tower of Independence Hall lamps came aglow. Tens of thousands of lights flashed on across the Ford complex. Fireworks colored the sky. An airplane carrying an illuminated sign, "Edison, 1879–1929," passed through the rain overhead. All over the world buildings blazed with Christmaslike decorations.

"And Edison said 'Let there be light!' " McNamee intoned.

When Edison and Hoover returned to Independence Hall, Edison, drained with fatigue and overcome with emotion, hesitated in the doorway and then collapsed onto a davenport in the hall. "I won't go in, I can't go in," he muttered.

While an apprehensive hum went through the hall, Mina brought him a glass of milk, whispered to him that millions of people were waiting for him, and slowly coaxed him back with her. After a lifetime of declining invitations to speak before an audience, he was to address the world. Mina handed him the speech. With a trembling voice he read into the microphone:

"I would be embarrassed at the honors that are being heaped on me on this unforgettable night were it not for the fact that in honoring me you are also honoring that vast army of thinkers and workers of the past without whom my work would have gone for nothing." His voice dropped lower and lower, and at the end trailed off into an almost incomprehensible whisper: "As to Henry Ford, words are inadequate to express my feelings. I can only say to you that, in the fullest and richest meaning of the term—he is my friend."

Exhausted, Edison slumped in his chair. Mina and the President's doctor led him from the hall, and he did not hear Hoover's concluding address.

A bleak and poignant reflection of the ceremony took place simultaneously at Menlo Park, where a hundred-foot wood scaffolding topped by a giant lightbulb now marked the site of the laboratory.* A few hundred people gathered outdoors, listened to the broadcast from Dearborn, and saw the bulb lighted at the appropriate moment.

Appearing like featured players in the road company of a hit drama were Dot, Tom, and Will.

Dot had returned to the United States in 1925. After living in Germany for more than thirty years she had been thoroughly Teutonized. While America was at war with Germany, Edison had continued to deposit $200 a month for her in a Swiss bank (a violation of the law against trading with the enemy by the chairman of the Naval Advisory Board), and she had accumulated a nest egg of $7000 in Switzerland. This enabled her to buy shoes, chocolate, and other luxuries unavailable to less-well-connected Germans.

Oscar achieved the rank of colonel, and in June, 1918, Dot wrote her father from Basel: "My old darling is getting old and grayer every day, but he is a dear fellow and our honeymoon will end it seems only when we part for the last time."

It was a time of last illusions. A year later Dot's "old darling" had taken a mistress, and Dot became involved in a messy divorce. She charged that Oscar was a drunkard who had allowed fellow officers to use their apartment for assignations, and had invited friends to ogle her in her bath. He, in turn, accused her of having had a half-dozen

* Several years later the tower was demolished in a storm, and a concrete structure was put up in its place.

lovers. "I shiver at the very thought of living with such a man!" she exclaimed. Later Oscar attempted to blackmail Edison by threatening to reveal some very unpleasant facts about Dot if Edison did not furnish him with money.

By 1924 Dot hated Germans and Germany. "Now the Germans dislike each other for they see how rotten they are at the core," she wrote Mina. "The dishonesty and immorality are appalling. Germany is preparing almost feverishly for another war. If they ever begin another war they ought to be expelled from Germany and made to live like the jews without a country."

Dot, in her fifties, now called Mina, seven years older than she, "mother."* When Dot arrived back in the United States, Edison bought her a bungalow on one and a half acres in Norwalk, Connecticut. She renewed her friendship with Tate and the other acquaintances of her childhood, kept a flat in New York, was a devotee of the theater and the opera, and continued to get into scrapes over her amours. Paying one man who tried to blackmail her thirty dollars, she Freudianized: "None of his treats had any success."

Will had enlisted as an overage mechanic in a tank retriever company in 1918, arrived on the Western Front the last day of the war, and had had a good time in France. After his return, he tried running a farm in Morristown, New Jersey, and operating a Ford tractor rental business. Neither endeavor was successful. In 1926 Edison agreed to build him a $12,000 house—about the same amount of money he had spent on homes for Charles and Madeleine—near Wilmington, Delaware. Will set up a ham radio transmitter, and Blanche, chronically dissatisfied, made friends with the horsey set. Before John Miller (as financial executive) called a halt, the $12,000 house expanded into a $28,000 villa. The place was not yet completed when Will fell seriously ill, and Mina, always good-hearted in such situations, went to minister to him. In gratitude, he sent her two doctor bills totaling $4000.

Tom remained all his life a muddled, ineffectual hypochondriac. Under the influence of Mina, his wife, Beatrice, developed religious fervor, and convinced Tom to be baptized and adopt the path of "wisdom, sacredness, and sunshine." Shortly after this transformation, in 1916, the child Beatrice was carrying was stillborn. Afterward she became harmlessly, but quite pronouncedly, dotty.

Tom, too, as the years progressed, teetered on the edge of psychosis. After the Ecometer Company went bankrupt, he puttered about the laboratory as an experimenter. He announced he had invented an automotive still that would make fuel from kerosene and chemicals while the car was in motion in the same way that a generator produced electricity. His various illnesses became more pronounced. When he

* This practice later led to confusion when Dot related stories, and her listeners did not realize that "mother" was Mina.

drank, he had wild swings of mood and terrible explosions of temper. "He has taken a new form of disease," Beatrice complained. "He becomes extremely morose, sulky, and very rough and unkind to me." In 1922 he was hospitalized with brain spasms. He looked upon his father as he would have upon a hated and unforgiving deity. He wrote the Edison Pioneers that they were "the Frame through which my glorious Father has so brilliantly shined. I know of no man who is a greater admirer of this Great Man—my dear illustrious Father than myself. As his oldest son and namesake I pledge my most lovable support to help Blaze a Glory which will never die in the everlasting self made works of my Father."

But after Edison returned from Florida in June of 1929, and he and Tom had a casual encounter in the lab, Tom bitterly told Dot: "His greetings were as cordial as I should expect I suppose—but it is hard indeed for me to become reconciled to his malice towards me—How happy I am that this feeling was not an inheritance of mine—for I seem to love everyone."

In actuality, Edison was simply indifferent to Tom, as he was in large measure to some of his other children. He looked upon them as he might upon unsuccessful experiments. When he returned to West Orange from the golden jubilee he seemed much better and was even jaunty—he playfully poked a cameraman in the ribs, slapped the backs of the train's engineer and fireman, pulled the brim of the cap down over the eye of his chauffeur, and twirled a new walking stick.

The company's business was not so frisky; and since Edison no longer seemed interested in the company, Charles kept most of the news from him. The phonograph industry had been damaged so badly by radio that, in the spring of 1929, even Victor found it unfeasible to continue as an independent company, and merged with RCA. On the day of the golden jubilee, Walter Miller, who had headed the production of Edison records for more than thirty years, issued a memorandum: "Stop all recording at once."

Manufacture of radio-phonographs continued for another year. But in December, 1930, Charles had to decide whether to pay $100,000 to RCA for a license the following year. The company was turning out only 2 percent of the five million radios sold annually in the United States. The line was unprofitable. Charles decided to retire Edison from the radio and amusement-phonograph fields. Thenceforward the company produced only the cylindrical business phonographs (dictating machines), which provided a steady if unspectacular income until they were made obsolete by the spread of tape recorders in the 1960s.

Six years of excellent battery sales had put the company in a sound financial position. The stock market collapsed the day Edison arrived back at West Orange from Dearborn, and two months later the company's financial expert warned: "We are in our opinion on the threshold of a major business depression."

In business affairs, Charles was conservative and cautious. "The old expense guillotine is running red with gore," he told his mother in June, 1930. "All plans I am making are based on the theory that things are going to get steadily worse." So as businessmen who were able to spend were unwilling, while those willing to spend were unable, the business projections became self-fulfilling prophecies.

Edison and Mina went to Fort Myers before Christmas in 1929, and remained there, much to the discomfort of Mina, who did not like the heat, until mid-June. Mina at times seemed as depressed as the economy. "Life is a strange problem and one wonders what it is all for anyway?" she wrote Charles. "It is beautiful here and we go on doing things to keep busy, but what is the use of it all?"

Edison had thoughts for only one thing. "I can't get my mind off the extremely complex rubber experiments," he told Meadowcroft. He would be all right one day, then sick to his stomach the next. He took nothing except milk, not even water. His kidneys had recuperated only partially from the attack of nephritis, and the diet aggravated their dysfunction. He still had his ulcer; he was diabetic; his entire internal apparatus was a wreck.

The doctors told him he must begin eating normally again and drink plenty of water. But Edison paid no more attention to them than he had to anyone else during his life—*he* knew what was good for himself.

In August, 1930, he was too ill to accept an invitation from President Hoover to the White House. By winter he was confined to a wheelchair. "Don't stop experimenting. It's good for long life!" he had once advised an acquaintance. In Florida a servant had to wheel him about, but he still spent his time in the laboratory. Occasionally he had himself taken to the dock, where he cast a line into the water. Sometimes he stayed up all night—once he picked up a book at one o'clock in the morning, got interested in it, and did not go to sleep until he had finished it.

His stomach was in constant turmoil. Mina attributed his difficulties to his frustration with the rubber experiments, which were not showing results. But his hands were so shaky he could no longer write a letter. When he returned to West Orange in mid-June, 1931, he was exhausted, and remained in bed for several days.

On June 23 he asked Charles about the company's finances, and Charles replied that they were excellent. Thomas A. Edison, Inc., had $5.3 million in Liberty Bonds and $700,000 in cash on hand.

Through July, Edison was in and out of bed. When he went to the laboratory, he missed Meadowcroft, whose mind had been affected by a near-fatal bout with pneumonia in January. He had difficulty staying awake at his desk. Ideas still came to him, and he jotted them down in disconnected fragments. His skin was the color of ash. He told Dr. Howe his machine would have to start functioning better or it might as well not function at all. On Thursday, July 30, he prepared a codicil to his will. On Saturday he was laboriously making his way about Glenmont when his legs suddenly gave way and he collapsed like a disjointed puppet.

His kidney function was a fraction of normal and his diabetes had become acute. Monday afternoon he seemed, nevertheless, much better. He sat up and asked to see the newspapers. Dr. Howe coaxed him to sip a little water by flavoring it with peppermint. By Friday he was joking and making caustic remarks again, and spent considerable time out of bed. The crisis was over. He asked to go for a ride, and was taken almost daily for a short trip in the car.

On August 21 he suffered a relapse. His uremia worsened. He lay in the large double bed in the sunny corner room where Mina had given birth to all three children. One of the first air conditioners in the world was installed, and Mina kept the room fragrant and colorful with flowers. Dr. Howe came every morning from eight thirty to nine o'clock. By the first week in September Edison's condition was critical. Periodically he was dizzy, and oxygen was administered to keep him from losing consciousness.

From time to time he roused himself enough for a drive around the spacious grounds. But there were as yet no miracle drugs or life-support equipment, so there could be only one outcome. Even Mina, who had largely shut her mind to his decline during the past few years, accepted the truth. Madeleine, Charles, and Theodore talked to her, and then held council with their uncle John.

"We all realize that before very long darling Mr. Edison will have to leave us," John Miller addressed his sister on September 17, and continued by telling her the family wanted "to have things well understood and arranged so that there will be as little confusion as posible when the time comes for your darling. In the twilight of the evening you will put your 'dearest one' into his bed of long rest and peace."

The three children and their spouses organized themselves into around-the-clock shifts of "Officers of the Day" and "Officers of the Night," so that someone was always at hand. They formed committees to take charge of every function, and composed an "official list" of ninety people to be invited to the funeral. Mina prayed, read the Bible, and cried a lot. She drew up, and kept revising, a list of music to be played, and bought a new grand piano for the house. John Miller ordered an $1875 bronze coffin. Madeleine sketched a shrub and floral display for the concrete vault, which Mina directed should be placed as close to the surface as practicable.

Gradually, more and more reporters clustered about the house. To keep them from overrunning the grounds, a press room was set up in the steel-and-concrete garage. Twenty telephones and eight telegraph lines were brought in. Edison was on stage as much as he had ever been during his most spectacular feats. In life he had rushed helter-skelter toward each climax, not knowing until the last moment whether the outcome would be triumph or disappointment. Now that he was no longer capable of exercising any control, the final drama was as organized as the funeral of a great king. Only *he* was not reconciled that this should be his farewell performance.

He had been seriously ill many times before and always recovered. His father had been given up for dead at the age of eighty-four, and had bounced back to laugh at the doctor. Every day Edison expected a turning point in his condition. He was convinced that soon he would begin to feel better. Only at the end of September did doubts enter his mind. Perhaps there would be no breakthrough following this struggle; no favorable solution to this experiment.

With the onset of October he had no more interest in his surroundings, his treatment, or the people who were in the room. He was unable to assimilate food, and was only semiconscious even when he was awake. On the morning of October 7 President Hoover called and asked to be notified daily of Edison's condition. Fifty reporters, including some women, camped in and about the garage, played cards, and fed the squirrels. Faith healers, concocters of homeopathic remedies, ministers who wanted Edison to repent, cranks, and fanatics all zeroed in on Glenmont.

On the night of Wednesday, October 14, Edison drifted into a coma. Friday, Miller Reese Hutchinson called: "Might I please have just one more look at him, because I worship him."

"Too late," Mina replied curtly.

All day Saturday Edison's pulse grew weaker and more rapid, his breathing shallower. In the second-floor bedroom and the adjacent sitting room Mina and the children awaited the inevitable hour. Beyond the wide expanse of lawn the red and yellow leaves of the deciduous trees painted surrealistic patterns amid the evergreens. A half moon rose over the rust-red walls and obelisklike chimneys of the house, bathing Glenmont in a soft light like a château tucked away in an isolated forest.

In the hallway outside the bedroom, the grandfather clock chimed each quarter hour. At 3:24 on the morning of Sunday, October 18, 1931—the fifty-second anniversary of Charley's death in Paris—Edison's life came to an end.

CHAPTER 42

Epitaph

For two days and nights Edison's body lay in state in the laboratory library. Thousands of people three abreast formed serpentine lines outside the gates and moved slowly forward in the autumn chill to view the flag-draped coffin. President and Mrs. Hoover and most of those who had been at the jubilee were there. Others came too: Calvin Coolidge and his wife, Andrew Carnegie, Dr. Karl Compton, and Hamilton McK. Twombly, Jr. A score or more of those whose lives had been intertwined with his appeared: Nellie Poyer, Alice Holzer, Batchelor's widow, Upton's two daughters, Lillian Gilliland, Aylsworth, Frank Dyer, Jehl, Sprague, Charles Dally, McCoy, Fred Ott, and Insull, whose utilities empire was on the verge of collapse. Henry Ford secreted himself behind a screen in an alcove. John Ott was missing—crippled and confined to bed, he had died after being informed of Edison's death, and his crutches were placed by Edison's coffin.

Only a year before, Edison had greeted Mina's entrance into the laboratory with an exultant shout: "I'm a heathen!" But Mina's minister had the last words over his coffin twice a day. On Wednesday, the twenty-first, his body was returned to Glenmont. The funeral ceremony, lasting nearly two hours, was held in the drawing room. After the playing of "I'll Take You Home Again, Kathleen," Beethoven's "Moonlight Sonata," Wagner's "Evening Star," Arensky's "Elegy," and several other selections, Edison was taken down the hill to be buried in Rosedale Cemetery.*

As soon as Edison's will was submitted for probate, it caused the smoldering resentment he had left behind among his children to erupt.

* Many years later his body was reinterred at Glenmont, and he and Mina lie in adjacent graves on the grounds.

During the 1920s he had given Mina about 14,000 shares, Charles 4000 shares, and Theodore 3000 shares of the 30,000 shares of Thomas A. Edison, Inc. His will directed that the remaining 9000 shares, valued at about $85 a share, be equally divided between Charles and Theodore. Additionally, his personal fortune consisted of $1,342,000 in United States bonds, $48,000 in railroad bonds, and $48,000 in cash. His portfolio contained 76,000 shares of 37 different defunct companies as diverse as the New Dunderland Company, the Consolidated Railway Telegraph Company, the Treasure Box Mining Company, the Sims-Edison Electric Torpedo Company, and the Architectural Concrete Company. (Had he, on the other hand, kept his stock in General Electric, it would have been valued at about $12 million, and he would have received $5.5 million in dividends before his death.) Everything else, including all real estate, he had previously given to Mina.

Eighty percent of the bond income was to be divided between Charles and Theodore, who were named executors of the estate. Each of the other four children was to receive 5 percent of the remainder. When Madeleine, whose husband had never been fullly accepted by Charles and Mina, learned that she had been lumped with Dot and Tom and Will, she was furious. She had given Edison his only grandchildren (four boys), and had been his staunchest defender in the press. She and Charles had been cool toward each other for years. Retaining an attorney, she charged that Charles had exercised undue influence over their father when he was no longer in full possession of his faculties, and initiated proceedings to break the will.

Tom, who was dominated by Charles, was not at all inclined to enter the contest. Dot and Will, however, came in on Madeleine's side. Will soon ran out of funds, and reached a settlement with Charles. Dot's pique continued to be directed primarily at Mina, and she wrote Henry Ford: "I thought Mrs. Edison would stop her unjust treatment of us after the notoriety given her because of Father's Will but I suppose until the end she will seek fame as she did wealth." When Dot discovered she had very little chance of gaining anything, she dropped out of the case.

Theodore, unique in the Edison family in that he cared little about money—he had complained to Mina while he was at MIT that she was sending him *too much*—wanted no part of the dispute, and after a time resigned as co-executor.

Theodore and Charles, in truth, agreed on very little. One of Charles's first acts after his father's death was to shut down the laboratory. Theodore, who did not want the workers cast into the unemployment lines in the depths of the depression, formed Calibron Products as a research firm to provide jobs for them. Theodore valued his privacy. Charles was politically inclined. After serving in various posts in the early years of the Roosevelt Administration, he was named assistant secretary of the navy in 1936, and subsequently moved up to the post of secretary for a year. A conservative Democrat, he served as governor of New Jersey from 1941 to 1944. Meanwhile, he brought in Wall Streeters to run the company. In 1957 Charles sold Thomas A. Edison Industries to the McGraw Electric Company, and it became McGraw-Edison. Today all the factory buildings at West Orange are gone, and the only remnant of the Edison company is in the hyphenated name.

Theodore played no part in the firm's management, but devoted himself to research. In 1947 he used $1.25 million of the inheritance from his mother to establish the Edison Industries Mutual Association to distribute shares in the company. Eventually, Theodore hoped, the workers would take over the company and divide the profits. Reserved and intellectual, he resembled his maternal grandfather, Lewis Miller. He was an early opponent of the Vietnam War, an advocate of population control, and an active environmentalist.*

Mina, placing much of the blame for the discord on John's influence over Madeleine, tried to make peace between her and Charles. "Please dearies," she requested Charles and his wife, "both help Madeleine. Put aside all bitterness and let us begin all over again. Madeleine has no strength to help her, no big strong character to advise with her. It takes great souls to love." Eventually, after several years, Charles and Madeleine reached an accommodation.

Mina, as sentimental as a mother with her baby's shoes, dream-walked through the year after Edison's death. "Oh it is hard and lonely without Dearie!" she cried. "When I close my eyes he comes before me. If one could only know just where he is and how he is. Striving to believe that he lives in bliss but not forgetting us."

She assuaged her grief and compensated for her sense of uselessness in no longer having Edison to care for by fashioning plans for a variety of memorials. She proposed a $10 million worldwide public subscription for a monument consisting of a dynamo crowned by a shaft of light reflected in a pool—a cross between the Jefferson and Washington memorials. Another monument, consisting of a Grecian tripod and brazier, was to be placed at the confluence of the Potomac and Anacostia rivers in Washington, D.C. For Fort Myers she planned a library patterned after the Taj Mahal.

"She would like to plaster the entire United States with memorials of all kinds," the eminently sensible and practical Madeleine told a friend of the family, "so that posterity will remember the debt that civilization owes to my father. I feel on the other hand that if posterity doesn't remember or give credit it is just too bad for posterity, but it doesn't detract any from my father. He did his work because he loved

* Theodore and Madeleine remain vigorous and alert. Charles died in 1969 and Dot in 1956. Tom and Will survived their father by only a few years. Tom killed himself in 1935, and Will died an excruciating death from cancer two years later.

it and not because he thought anybody was going to name public works after him."

Mina had a common problem of the very wealthy—a disassociation from the struggles of the not-so-well-to-do and the problems of the nation. In the midst of the worst depression the world had ever known, people were not lining up to contribute to memorials.

"My mother never in her whole life has known what it is really to *need* money," Madeleine commented. "Consequently I don't think she has the least idea of what things cost or the relation of the ideal in her head to the practical translation of it in dollars and cents."

So Mina daydreamed of memorials. But, like a mental pilgrim to Jerusalem, she was satisfied with the vision, fearing perhaps that reality might be disappointing. In 1935 she became reacquainted with Edward Everett Hughes, whom she had known as a girl.

"A new love claims two souls," she wrote Charles from Chautauqua. "There has been a silver thread all these years weaving itself into the fabric of time and an attachment of childhood and youth has been renewed in the autumn time of life." A few months later, at the age of sixty-nine, she remarried. She was happy. She had found someone else to look after.

CHAPTER 43

Legend and Legacy

Why Edison?

How did the offspring of such an ordinary family become the most extraordinary inventor in history?

Had Charles Dickens met him as a trainboy, he would have seen little more in Al than in any other ambitious American youth, a character far too commonplace to be the subject of a novel. As for Horatio Alger, he would have noted Al's lack of thrift, his disbelief in religion, his uncouthness, and some of his other less-than-admirable qualities, and concluded he was destined to come to no good end.

It was, in a way, typical of the contradictions in Edison's personality and life that, though he was the great-grandson of a Tory who had almost lost his life in the American Revolution, he took Tom Paine for his idol. Rebellion was one of his most notable early characteristics —rebellion against his disciplinarian mother, against a stern and unforgiving church, against a dull and rigid school. From boyhood to old age he could not bear to have anyone tell him what to do, but remained undisciplined and iconoclastic.

Since he was, in effect, an only child, he was forced to rely on his imagination for much of his amusement. His isolation was accentuated by his deteriorating hearing and his awkwardness. Early in life he was the focus of attention of three women, and as he grew older he was to be self-centered, demanding, and impatient. Before becoming a train hawker, he was a bright and lonely misfit, who in today's society would have been probed and tested, then pushed, pulled, kneaded, and reshaped to make him conform to "the norm."

The job on the Grand Trunk Railroad set him free to develop his aggressiveness, initiative, and ingenuity. Coming from a home in which expectations were always high even if fulfillment was low, he was sublimely optimistic, and this optimism developed into supreme self-confidence. In no other land or period in the history of the world did such an explosively expanding economy scatter the seeds of opportunity everywhere, and a perceptive youth had but to stoop and pick them up.

Had he been a well-rounded person of more physical talent, had he had more success in school or as a telegrapher (not to speak of as a locomotive engineer!), he might have felt less pressure to prove himself, and turned out to be just another nineteenth-century middle American. "If I had not had so much ambition and had not tried to do so many things, I probably would have been happier, but less useful," he remarked shortly before his death.

His poor hearing forced him to rely more on his sense of sight, provided incentive for him to become an excellent reader, and reduced the ordinary distractions and disturbances of the world. When he withdrew into the semisilent universe of his mind, he wove imagination, ingenuity, and technical knowledge into new configurations. "The man who doesn't make up his mind to cultivate the habit of thinking misses the greatest pleasures in life," Edison said. "The brain that isn't used rusts." He attributed the cause of people's troubles to one source: "It is all because they won't think, won't think!" Every room in the laboratory contained the quotation from Sir Joshua Reynolds: "There is no expedient to which a man will not resort to avoid the real labor of thinking."

"My business is *thinking,*" Edison declared.

But Edison was not only a thinker; he was a visionary. And as such his thoughts were often unorthodox. His experiences with conventional institutions and accepted modes of doing things were of such negative character that he was convinced there must be better ways. This exposed him, like most searchers for new paths, to the criticism of the establishment and the derision of his peers. When his critics turned out to be wrong a significant percentage of the time, his faith in his own certitude and his exclusion of all opinion that did not conform to his ideas became ever more pronounced.

From an early age, Edison had an adversary relationship with many of the people with whom he came into contact. He irritated them by upsetting the established order with his innovative sallies. They could not understand why, if an operation or mode of proceeding was good enough for the rest of the world, it was not good enough for Edison. Edison's brashness and disinclination to be diplomatic aggravated his poor relations with people in authoritative positions—had he been forced to make a living by working on a job he would have fared badly. He seldom cultivated friendships, but looked upon acquaintances as dispensable, to be used for his own advantage and advancement, then forgotten. He was what Herbert Hoover extolled, "the rugged individualist." His ethics and his standards were his own; and when they conflicted with his drive for success, they could be overridden.

The Edison who came to New York was an ambitious, aggressive, rebellious, single-minded, self-centered, imaginative, creatively intrepid youth with a half-dozen years of experience in the country's fastest-growing industry. He invented to make money, and to prove himself to his critics. Once he had the money, he did not spend it on drink, because alcohol disagreed with him; he did not spend it on girls, because he was shy and not a womanizer; he did not speculate on Wall Street, because one of his earliest and most profound experiences in New York was watching what happened to those who did; he did not spend it on luxuries for himself, because they meant nothing to him; he did not save it up, because he lacked any inclination for thrift. Instead, he reinvested it in his business.

"I always wanted to obtain money to go on inventing," he said. And however much he wasted, piece by piece he built up his inventive facilities. Although the Menlo Park laboratory was small and primitive by today's standards, in 1876 it was unique. Edison was the only inventor with his own fully equipped laboratory, the only one who could afford to maintain a staff. Neither Michael Faraday nor Joseph Henry had facilities comparable to Edison's. Alexander Graham Bell's laboratory was rudimentary. William Sawyer in New York and Joseph Swan in England were competing against a well-ordered if not well-organized enterprise.

At a time when modern science was still in a natal stage, inventors were forced to do nearly everything for themselves. Obviously, a successful operation necessitated significant wherewithal, which, among inventors, only Edison possessed and was willing to spend. With Batchelor and his other associates Edison composed an inventive *team*. If Edison wanted to experiment with an alloy, he usually had to make it himself. When he needed an efficient vacuum pump, he had to design it, construct it, then redesign and reconstruct it over and over. If electrical measuring instruments were required, Upton had to devise them.

Edison was able to establish not only continuity, so that discoveries made in the development of one invention could be applied to another, but scope. The phonograph was an outgrowth of multiple lines of investigation. Edison sustained the experiments on the incandescent light for over a year without coming close to obtaining a lamp of practical value, yet in the end triumphed and produced not only a commercial light but a generator and a vacuum pump considerably in advance of their time.

Since Edison did not regard an invention as an end in itself but rather as one more discovery in a never-ending search, invention followed invention like the steps on a ladder. The amazement with which he was regarded, even early in his career, was epitomized by the *London Standard* in October, 1878:

"Why a man should tear himself to pieces to concoct a string of inventions is a problem which may interest the psychologist. It seems as inevitable for Mr. Edison to invent as for a fish to swim. It is his nature, or it may be described as an all-devouring passion, which nothing can satiate. We appear to behold a union between the restless spirit of American enterprise and the scientific genius of the present age. Mr. Edison invents with the same impetuous ardor that his countrymen speculate. There is the same disdain of precedent and the same determination to 'go ahead.' "

His first device, in Indianapolis, was conceived to help him learn telegraphy—he rigged two old Morse embossing registers so that incoming messages could be played back at a slower speed. When he modified the apparatus so that the tapes would activate a sounder key and could be used for transmitting, it became a **REPEATER.**

He next entered into competition with a number of other telegraphers to invent machinery that employed electrical impulses not to inscribe dots and dashes on a piece of paper but instead translate signals into roman characters. These devices were the **PRINTERS** for the stock-ticker system and private telegraphy.

So that this procedure could be reversed, and roman characters could be encoded for transmission on the automatic telegraph system, Edison perfected a **PERFORATOR.** From the perforator he progressed into the field of automatic telegraphy, which stimulated him to study chemistry. While working on various devices for automatic telegraphy, he discovered the principle of the **ELECTROMOTOGRAPH.** Experiments with chemical solutions engendered the **MIMEOGRAPH**, and as an adjunct for this he designed the **ELECTRIC PEN.**

In addition to his work in printing and automatic telegraphy, Edison was heavily involved in multiplex telegraphy. Automatic and multiplex telegraphy were different technologies employed for the same end: the more intensive use of the wires. Edison's **QUADRUPLEX** enabled one wire to handle four messages simultaneously.

When multiplexing turned to acoustics, and Bell invented the telephone, Edison converted the electromotograph into the **MUSICAL TELEPHONE**, and it ultimately became the **CHALK RECEIVER**, or **LOUDSPEAKING TELEPHONE.** As he labored along similar principles to develop an autographic copying press, an autographic (or facsimile) embossing telegraph, and a telephone speaker, he produced, instead, the **PHONOGRAPH.**

Edison's automatic telegraph experiments with carbon resulted in his discovery that the conductivity of the element varied precisely according to the pressure it was under, and he applied the principle to the development of the **CARBON-BUTTON TELEPHONE TRANSMITTER.**

The **TASIMETER**, embodying the same principle, followed. This instrument led to his attempt to design an incandescent lamp with an automatic cutoff. Although not one element of the original concept remained by the time Edison brought the **INCANDESCENT LAMP** to realization, his work completely altered the thrust of development in electricity.

As an integral part of the system he constructed the most powerful **GENERATOR** yet devised, then initiated plans to convert the generator into an **ELECTRIC MOTOR**, and used the motor to drive machinery and an **ELECTRIC LOCOMOTIVE.** Since he still expected he would need great quantities of platinum for an incandescent light, he employed the generator in conjunction with a magnet for an **ORE SEPARATOR**, which was later modified for iron mining. The offshoot of the ore milling process was the **CEMENT WORKS.**

In order to deliver electric power, Edison had to design a **DISTRIBUTING SYSTEM**, as well as numerous ancillary items such as switches, fuses, sockets, and insulation. Some of Edison's experiments with telegraphy, and the discovery of the **EDISON EFFECT**, laid the foundation for radio.

When Edison gave his attention once more to the phonograph, he conceived the idea of synchronizing the phonograph with a zootropic device. From these experiments the **KINETOSCOPE** and **MOTION PICTURES** emerged.

After Edison's interest in electric transportation turned to self-propelled vehicles, he developed his final major invention, the **ALKALINE STORAGE BATTERY.**

No other inventor has approached the number of patents issued to him singly or jointly—1093—and they exhibit a notable unity:

Telegraph	150	Ore Separator	62
Electric Pen and Mimeograph	5	Cement	40
Telephone	34	Motion Pictures	9
Phonograph	195	Battery	141
Electric Light and Power	389	Automobile	8
Railroad	25	Miscellaneous*	35

Although Edison had no more than four years of schooling—in a day when few but privileged children went beyond grade school—he was not uneducated. He emerged from his years as a newspaperboy and telegrapher with a wide span of general knowledge. His on-the-job training in telegraphy, supplemented by his inquisitiveness and experimentation, gave him an excellent grounding in practical electricity. When,

* Among the miscellaneous patents are three for typewriters, one for vacuum preservation, one for an auto giro (a cross between a helicopter and an airplane, having a rotor for lift and a propeller for forward motion), three for chemicals, three for military projectiles, two for radio, and one for rubber.

subsequently, he learned chemistry, he developed into the outstanding empirical chemist-electrician of his age. Time and again his interdisciplinary self-education proved the key to success.

Complementing his expertise as a chemist and an electrician was his ability as a promoter. He had picked up from his father the art of scheming. From his association with newspapermen he learned the importance of the press. He was not only a good inventor, but always managed to convince potential backers he was better than he was; and when he projected himself into the newspapers, he was never short of superlative. He proceeded on the well-established theory that successes will be remembered, while failures are forgotten. He acted on the basis that a promise can bring practical advantage, so it is self-defeating to worry whether the promise can be kept. The genesis of the quadruplex, the carbon-button telephone transmitter, and the phonograph made the disappointment of other, unfulfilled expectations seem insignificant. Every success augmented the mystique growing up about him, and added to his standing with the financial community—from the time that Marshall Lefferts and George Harrington began to back him, his ties with Wall Street made him unique as an inventor.

In keeping with his background and experience, Edison's inventions were distinguished by two primary characteristics.

They were *practical;* that is, they were intended to earn money. "A scientific man busies himself with theory. He is absolutely impractical. An inventor is essentially practical," Edison asserted. "Anything that won't sell, I don't want to invent. Its sale is proof of utility, and utility is success." Although he was occasionally diverted by phenomena like "etheric force," he did not pursue them—he could not, in truth, afford to pursue them. When the phonograph proved inapplicable for recording telephone conversations and sales of the machines lagged, Edison lost interest until Tainter threatened to transform the invention into a commercial instrument. Although he was a pioneer in wireless transmission, he did not follow up the experiments after the rather unsuccessful Grasshopper operation or renew them after Marconi demonstrated long-distance transmission was practicable, because he failed to perceive that any substantial revenues could be derived. Since, for a long time, motion pictures appeared to have only limited profitability, Edison during his entire career took out only nine patents in the field. In comparison, he obtained sixty-two patents for ore separation, which he expected to be the source of fabulous wealth.

In addition to being practical, his inventions were *topical.* He worked either in the mainstream of technology, as with the telegraph, or jumped into a developing field, as with the telephone, the incandescent light, the X-ray, the electric automobile, and, at the very last, the domestic production of rubber. When he himself was ahead of everyone else, as with the phonograph, wireless telegraphy, and the kinetoscope, he paused until commerce caught up to him.

"Verily the pioneer has to get his justice in the same way that florists get bouquets from Century Plants," he remarked. And Edison wanted to get his recognition and money at once, not in a hundred years.

By dealing in the practical and topical, he was able to interest men whose motive was profit, and to obtain backing. His contacts, his ability to raise money, his exploitation of people, his ambition, his imagination, his dedication—these were the ingredients of his success. There was one more, perhaps the most important of all: his persistence.

Conversely, many of the elements that were the ingredients of his success as an inventor were the cause of his failure as a businessman.

His disinterest in tailoring his products to customers' preferences and demands, his disdain of economics, his jealousy of the inventions and products of others, his stubborn assertion of his own correctness against all contrary opinion, his insistence on being different from everybody else, his unwillingness to cooperate and compromise, his impatience on the one hand and tendency to procrastinate on the other, all contributed to his lack of commercial success.

Despite his talent for press agentry, he understood little of the psychology of dealing with the public. His privateering approach, picked up from Orton and Gould, alienated people and was self-defeating when he destroyed the confidence and good will of men with whom he had to conduct business. His persistence in continuing failing ventures after all reasonable probability of their success was exhausted was a blueprint for bankruptcy.

Not only was it virtually impossible for Edison to reach a disagreeable decision, it was difficult for him to come to *any* decision. The process of arriving at a logical choice was something he had not learned in his self-education. That was why in experimenting he tried whatever seemed reasonable. Either a mechanism or a chemical reaction would work or it would not. *It* made the decision. Not he.

As his fame increased, there was a surge of interest in his background. Edison, however, was close-mouthed about his personal life. When he rattled off anecdotes, he used them as a means of defense against too close an inquiry into his family relationships and business transactions. Men like Mac Kenzie strove to expand the importance of their association with him, and told exaggerated tales. George Bliss, using largely thirdhand material, was the first to attempt to put together an account of Edison's life, which he presented before the Chicago Electric Society early in 1878. Soon afterward, he reported chagrinedly to Edison:

"Ashley writes me that my article is a romance and all that I have ever heard about you isn't so." A few months later, nevertheless, a Chicago publisher, Rhodes & McClure, expanded on the Bliss biography in a slim volume. Edison was credited with such feats as using the whistle on a locomotive to send Morse-code messages across the St. Clair River, a tale that was appropriated from the 1860 visit to the United States by the Prince of Wales, when the prince's guide, Colonel Wilson, had used two locomotives to signal back and forth across the Mississippi. Edison's experience at Stratford was embellished by grafting on a portion of another telegrapher's story of his service on the Grand Trunk Railroad, an account that appeared in *The Operator* under the title of "Brian Born's Hard Luck" adjacent to Bliss's life of Edison. Yet Bliss's "romance" and the material from the McClure book were treated as source material in almost every subsequent book about Edison.

"Edison never went to school over two months in his life," Bliss said, a statement Edison himself topped a quarter century later when he yelled in exasperation: "School? I've never been to school a day in my life! D'you think I would have amounted to anything if I had?"

Yet Edison had not only gone to three schools at Fort Gratiot and Port Huron but, when his knowledge of chemistry was inadequate for his needs, had attended classes at Cooper Union. It was, indeed, a time when it was fashionable to proclaim that one was self-educated and uncorrupted by schooling, and men with a great deal more education than Edison made protestations similar to his.

Following the revival of the phonograph, George P. Lathrop obtained Edison's permission to talk to a friend, Samuel Clemens (Mark Twain), about writing a biography of the inventor.

"Clemens responded very favorably to my suggestions about the book, but seemed to think it important to have it put in autobiographical form," Lathrop wrote Tate. Unfortunately, Lathrop noted, "Edison has more than once stated to me his decided aversion to autobiography."

Clemens visited the laboratory and tried out the phonograph, and he and Edison regaled each other with stories. But the project broke down over whose name was to appear as the author's. Edison was fundamentally honest. He preferred to do the "honorable" thing—in 1879 Hilborne Roosevelt told his brother: "Edison is one of the most honorable men I ever met and when he is a friend to anyone he will never go back on him." But Edison's ideals were always being compromised by his practicality. Consequently, as with the Edison Speaking Phonograph Company and the North American Phonograph Company, he rationalized and did not what was honorable, but what was expedient. Long before Freud popularized the id and the ego, Edison had his conflicts between "Mind Number 1" and "Mind Number 2." In his private life "Mind Number 1" generally prevailed. In his professional life "Mind Number 2" and self-interest dominated. During the nineteenth century commerce and finance were widely regarded as a poker game in which a person pitted his wits and money against the other fellow's; and if he lost either or both, it was unfortunate but no cause for complaint. To Edison, people who invested in his projects were speculators, and the fact that he was able to convince them that the projects were sure things and not gambles was part of the game.

Still, in a book all of this would require explanation. If someone else wrote about the events, they could be obscured and varnished. But Edison did not want to sign his own name to the account.

In 1893–94 William K. L. Dickson and his stepsister wrote a brief, sycophantic book in which Dickson placed himself in Edison's service several years before he had actually arrived. From time to time Edison was featured in other publications, such as *Four American Inventors,* which led him to comment: "The statements are correct and all right although the taffy is rather strong." In late 1907 Hearst's *Cosmopolitan* magazine urged Edison to authorize a book and its serialization.

"When I go into senile decay I may consider the autobiographical scheme but as long as I can put in 18 hours daily I don't want to waste any time on it," Edison replied.

Edison responded in a similar vein to Thomas C. Martin, editor of the *Electrical World,* who was acting as an intermediary for *The Century* magazine.

A few months later, however, in February, 1908, Johnny Randolph committed suicide, and Edison once again became seriously ill with a mastoid infection. The New York Phonograph Company triumphed, and Edison cut production sharply at the Phonograph Works. He was struggling with the battery. The Dunderland venture was heading for bankruptcy, and the Portland cement plant was in trouble. Frank Dyer thought Edison's image could stand polishing. After Edison returned from Fort Myers, Dyer persuaded him to submit to an official biography.

Dyer and Martin were, ostensibly, to be the authors of the book, but the wispy William Meadowcroft, who had joined Eaton's law firm as a clerk in 1875 and had written the best-selling *The ABC of Electricity,* was put on the laboratory payroll to do most of the work.

Meadowcroft and Martin struggled through the winter of 1908–1909. Millions of letters, notebooks, clippings, documents, and papers of all kinds chronicling Edison's career existed. But, hopelessly jumbled, they were packed away in boxes, crates, trunks, and cabinets scattered about the laboratory, the factory, and Glenmont. Much of the material was inaccessible, and did not, in any case, contain the kind of information to be presented in a staid, official biography. When Dyer saw the results of Meadowcroft's and Martin's work, he told Edison:

"The book ought to be dignified, but as you have often told me you never became thoroughly interested in a subject until it looked pretty hopeless, I hope you will feel the same way toward the book in its present condition and help us out. You are the only one who can inject the spark of life into the apparent corpse."

Sitting down with Meadowcroft, Edison related humorous stories and impressions of his career. These were melded into the book, which became a strange pastiche. Edison told stories primarily for their effect, not their accuracy. He had little interest in dates, a bad memory for figures, a great capacity for generalization, a history of inconsistency, and a penchant for exaggeration.

"My refuge was the Detroit Public Library.* I didn't read a few books. I read the library," he claimed, although in his correspondence he admitted: "I started to do it but gave it up after reading about ten books that were pretty dry reading."

Edison's fanciful account of how he had started at Western Union led P. H. Shaughness, who had been there with him, to comment: "You can invent history as easy as other things. Now that Mark Twain has retired as humorist you are in line of promotion."

Edison recounted that when he had sold his stock ticker to Marshall Lefferts he had hoped to get $5000, but didn't dare to ask for such a sum, and had suggested Lefferts make an offer. Thereupon, Lefferts had said: "How would $40,000 strike you?" Edison asserted it had stricken him almost into unconsciousness, but he had revived enough to take the check to the bank and stuff $40,000 in bills into his pockets. The tale led Reiff to laugh: "You are made to say that Lefferts gave you $40,000 when you only expected $5000 and Orton offered you $100,000 for all your inventions. Of course it seemed funny to me that Lefferts and Orton were throwing $40,000 and $100,000 around so recklessly."

According to the biography, Edison had decided in 1877 (long before he had in fact turned his thoughts to the development of an incandescent lamp) that carbon was the substance he was seeking. "On October 21, 1879, after many patient trials, he carbonized a piece of cotton sewing thread bent into a loop or horseshoe form. This lamp lighted up brightly to incandescence and maintained its integrity for over forty hours, and lo! the practical incandescent lamp was born." Moving ahead to December 21, the book reported "some hundreds of these paper-carbon lamps had been made and put into actual use," thus multiplying the true number by about ten. The bamboo filament was supposedly discovered "one day in the early part of 1880," advancing the actual date by about six months.

Talking about the mimeograph, Edison said: "Toward the latter part of 1875 in the Newark shop I invented a device for multiplying copies of letters, which I sold to Mr. A. B. Dick of Chicago," thereby skipping entirely over the dozen years the rights had been held by the Gillilands and George Bliss. (Gilliland was, of course, unmentionable.)

Referring to Wilber's embezzlement of $1300 in patent fees, Edison

* The Detroit Public Library actually was not yet then in existence. Edison went to the library of the Young Men's Society.

elaborated that Wilber (whose name he did not mention) had sold seventy-eight of his inventions to other people, who had thereupon patented the inventions themselves, an imaginary occurrence that, had it been true, would have been the biggest patent scandal in history and brought on a lawsuit of gigantic proportions.

Edison insisted that inventing per se had not been profitable. "Whatever money I have made has been made in manufacturing and not selling the patents," he declared. This statement overlooked no more than a million and a half dollars in cash, not to speak of remuneration in stock. The sums ranged from such relatively small amounts as $5000 for his Polyform concoction to $47,000 for automatic telegraph rights ($17,000 from England and $30,000 from Gould), $50,000 for the quadruplex, $100,000 each for the telephone transmitter and the electromotograph, $175,000 for British telephone rights, $70,000 (plus $330,000 in stock) for British ore-milling rights, and $435,000 from Lippincott (actually received) for phonograph-marketing rights. The Bell Telephone Company paid him $130,000 over a span of more than twenty years for an option on future telephone inventions, though Edison never produced any. He repeatedly received sizable sums for marketing rights to various inventions in Europe, Central and South America, Australia, and Asia. In many cases, such as the electric light, the kinetoscope, and the battery, Edison was not interested in selling the rights; and in the instance of the phonograph he got them back.

Though he made a major contribution to the development of the stock ticker, he never tired of playing the innocent on Wall Street, and of casting himself in the role of victim. He was an actor, and assumed whatever part he thought would gain him the sympathy and plaudits of his audience. He described how he had paid $10,000 (expanded from the actual sum of $6200) to C. E. Chinnock to put the Pearl Street station on a commercial footing, and how, when he had requested the Light Company directors to reimburse him, they had said "They were sorry. Wall Street sorry." In reality, Edison had made the deal entirely on his own initiative, and had received a settlement of $67,000, in which the Chinnock claim was included.

Discussing the electric-railway project of 1881, the authors became entangled in a welter of confusion. "Mr. Edison is authority for the statement that Mr. Villard advanced between $35,000 and $40,000, and that the work done was very satisfactory; but it did not end at that time in practical results as the Northern Pacific went into the hands of a receiver. Mr. Insull states that the money advanced was treated by Mr. Edison as a personal loan and repaid to Mr. Villard." The sum Villard had actually advanced in 1881 was $12,000, and Edison, on failing to perform the contract, had been forced to return it or lose the Light Company stock he had put up as collateral. The $40,000 referred to was part of the $250,000 the North American Company, Villard's umbrella organization, was supposed to provide in 1889–90 for the low-voltage-electric-railroad project that was, after a few months, abandoned.*

Such distortions abounded in the two volumes, which were intended not so much as a biography as a business encomium.† (In the nearly 1000 pages, one sentence was devoted to Mary and her children, one paragraph to Mina and her children.) Failure was explained away. The iron-ore venture, it was said, would have succeeded except for the uncontrollable misfortune that the price of ore fell from $6.50 to $3.50 a ton. The cement plant was treated as an innovative success, and the concrete house described as an accomplished fact instead of a drawing-board nightmare. While the book spoke of "Edison's conception of the workingman's ideal house," a 2250-square-foot structure to be built for $1200, the laborers at the cement plant were living in as miserable a shantytown of packing crates, discarded tin, and general jetsam as ever existed in America.

Since Reiff's and Edison's suit against the Gould estate was still before the court, Edison referred to the automatic telegraph and the quadruplex only in generalities, made Gould out to be the villain, and pretended he had had no more contact with the financier after 1875.

Edison related the entire history of his electric business from 1886 to 1892 as follows: "At these new works our orders were far in excess of our capital to handle the business and both Mr. Insull and I were afraid we might get into trouble for lack of money. Mr. Insull was then my business manager, and running the whole thing; and therefore, when Mr. Henry Villard and his syndicate offered to buy us out, we concluded it was better to be sure than sorry; so we sold out for a large sum."

Subsequent writers groping for a more rational explanation theorized, without facts, that Edison had been the victim of a Wall Street plot, and that the sale had been precipitated by Insull's profligacy.

As time progressed, the accounts of Edison's life and inventions became ever more exaggerated and distorted. Everyone tried to place his relationship with Edison in the most favorable light. Henry Ford wrote: "After that first meeting in 1896, I saw him again two or three years later in his laboratory in West Orange, New Jersey. I had the thought of finding a starting battery which would give enough power to enable us to combine a starter and a generator in one motor unit." Ford thus predated his relationship with Edison by more than a dozen years.

Rosanoff ended his lengthy article on Edison not by being fired after failing to produce anything to Edison's satisfaction, but by triumphantly discovering the sought-for wax.

* Edison also negotiated a $40,000 loan from Villard in the late 1890s for the ore mill.

† The campaign to glorify Edison even included touching up a photo of the electric locomotive to show Edison instead of Charles Hughes at the controls.

Edison himself was amused by the mythology growing up about him. "This is my 998th birthday," he remarked in the late 1920s. "It's the newspapers, bless 'em. We've been amusing ourselves at home by collecting these yarns and calculating how old I should have to be to live through all the adventures they have set me down for. When we quit last night I was only 970, but the bureau boosted me 28 years this morning."

An incurable showoff, Edison was, throughout his life, his own best creator of fiction. To himself he admitted: "I am afraid I am like the parrot and the dog. I talk too much, more than I perform." But he was never able to restrain himself from popping off, or from discussing grandiose projects as if all that was required to materialize them was a snap of the imagination.

Since he was sensitive about his early slow physical development, he took pains to show himself to be "manly." He suppressed emotions and sympathy, and made light of tragedy. Recalling an incident when a visitor's poodle was swept up by the machinery at Goerck Street, he related: "The dog was whirled around 40 or 50 times, and a little piece of leather came out—and the ladies fainted." No weakness was to be admitted. Edison always maintained that while a whole ship's company might gasp and heave, he would sit unaffected and puff away on his cigars. Repeatedly sick and chronically suffering with his stomach, he claimed he had only one serious illness in his entire life.

The most pervasive legend was that he slept only a few hours out of every twenty-four. It was, in reality, a smokescreen to divert attention from his habit of nodding off. While in his younger years there were stretches of several days when he drove himself and others past the point of exhaustion, he always managed to catch up on his sleep afterward.

Tate related: "His genius for sleep equalled his genius for invention. He could go to sleep anywhere, any time, on anything." He and his secretary once went to the beach for the weekend, and Edison slept thirty-six hours straight with only a two-hour interruption for a dinner of steak, potatoes, and apple pie. It was common for him to work a day and a night, taking occasional naps at the laboratory, then go to bed at Glenmont for eighteen hours.

It was another example of the division between the ideal and the practical. Ideally, Edison contended: "Sleep is an acquired habit. Cells don't sleep. Fish swim about in the water all night; they don't sleep. Even a horse don't sleep, he just stands still and rests. A man don't need any sleep." In practice, Edison did like everyone else, and went to sleep.

The man whose relationship with Edison turned out to be the most curious and ironic was Francis Jehl. Over the years, Jehl grew more and more resentful of Edison because, he felt, Edison had never fulfilled the "blue sky" promises he made at Menlo Park. In 1913, Jehl attacked Edison in a lengthy letter:

"He would never hesitate to spend large sums of money when chasing Mammon and the laurel wreath. He was always seemingly affectionate, but only to those who were of use or benefit to him in satisfying his greed and lust. In appearance he could well pose as a profound thinking man, that seemed guileless, pure and natural, but that was only a mask. Mind, I don't say that he is not a genius, for that he is; when an abnormal man can find such abnormal ways and means to make his name known all over the world with such rocketlike swiftness, and accumulate such wealth with such little real knowledge, a man that cannot solve a simple equation, I say, such a man is a genius—or let us use a more popular word—a wizard."

A few years later, Jehl, whose letter was as one-sided as other accounts of Edison were exaggerated, returned to the United States with $300 given him by Edison. After he was hired by Henry Ford, Jehl inflated his own importance in the development of the incandescent light by moving the role he had played in the making of the bamboo filament (when Batchelor was in Europe) forward to the time of the realization of the horseshoe filament. He became a key figure in the lionizing of Edison at the golden jubilee, and contributed substantially to the apotheosizing of the inventor.*

Thus, layer by layer the legends and myths obscured Edison's character, until he was like a magnificent master vanished beneath the encrustations of fiction. As the cynical, knowledgeable, cussing, ruthless, hard-driving, revolutionary, willful, and visionary Al Edison disappeared he was replaced by an Olympian and omniscient genial genius.† Yet in the process Edison was not only deprived of his human qualities and turned into a kind of Greco-Roman demigod, a singular once-in-a-millennium phenomenon whom ordinary mortals might only regard in stupefied wonder, but his achievements and his legacy were also overshadowed and diminished.

Edison himself never wavered in his assertion that he was not a wizard or a genius—in fact, he despised the designation. When an acquaintance once referred to his "Godlike genius," Edison snorted: "Godlike nothing! Sticking to it is the genius!"

"Any other bright-minded fellow can accomplish just as much," he proclaimed, "if he will stick like hell and remember nothing that's any good works by itself. You got to *make* the damn thing work."

Edison's talent was not unlimited. Its boundaries were quite well de-

* In his *Menlo Park Reminiscences,* published by Ford in the late 1930s, Jehl had only praise for Edison. A single, bland, between-the-lines remark hinted at his 1913 feelings.

† In 1940 MGM released a syrupy movie, *Young Tom Edison*, thereby creating an entirely imaginary figure, since there had never been a young Tom, only a young Al.

fined. Even in electricity and chemistry, theory and abstractions were beyond his grasp. When he ventured into other fields, such as engineering and mining, the results could be disastrous. He invented best only when his own ingenuity was complemented by the skills of his collaborators.

Jehl said: "Edison is in reality a collective noun and means the work of many men."

Batchelor was his most important co-inventor. From 1870 to 1880 there was little Edison did that Batchelor was not involved in; and in such inventions as the mimeograph and the telephone Batchelor probably played the principal role. Jim Adams and Charley Edison made important contributions. Upton's work was essential to the development of the lighting system. At the West Orange laboratory, Aylsworth was the key figure in the continuing improvement of phonograph recording and the production of the alkaline storage battery. Dickson did most of the experimenting on the kinetoscope and the kinetograph.

Edison's character would stand in better light had he been more generous in sharing the credit. Yet it was precisely his need for attention, praise, and justification that drove him on. When writers lacked the energy and perspicacity to look beyond one man, he was not charitable enough to point their error out to them and deprive himself of part of the glory. Only at the golden jubilee did he allude to the conflict in his mind:

"I would be embarrassed at the honors that are being heaped on me on this unforgettable night were it not for the fact that in honoring me you are also honoring that vast army of thinkers and workers of the past without whom my work would have gone for nothing."

It was, nevertheless, unquestionably Edison who provided the driving force and made the work of the others possible. Batchelor, though he had a mind fully as lively as Edison's, produced practically nothing individually. Adams would never have been anything but a drunken sailor had he not stumbled into Edison's laboratory in Newark. Charley dissipated and died as soon as he was able to duck out from under Edison's influence. Upton produced only under Edison's leadership and spur. After leaving Edison, Dickson spent forty years in the motion-picture industry and in his small laboratory in England without developing much.

Edison's accomplishments become even more astonishing when compared to those of other leading inventors. James Watt's fame rests on taking an existing machine and perfecting it. Robert Fulton adapted the steam engine to ships (but this was by no means an original concept and he had considerable competition), promoted canal building, and was a pioneer in experimenting with submarines and torpedoes. George Stephenson produced the first practical railroad, but he was simply carrying the work of others to conclusion. Samuel Morse, like Fulton a successful painter, was responsible for one remarkable invention, the telegraph. Cyrus McCormick revolutionized agriculture, but did not venture beyond that field. Charles Goodyear hit upon the process of vulcanizing rubber accidentally. Alexander Graham Bell, after proving that a continuous current was the secret to telephonic transmission, never developed another significant invention. Wilbur and Orville Wright were the first to fly a powered heavier-than-air machine, but were preceded by the near misses of several other men. Guglielmo Marconi took up the theories and experiments of a number of men, among them Edison and Heinrich Hertz, and carried them to a practical conclusion. Henry Ford, who was the world's most ingenious manufacturer and the epitome of the practicality that Edison always aimed for but never achieved, could scarcely be called an inventor at all. Only Michael Faraday, who greatly advanced the sciences of chemistry and metallurgy, was the principal discoverer of the interrelationship between magnetism and electricity, laid the foundation for the generator, made the first transformer, and provided a clue to the existence of the electron, rivaled Edison in diversity.

Edison's inventions are like the building blocks of an obelisk rising out of the history of the world. No other man has ever been responsible for striking the spring of so much wealth, nor had such influence on the lives of so many people.*

The mechanism of today's telephone speaker remains essentially the same as that of the carbon-button transmitter of 1878. The microphone is merely an adaptation of the carbon button. The "Edison effect" globe of 1884 laid the basis for the radio vacuum tube. The alkaline battery was an important step forward in the technology of stored electricity. The mutation of the zootrope from a novelty into a commercial instrument was initiated by Edison, and it was at West Orange that the world's first motion-picture studio was established.

The phonograph, for which the kinetoscope was intended as an adjunct, was Edison's most novel and revolutionary invention. Although his belief that a recording and reproducing device would prove valuable in telephone transmission was erroneous, the three different modes with which he experimented proved to be startlingly perspicacious. The disk remains the phonograph of today. The cylinder continues to be used in such dictating machines as are still in operation. The paper band had some similarity to magnetic tape—the underlying idea was comparable even if the technologies are a century apart.

* Motion pictures are a $2 billion industry in the United States; phonograph and recording, $1.5 billion. The electric industry generates nearly 2 trillion kilowatts of power and grosses almost $150 billion a year. Sales of household appliances, washers and air conditioners, stoves and refrigerators, radio and television sets account for more than $15 billion annually. General Electric has assets of $10 billion and sales of $13 billion. The A. B. Dick Company, which began by marketing the mimeograph, grossed $281 million in 1975.

Yet quadruplex, telephone transmitter, phonograph, kinetoscope—all pale before Edison's achievement in electric light and power. The incandescent lamp was not so much an invention as the marker of a technological and social revolution. To bring the incandescent light to commercial fruition, Edison had not only to develop a lamp but to produce a generator of far greater capacity than previously conceived, and to design a system of electric distribution. Together with light, he endeavored to bring power to the world—when, early in the fall of 1879, he appeared to have reached a dead end with the lamp, he planned to construct power stations for electric motors.

When Edison launched his quest in 1878, only a few homes, dependent on candles, kerosene lamps, and gas lights, had the illumination of 25 candlepower considered the minimum necessary for reading. Public transportation consisted of horse cars and trains pulled by steam locomotives. The limited power of steam elevators held building heights to a few stories. Dentistry consisted mostly of pulling teeth, since the slow, foot-powered drills made filling cavities excruciating experiences for patients. Factories, dependent on coal and water, were concentrated along waterways, blackened cities with their pollution, and only infrequently operated at night.

Between 1878 and 1882 Edison constructed the prototype for the entire electric light and power industry. The first skyscraper, using electric elevators, was erected in Chicago in 1885. During the next two decades, electric traction systems were built in every major city, and the first electric locomotives came into use. Motors combined with machines and tools increased industrial productivity and lightened household tasks. Electricity revolutionized the factory. The path was cleared for modern medicine and dentistry. Business became more and more dependent upon power; and Edison's early oddity, the electric typewriter, appeared in offices. Home life was transformed because night was no longer a time of darkness—today one house benefits from more candlepower than an entire village in the nineteenth century.

Edison threw open the door through which not only he but a host of scientists and inventors rushed to make discovery after discovery. One hundred years after Menlo Park, Edison's vision of a universe run by electricity has been fulfilled. Blackouts, when they occur, bring life almost to a standstill. It is possible to conceive of a world without automobiles, airplanes, and natural gas, but to revert to an age without electricity would cripple civilization.

Edison was a scientific explorer who discovered new continents where prevailing opinion held that none existed. If sometimes he sailed off the edge of the world, that was part of the risk of exploration. During his career he repeatedly exhibited the amorality of a conqueror; but his conquests were dedicated to progress. When Mina, and other people, tried to place a halo on his head they failed to understand that had he been philanthropic and saintly he could never have accomplished what he did. It is remarkable that a man who during his life had recurrent serious illnesses and took extended vacations could compile such a monumental record.

What if there had been no Edison? Would not another man or other men have made the same discoveries? Eventually, in all likelihood, yes. But Edison's enormous impact was due to the uniqueness of his character and his particular approach to life. He was a scientific primitive with talent enough to push ahead, but not the education or sophistication to know that he was, presumably, attempting the impossible. He was the epitome of the practical inventor. At Menlo Park he established and equipped—without conscious intent—the forerunner of the industrial research laboratory. He poured into research and development more money than his enterprises yielded—his ambition for wealth was repeatedly subverted by his passion for invention.

He was inimitable in his time, and it is probable that he advanced the art of electricity by at least a generation. As for the phonograph, the discovery that sound could be recorded and reproduced was such a rare combination of insight, imagination, and good fortune that it is futile to speculate when someone else might have hit upon the secret.

Edison succeeded, where others failed or never tried, because it was his nature to dare. Moses Farmer, who seemingly had every advantage over Edison, started everything and dropped everything. He lacked the drive and the great vision, he was unwilling to gamble, he looked for a safe nest. Men like Upton and Batchelor didn't have Edison's audacity, his hunger for recognition, his ability to inspire confidence, or the one-sidedness that led Edison to live for his work and by his work. Edison himself recognized that he was an anomaly. "Don't touch this business," he warned an aspiring inventor. "Not one man in a thousand ever succeeds in it."

Edison succeeded because he was an eternal optimist who would not let himself or others consider the possibility of failure; because he was an unconventional thinker, who accumulated the resources that enabled him to transform his ideas into reality; because he charged ahead when others hung back; because he demolished the opposition and bowled over impediments. A child of the rough-and-ready universe of the Industrial Revolution, where many failed but a few succeeded spectacularly, he was the product of a unique conjunction of talent, ambition, and opportunity. There was never anyone like him before. And, in the hundred years since, the world has changed so radically it is highly improbable that there will ever be anyone like him again.

APPENDIXES

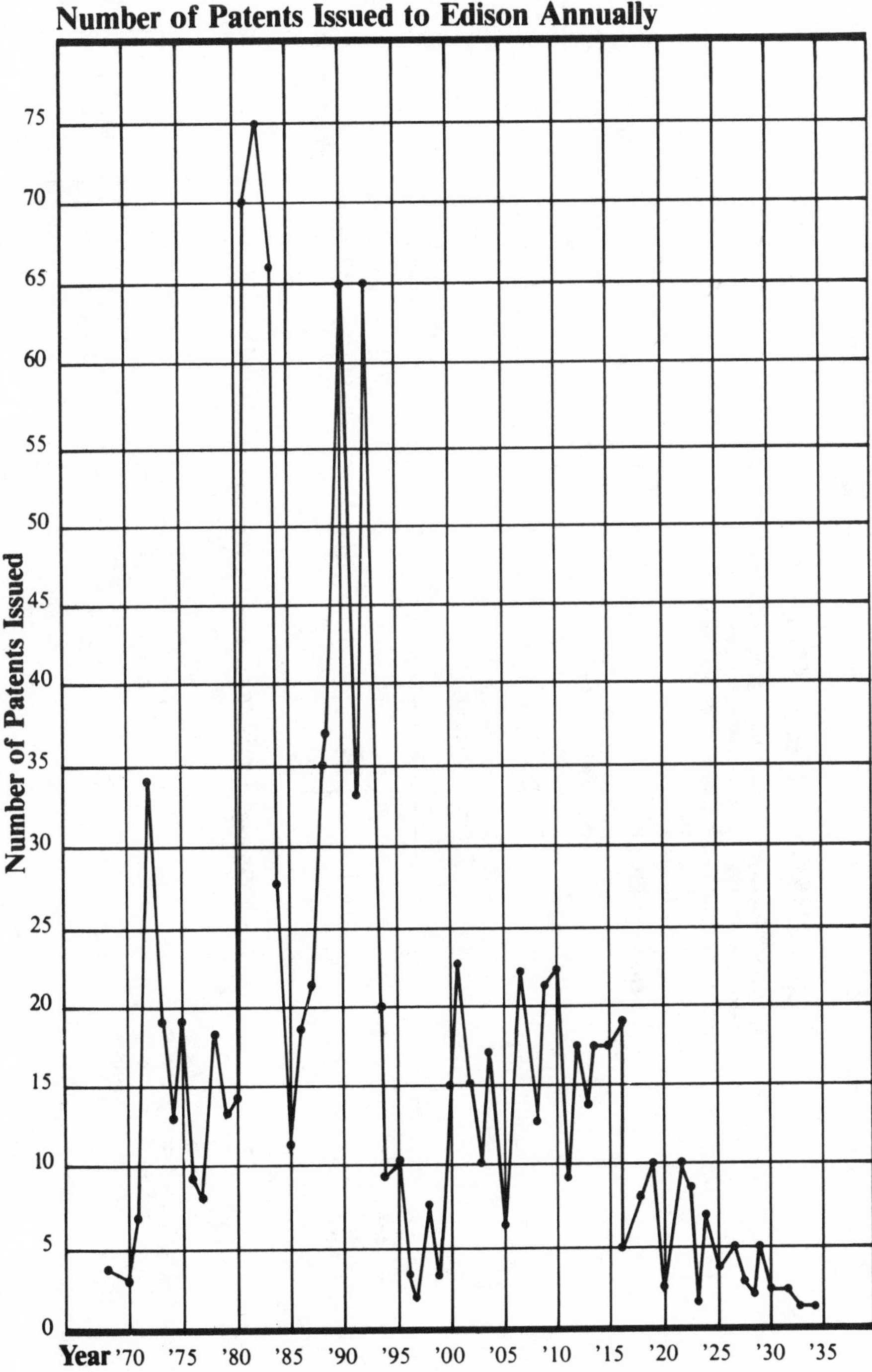
Number of Patents Issued to Edison Annually
Number of Patents Issued
75
70
65
60
55
50
45
40
35
30
25
20
15
10
5
0
Year '70 '75 '80 '85 '90 '95 '00 '05 '10 '15 '20 '25 '30 '35

AN EDISON CHRONOLOGY

The Chronology demonstrates how Edison throughout hie life carried multiple projects forward concurrently. At various points projects divide and merge, but there is a continuity present from beginning to end.

October, 1869 Pope, Ashley, and Edison partnership to produce stock ticker and private telegraphy equipment

April, 1870 Firm merged with Gold & Stock Company

October, 1870 Edison signs new agreement with Marshall Lefferts of Gold & Stock Company to furnish stock printers

January, 1871 Edison approached by Western Union re: fashioning of duplex

February, 1873 Edison experiments on duplex for Western Union

August–September, 1870 Edison signs agreement with Daniel Craig and George Harrington to provide auxiliary equipment for Automatic Telegraph; forms American Telegraph Works

April, 1871 Edison signs new contract with Harrington and sets to work on developing own Automatic Telegraphy system

April, 1870 Edison and Unger partnership to manufacture telegraphy equipment

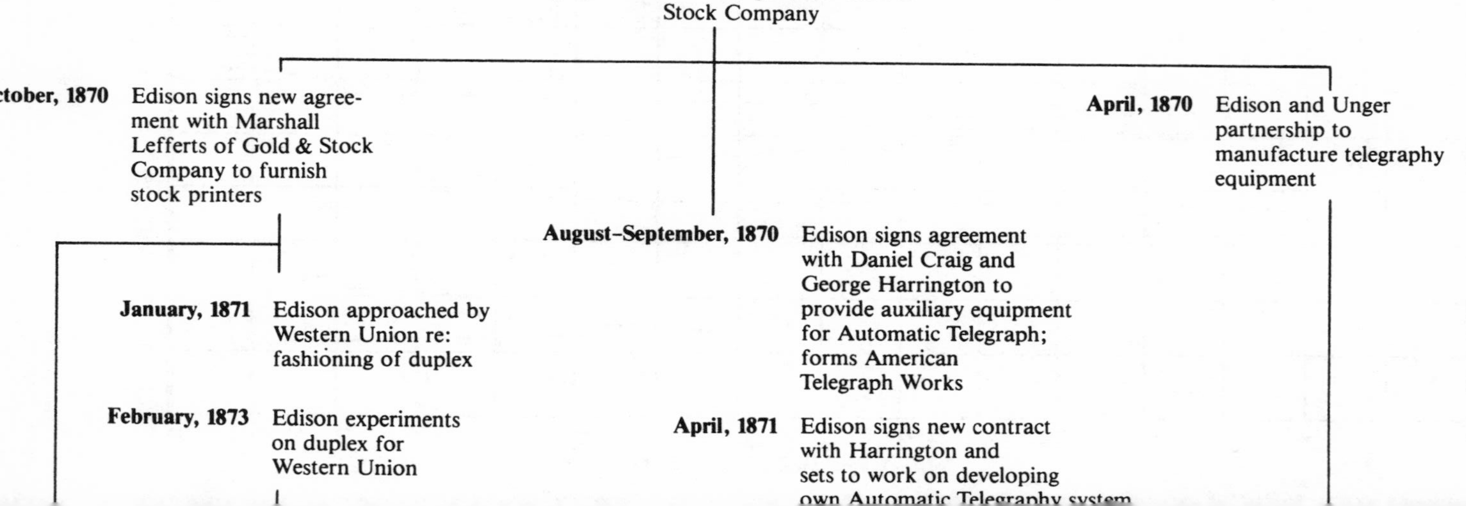

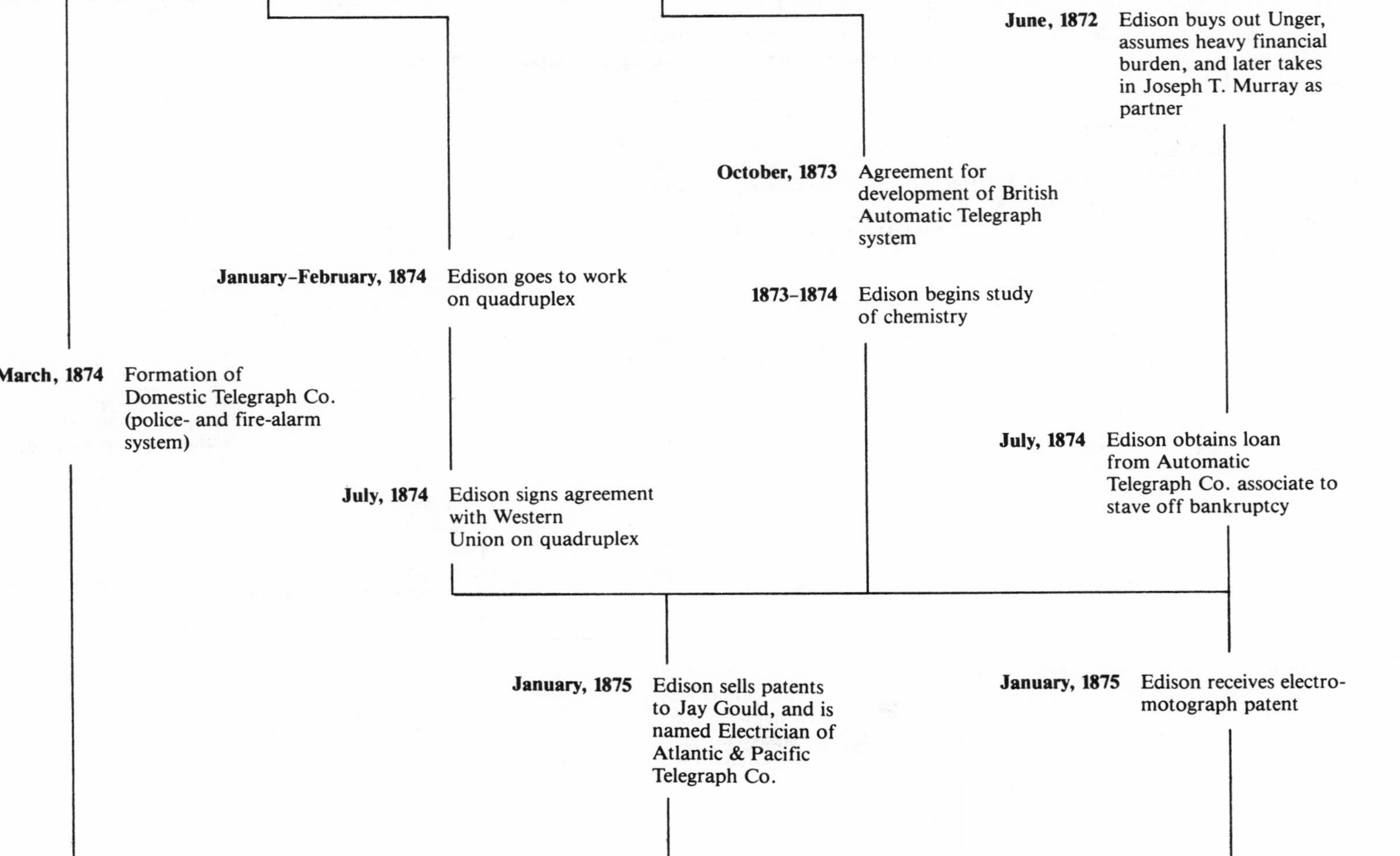

June, 1872
Edison buys out Unger, assumes heavy financial burden, and later takes in Joseph T. Murray as partner
October, 1873
Agreement for development of British Automatic Telegraph system
1873–1874
Edison begins study of chemistry
January–February, 1874
Edison goes to work on quadruplex
March, 1874
Formation of Domestic Telegraph Co. (police- and fire-alarm system)
July, 1874
Edison obtains loan from Automatic Telegraph Co. associate to stave off bankruptcy
July, 1874
Edison signs agreement with Western Union on quadruplex
January, 1875
Edison sells patents to Jay Gould, and is named Electrician of Atlantic & Pacific Telegraph Co.
January, 1875
Edison receives electro-motograph patent

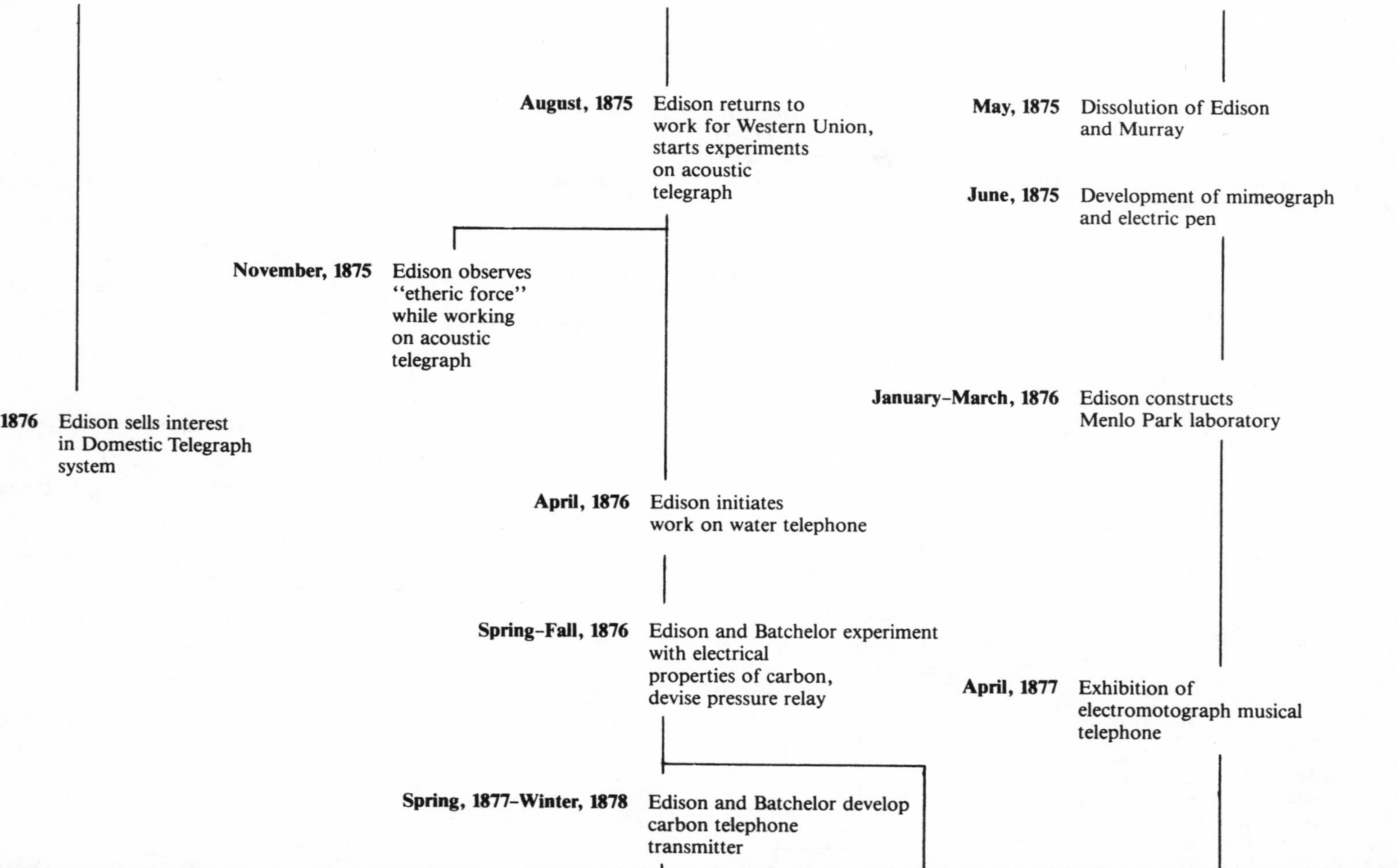
1876 Edison sells interest in Domestic Telegraph system
August, 1875 Edison returns to work for Western Union, starts experiments on acoustic telegraph
November, 1875 Edison observes "etheric force" while working on acoustic telegraph
April, 1876 Edison initiates work on water telephone
Spring–Fall, 1876 Edison and Batchelor experiment with electrical properties of carbon, devise pressure relay
Spring, 1877–Winter, 1878 Edison and Batchelor develop carbon telephone transmitter
May, 1875 Dissolution of Edison and Murray
June, 1875 Development of mimeograph and electric pen
January–March, 1876 Edison constructs Menlo Park laboratory
April, 1877 Exhibition of electromotograph musical telephone

June, 1878 Edison builds tasimeter, based on electrical properties of carbon

September, 1878 Edison, employing tasimeter principle, initiates work on incandescent illumination

November, 1878 Incorporation of Electric Light Co.

June–July, 1879 Upton and Edison design commercial dynamo

October–November, 1879 Evolution of the carbon incandescent lamp

April, 1880 Formation of Lamp Manufacturing Co.

Summer–Fall, 1880 Development of system of electrical distribution

December, 1880 Incorporation of Electric Illuminating Co. of New York

September, 1882 Pearl Street station goes into operation

1882 Observation of Edison effect lays the basis for the radio tube

November, 1879 Bell and Edison inventions combined to establish basis of modern telephone

December, 1879 Formation of Edison Magnetic Ore Milling Co.

Fall, 1882 Ore Milling Co. lapses into dormancy after unsuccessful trials

May, 1880 Testing of electric railroad

Fall, 1882 Improved electric railroad fails commercial tests

Spring, 1883 Merger of Edison and Field electric railroad companies

July–August, 1877 Edison discovers principle of sound recording while experimenting with telephone and telegraph repeater

December, 1877 Edison designs phonograph

April, 1878 Formation of Edison Speaking Phonograph Co.

Fall, 1878 Tinfoil phonograph proves commercially impractical

January, 1879 Development of electromotograph speaking telephone receiver

1881 Edison moves operations from Menlo Park to Manhattan

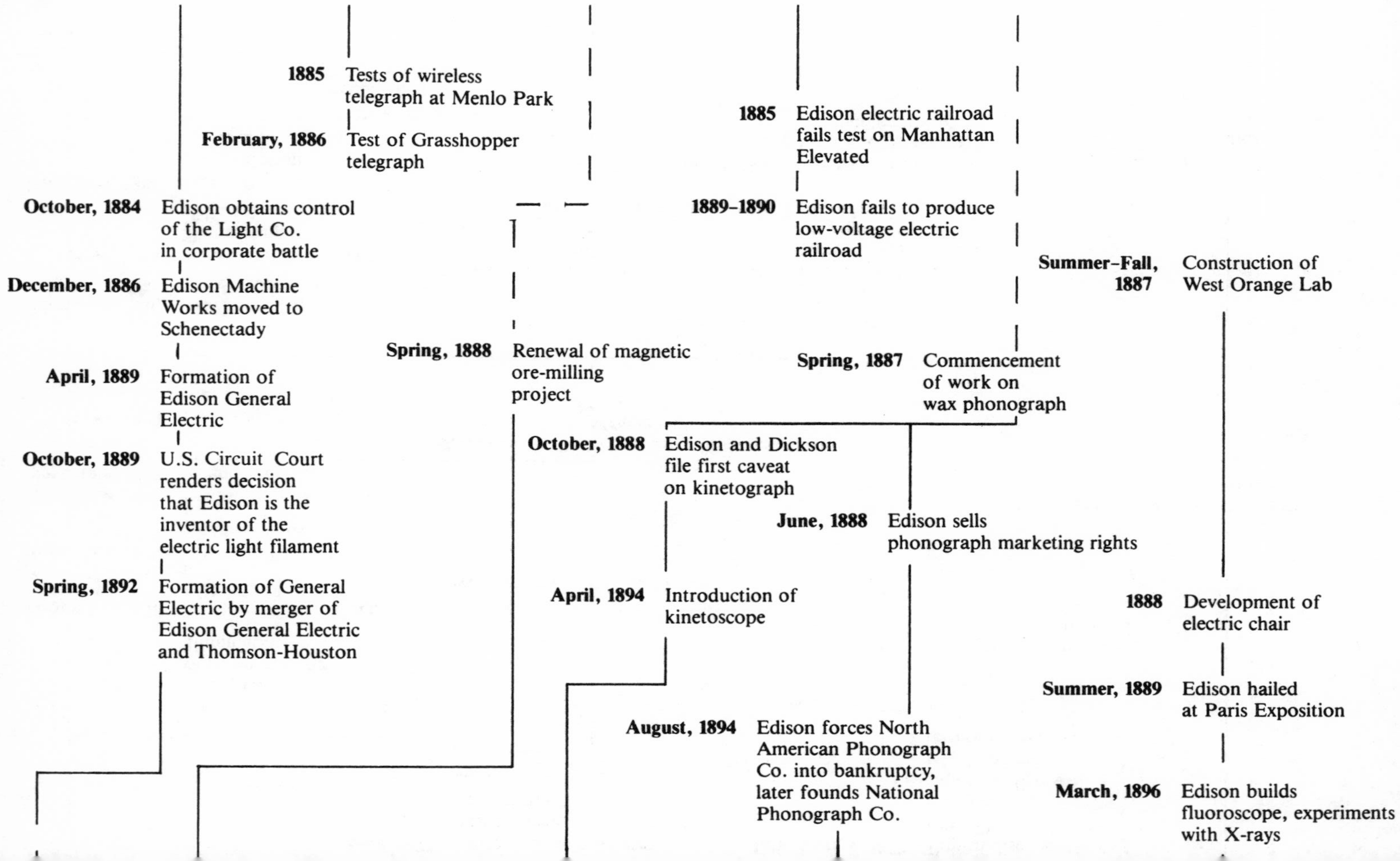

1885 Tests of wireless telegraph at Menlo Park

February, 1886 Test of Grasshopper telegraph

October, 1884 Edison obtains control of the Light Co. in corporate battle

December, 1886 Edison Machine Works moved to Schenectady

April, 1889 Formation of Edison General Electric

October, 1889 U.S. Circuit Court renders decision that Edison is the inventor of the electric light filament

Spring, 1892 Formation of General Electric by merger of Edison General Electric and Thomson-Houston

Spring, 1888 Renewal of magnetic ore-milling project

1885 Edison electric railroad fails test on Manhattan Elevated

1889–1890 Edison fails to produce low-voltage electric railroad

Spring, 1887 Commencement of work on wax phonograph

October, 1888 Edison and Dickson file first caveat on kinetograph

June, 1888 Edison sells phonograph marketing rights

April, 1894 Introduction of kinetoscope

August, 1894 Edison forces North American Phonograph Co. into bankruptcy, later founds National Phonograph Co.

Summer–Fall, 1887 Construction of West Orange Lab

1888 Development of electric chair

Summer, 1889 Edison hailed at Paris Exposition

March, 1896 Edison builds fluoroscope, experiments with X-rays

Summer, 1900 Final abandonment of magnetic ore-milling project

July, 1902 Portland cement plant goes into operation

October, 1929 Golden jubilee of incandescent light at Dearborn

Winter, 1913 Edison attempt to introduce talking motion pictures fails

February, 1918 Closing of motion-picture studio

December, 1909 Edison initiates work on disk phonograph

December, 1930 End of phonograph production

1899 Edison starts developing alkaline battery for electric automobiles

1902 Edison secretly initiates work on electric vehicles

January, 1912 Edison agrees to design self-starter for Model T

October, 1928 Edison receives special congressional medal

Reference Guide

The Reference Guide is intended to assist readers in quickly identifying persons who appear and reappear in the book. People are listed according to their relationship with Edison and the chapter in which they are introduced. A cross-reference is provided in the Index. (Persons of lesser importance in the narrative are included in the Index only.)

Edison's Family

	Chapter Introduced
Nancy Elliott Edison—mother of Thomas Alva Edison.	1
Sam Edison, Jr.—father of Thomas Alva Edison.	1
William Pitt Edison—Thomas Alva Edison's older brother, who failed in whatever venture he undertook.	1
Charley Edison—talented, profligate son of William Pitt Edison; he became an assistant in the Menlo Park laboratory.	8
Mary Sharlow—mistress of Sam Edison, Jr., and mother of three of his children.	8
Mary Stilwell—the teenaged first wife of Thomas Alva Edison.	5
Alice Stilwell Holzer—older sister of Mary Stilwell.	5
Charles Stilwell—younger brother of Mary Stilwell.	5
William Holzer—a glassblower who married Alice Stilwell and became an executive in Edison's lamp factory.	14

	Chapter Introduced
Margaret Stilwell (Grammach)—Mary Stilwell's mother	19
Dot (Marion), Tom, and Will—Edison's and Mary's children, born in 1873, 1876, and 1878.	11
Mina Miller—young and attractive second wife of Thomas Alva Edison.	19
Lewis Miller—agricultural-machinery manufacturer and father of Mina Miller.	19
Mary Valinda Miller—devoutly religious mother of Mina Miller.	19
John Miller—the younger brother of Mina Miller, he was subsequently an executive with Thomas A. Edison, Inc.	20
Theodore Miller—Mina's brother killed in the Spanish-American War.	20
Mary Emily Miller—younger sister and confidante of Mina.	20
Madeleine, Charles, and Theodore—Edison's and Mina's children, born in 1888, 1890, and 1898	23, 26, 32

Friends of Early Days

Jim Clancy—a neighboring youth at Fort Gratiot who helped Edison practice telegraphy and on occasion lent him money.	1
James Mac Kenzie—telegrapher who taught Edison the trade at Mount Clemens, Michigan.	1
Ezra Gilliland—young telegrapher who was Edison's friend, and his associate off and on for twenty-five years.	2
Milt Adams—telegrapher responsible for getting Edison a job in Boston.	2

Edison's Associates in Boston, 1868–69

Sam Ropes—telegrapher through whom Edison obtained his first contract for an invention.	2
George F. Milliken—the manager of Western Union's Boston office, he introduced Edison to Frank Pope.	3
Charles Williams—proprietor of a famous scientific-instrument shop on Court Street.	3

	Chapter Introduced
Frank Hanaford—Edison's partner in his stock-ticker and private telegraphy ventures.	3
D. C. "Bob" Roberts—telegrapher associated with Edison in some of his early inventions.	3
E. Baker Welch—Cambridge businessman who provided money for the development of Edison's duplex.	3

Edison's Associates in New York, 1869–70

Frank Pope—one of the most respected electrical engineers in the telegraph industry, he helped Edison establish himself in New York.	3
James Ashley—Frank Pope's associate and the editor of the *Telegrapher*.	3
Samuel S. Laws—proprietor of the Laws Reporting Telegraph, a stock-ticker system; he commissioned several inventions from Edison.	3
Marshall Lefferts—president of the Gold & Stock Company, he took a paternal interest in Edison and arranged a manufacturing partnership in Newark for him.	4

Friends and Associates Dating from the Newark Period, 1870–75

Daniel H. Craig—the former manager of the Associated Press and wily owner of the National Telegraph Company, he introduced Edison to the field of automatic telegraphy.	4
George Harrington—former assistant secretary of the treasury, who backed Edison in the organization of the American Telegraph Works.	4
Josiah Reiff—an associate of George Harrington, he became Edison's lifelong friend.	4
Uriah Painter—Washington wheeler-dealer tied to Harrington and Reiff, he also came to represent Edison in the capital.	4
General William Palmer—builder of the Kansas-Pacific Railroad, he was interested in acquiring Edison's automatic telegraph system.	4

	Chapter Introduced
Edward H. Johnson—General Palmer's telegraphy expert and a lifelong business associate of Edison.	6
Colonel George Gouraud—London financial representative for George Harrington and General Palmer, he was subsequently associated with Edison in the marketing of the telephone and phonograph.	6

Edison's Partners and Employees Dating from the Newark Period

Charles Batchelor—Edison's principal partner and co-inventor.	5
John Kruesi—an excellent, Swiss-born machinist who made the models for all of Edison's inventions at Newark and Menlo Park.	5
John and Fred Ott—two young mechanics who joined Edison in the early 1870s and remained with him until his death.	5
Jim Adams—a tubercular sailor who wandered into Edison's shop and assumed an important role as his associate.	7

Manufacturing Partners

William Unger—machine-shop operator associated with Edison from 1870 to 1872.	4
Joseph T. Murray—an amiable man of limited talent who became Edison's junior partner in 1872.	6

Western Union Executives

Tracy Edson—vice-president of the Gold & Stock Company, which became a Western Union subsidiary; he introduced Edison to William Orton.	4
William Orton—the sharp-dealing president of Western Union who encouraged Edison to develop his duplex and quadruplex.	3
George Prescott—the Western Union electrician, an important figure in Edison's dealings with the company.	6

	Chapter Introduced
Thomas T. Eckert—general manager of Western Union's Eastern Division, he was a key figure in machinations revolving about Edison.	6

Friends and Associates Dating from the Menlo Park Period, 1876–82

William Carman—the Electric Pen Company accountant who introduced Edison to Menlo Park.	7
Dr. Leslie Ward—a Newark physician who was one of the founders of the Prudential Life Insurance Company of America, and Edison's family physician for ten years.	8
George F. Barker—a University of Pennsylvania professor and Edison's champion in academic circles.	10
Reverend John Heyl Vincent—a prominent Methodist clergyman who lived near Menlo Park and met Edison after the invention of the phonograph. Edison became his son's rival for the hand of Mina Miller.	10
Edwin Fox—a friend of Edison who frequently contributed articles to the *New York Herald* and wrote a lengthy story on the incandescent light.	11

Edison's Secretaries

Stockton L. Griffin—a moustachioed telegrapher who became Edison's first personal secretary in 1878.	10
Samuel Insull—an ambitious, brilliant Englishman who replaced Griffin in 1881 and within a few years worked himself into the key executive position in Edison's enterprises.	16
Alfred O. Tate—a Canadian who replaced Insull as secretary and became an executive with Edison's North American Phonograph Company in 1892.	21
Johnny Randolph—a kind, considerate man liked by everyone, he held the position of secretary from 1893 until his death in 1908.	24
William H. Meadowcroft—a wispy, myopic author of two popular books on electricity, he was Edison's biographer as well as his secretary from 1909 to 1931.	37

Chapter Introduced

Edison's Patent Solicitors and Attorneys

Lemuel Serrell—Edison's patent solicitor from 1870 until the middle of the decade. 6

Grosvenor P. Lowrey—patrician and witty attorney for Western Union, he became Edison's attorney and friend after the quadruplex trial in 1877. 7

Benjamin Butler—a man with an authoritative personality who gained the nickname Bloody Ben during his command of Union Troops in New Orleans; he was Edison's attorney in the quadruplex trial. 8

Zenas F. Wilber—a well-connected, corrupt former examiner in the patent office, he served as Edison's patent solicitor in Washington, D.C., for several years. 8

Sherbourne Eaton—a noted New York lawyer who at times was Edison's personal attorney as well as the president of the Edison Electric Light Company. 16

Colonel George Dyer—onetime partner of Zenas F. Wilber, he took over as Edison's patent solicitor when Wilber was discovered to have embezzled $1300. 17

John Tomlinson—a young Columbia University graduate who was Edison's attorney in the Lucy Seyfert suit and played a principal role in the Edison-Gilliland controversy. 18

Richard Dyer—the son of Colonel George Dyer, he assumed the position of Edison's patent solicitor and attorney in the 1890s. 25

Frank Dyer—younger brother of Richard Dyer and successor to the position of Edison's attorney. 28

"Judge" Howard W. Hayes—a New Jersey lawyer and "fixer" who was retained by Edison to work out a settlement in the bankruptcy of the North American Phonograph Company. 28

Associates in the Electric Pen and Mimeograph

Robert Gilliland—father of Ezra Gilliland and first manufacturer of the electric pen and mimeograph. 2

	Chapter Introduced
George Bliss—manager of Western Electric, he acquired marketing rights to the electric pen and mimeograph, and wrote a highly imaginative and overblown account of Edison's early life.	8
A. B. Dick—Midwest lumber manufacturer who later took over the mimeograph and established a multimillion-dollar business-machine company.	21

Associates and Employees in the Phonograph Field

Gardiner Hubbard—member of a prominent and wealthy Boston family and the father-in-law of Alexander Graham Bell, he was one of the principal stockholders in the Edison Speaking Phonograph Company.	8
Hilborne Roosevelt—an organ manufacturer and the cousin of Theodore Roosevelt; he was one of the investors in the first phonograph company.	10
Charles Cheever—wealthy New Yorker crippled in infancy, he invested in a number of Edison enterprises.	10
Jesse Seligman—New York banker who formed the Edison United Phonograph Company to market the invention in Europe.	23
Jesse Lippincott—glass manufacturer who tried to form a recording-machine monopoly but went bankrupt instead.	23
Jonas W. Aylsworth—a chemist who became Edison's indispensable assistant in the development of the phonograph and alkaline storage battery.	23
William Gilmore—hard-driving head of Edison's phonograph enterprise and motion-picture division from 1894 to 1908.	28

Inventors and Rivals in the Phonograph Industry

Charles Tainter—the principal inventor of the graphophone, an improved version of the phonograph.	21
Emile Berliner—the designer of the disk phonograph.	28

	Chapter Introduced
Edward D. Easton—president of the American Graphophone and Columbia Phonograph companies, he carried on and ultimately won a fourteen-year legal battle against Edison.	28
Eldridge Johnson—president of the Victor Talking Machine Company and a friend of Edison.	28

Associates and Employees in Electric Lighting

Francis Upton—a brilliant graduate of Bowdoin and Princeton who joined Edison at Menlo Park and was responsible for the theoretical work in electric light and power development.	11
Frank McLaughlin—a friend of Edison's telegraphy days who went to Canada and California to hunt platinum for the inventor.	12
Ludwig Böhm—German-born glassblower who constructed the vacuum pumps and the globes for incandescent light.	12
J. C. Henderson—chief engineer for financier Henry Villard.	12
Francis Jehl—young laboratory assistant who operated the vacuum pump and was instrumental in the development of the bamboo filament.	12
Sigmund Bergmann—immigrant youth who worked for Edison in Newark and ultimately became one of the leading electrical-equipment manufacturers in the world.	14
Charles Clarke—a Princeton graduate who played a major role in establishing the world's first electric light and power system in New York.	14
William Hammer—a youth who had a knack for showmanship and set up many of Edison's lighting exhibitions.	15
Charles Dean—hard-nosed foreman of Edison's machine shop.	15
Frank Sprague—electrical engineer employed by Edison to supervise small-town installations; he subsequently built the first comprehensive electric traction system in America.	17
Nikola Tesla—a somewhat mad Croatian genius who was one of the leading inventors in the field of alternating current.	17

	Chapter Introduced
Arthur E. Kennelly—Edison's electrical expert during the early days of the West Orange laboratory.	22
Harold P. Brown—a New York electrical engineer principally responsible for the electric-chair experiments at Edison's laboratory.	22

Inventors and Rivals in the Electric-lighting Field

Elihu Thomson—Philadelphia inventor who founded Thomson-Houston, an arc-light company that ultimately merged with Edison General Electric.	8
Henry Draper—noted scientist, whose father's remark drew Edison into the investigation of incandescent lighting.	10
William Wallace—manufacturer whose factory Edison visited to inspect the arc-light installation.	11
Joseph Swan—British inventor working on an incandescent lamp concurrently with Edison.	11
William Sawyer—New York inventor and bitter rival of Edison.	11
Hiram S. Maxim—prolific inventor who was hired by the United States Electric Light Company to produce an incandescent light in competition with Edison.	14
George Westinghouse—air-brake manufacturer who established one of the three major firms in electric light and power in the United States.	22

Acquaintances, Associates, and Rivals in Motion Pictures

Thomas Lombard—manager of the North American Phonograph Company, he and Alfred Tate introduced the kinetoscope to the public.	28
Eadweard Muybridge—the developer of sequential photography, whose lecture at the West Orange laboratory gave Edison the idea that ultimately resulted in the development of motion pictures.	29
William K. L. Dickson—the Edison assistant who conducted the experiments leading to the production of the kinetoscope and the kinetograph.	29
Thomas Armat—the inventor of the motion-picture projector.	29

	Chapter Introduced

Friends, Associates, and Employees of Later Years

Walter Mallory—a Chicago iron manufacturer who joined Edison in his ore-mining venture and later headed the Cement Works.	20
Joseph McCoy—Edison's private investigator.	28
William C. Anderson—manufacturer of the "Detroit Electric" automobile, he reintroduced Edison to Henry Ford.	35
Henry Ford—Detroit auto manufacturer who became a strange combination of godson-godfather to Edison.	35
Miller Reese Hutchinson—talented inventor with ambition to be Edison's heir to the laboratory.	35
Harvey Firestone—tire manufacturer and frequent companion of Edison after World War I.	39
Stephen Mambert—capable efficiency expert who placed Thomas A. Edison, Inc., on sound financial footing for the first time but was eased out by Charles Edison.	39

Financiers

Jay Gould—lifelong rival of the Vanderbilts, he purchased Edison's telegraph patents in 1875.	6
Hamilton McK. Twombly—a patrician Bostonian who married one of Commodore Vanderbilt's nine daughters and became an important backer of Edison.	10
J. P. Morgan—New York banker who met Edison at the time of the development of the incandescent light.	11
Henry Villard—flamboyant promoter and president of the Northern Pacific Railroad, he organized the formation of Edison General Electric.	12
Charles Coffin—president of the Thomson-Houston Company and first head of General Electric.	26

Inventors

	Chapter Introduced
Moses Farmer—talented telegraphic and electrical inventor who failed to carry his ideas to commercial fruition.	3
Alexander Bain—inventor of the first automatic telegraph system.	4
George Little—developer of an automatic telegraph system for which Edison was hired to construct ancillary equipment.	4
Elisha Gray—Midwestern inventor whose experiments with acoustic telegraphy led to the development of the telephone.	7
Alexander Graham Bell—discoverer of the principle of the continuous, undulatory current, which was the basis for the telephone.	8
Stephen Field—nephew of Cyrus Field, layer of the Atlantic cable, he was an associate of Edison in a number of enterprises.	10
Sir William Thomson (Lord Kelvin)—noted British scientist who became a friend of Edison.	10
David Hughes—British-American inventor who set off an international quarrel between himself and Edison when he claimed credit for inventing the microphone.	10
Guglielmo Marconi—developer of wireless telegraphy, for which Edison possessed one of the basic patents.	30
J. Ambrose Fleming—British scientist who built upon the principle of the "Edison effect" to develop the radio vacuum tube.	30
Reginald Fessenden—a former Edison employee who transmitted the first voice message across the Atlantic.	30

Miscellaneous

Jim Symington—the peppery companion of Edison's father, Sam.	1
William and Lucy Seyfert—Philadelphia banking couple who lent Edison a substantial sum in 1874. Lucy Seyfert later filed suit for repayment.	6

	Chapter Introduced
William H. Preece—electrician of the British Post Office, he was alternately a friend and adversary of Edison.	9
George P. Lathrop—author and son-in-law of Nathaniel Hawthorne, he proposed the writing of a science-fiction book to Edison.	30
Thomas C. Martin—noted editor of electrical journals who worked on an authorized biography of Edison.	31

Source Notes

Unless otherwise indicated, all references listed are located at the Edison National Historic Site, West Orange, New Jersey.

Other principal sources are identified as follows:

Baker Library—	Henry Villard Collection, Baker Library, Harvard University
C. E. Fund—	Charles Edison Fund, Orange, New Jersey
Ford Archives—	the Ford Archives in the Henry Ford Museum, Dearborn, Michigan
Hammer Collection—	the Hammer Collection in the Museum of the History of Technology, Smithsonian Institution, Washington, D.C.
Tannahill Library—	Henry Ford Museum, Dearborn, Michigan

Miscellaneous papers not otherwise identified are listed by file (indicated in italics) and date. Thus, *Telegraph, Phonoplex,* Nov. 23, 1885, shows the item is held in the specified vault file at the Edison National Historic Site.

The Upton letters, unless indicated as being in the Hammer Collection, are in the Upton Papers, Upton Historical Reference File, Edison National Historic Site.

Abbreviations used throughout the Source Notes:

HRF— Historical Reference File in the Main Laboratory office, West Orange
N— Laboratory notebook in the series located in the inner vault
PN— Edison pocket notebook
Ed.— Thomas Alva Edison
Batch.—Batchelor
Quad.—quadruplex

Chapter 1. A Bushel of Wheat

Page Identification—Source

3 him Rinkey—Marion Edison Page to Ed., April 25, 1878

3 sixty pounds—George P. Lathrop, "Edison's Father," in *Once a Week,* Jan. 20, 1894

3 Presbyterian Church—Luke Stow to Ed., Aug. 31, 1887

4 the participants—Dr. Buckner to Ed., Feb. 18, 1891

4 for hours—James Symington to Ed., Feb. 23, 1891

4 gambling alone—Lathrop, op. cit.

4 this lover—Marion Edison Page to Mina Edison, July 22, 1892; Ed. note for Henry Ford, Box 1, Ford Archives

5 the covers—Meadowcroft Compilation of Edison Notes, Jan. 9, 1920; Ed. Notes for Dyer & Martin, Book 2

6 the boiler—P. L. Hubbard to Ed., Dec. 23, 1911; *History of St. Clair County, Mich.,* p. 616, in Burton Collection, Detroit Public Library

6 the fort—Carrie Farrand Ballentine to Ed., Mar. 3, 1911. (The house was originally built by Edgar Jenkins, an eccentric early settler at the fort. He sold it to B. C. Farrand, the father of Carrie Farrand Ballentine, who in turn sold it to Sam Edison.)

6 the school—William Brewster to Ed., May 10, 1883, and Dec. 22, 1891

7 pilfer from it—Ed. Diary, p. 44

7 and buttocks—Ed. Notes, op. cit.

7 science book—Ed. note for Ford, Box 1, Ford Archives

7 parents' pantries—"Sunday 1" interview in *Ed. Personal,* 1890; Willis D. Engle to Friend Al, July 24, 1874

7 he enjoyed—Ed. to P. L. Hubbard, Nov. 9, 1893

7 and the like—George Bliss, "Life of Edison," *Chicago Tribune,* Apr. 8, 1878

7 on Alva—*Norwalk Reflector,* Mar. 10, 1896

8 little addled—Lathrop, op. cit.

8 many books—Symington to Ed., Mar. 14, 1878

8 feet apart—"Sunday 1," op. cit.

8 Detroit Schottisch—The Rev. G. B. Engle to Ed., Aug. 13, 1885; Mary Engle to Ed., Oct. 4, 1886

8 confide in her—Caroline Ballentine, "The True Story of Edison's Childhood and Boyhood," *Michigan History,* Jan., 1920, p. 183

9 the train—Lathrop, op. cit.

9 his life—*The Diary and Sundry Observations of Thomas Alva Edison,* Dagobert D. Runes, ed., New York, 1948, p. 58

9 hours later—Grand Trunk Railroad timetable, June, 1862, Tannahill Lib.

9 eleven o'clock—Ed. to Friend Willie, Aug. 10, 1862, Tannahill Lib.

10 on its own—Meadowcroft, op. cit.

10 Paine's path—*The Diary and Sundry Observations,* op. cit., p. 154

10 full of chemicals—Ed. Notes, op. cit.

10 encyclopedic enterprise (fn)—Ed. membership card No. 33, Detroit Young Men's Society, Sept. 15, 1862, in Tannahill Lib.; Ed. to James Swift, Dec. 2, 1908

11 out the door—*The Operator,* May 15, 1878; Ed. Notes, op. cit.

Page	*Identification—Source*
11	ice house—Henry Barnett to Ed., Mar. 11, 1898; James Clancy to Ed., Mar 3, 1889
11	mother's stove—"Sunday 1," op. cit.
11	*Paul Pry*—Meadowcroft, op. cit.
12	he could not—William D. Wright to Ed., Oct. 10, 1878, and June 29, 1910
12	on consignment—Meadowcroft, op. cit.
12	quarter each—Meadowcroft, op. cit.
12	and forlorn (fn)—Wright to Ed., June 29, 1910
13	he recalled—Meadowcroft, op. cit.; Ed. Notes, Book 1, op. cit.
13	turkey shoot—Meadowcroft, op. cit.
14	caustic comments—"Sunday 1," op. cit.; Ed. Notes, Book 2, op. cit.
14	Morse code—Mrs. M. J. Powers to Ed., Feb. 18, 1880
14	Port Huron—Ann Mac Kenzie to Ed., Jan. 4, 1912
14	about it (fn)—WTL letter in *Detroit Free Press,* Dec. 7, 1889
15	study telegraphy—J. Carle Barron to Ed., June 23, 1878
15	he craved—Ibid; *James Mac Kenzie* HRF
15	over the line—Clancy to Ed., Feb. 22, 1878; Wright to Ed., June 29, 1910; Meadowcroft, op. cit.
15	send well—Meadowcroft, op. cit.
15	shredded paper—P. L. Hubbard to Ed., Jan. 7, 1909, and Dec. 23, 1911
16	military telegrapher—John F. Schultz to Ed., Jan. 19, 1905

Chapter 2. A Drifter and a Dreamer

17	3500 citizens—R. Thomas monograph, Perth County Historical Archives, Stratford, Ontario
17	on the line—G. E. Mitchel to Ed., Jan. 22, 1909; Margaret Colclaugh to Ed., Oct. 8, 1906; Buffalo–Lake Huron Time Schedule, 1862
17	to "CAUTION"—Buffalo–Lake Huron Time Schedule, 1862
18	one boxcar—*Stratford Herald,* Dec. 18, 1863
18	outer office—Meadowcroft, op. cit.; Ed. Notes, Book 1, op. cit.
18	Port Huron—Sam Edison to Manager of Grand Trunk Railroad, Canada, Sept. 10, 1884, in *Ed. Personal,* 1911
18	small workshop—Menlo Park Scrapbook, *Cincinnati Commercial* interview with Ezra Gilliland (no date)
19	telegraph office—Meadowcroft, op. cit.
19	the operation—Ibid.; Harriet H. Hayes to Ed., Oct. 26, 1911
20	to St. Louis—Meadowcroft, op. cit.; George Parmalee to Ed., Jan 7, 1891; Ed Parmalee to Ed., Mar. 9, 1891; Harriet H. Hayes to Ed., Oct. 26, 1911
20	tobacco incessantly—Ed. response to "Longevity Inquiry," 1930
20	the theater—J. W. Doan to Ed., Jan. 24, 1880; E. W. Clowes to Ed., Feb. 3, 1880; Nat Hyams to Ed., Feb. 11, 1912
20	would copy—Meadowcroft, op. cit.
20	the Ohio—George Lockwood to Ed., Nov. 28, 1911
20	on tragedy (fn)—Jot Spencer to Ed., April 8, 1878
21	did it—Meadowcroft, op. cit.
21	wire nights—Ibid.

Page Identification—Source

21 telegraph office—Irene Schloss to Ed., Jan. 31, 1911; Bates to Ed. Dec. 12, 1912
22 all night—Bates to Ed., Dec. 12, 1912; Meadowcroft, op. cit.
22 the office—Recollections of George "Fatty" Stewart, *New York Sun,* March 10, 1878
22 in Arkansas—George Donegan to Ed., Feb. 21, 1912
22 fall of 1865—*The Operator,* May 15, 1878; Meadowcroft, op. cit.
22 was hired—F. L. Dyer and T. C. Martin, *Edison: His Life and Inventions,* New York, 1910, p. 79
22 "Poverty Flat"—Western Union roster of Louisville office in *Ed. Personal, 1911;* E. W. Clowes to Ed., Jan. 21, 1912
23 an abstainer—Meadowcroft, op. cit.
23 take press—*Boston Globe,* May 28, 1878
23 interested him—Meadowcroft, op. cit.
23 Edison replied—John A. Cassell to Ed., June 1, 1889
23 his ideas—C. J. McGuire to Ed., Feb. 12, 1912
24 fellow operators—Meadowcroft, op. cit.; Ed. note on Southwest Telegraph Co. blank, Tannahill Lib.
24 South America—Lawler to Ed., Mar. 2, 1908; unidentified newspaper clipping, Dec. 27, 1905
24 to Louisville—Clancy to Ed., Feb. 13 and Mar. 30, 1889
24 one day—Meadowcroft, op. cit.
24 Mexican port (fn)—*New York Times,* Jan. 24, 1892
25 was fired—Ibid; *The Operator*, May 15, 1878; *Boston Globe*, May 28, 1878
25 small sum—$250—Samuel W. Ropes contract with Ed., Joel Hills, and William E. Plummer, Jan. 1, 1867, in Tannahill Lib.
25 Mechanics' Library—Meadowcroft, op. cit.; Menlo Park Scrapbook, *Cincinnati Commercial* interview with Ezra Gilliland
25 and tools—Meadowcroft, op. cit.
25 high voltage—Lawler to Ed., Mar. 2, 1908
26 going on—Meadowcroft, op. cit.; Dyer & Martin, op. cit., p. 92
26 in Cincinnati—*The Operator,* May 15, 1878
26 Ruhmkorff Coil—Mary Richardson French to Ed. in Main Laboratory Photo File

Chapter 3. The Walking Churchyard

27 Western Railroad—Clancy to Ed., Feb. 13 and Mar. 30, 1889
28 night shift—Recollections of George "Fatty" Stewart, op. cit.
28 the fraternity—*The Operator,* May 15, 1878; G. M. Shaw, "Sketch of Edison," *Popular Science Monthly,* Aug. 1878
28 other foot—Ed. Notes, Book 1, op. cit.
29 in 1866—James Reid, *The Telegraph in America,* New York, 1877, p. 656
29 open flame—Meadowcroft, op. cit.
29 but two-toned—Recollections of George "Fatty" Stewart, op. cit.; *Boston Globe,* May 28, 1878; Ed. Notes, Book 1, op. cit.
30 the Bible—*Boston Globe,* May 28, 1878
30 was born—Reid, op. cit., pp. 602–605
30 gold exchange—Ed. inventory, Feb. 13, 1869, in Tannahill Lib.

Page *Identification—Source*

31 Rome howl!—D. C. Roberts to Ed., May 28, 1877, June 6, 1890, and Jan. 20, 1911
31 Refining Company—D. N. Skillings & Co., and Continental Sugar Refining Co. to Ed. and Hanaford, May 5, 1869
31 summed up—John E. Clarke to Ed., Dec. 11, 1880
31 fellow telegraphers—Ed. Notes, Book 1, op. cit.; Meadowcroft, op. cit.
31 objected to "B.C."—Ibid.
31 their talk—Recollections of George "Fatty" Stewart, op. cit.
32 capricious inventions—Reid, op. cit., p. 564
32 a child—Reid, op. cit., p. 640
32 his ability—George L. Newton to Ed., May 7, 1878
32 the device—Boston, April 7, 1869, in *Teleg. Duplex, 1875;* Ed. Notes, Book 1, op. cit.
32 his acquaintances—Ed. to Hanaford, July 26, 1869
33 before I die—Ibid.
33 Stock Company—Reid, op. cit., p. 605
34 other things—Ed. to Hanaford, Sept. 17, 1869
34 New York—*Telegrapher,* Oct. 1, 1869; Reid, op. cit., p. 621
34 after midnight—Ibid.
34 walking churchyard—Ed. to Hanaford, Jan. or Feb., 1870
34 the premises (fn)—Ed. Notes, Book 1, op. cit.

Chapter 4. The Proliferating Inventor

35 in September—Memorandum of Agreement between Gold & Stock Co., Frank L. Pope, James N. Ashley, & Thomas A. Edison, April 18, 1870
35 out to him—Ed. Notes, Book 1, op. cit.
36 pocket money—Ed. Accounts, May 7, 1870
36 of science—Ed. to Hanaford, June 10, 1870
36 newspaper techniques—Reid, op. cit., p. 400
37 Telegraph Company—*Telegraph, Automatic,* May 3, 1870
37 better treatment—Craig to Ed., Aug. 12, 1870
38 the thing—Ibid.
38 writing machine—Craig to Ed., Aug. 12, 1870
38 the invention—Memorandum of Agreement between Edison & Craig, Aug. 17, 1870
38 winning ticket—*The Operator,* Nov. 15, 1874
38 little response—Quadruplex Case, vol. 70, p. 39
38 remained empty (fn)—Reid, op. cit., p. 375
39 wheeler-dealer—Painter obituary in *Phono., General,* 1899
40 in stock—Quad. Case, vol. 70, pp. 41 et seq.
40 $1,350 a year—Statement of George Harrington of his relations with Thomas A. Edison, E 5796-2
40 heads together—Craig to Ed., Aug. 19, 1870
40 patent solicitor—*Amer. Teleg. Works,* Oct. 3, 1870
40 may devise—Lefferts draft of Ed. memo for Gold & Stock Co., Oct. 19, 1870
40 $1.25 million—Ibid.
41 $1000 a year—Ed. lease from Gould Machine Co., Oct. 1, 1870

Page *Identification—Source*

41 on the thirty-first—Amer. Teleg. Works account books, Sept.–Dec. 1870
41 did not work—Ed. to Harrington, Dec. 7, 1870
42 *more sure*—Ed. to Craig, Dec. 7, 1870
42 practically empty—PN 70-10-31
42 new *vim*—Craig to Ed., Jan. 12, 1871
42 doubt it—Craig to Ed., Feb. 13, 1871
42 financial news—Reid, op. cit., p. 609
42 Western Union—Edson to Ed., Jan. 12 and 13, 1871

Chapter 5. A Tool of Wall Street

44 into models—*John Kruesi* HRF
44 over $30,000—Statement of George Harrington, op. cit.
45 too hot—Craig to Ed., Mar. 16, 1871
45 be tremendous—Ed. to Harrington, spring 1871
45 mechanical printers—Statement of George Harrington, op. cit.
45 health and mind—Ed. to Harrington, July 22, 1871
45 brained capitalist—Ed. Lab. Book, E 1676, July 29, 1871
45 Little's machinery—Craig to Ed., Oct. 6, 1871
45 the company—Ed. to Craig, Oct. 3, 1871
45 be transmitted—Quad. Case, vol. 70, p. 320
46 Wall Street news—Reid, op. cit., p. 609
46 to date—PN 71-09-06
46 Wall Street concrn—*John Ott* and *Fred Ott* HRF; P. J. Boorum to Ed., Aug. 29, 1878
47 to death—*The Operator,* July 1, 1878; Associated Press interview with Ed., Feb. 12, 1926
48 worth a Damn! !—Ed. Lab. Book, E 1676, Feb. 1, 1872
48 feathered hats—PN 72-05-01
48 twenty-year-old—Symington to Ed., May 14, 1878
48 her father—Sam Edison to Mary Edison, c. 1877
48 avoid bankruptcy—Pitt Edison to Ed., July 12, 1871
48 with $3100—Quad. Case, vol. 70, p. 704
49 promissory note—*Edison & Unger,* Jan. 1 and July 3, 1872

Chapter 6. The Night of Suspense

50 devoted to him—*Boston Globe,* July 6, 1902
51 the paper—Patent No. 133,841, Dec. 10, 1872
51 fix this—*Edison & Murray,* Oct. 9, 1872
51 $2000 a year—Ed.–Reiff contract, Nov. 5, 1872; Quad. Case, vol. 70, p. 706
51 to work—Testimony of E. H. Johnson in *George Harrington, et al.,* v. *Atlantic & Pacific Teleg. Co.*
52 my fingernail—Ed. Lab. Book, E 1679
52 in New York—Mary Childs Nerney, *Thomas A. Edison: A Modern Olympian,* New York, 1934, p. 262
52 aurichloride of sodium—Ed. Lab. Book, E 1679
52 anything in use—Testimony of E. H. Johnson, op. cit.
52 *Richard III*—Ed. Lab. Book, E 1679, Feb. 9, 1873

Page	*Identification—Source*
53	had bought—Quad. Case, vol. 71, p. 108–17 and 126
53	same time—Quad. Case, vol. 71, p. 125
53	do next—Quad. Case, vol. 71, p. 126
54	April 23—Quad. Case, vol. 71, Joseph Murray diary
55	Atlantic cable—Gouraud to Reiff, April, 1874
55	laborynthine *complexity*—Johnson to Ed., June 11, 1873
55	were inadequate—Quad. Case, vol. 71, p. 126
55	the sum—Murray to Ed., June 12, 1873
55	the other—Johnson to Ed., June 13, 1873
56	for transcription—Uriah Painter to Ed., Dec. 5 and 15, 1880
56	for twenty years—*Teleg., Automatic,* Oct. 1, 1873
56	trotting along—William Pitt Edison draw on Mechanics National Bank, Dec. 24, 1873
57	its weapon—Quad. Case, vol. 70, Exhibits, p. 6
57	said sarcastically—Western Union reply to Postmaster General Creswell, Dec. 6, 1873
57	one wire—*Telegraph,* Jan. 27, 1874
58	used to it—Quad. Case, vol. 71, p. 127
58	extraordinary feat!—Ed. Lab. Book, E 1679, Feb. 3, 1874
58	be president—Ed. agreement with Tracy Edson, Mar. 26, 1874
58	nervous diseases—Batch. notebook
58	not working—James W. Brown to Ed., Apr. 12, 1874
58	you desire—Reiff to Ed., April, 1874
59	further progress—Gouraud to Reiff, April, 1874
59	than Morse—Reiff to Ed., May, 1874
59	and duplex—Quad. Case, vol. 71
59	divided equally—Quad. Case, vol. 71, p. 127
59	the floor—William M. Allison to Ed., Jan. 3, 1888
60	said unhappily—Quad. Case, vol. 70, pp. 446 and 718; *Daniel H. Craig & James B. Brown* v. *George Little, et. al.,* p. 81
60	word on it—A. E. Sink in *Telegraphic Age,* Sept. 16, 1895
61	the interest—Quad. Case, vol. 70, pp. 448 and 451
61	$3000 advance—Quad. Case, vol. 70, p. 706
61	from them—Quad. Case, vol. 70, Ed. testimony
62	agreeing to it—Quad. Case, vol. 70, pp. 448–449
62	he declared—*New York Times,* July 9, 1874
62	his stomach—Quad. Case, vol. 70, p. 438; Affidavit of Thomas A. Edison, April 27, 1875 (E 5796-1)
62	loss of $100,000—Harrington to Ed., July 9, 1874
62	the letter—Quad. Case, vol. 70, p. 431
63	home tonight—Telegram of Sept. 3, 1874
63	charged more—Ed. Lab. Book, E 1677, July, 1874
63	of late—Johnson to Reiff, Aug. 5, 1874
63	Prescott agreed—Affidavit of Thomas A. Edison, op. cit.; Quad. Case, vol. 71, testimony of Lemuel Serrell
63	pull together—Reiff to Ed., Sept. 11, 1874
64	and fortune—*Telegrapher,* Sept. 26, 1874
64	to advance—Quad. Case, vol. 73, p. 36
64	would talk—Orton to Prescott and Ed., Dec. 17, 1874; Affidavit of Thomas A. Edison, op. cit.
64	wait two—Quad. Case, vol. 71, p. 142; Affidavit of Thomas A. Edison, op. cit.
65	and lower—Affidavit of Thomas A. Edison, op. cit.

Page *Identification—Source*

65 his return—Quad. Case, vol. 71, p. 250
65 by Eckert—Quad. Case, vol. 71, p. 288
66 telegraphy inventions—Quad. Case, vol. 70, pp. 468–69 and 691
66 current market—Quad. Case, vol. 71, Agreement of Dec. 30, 1874, between Jay Gould, Josiah Reiff, & Jonathan McManus; vol. 70, pp. 351, 489–90
66 the road—Edison Ledger Book, PN 75-01-05, 1875
67 New York—Quad. Case, vol. 71, pp. 177 et seq.
67 notify Edison—Quad. Case, vol. 71, p. 265
67 boots on—Quad. Case, vol. 71, pp. 703–708

Chapter 7. The Tangled Web

68 to Gould—Master's Report, Quad. Case, U.S. Circuit Court for the Southern District of New York, Jan. 30, 1909
68 per minute (fn)—Quad. Case, vol. 70, p. 646
69 developed frequently—Quad. Case, vol. 70, p. 764
69 system economic—Master's Report, op. cit.
69 recuperate in Europe—Quad. Case, vol. 70, p. 631
69 filed suit—Master's Report, op. cit.
70 he concluded—Ed. to Gould, July 26, 1875
71 was applied—*Telegraph, General:* Ed. rough draft of Oct. 7, 1878; Speaking Telegraph Interference, vol. 1, testimony of Robert Spice
71 January, 1875—Patent No. 158,787, Jan. 19, 1875
71 ad infinitum—Ed. Lab. Book, E 1676, July 24, 1874, et seq.
71 on oxygen—Ed. Lab. Book, E 1676, Mar. 10 and May 10, 1875
72 home nights—Ed. Lab. Book, E 1676, June 2–10, 1875
72 white pitcher—Charles Palance to Ed., May 28, 1909
72 My Mother—Ed. Lab. Book, E 1678
72 Edison's shop (fn)—*New York Commercial Advertiser,* Feb. 15, 1877
74 one cereal—Ed. Lab. Books, E 1679 and E 1692
74 in June—Ed. Lab. Book, E 1676, June 1, 1875
74 his claim (fn)—Gilliland to Ed., Feb. 22, 1875
75 he suggested—Ed. handwritten note, Aug., 1875
75 royalty offer—Prescott to Ed., Aug. 10, 1875
75 was delighted—Quad. Case, vol. 70, Exhibits, p. 24
75 electrified wire—E.H. Johnson, "The Telephone: Its Origin and Development," in *The Telephone Handbook,* New York, 1877
75 over wires (fn)—Francis Jehl, *Menlo Park Reminiscences,* Dearborn, 1936-1941, p. 101
76 New York—*The Operator,* Oct. 1, 1875
76 acoustic telegraph—Ed. deposition, Speaking Telegraph Interference, vol. 1, Nov. 8, 1880
76 a profit—Edson to Ed., June 4, 1875
76 he wired—Gouraud to Reiff, Sept. 18, 1875
76 for $16,500 (fn)—*Telegraph, General,* Apr. 5, 1876
77 to Batchelor—*Bennington* (Vermont) *Banner,* Nov. 14, 1929
77 optical nerves—Ed. to A. B. Chandler, Nov. 11, 1875
77 eminently unsatisfactory—Ed. deposition, op. cit.

Page *Identification—Source*

77 the said Orton—Edison–Western Union agreement, Dec. 14, 1875

78 Port Huron—Deed from George Goodyear to Ed., Dec. 29, 1875, in *Menlo Park,* 1924; mortgage, Ed. and Mary to George Goodyear, in *Insull,* July 2, 1890; Newark expense book, and documents of Jan. 3 and 20, Feb. 7, and Mar. 20, 1876, in Box 1, Ford Archives

Chapter 8. A Remarkable Kaleidoscopic Brain

79 civilised world—Sam Edison note, 1871

79 perfectly insane—Cousin Jake to Ed., Apr. 13, 1874

79 manage this—Symington to Ed., May 14, 1878

79 was seventy-eight—Ed. Nat. Hist. Site, No. 1792, *Mary Sharlow*

80 *unknown force*—Ed. Lab. Book, oversized vol. 1, Nov. 22, 1875

81 telegraphic transmission—Batch. note excerpts #9, "A New Force," Nov.–Dec., 1875; Ed. Lab. Book, oversized vol. 1, Nov. 22, 1875–Jan. 7, 1876; *The Operator*, Jan., 1876

81 induced electricity—*Journal of Franklin Institute,* May 10, 1876

81 on the other—Letter in *Scientific American,* Jan. 22, 1876

81 acoustic telegraph—Batch. note excerpts, Dec. 3 and 26, 1875

81 he shouldn't—Speaking Telegraph Interference, vol. 1, testimony of Reiff

81 your traps—Orton to Ed., Jan., 1876

81 telegraph books—Ed. to Orton, Feb. 19, 1876

81 electric circuit—Robert V. Bruce, *Bell,* Boston, 1973, p. 169

82 to litigation—Bruce, op. cit., p. 141

82 the voice—Ibid., pp. 164–65

83 transmitting speech—*New York World,* Mar. 22, 1888, and Aug. 23, 1889; Electric Light Catalog No. 1085: Undated *New York World* article, "Patent Office Frauds"; Bruce, op. cit., pp. 166–73

83 his application—Affidavit of Zenas Fisk Wilber, Apr. 6, 1886, published in *New York Herald,* May 22, 1886

83 and came—Bruce, op. cit., pp. 180–81

83 to see it (fn)—Ibid., p. 252

84 and Cambridge—Bruce, op. cit., p. 204

84 to answer—Gould to Ed., Jan. 26, 1877

84 not me—Ed. to Gould, Jan. 20, 1877

84 make it walk—Painter to Ed., Jan. 10 and 31, 1877

85 layers of cloth—Ed. Lab. Book, vol. 5, Nov. 11, 1876; Ed. Experimental Records, vol. 4, Oct. 22, 1876

85 new concern—Ed. Experimental Records, vol. 5, p. 7; Am. Novelty Co., Dec., 1876; *E. H. Johnson* HRF

85 work on—Ed. Patent No. 295,990, filed Dec. 4, 1878

85 debt of $3700—*Personal Bills,* 1877

86 in Chicago—Ed. Electric Pen & Duplicating Press, Dec. 27, 1876

86 but evaporated (fn)—*Electric Pen,* 1877; Jim Adams to Batch., July 22, 1878

87 Victorian geegaws—New Jersey Supreme Court, *Lucy Seyfert* v. *Thomas A. Edison,* sheriff's inventory, May 13, 1884

87 old newspapers—Charles Hughes statement, June 19, 1907

Page *Identification—Source*

87 induced insomnia—Alfred O. Tate, *Edison's Open Door,* New York, 1938
87 guest room—David Marshall, *Recollections of Edison*, Boston, 1931
87 decapitate himself—Tate, op. cit., p. 145
87 bath a week—Ed. response to "Longevity Inquiry," 1930
87 he recorded—Experimental Records, vol. 4, May 26, 1877
88 them before—Experimental Records, vol. 4, Dec. 10, 1876
88 his eyes—Meadowcroft, op. cit.
88 person's face—*Science News,* Nov. 15, 1880; "The Great Inventor's Triumph in Electric Illumination," *New York Herald,* Dec. 21, 1879
89 the laboratory—"The Great Inventor's Triumph," op. cit.; Meadowcroft, op. cit.
89 law firms—*Port Huron & Sarnia Street Railroad,* 1876–77
89 secretary of war—Painter to Ed., Jan. 15 and Feb. 7, 1877
89 horsecar business—Batch. Diary, Feb. 27 and Mar. 1, 1877
89 to suffer—*Teleg., Acoustic,* Dec. 6, 1876
89 five years—Lefferts to Ed., April 25, 1876
90 Western Union—Batch. Scrapbook, No. 1-867, Feb. 20, 1877
90 upon cables—Ed.–Western Union agreement of Mar. 22, 1877; Batch. Diary, Mar. 22, 1877
90 come out—Batch. Diary, Apr. 26, 1877; Norman C. Miller notes to Ed. in *Telegraph, Patents,* 1877; Orton to Ed., May 3, 1877
90 never made (fn)—Ed. memo for Orton, Mar. 22, 1877; Reiff to Ed., Apr. 12, 1894
91 their ranks—Johnson to Ed., May 10, 1877
91 be fixed—Reiff to Ed., Apr. 12 and June 4, 1877
91 receiving nothing—Quad. Case, vol. 73, p. 33
91 he confessed—Quad. Case, vol. 70, p. 330
91 and quadruplicity—*Telegrapher,* Mar. 25, 1876
91 any of them—*New York World,* Jan. 12, 1878
91 community today—Quad. Case, vol. 73, p. 58–60
92 two companies (fn)—Batch. Scrapbook, No. 1-867, Aug. 15 and 20, 1877

Chapter 9. The Speaking Telegraph

93 was under—*Telephone,* 1880; Ed. note on "Interference M"
93 commercially impractical—Batch. Notebook excerpts on Speaking Telegraph, Feb. 12, 1877
93 of plumbago—Batch. Notebook excerpts on Speaking Telegraph, Mar. 18, 1877
94 miles away—Johnson, op. cit.; *Telegraphic Journal,* Sept. 15, 1878
94 on May 23—Batch. Diary, May 23, 1877
94 The River—Johnson, op. cit.; *Johnson* HRF
94 Bell's Monopoly—Ed. to P. A. Dowd, May 14, 1877, Ford Archives
94 some weeks—*Telephone,* May 27, 1877
95 the other—Batch. Notebook excerpts, June 6, 1877
95 or Gelatin—Speaking Telegraph, Edison Exhibits 113-11, 164-11, and 136-11; note on Sextuplex Telegraph, June 25, 1877

Page *Identification—Source*

95 You Bet—Batch. Notebook excerpts, June 6, 1877
95 was fastened—Speaking Telegraph, Ed. Exhibit 7-12, July 11, 1877
95 early fall—*The Engineer,* Nov. 29, 1878
95 other substances—Speaking Telegraph, Edison Exhibits 93-12, Aug. 7, 1877, and 38-12, July 20, 1877
96 number of times—Batch. Diary, May 18, June 25, July 3, 1877
96 to Orton—Batch. Diary, Aug. 2 and 20, 1877
96 them together—Speaking Telegraph, Ed. Exhibits 82-12, Aug. 5, 1877, and 94-12, Aug. 7, 1877; Batch. Diary, July 30, 1877
96 per day—Ed. to T. B. A. David, Aug. 29, 1877
96 South Street—Batch. Diary, Aug. 30, 1877
96 the embosser—Ed. memo to Orton, Aug. 31, 1877

Chapter 10. The Sound Writer

97 Western Union operations—Ed. Oversized Lab. Notebooks and sketches: Edison Embossing Telegraph, June 26, 1877; Embossing Transmitter, July 5, 1877; Speaking Telephone, July 17, 1877; Translating Embosser, Sept. 8, 1877; Autographic Telegraph, Dec. 3, 1877; vol. 2, E 173-203, June 29, 1877; Experimental Records, vol. 5, Jan. 9, 1877
97 he hypothesized—Speaking Telegraph, Ed. Exhibit 179-11, June 24, 1877
98 the line—Ed. oversized Lab Book, Dec. 3, 1877, Autographic Telegraph
98 upper-crust Republicans—Ed. to Butler, Oct. 13, 1877
98 spring contact—Ed. oversized Lab Book, vol. 2, E 173-203, July 5, 1877
99 murmuring sound—Speaking Telegraph, Ed. Exhibit 25-12
99 original "Halloo!"—George P. Lathrop, "Talks with Edison," *Harper's,* Feb., 1890, p. 428
100 voice perfectly—Speaking Telegraph, Ed. Exhibit 25-12, July 18, 1877
100 Johnson closed—Johnson to Ed., July 17, 1877
100 talking Telegraph—*Phonograph,* oversized vol. 13, Aug. 12, 1877, and Speaking Telegraph, Ed. Exhibit 115-12, Aug. 17, 1877
100 the membrane—*Phonograph,* oversized vol. 13, Sept. 7 and 21, 1877
104 another contract—Telegraph, Quad., Sept. 18 and 19, 1877
104 at all—Butler to Ed., Oct. 23, 1877
105 to cloth—Speaking Telegraph, Ed. Exhibits 163-12 and 179-12, Sept. 20 and 24, 1877
105 the lab—Batch. Diary, Aug. 18 and 25, 1877
105 could do—*Telephone, General,* 1878, Ed. rough draft of article
105 was miraculous—Batch. Notebook excerpt, Nov. 9, 1877
105 practical transmitter—Batch. Notebook excerpt, Nov. 24, 1877; Speaking Telegraph, Ed. Exhibit 103-13, Nov. 12, 1877
105 of telephones—Batch. Scrapbook, Dec. 1, 1877
106 New York—*Scientific American,* Nov. 17, 1877
106 band of paper—Speaking Telegraph, Ed. Exhibits 54-13, Nov. 5, 1877; and 86-13, Nov. 10, 1877; *Phonograph,* oversized vol. 13, Dec. 3, 1877

Page Identification—Source

106 on and on—*Phonograph,* Ed. handwritten manuscript, Nov. 23, 1877
106 machine it—*Phonograph,* oversized vol. 13, Nov. 29 and Dec. 3, 1877; Batch. Diary, Dec. 4, 1877
107 said glumly—*New York Sun,* Feb. 22, 1878; William Carman account in *Newark News,* Jan. 26, 1902
107 *Scientific American*—Batch. Diary, Dec. 6 and 7, 1877
107 for them—*Scientific American,* Dec. 22, 1877
107 Faradization. Ahem!—*Phonograph,* oversized vol. 17, No. 1-30, 1877
107 Parlor Phonograph—*Phonograph,* oversized vol. 13, Dec. 11, 1877
107 Western Union building—*New York Sun,* Jan. 2, 1878
107 absolutely perfect (fn)—Ed. to Frank Foell, Dec. 29, 1877, Tannahill Lib.
108 represented to be—Ed. Notes, Book 2, op. cit.
108 of August—Vincent to Ed., Feb. 25, 1878
108 advertising messages—Ed. Agreement with Daniel M. Somers & Henry J. Davis, Jan. 7, 1878
108 to clocks—Batch. Diary, Jan. 29, 1878
108 in recording—Ed. to Alfred Mayer, Feb. 12, 1878; Patent No. 201,760
108 New York harbor—*New York Sun,* Feb. 22, 1878; *Charleston Journal of Commerce,* Feb. 26, 1878
108 her calf—Edison's Desk Box ¾
109 the East—*Phonograph,* oversized vol. 17, Jan. 8 and May 15, 1878; Ed. sketches for caveat, Feb. 3, 1878; Phono. device for Caveat No. 2, Mar. 5, 1878
109 a window—*Washington Post, American Union,* and *Washington Star,* Apr. 19, 1878
109 big telescope—*Washington Star,* Apr. 19, 1878
109 him $250—Wilber to Ed., Mar. 31, 1878
109 being feted—Ed. Notes, Book 2, op. cit.
109 without him—*Phonograph, General,* 1895, "Edison on Inventions"
110 clever ventriloquist—Jehl, op. cit., p. 273
110 is ridiculous—*New York Sun,* Apr. 29, 1878
110 patent office—*Washington Star,* Apr. 19, 1878
110 disk machines—*Telephone, England,* Mar. 7, 1878
110 the concern—Ed. Speaking Phonograph Co., agreement with Hilborne Roosevelt, Charles A. Cheever, et. al.; Painter to Ed., Apr. 22, 1878
110 Western Union—Batch. Diary, Dec. 27, 1877
110 seventeen years—Painter to Ed., Mar. 24, 1878
110 on others—Speaking Telephone Interference, vol. 1, testimony of Henry Bentley
110 them working—*Telephone, England,* Jan. 27, 1880
110 electromotograph receiver (fn)—Meadowcroft, op. cit.
110 phonograph experiments (fn)—*Phonograph, General,* May, 1878; Batch. Diary, Jan. 22, 1878
111 same room—*Telephone, General,* 1878, Ed. rough draft of article; Batch. Notebook excerpt, Apr. 2, 1878; Speaking Telephone Interference, vol. 1, testimony of Henry Bentley; *Journal of the Telegraph,* Apr. 16, 1878

Page *Identification—Source*

111 Hamilton McK. Twombly—*New York Times,* Apr. 23, 1878
111 over $900—Ed.–Western Union contract of May 31, 1878
112 the world—Barker to Ed., Nov. 20, 1877
112 you delay—Reiff to Ed., Jan. 8, 1878
112 Batchelor's motto—Inscription on bound volume of patents presented to Ed. by Batch.
112 two rabbits—Batch. Diary, Jan. 1, Feb. 2, and Dec. 9, 1877
112 the laboratory—*New York Sun,* Apr. 29, 1878
113 to function—*The Diary,* op. cit., p. 51
113 the catfish—*New York Sun,* Apr. 29, 1878
113 Napoleon of Science—*New York Sun,* Mar. 10, 1878
113 Common Earth—*New York Daily Graphic,* Apr. 1, 1878
113 feet away—*Boston Evening Traveler,* May 23, 1878
113 got a cent—*New York Daily Graphic,* June 6, 1878
113 an earthquake—*Washington Post,* May 20, 1878
113 their city—*Phonograph,* Apr. 25, 1878
113 railroad passes—Roosevelt to Ed., June 14, 1878
113 the current—Reiff to Ed., Aug. 7, 1877
114 credit whatever—*New York Daily Graphic,* May 31, 1878; *New York Daily Tribune,* July 13, 1878
114 and "perfidy"—*Telegraph, General,* July 19 and 30, 1878; Oct. 7, 1878
114 a reporter—*New York World,* June 27, 1878
114 them five dollars (fn)—Mrs. A. D. Coburn to Ed., May 13, June 23, and July 25, 1878
115 in all—Batch. Order Book, June 8, 1878
115 Edison's priority—Sir William Thomson to Sir Henry Thompson, June 7, 1878; *New York Herald,* June 27, 1878; Speaking Telephone Interference, vol. 1, testimony of E. H. Johnson, April 27, 28, and 29, 1881
115 his accusations—*New York Tribune,* Aug. 12, 1878
115 quiet down—Henry Knapp to Ed., June 23, 1878
115 heat-measuring device—Langley to Ed., June 4, 1878
116 by the stars—*Washington Star,* Apr. 19, 1878
116 sun's corona—*New York Tribune,* June 8, 1878
116 on the stars—*Boston Evening Traveler,* May 23, 1878
116 elevated railroad—*New York Tribune,* July 3, 1878

Chapter 11. The Napoleon of Invention

117 and blood—W. K. L. Dickson, *The Life and Inventions of Edison,* London, 1894, pp. 233–35
117 Station platform—*Daily Graphic,* July 19, 1878
117 in Schenectady—*Kruesi* HRF
118 dizzy spin—Ed. Notes, Book 1, op. cit.
118 of icebergs—*New York Herald,* July 30, 1878; John A. Eddy, "Thomas A. Edison and Infra-Red Astronomy," in *Journal for the History of Astronomy,* 1972, pp. 165–72
119 their own—*Telegraphic Journal,* June 1, 1878
120 home illumination—Moses Farmer statement of Oct. 30, 1878, in *Ed. El. Light Co.* v. *U.S. El. Light Co.,* pp. 2185–87

Page	*Identification—Source*
120	be maintained—*Philadelphia Magazine,* vol. 30, 1847, p. 45
121	bighorn sheep—Barker to Ed., Oct. 20, 1878
121	may desire—letter of J. J. Dickey, July 17, 1878
121	light business—Ed. note on Dr. Alexander Schweichler letter, May 24, 1878
121	and tunneling—*New York Graphic,* Aug. 28, 1878
121	August 20—Griffin to Ed. on telegraph, Aug. 5 and 13, 1878
122	self-evident—Griffin to Ed., Aug. 16, and Ed. to Griffin, Aug. 19, 1878, on telegraph
122	the exposition—*New York Tribune,* Aug. 27, 1878
122	a day—Recollections of Marion Edison Oeser, Mar. 1956
122	Dot and Dash—*New York Graphic,* Aug. 28, 1878; *New York Sun,* Aug. 29, 1878
122	telephone apparatus—*Telegraph, General,* Sept. 5, 1878
123	heated air—El. Light sketches for "Electric Draft Lamp," Aug. 29 and Sept. 9, 1878
123	September 8—*El. Light,* Sept. 2, 5, and 6, 1878
123	of murder—*New York Mail,* Sept. 10, 1878
123	right track—Ed. Lab. Book, vol. 5, Sept. 8, 1878
124	his features—*Scientific American,* July 13, 1878
124	shared it—*New York Sun,* Oct. 20, 1878
124	big bonanza—Ed. to Wallace, Sept. 13, 1878
124	electric lighting—Caveat for El. Light Spirals, Sept. 13, 1878
124	prevent oxidation—Caveat No. 2, Sept. 25, 1878
124	Caveat 5—Ed. Lab. Book, vol. 1, copied from El. Light Records by William Carman, Caveat 3, Oct. 3, 1878, Caveat 4, undated, Caveat 5, Oct. 16, 1878; see also Electric Light Subdivision, oversized vol. Sept.–Oct., 1878; Ed. Patents No. 214,636, Thermostatic Shunt, filed Oct. 5, 1878, and No. 214,637, Thermostatic Regulator, filed Nov. 18, 1878
124	much cheaper—*New York Sun,* Oct. 20, 1878
124	into movement—H. S. Olcott to Ed., Apr. 4, 1878
125	gaslight companies—Ed. to Theodore Puskas, Oct. 5 and 14, 1878; Ed. to Gouraud, Oct. 8, 1878
125	incandescent light—*Electric Light,* Sept. 24, 1878
125	came out—*Electric Light,* Sept. 26. 1878
125	"good news"—*Electric Light,* Sept. 30, 1878
125	reduced cost—*Electric Light,* Lowrey memo to Twombly, Oct. 1, 1878
125	for you—Lowrey to Ed., Oct. 2, 1878
125	for Edison—Lowrey to Ed., Oct. 12, 1878
125	Menlo Park—*Ed. Personal Bills,* 1878–79; *Menlo Park Lab Accounts,* 1879
126	expanded shop—Ed. to Puskas, Nov. 13, 1878; Lowrey to Ed., Dec. 23, 1878
126	two days—*Electric Light,* Oct. 5, 1878
126	my mind—*New York Herald,* Oct. 12, 1878
126	500 horsepower engine—*New York Sun,* Oct. 20, 1878; *Telegraphic Journal,* Oct. 15, 1878
127	taken seriously—*Telegraphic Journal,* Oct. 15, 1878
127	to bed—*Electric Light,* Sept. 24, Oct. 1 and 23, 1878
127	his head—*New York Sun,* Oct. 20, 1878

Page	*Identification—Source*
127	bad cold—*Electric Light,* Oct. 19, 1878
127	the blink—Ed. to builder, Oct. 28, 1878
127	far behind—*Scientific American,* Nov. 16, 1878; *New York Times,* Oct. 30, 1878; *Electric Light,* Oct. 30, 1878
128	see anyone—*Electric Light,* Oct. 24, 1878
128	the organization—Griffin to Lowrey—Nov. 1, 1878
128	was unthinkable—Ibid.
128	absolutely complete—Lowrey to Ed., Nov. 2, 1878
128	has been done—*Ed. El. Light Co.,* v. *U.S. El. Light Co.,* U.S. Circuit Court for the Southern District of New York, vol. 4, p. 2602
128	been accomplished—Lowrey to Ed., Oct. 31, 1878
128	New England family—Ed. to H. R. Butler, Nov. 12, 1878
128	incandescent development—Upton Historical Reference file, Nov. 13, 1878, and letter of Nov. 7, 1878
128	November 12—Lowrey to Ed., Oct. 31 and Nov. 7, 1878
128	a magazine—Ed. Lab Book, Electric Light, vol. 1, p. 208; Patent No. 224,329
128	their support—*Electric Light,* Nov. 13, 1878
129	sum of $30,000—Ed. Lab Book, El. Light, vol. 1, p. 208, Nov. 15, 1878; Lowrey to Ed., Nov. 13, 1878; *Electric Light,* Nov. 15, 1878; Ed. El. Light Co., op. cit., p. 2348
129	J. Hood Wright—*Electric Light,* Oct. 25, 1878
129	of Europe—Ed. to Puskas, Nov. 13, 1878; *Electric Light,* Dec. 2, 1879; *Electric Light, England,* June, 1879
129	only trifles—Rothschild to Belmont, Oct. 25, 1878
129	international finance—*Electric Light,* Dec. 2 and 5, 1878
129	he said—*New York Herald,* Dec. 3, 1878
129	$100,000 to $125,000—*New York Sun,* Nov. 25, 1878
129	nineteenth century—*London Times,* Oct. 12, 1878
130	and dynamics—*New York Sun,* Oct. 20, 1878; *The Electrician,* Nov. 23, 1878
130	to obtain—*New York Herald,* Dec. 3, 1878
130	the heat—Ed. Lab Book, Electric Light, vol. 1, "Electric Light Law," Nov. 15, 1878
130	of electricity—Francis Jehl letter of April 22, 1913, copied by William Hammer, Mar. 15, 1929, in Hammer Collection
130	bluestone battery (fn)—Ibid.
131	550 volts pressure—John W. Howell, *History of the Incandescent Lamp,* Schenectady, 1927, p. 52
132	not explain—Ed. El Light Co., op. cit., vol. 4, pp. 2571 and 2578
132	*high* resistance—N 8-12-15.2, pp. 11-13, and No. 31, p. 73
133	an impossibility—Ed. El. Light Co., op. cit., vol. 6, p. 4109
133	from experimenting—Ibid., vol. 4, p. 2571
134	the hill—Francis Upton Papers, Dec. 12, 1878, in *Upton* HRF
134	midnight suppers—*New York Herald,* Dec. 3 and 11, 1878
134	of success—Jehl letter, op. cit.
134	pencil and paper—N 28-11-28 and N 78-12-20.2
134	his notebook—N 78-12-20.3
135	been filed—N 78-11-16 and 79-00-01; *Electric Light,* Dickerson to Lowrey, Dec. 11, 1878
135	other on—Jehl letter, op. cit.

Page	*Identification—Source*
135	investors' anxiety—Lowrey to Ed., Dec. 23, 1878
135	for dictating—Ed. note on Painter letter, Jan. 9, 1879
135	the phonograph—*Phonograph,* Oct. 12–14, 1878, and Jan. 21, 1879
135	its development (fn)—*Electric Light,* 1879, note referring to Marcus patent of Oct. 20, 1878; Robert Conot, *American Odyssey,* New York, 1974, p. 111
136	or utility—*Phonograph Accounts,* 1878; *Phonograph,* 1880 and 1883
136	telephone instrument—Johnson to Ed., Apr. 10, 1878
136	problem whipped—Speaking Telephone Interference, vol. 1, p. 80
136	the mail—Letter Book E 3440, p. 130, Mar. 4, 1877
136	calmed down—D. Murray to Ed., Oct. 22, 1879; Ed. to Puskas, June 2, 1879
136	system working—*History of St. Clair County,* op. cit., pp. 625–26
136	famous uncle—Simeon O. Edison to Ed., Dec. 11, 1879
136	strontium drum—Ed. to S. W. Stratton, Nov. 24, 1908
136	feet away—Speaking Telephone Interference, vol. 1, p. 80; Batch. Notebook 1304, May 18 and 19 and June 25, 1879; Ed. Lab. Book, vol. 4, Electromotograph Receiver, June 19 to Oct. 13, 1878
137	our discontent—N 78-12-02, Dec. 28, 1878
137	back wages—*Telephone, Foreign,* Dec. 15, 1878
137	vivid incandescence—Caveat No. 6, Jan. 13, 1879; N 78-12-15.1, Jan. 9, 1879; see also sketches in oversized Electric Light vol., Jan. 13, 1879
137	in again—Ed. to Puskas, Jan. 3, 1879
137	3000° F.—Lowrey to Ed., Feb. 10, 1879
137	to bed—Barker to Ed., Jan. 17, 1879
137	psychological malaise—James A. Crowell, "What It Means to be Married to a Genius," *American Magazine,* Feb., 1930
137	to remedy—Ed. to Puskas, Jan. 3 and 28, 1879
137	a year—*New York Herald,* Jan. 30, 1879
138	now knows—*New York Graphic,* Feb. 6, 1879
138	of gas—*English Mechanic,* Feb. 21, 1879
138	do everything—quoted in *Telegraphic Journal,* July 15, 1880
138	really absurd—*Telegraphic Journal,* Feb. 1, 1879
138	join in it—Lowrey to Ed., Jan. 25, 1879
138	tungsten, et al.—Jehl, *Reminiscences,* op. cit., p. 238
138	fell asleep—N 79-01-19, Jan. 19–27, 1879

Chapter 12 A New Light to the World

139	and Alice—*Ed. Personal,* Jan. 8, 22, and 28, 1879; Reiff to Ed., Feb. 11, 1879
139	been found—N 78-12-20.3
140	the globe—N 78-12-15.1, Feb. 10, 1879, and N 79-01-01, Feb. 12, 1879
140	borrow it—Ed. telegrams to professors George Barker and Henry Morton, Jan. 22, 1879; Francis Jehl Memo to William Hammer, April, 1881, in Edisonia, Hammer Collection
140	ten minutes—N 78-11-22, Feb. 8, 1879

Page	*Identification—Source*
140	radio tube—N 78-12-31, Feb. 12, 1879
140	platinum increased—N 78-12-31, Feb. 6, 1879
140	the current—N 31, pp. 51–69; Caveat of Mar. 17, 1879; El. light Patent No. 227,229 sent to Serrell Mar. 10 and filed Apr. 12, 1879
141	March 16—N 78-11-21, Mar. 16, 1879
141	commercial standpoint—Batch. Notebook 1304, Mar. 18, 1879
141	lamps simultaneously—N 79-02-15.2, Provisional Specification for a New English Patent
142	stopping point—*New York Times,* Oct. 19, 1931
142	with nitrogen—El. Light oversized vol. 2, Feb. 20, 1879
142	five hours—*Consolidated Light Co.* v. *McKeesport Light Co.,* U.S. Circuit Court for the Western District of Pennsylvania, vol. 1, pp. 757–59
142	certainly achieved—Ed. to Puskas, Nov. 13, 1878
142	for invention—Upton to father, Apr. 27, 1879, Hammer Collection
142	fairly steady— N 78-12-04.2; El. Light oversized vol. 2, sketch of Apr. 19, 1879; Francis R. Upton, "Edison's Electric Light," in *Scribner's,* Feb., 1880
142	same design—Barker to Ed., Nov. 30, 1880
142	greatly increased—N 79-02-15.2, Improvements in Electric Lighting and in Apparatus for Developing Electric Currents, Edison's British patent application No. 2402, Nov. 25, 1879; Batch. Order Book, Feb. 22, Mar. 13, Apr. 12, 1879; Batch. Notebook 1304, Dec. 18 and 25, 1878
143	another dynamo—N 31, p. 81; Edison Electric Light Patent No. 227,229; N 79-02-15.2, "Provisional Specification for a New English Patent"
143	the age—J. C. Henderson to Villard, Apr. 15, 1879
143	Electric Light—*New York Herald,* Apr. 27, 1879
144	enormous results—DeLong to Ed., Apr. 21, 1879; Jerome Collins to DeLong, May 5 and 14, 1879; "Accounts of Outfitting Arctic Steamer Jeannette," National Archives, Washington, D.C.
144	sent away—Upton to father, May 24, 1879, Hammer Collection
144	pillarlike magnets—Ed. Patent No. 219,393, "Dynamo-Electric Machine"
144	1100 pounds—Jehl letter of April 22, 1913, op. cit.; *New York Times,* July 7, 1879; *Scientific American,* Oct. 18, 1879; Jehl, *Reminiscences,* op. cit., p. 312
144	80 per cent—*Scientific American,* July 26, 1879; N 79-02-15.2, Improvements in Electric Lighting, op. cit.
144	the lab—N 79-01-14
144	De Long, perished (fn)—*The Voyage of the Jeannette, Journal of Lieutenant Commander George DeLong,* Boston, 1884
144	their own—Batch. Notebook 1304, Feb. 23 and Mar. 23, 1879; Ed. to T. L. Clingman, Sept. 8, 1879
144	in stock—*El. Light Filament,* Apr. 2 and May 5, 1879
144	the knees—N 22, Apr. 1, 1879
144	the globes—N 78-11-22 and 78-12-31; Ed. El. Light Co., op. cit., vol. 5, p. 3025
145	curled edge—*New York Sun,* July 7, 1879; John Allyn to Ed.,

Page *Identification—Source*

July 30 and Dec. 6, 1879; *Telegraph, General,* 1879; Milt Adams to Ed., Sept. 22, 1879; *Ed. El. Light Co.,* July 14 and 15, 1879; N 78-11-22, p. 149; *Ed. Ore Milling Co.–Platinum Search,* 1879

145 Haid reported—*New York Sun,* July 7, 1879

145 public demonstration—Patent No. 221,957, "Improvement in Telephones," filed Mar. 31, 1879

145 would do—Ed. to Bliss, Feb. 27, 1878

146 six instruments—*New York Herald,* Feb. 27, 1879

146 he is here—Gouraud to Ed., Apr. 10, 1879

146 telephone business—*London Times,* Mar. 17, 1879

146 England hum—George Bernard Shaw, preface to *The Irrational Knot*

146 in Paris—Charley Edison cable, May 5, 1879

146 Dutch clock—Charley to Ed., May 15, 1879

146 home immediately—Ed. to Charley, May 12, 1879

146 uncover it (fn)—Pitt Edison to Ed., Feb. 17, 1879

147 $25,000 in hand—Arnold White to Ed., May 13, 1879; *Telegraph, England,* May 23, 1879

147 five hundred books—Upton to his father, May 24, 1879

147 carbon-button transmitter—Batch. Notebook 1304, May 18 and 19, June 25, and July 3, 1879

147 more attractive—Johnson to Ed., May 22, 1879; Olin D. Russell to Ed., May 26, 1879

147 in New York—N 79-06-12; *Telegraph, General,* 1879, Ed. draft for patent application; *New York Sun,* July 7, 1879

147 them out—Ed. to George Walker, July 16, 1879

147 carbon buttons—N 79-07-07.2

148 half as loud—N 79-06-12, July 7, 9, and 10, 1879

148 a second—Ed. to Tracy Edson, July 7, 1879

148 Roman candle—N 79-02-15.2, Provisional Specification for a New English Patent

148 was impossible—Upton to his father, Apr. 27, 1879

148 previous strength—N 79-04-03; N 79-02-15.2, Provisional Specification for a New English Patent; Caveat of July 23, 1879

149 worse than useless—Batch. Notebook 1304, July 31 to Sept. 9, 1879; N 79-07-31, July 31 to Aug. 22, 1879; Ed. to T. L. Clingman, Sept. 8, 1879; Howell, op. cit., pp. 54–55

149 Edison liberally—*New York Tribune,* Sept. 1, 1879; *New York Times,* Sept. 3, 1879

149 profit go—Upton to his father, June 22 and 29 and July 6, 1879

149 their wares—N 79-10-18, Oct. 18, 1879

149 running motors—*Electricity, Motors,* Sept. 8, 1879

149 prohibitive figure—N 79-07-07.1, July 10, 1879

149 he declared—Ed. handwritten prospectus in *Ed. Ore Milling Co.,* 1879

149 platinum and gold—Correspondence with G. W. Carleton, June 21, 1879

150 be formed—Ed. handwritten prospectus, op. cit.; *Ed. Ore Milling Co.,* July 24, 1879, et seq.

150 awhile longer—Lowrey to Ed., Sept. 10 and Oct. 16 and 24, 1879; *Ed. El. Light Co.,* Sept. 17, 1879

Page Identification—Source

150 a belcher—*Telegraph, General,* Oct. 8, 1879; *Telegraph Agreements,* 1880; N 79-09-18, p. 93
150 varied widely—N 79-04-09; N 79-09-18
151 be obtained—Jehl letter of April 22, 1913, op. cit.
151 required sealing—Jehl memo to Hammer, op. cit.
151 *New York Herald*—Ed. El. Light Co. trial, op. cit., vol. 5, pp. 3420 and 3251
151 at Heidelberg—Jehl, *Reminiscences,* op. cit., pp. 324–25
151 the Sprengel—Ed. El. Light Co. trial, op. cit., p. 3029
151 green phosphorescence—N 79-08-22 and N 79-07-31, Sept., 1879
152 Upton enthused—N 79-08-22, Sept. 26, 1879
153 the apparatus—N 79-08-22, Oct. 1 and 2, 1879
153 it before—N 79-09-20, Oct. 3, 1879; Ed. El. Light Co. trial, op. cit., pp. 3013–14, 3026
153 platinum wire—N 79-08-22, Oct. 8, 1879; N 79-07-31, Sept.–Oct., 1879
153 the time—N 79-04-21, Oct. 11, 1879
153 with platina—N 79-04-09, p. 45
154 ohms resistance—N 79-08-22, Oct. 13, 1879
154 to silicon—N 79-08-22; N 79-10-18; N 79-09-18, Sept. 18–Oct. 4; N 79-02-15.2, Oct. 17 and 20, 1879; N 79-06-16.1, Aug. 8–15, 1879; N 79-07-07.2, Aug. 12, 1879; Francis Upton, "Edison's Electric Light," *Scribner's,* Feb., 1880
154 incandescent light—*Electric Light,* May 13, 1878
154 months before—N 79-07-12, p. 5
154 both poles—N 79-08-22, Oct. 11 and 13, 1879
154 Lord only knows!—Ed. to Johnson, Oct. 10, 1879
154 the pump—N 79-08-22, Oct. 13, 1879
154 high-resistance mixture—N 79-02-15.2, Oct. 17 and 20, 1879
155 sticks resumed—N 79-08-22, Oct. 17, 1879
155 Paris–Brussels link—Charley to Ed., June 26, 1879
155 I said so—Nelly Edison to Ed., June 29, 1879, and Ed. response
155 were sous—*Charles Edison* file, May 8, 1879
156 October 18—D. Murray to Ed., Oct. 22, 1879
156 Port Huron—*Charles Edison,* Oct. 20, 1879, et seq.; B. H. Wilton to Ed., Oct. 23, 1879; Ed. to Wilton, Oct. 27 and Nov. 3, 1879
156 a scandal—Bailey to Ed., Mar. 8, 1880
156 all safe—D. Murray to Ed. in *Charles Edison,* 1879
156 the boxes—Batch. to Johnson, Dec. 7, 1879
156 he philosophized—*The Diary and Sundry Observations,* op. cit., p. 58
157 glass tubes—Marshall, op. cit.; Jehl, *Reminiscences,* op. cit., vol. 1, pp. 352–53
157 2 ohms—N 79-08-22, Oct. 19, 1879
157 flashed out—N 79-07-31, Oct. 7 and 21, 1879; Lab Book, E 173-203, Case 242; N 79-08-22, Oct. 22 and undated, pp. 175–77
157 113 ohms—N 79-07-31, Oct. 22, 1879
160 glass cracked—N 79-07-31, Oct. 21–28, 1879
160 local tavern—Jehl, *Reminiscences,* op. cit., pp. 516–18
160 he exploded—N 79-07-31, Oct. 28, 1879
160 by incandescence—Ed. caveat of Oct. 28, 1879
160 a bobbin—Ed. Patent No. 223,898, "Electric Lamp"

Page	*Identification—Source*
161	the world—Upton to his father, Oct. 19, 1879, Hammer Collection
161	string wires—*El. Light,* Nov. 4, 1879; Ed. to Norvin Green, Nov. 4, 1879
161	roll in—Upton to his father, Nov. 2 and 9, 1879
161	the threads—N 79-07-31, Lamp No. 29
161	sixteen hours—N 79-08-22, Nov. 12–17, 1879
161	wished it for—Upton to his father, Nov. 22, 1879
161	failed completely—N 79-08-22; N 79-07-31
161	November 25—Ed. British Patent Application No. 2402, op. cit.
161	tried before—N 79-07-31, Nov. 28, 1879
161	Edison wrote—Ed. to Henry Cox, Dec. 1, 1879
162	December 3—Griffin to Bailey, Dec. 2, 1879
162	white carpet—N 79-07-31, Dec. 3, 1879; Mary Holzer to Norman Speiden, June 5, 1956
162	for Edison—Lowrey to Ed., Nov. 13, 1879
162	electric machines—*Scientific American,* Oct. 15 and Dec. 6, 1879
162	as lights—*New York Sun,* Oct. 20, 1878
162	the lights—N 79-07-31, Dec. 3, 1879
162	immense success—Lowrey to Ed., Dec. 5, 1879
162	the country—*Ed. Light Exhibit,* 1879
162	for $10,000—*Ed. El. Light Co.,* Dec. 5, 1879
162	Wall Street—Lowrey to Ed., Dec. 5, 1879
163	be resumed—Johnson to Batch., Nov. 27, 1879, and Johnson to Ed., Dec. 2, 1879
163	$25,000 due—James Banker to Ed., Nov. 29, 1879
163	Edison crowed—Ed. to Frank McLaughlin, Nov. 11, 1879
163	in California—*California Weekly Mercury, Nov.* 14, 1879
163	with electricity—*London Weekly Dispatch,* Oct. 25, 1878
163	the 260th—N 79-07-31, Dec. 5–19, 1879
163	go with it—U.S. Caveat and English Provisional Protection, Dec. 19, 1879
163	platinum lamp—Ed. British Patent Application No. 2402, op. cit.
163	western trip—Fox to Ed., May 3, 1879
163	taken notes—*El. Light Articles,* Mar. 3 and Aug. 13, 1879
164	"The Triumph of the Electric Light"—*New York Herald,* Apr. 27, 1879
164	an afterthought—*New York Herald,* Dec. 21, 1879
164	were ready—*El. Light Articles,* Joel Cook to *London Times,* Dec. 28, 1879
164	us backwards—Du Moncel letter in *La Lumière Electrique,* Jan. 1, 1880
164	the public—*New York Sun,* Dec. 22, 1879
165	the World—*Puck,* New Year's issue, 1880

Chapter 13. A Wide-Awake Wizard

166	through unscathed—Lab. Book, E 173-212, Jan. 3, 1880
166	you see—H. R. Fraser, *The Old Man,* unpublished manuscript, pp. 178–79
166	40 hours—Ed. El. Light Co. trial, op. cit., vol. 5, pp. 3023–24

Page *Identification—Source*

166 550 hours—Upton to his father, Jan. 25, 1880
166 Upton *Immortal!*—N 79-04-03
167 enormous profits—Villard to Stern, Feb., 1880, Baker Lib.
167 is practicable—Stern to Villard, Feb. 9 and 16, 1880, Baker Lib.
167 down to $1500—Upton to his father, Dec. 28, 1879, and Jan. 25, 1880; Goddard to Ed., Feb. 2, 1880
167 twenty minutes—*New York Sun,* Dec. 22, 1879, and Jan. 4, 1880; *New York World,* Dec. 24, 1879
167 every demonstration—Consolidated Electric Light Co. trial, op. cit., vol. 1, pp. 35–36 ff
167 the lamp—*El. Light Articles,* Joel Cook to *London Times,* Dec. 28, 1879, and *Times* editorial, Dec. 31, 1879
167 his head—*Saturday Review,* Jan. 10, 1880
168 aggressive enemy—Ed. to Theodore Waterhouse, Jan. 29, 1880
168 over reported—Jim Seymour to Ed., Apr. 11, 1880
168 new company—*Telephone Agreements,* May 13, 1880
168 Holcombe v. Bell (fn)—Speaking Telephone Interference, vol. 1
169 the patents—Bruce, op. cit., pp. 268–71
169 $6,000 yearly—*Western Union Agreements,* Ed. undated note to John Tomlinson
169 low temperature—*Telephone, General,* Ed. draft for caveat, June, 1879
169 be replaced—Vail to Ed., Dec. 18, 1879
169 a century—Ed. to Vail, Dec. 21, 1879
169 Prescott $100,000—*Telegraph, Western Union,* Jan. 2, 1880
169 limited to $3,000—Contract of May 12, 1879
169 Bell's patent (fn)—*New York World,* Mar. 22, 1888
170 positive value—Ed. note on letter from Norvin Green, Mar. 12, 1880
170 Edison snorted—Green to Ed., Mar. 12, 1880, and Ed. marginal notes
170 with Gould—*Electromotograph,* Ed. telegram to Green, 1880; Gould to Ed., Mar. 17, 1880
170 twelve years—Green to Ed., Mar. 22, 1880; Western Union agreement of Apr. 3, 1880
170 to $175,000—*Telephone, Europe,* 1880, Ed. trust set up through George Renshaw
170 another $14,000—*Telegraph,* Jan. 17, 1888
170 $13,000 to $45,000—PN, Oct., 1878, to Dec., 1879
170 New York—*El. Light, SS Columbia,* Apr. 16, 1880
171 them in—Jehl, *Reminiscences,* op. cit., pp. 558–63
171 held aboard—*New York Herald,* Apr. 29, 1880; Upton to his father, May 9, 1880
171 West Coast—Ed. El. Light Co. trial, op. cit., vol. 5, pp. 3015, 3244
171 Washington, D.C. (fn)—*El. Light,* Dec. 31, 1929
172 berserk camel—Marshall, op. cit.; *Electric Railroad,* May, 1880 (in 1882 file); Jehl, *Reminiscences,* op. cit., p. 859; Charles Hughes statement of June 19, 1907
172 "A daisy!"—Lowrey to Mrs. Lowrey, June 5, 1880
172 the tracks—W. W. Riley to Ed., May 28, 1880; Lowrey to Ed., Aug. 7, 1880; Butler to Ed., July 7, 1880; Gouraud to Ed., Aug. 20, 1880; *Electric Railroad Accounts,* 1880

Page *Identification—Source*

172 the project (fn)—Gouraud to Ed., Sept. 9 and 11, 1880; *Vacuum Freezing,* 1880; *Fruit Preservation,* 1881; Jehl, *Reminiscences,* op. cit., p. 534

Chapter 14. The Giver of Light

173 standard bulbs—Ed. El. Light Co. trial, op. cit., vol. 4, pp. 2188, 4227, 4236; *Engineering,* May 14, 1880
173 1200 candlepower—John W. Urquart, *Electric Light,* London, 1893, p. 345
173 Robert Gilliland—*Ed. Lamp Works,* Apr. 24, 1880
173 green phosphorescence—Ed. El. Light Co. trial, op. cit., vol. 4, p. 4228; Jehl, *Reminiscences,* p. 338; N 80-07-05, Aug. 29, 1880
174 poisonous atmsophere—*Ed. Lamp Works,* undated Upton memo
174 to Europe—Batch. Scrapbook, July 21, 1880
174 to glow—N 80-07-05, July 9, 10, and 12, 1880
174 Edison wired—Segredor to Ed., Sept. 8 and 28, 1880; George Price to Ed., Sept. 17, 1880; V. F. Butler to Ed., Oct. 27, 1880; Ed. to Butler, Nov. 4, 1880
175 the filament—*Engineering News,* Dec. 25, 1880; *Ed. Lamp Works,* Upton letter of Mar. 3, 1881; Jehl, *Reminiscences,* op. cit., p. 865
175 than paper—Upton to Ed., May 4, 1881
175 nailed shut—Serrell to Upton, Apr. 25, 1881
175 tonight at—N 79-10-18, p. 36; Jehl, *Reminiscences,* op. cit., p. 512
175 thunder dramas—Jehl, *Reminiscences,* op. cit., pp. 500–503
175 shouted back—Ibid., p. 504
176 the phonographs—Bergmann to Ed., Apr. 1 and June 8, 1880, and Nov. 20, 1888
176 intermediate points—Upton to his father, May 9, 1880
176 full-scale test—William Andrews account in *John Kruesi* HRF; Batch. Scrapbook, Aug. 9, 1880
176 two weeks—Andrews account, op. cit.; Jehl, *Reminiscences,* op. cit., pp. 720–23
176 pine tar—N 80-07-05, Sept. 1–13, 1880
177 an hour—Andrews account, op. cit.
177 junction box—Marshall, op. cit.
177 the fields—Andrews account, op. cit.
177 lamps connected—Charles Clarke to S. B. Eaton, Apr. 11, 1881; Marshall, op. cit.
177 the powerhouse—Jehl, *Reminiscences,* op. cit., p. 591
177 the conductors—*Ed. El. Light Co.,* Dec. 2, 1880; Ed. to Tracy Edson, Nov. 20, 1880
177 large-scale production—Ed. to Weeks, Nov. 20, 1880; N 79-02-20.2; Jehl, *Reminiscences,* op. cit., p.856
177 any good—Marshall, op. cit.
178 in February—Recollections of Marion Edison Oeser, Mar., 1956; Mary Holzer to Norman Speiden, June 5, 1956; Marshall, op. cit.

Page	*Identification—Source*
178	close pals—Jehl, *Reminiscences,* op. cit., pp. 495–96, 513
178	the shed—Ibid., p. 516
178	the butter!—Hammer Reminiscences in Hammer Writings on Edison, Hammer Collection
178	Böhm left—Böhm to Ed., Oct. 14 and 20, 1880; Ed. to Böhm, Oct. 21, 1880
179	new employer—Wilber to Ed., Jan. 10, 1881
179	to him—Jehl, *Reminiscences,* op. cit., p. 611.
179	of piracy—Ibid., pp. 708–709
179	it greatly—*Illustrated Science News,* Dec. 21, 1880
179	had noted—Bessemer to Ed., Jan. 12, 1880
179	he rationalized—Ed. note on letter from A. B. Hutchins, Nov. 29, 1880
179	of mercury—Hammer to John Lieb, Apr. 30, 1925, Hammer Collection
179	seen treated—Upton to Ed., May 4, 1881
180	night and day—Edwin Fox to Ed., Oct. 22, 1880; *New York Tribune,* Nov. 25, 1880
180	20 years ago—Barker to Ed., Nov. 9 and 26, 1880; *New York Post,* Nov. 22, 1880
180	is forgotten—*Illustrated Science News,* Dec. 21, 1880
180	the woods—Lowrey to Ed., Nov. 28, 1880
181	streams of fire—Sarah Bernhardt, *Memories of My Life,* New York, 1907, pp. 392–95
181	giver of light—Jehl, *Reminiscences,* op. cit., pp. 770–75

Chapter 15. Edison Asked Only for the Earth

182	attended meetings—*Ed. El. Light Co.,* Dec. 16 and 17, 1880
182	his regrets—Batch. Scrapbook, Dec. 28, 1880
182	was plotted—Hammer to John Lieb, April 30, 1925, Hammer Collection
183	the city—*Ed. El. Light Co.,* Dec. 20, 1880; Batch. Scrapbook, Dec. 20, 1880; *Newark Advertiser,* Jan. 10, 1881; Jehl, *Reminiscences,* op. cit., pp. 777–78
183	Lowrey's face—Lizzie Upton letter, Dec. 27, 1880; Lowrey to Ed., Dec. 31, 1880; *Menlo Park Trial Balances,* Dec., 1880; Villard Papers, Baker Library, Oct. 27, 1880
183	successful completion—Ed. El. Light Co. Agreements, Jan. 12, 1881
184	property tax—*New York Herald,* Jan. 21, 1881
184	the earth—*Forty Years of Edison Service,* New York Edison Co., 1922
184	to begin—*Newark Advertiser,* Jan. 10, 1881; Batch. Scrapbook, April 21 and 30, 1881; *Ed. Illuminating Co. of New York,* Apr. 19, 1881
184	the façade—Batch. Scrapbook, Mar. 1, 1881
184	Until Morning—Hammer Writings on Edison; Hammer to John W. Lieb, April 30, 1925, Hammer Collection; Jehl, *Reminiscences,* op. cit., p. 516

Page	*Identification—Source*
185	of perpetuating—Jehl, *Reminiscences,* op. cit., p. 519
185	from oxidizing—Hammer to John W. Lieb, Apr. 30, 1925, Hammer Collection; Jehl, *Reminiscences,* op. cit., p. 860
185	this afternoon—Jehl, *Reminiscences,* op. cit., p. 862
185	successful failure—Jehl, *Reminiscences,* op. cit., p. 868; John W. Hammond, *Men and Volts,* New York, 1941, pp. 44–45
186	Goerck Street—Edison Companies, Box 1, B Group, Hammer Collection; Jehl, *Reminiscences,* op. cit., pp. 878–79
186	just right—Upton to Ed., Mar. 7, 1881; Ana Kirsten to Mina Edison, Nov. 26, 1929, in *Golden Jubilee* file
186	3 cents—Ed. Agreement with Ed. El. Light Co., Mar. 8, 1881
186	bargain price—*Ed. Lamp Co.,* May 4 and 9, 1881; Jehl, *Reminiscences,* op. cit., p. 814
186	its payments—*Ed. Lamp Co.,* Dec. 20, 1881; Ed. note of Feb. 4, 1881; Balance Sheet, 1881–87
187	of visitors—Jehl, *Reminiscences,* op. cit., p. 885; Batch. Scrapbook, Feb. 8, 1881; *New York World,* Apr. 5, 1881; *Commercial Advertiser,* May 7, 1881; Johnson to Ed., Oct. 22, 1881
187	called "electroliers"—*Forty Years of Edison Service,* op. cit.; Jehl, *Reminiscences,* op. cit., p. 753
188	Light Company—*Ed. Co. for Isolated Lighting,* Nov. 11, 1881; Isolated Lighting Plants in the U.S., 1882
188	manufacturing company—Walter L. Welch, *Charles Batchelor,* Syracuse University, 1972, p. 46
188	the board—Ed. to S. B. Eaton, Sept. 13, 1881; Calvin Goddard to Ed., Oct. 1, 1881
189	top award—*William J. Hammer* HRF; Meadowcroft, op. cit.; *Ed. El. Light Co. of Europe,* Oct. 23, 1881; Batch. Scrapbook, Nov. 3, 1881
189	for lying—Batch. Recollections of Ed., Nov. 2, 1909
189	was abandoned—H. Patterson to Henry Shields, Apr. 16, 1882, Correspondence Box 1879–1899, Hammer Collection; Batch. Recollections of Edison, Nov. 2, 1909; Johnson to Upton, Nov. 11 and Dec. 30, 1881
190	anyone else—Johnson to Ed., Oct. 22, 1881
190	us fully—Puleston to Ed., Jan. 24, 1880
190	as yourselves—Ed. to Puleston, Apr. 28, 1880
190	and legally—Puleston to Ed., May 2, 1880
191	Fleet Street—*London Electrician,* Apr. 22, 1882
191	more lovely?—Johnson to Ed., Oct. 22, 1881
191	be bygones—Ed. to Preece, Jan. 23, 1882; Preece to Ed., Feb. 10, 1882
191	were connected—Deposition of William Hammer, July 8, 1915, in *Albert Moritz* v. *Ed. Elect. Illuminating Co. of Brooklyn,* Early Edison Companies Box, Hammer Collection
191	central station (fn)—Johnson to Ed., Nov. 19, 1881
192	challenging Edison—*London Electrician,* Mar. 17, April 22, and June 19, 1882; *Ed. El. Light Co. of Europe,* Jan. 23 and 27, 1882; *William Hammer* HRF
192	he lives—Johnson Statement, Sept. 15, 1881
192	*London Standard* noted—*London Standard,* Jan. 20, 1881

Chapter 16. The Magic Touch

Page Identification—Source

194 Edison's business—Insull Notes for Meadowcroft, Feb., 1909

194 at $160,000—Ed. calculations in *Upton* HRF

194 at night—John W. Howell, *Lamps for a Brighter America,* Schenectady, 1927, p. 71; El. Light Notebook 173-154, 1880; *Commercial Advertiser,* May 7, 1881; Ed. rough notes of Dec., 1879, in *Upton* HRF; Johnson Statement of Sept. 15, 1881; Ed. to Johnson, Sept., 1888

195 lighting companies—Johnson Statement of Sept. 15, 1881; *American Journal of Science and Arts,* vol. 17, 1879, pp. 65–66; *New York Sun,* Aug. 11 and 21, 1881

195 "undertakers' wire"—*Commercial Advertiser,* May 7, 1881; Jehl, *Reminiscences,* op. cit., p. 896

195 dollars a day—*Ed. El. Light Co.,* Dec. 2, 1880

195 be required—Jehl, *Reminiscences,* op. cit., pp. 545–46, 732–33

195 first imagined—Johnson to Ed., Dec. 31, 1881, and Ed. reply

195 and replaced—Ed. handwritten notes, 1881

196 coal mine—*Forty Years of Edison Service,* op. cit.

196 never attended—Recollections of Marion Edison Oeser, Mar., 1956

196 other jewelry—Ibid.

196 western trip—J. W. Wexel to Ed., Dec. 16, 1878

196 electric development—Recollections of Marion Edison Oeser, Mar., 1956

196 his parents—Ibid.

196 no mistake—Ed. to Sam Edison, Oct. 21, 1876, Tannahill Lib.

197 his indigestion—Recollections of Marion Edison Oeser, Mar., 1956

197 her mother—Marshall, op. cit.; Marion Edison Oeser to Norman Speiden, May 23, 1947

197 near Trenton—*Ed. Personal,* Sept. 20, 1882

197 Menlo Park—Jehl, *Reminiscences,* op. cit., p. 514

197 nothing better—Dr. Ward to Ed., Jan., 1882

198 the engines—*Forty Years of Edison Service,* op. cit., p. 56; Meadowcroft, op. cit.; *Ed. El. Light Co.,* 1885

198 each pie—Meadowcroft, op. cit.

198 played on him—Dickson, op. cit., p. 239

198 snide remarks—Lowrey to Ed., Aug. 9, 1882; Report of S. B. Eaton, June 10, 1882; Eaton to Batch., Sept. 28, 1882; *Ed. El. Light Co.,* Statement of Board of Trustees, 1885; *Boston Herald,* Jan. 28, 1883

198 lamp capacity—*Ed. El. Light, Pearl Street,* 1882

198 I promised—*New York Sun* and *New York Herald,* Sept. 5, 1882

198 in place—*New York Sun,* Sept. 3, 1912

198 exceeded 1300—*Ed. El. Light, Pearl Street,* Nov. 4, 9, and 10, 1882; Report to Stockholders, Oct. 24, 1882

198 Elihu Thomson—*Ed. El. Light, Pearl Street,* Reports of Nov. 9 and 10, 1882; Ed. El. Light Co. trial, vol. 4, pp. 2157–58; *Bulletin of the Edison Electric Light Company,* Oct. 31, 1883

199 River ferries—Ed. to Johnson, Aug. 19, 1881; *S. B. Eaton* HRF

199 $500,000 investment—*Edison Company for Isolated Lighting,* Iso-

Page Identification—Source

lated Plants Installed in the U.S., 1882; Ed. El. Light Co. trial, pp. 2160–65, 3858–59; Eaton to Batch., Sept. 28, 1882
199 the money—*El. Light,* Sept. 28, 1882
199 to Avenue B—Bergmann to Ed., Nov. 20, 1888
199 new factory—Bergmann to Ed., Sept. 2, 1882, and Nov. 20, 1888
200 25 Gramercy Park—*Ed. Personal, Gramercy Park,* 1883–84
200 women do—Eaton to Batch., Sept. 28, 1882
200 for $500—Upton to Insull, Oct. 18, 1882; Upton to Ed., Apr. 27, 1882
200 out annually—*New York World,* Jan. 1, 1892; *Ed. Lamp Co.* Memo of Lamps Manufactured and Sold, 1888
200 Electrical Engineers (fn)—Ralph Pope to Ed., Apr. 18, 1910
201 were useless—*Electric Motors,* Dec. 30, 1882
201 same experience—Ed. El. Light Co. trial, vol. 4, p. 2553
201 the wind—Edison's Canadian Affidavit of Nov., 1888, in *Royal Electric Co.* v. *Ed. El. Light Co.,* pp. 2623–25
201 the Rothschilds—Upton to Ed., July 31, 1882

Chapter 17. A Jackknife and a Bean Pot

202 Dr. Haid—Proposed Plan of Reorganizing the Ed. Ore Milling Co., Jan., 1886
202 light now—Jehl, *Reminiscences,* op. cit., p. 530
203 the Midwest—*Engineering and Mining Journal,* July 16, 1881
203 four dollars per ton—*Ore Milling,* Annual Report, Jan. 16, 1883
203 from Sweden—*Chicago Tribune,* June 21, 1881
203 the sea—*Ore Milling,* June 22, 1881; *The Hampton Beach,* June 24, 1972
203 this district—W. J. Menzies to Ed., Nov. 9, 1882
203 been done—*Ore Milling,* Annual Report, Jan., 1884; George Fitton to Ed., Feb. 6, 1882
204 a railroad—Agreement of Sept. 14, 1881, between Ed. & Villard, in *Electric Railroad,* 1886; *Electric Railroad,* Memo of Agreement, Mar. 2, 1882; Edison Deposition of Sept. 12, 1898, Box 1, B Group, Hammer Collection
204 wet weather—*The Railway Age,* Sept. 21, 1882
204 the engineer—Marshall, op. cit.
204 be in it—Hughes, op. cit.
205 Company shares—*Electric Railroad,* July 1, 1882; *Ed. El. Light Co.,* Report to Stockholders, Oct. 24, 1882
205 abandon horsecars—R. M. Galloway to Ed., May 17, 1882; W. C. Baker, Jr., to Ed., July 5 and 28, 1882; Austin Corbin to Eggisto Fabbri, Aug. 25, 1882, in *Railroad,* 1898; Louis Glass to Ed., Aug. 13, 1882; Erastus Wyman to Ed., Dec. 14, 1882
205 excellent one—Judge William Kelly to Ed., July 22, 1882
205 the groundwork—Ed. contract with Biedermann and Havemeyer, June 19, 1882, Box 1, B Group, Hammer Collection; *Electric Railroad,* Sept. 1, 1882, in *Railroad,* 1898; Ed. to C. W. Havemeyer, Sept. 1, 1882; Ed. to Drexel-Morgan, Sept. 18, 1882
206 the matter—Edward Fox to Ed., Dec. 14, 1882; Meadowcroft, op. cit.
206 follow it up—Draper to Ed., May 22, 1882

Page *Identification—Source*

206 electric locomotive—*Railway Age,* Sept. 21, 1882; Dyer and Martin, op. cit., p. 459
206 and Application—Jehl, *Reminiscences,* op. cit., p. 857
206 carbon telephone—Sprague to Ed., May 26, 1878
207 New York—Johnson to Ed., Apr. 11 and May 18, 1883; Jehl, *Reminiscences,* op. cit., p. 857
207 at Chicago—*Electric Railroad,* Feb. 1 and Apr. 20 and 22, 1883; Ed. El. Light Co. trial, vol. 4, p. 2159; Hammond, op. cit., p. 77
207 as caretaker—*Menlo Park,* June 18, Aug. 30, and Nov. 26, 1883
207 sold them—*Iron Age,* July, 1884
207 few hours—Address of Philip A. Lange to Engineers Club of Manchester, England, Mar. 15, 1907; Harriet Sprague, *Frank J. Sprague and the Edison Myth,* New York, 1947, pp. 12–13
208 two thirds—*Central Stations,* Aug. 7, 1883
208 Edison's patent—Sprague to Ed., Mar. 2, 1883
208 with gas—*Ed. El. Light Co.,* Feb. 23, 1883; *Boston Advertiser,* June 1, 1885
208 of Providence—Louis Glass to Ed., July 16, 1910
208 the balls—Gardiner Sims to Ed., Sept. 5, 1883, and Ed. comment on letter
208 gas fixtures—*William S. Andrews* HRF
208 with light—*El. Light,* Sept. 9, 1883; *El. Light, Sunbury,* 1916
209 lightning rods—*El. Light, Sunbury,* 1916
209 water power—General Electric Co., *Historical Notes,* p. 20
209 improved little—Harry L. Keefer, Sunbury, to Ed., Feb. 2, 1916; *El. Light Defects,* 1883; *Andrews* HRF
209 be plenty—Eaton to Insull, Feb. 18, 1884; Ed. to Sprague, Jan. 9, 1884; Ed. to Insull, Mar. 8, 1884; *Central Stations, Sunbury,* Mar. 29, 1884; Letter of Frank Marr, Sunbury, Dec. 6, 1883
209 owners wrote—*Central Stations,* letter of C. B. Storey, Apr. 6, 1885
209 not corrected—*Central Stations,* letter of W. L. Garrison, Feb. 3, 1885
209 all my life—A. Stuart Grey to Insull, *Central Stations,* 1885
210 Edison told him—*Frank Sprague* HRF, Apr. 24, 1884
210 Motor Company—Correspondence, 1884, in Hammer Collection
210 happy family—Ed. to Johnson, twelve-page handwritten statement, in *Phonograph,* 1888
210 Sprague motor—C. H. Coster, Report of Special Committee of One, May 22, 1885
210 Jersey City inventor—Batch. Diary, Apr. 26, 1886
210 so quickly—*New York Mail & Express,* May 30, 1886; *Patent Suits,* 173-159, June 26, 1885; *Electric Railway Co. of the U.S.,* E 173-157, Oct. 1–Dec. 29, 1885
210 popular imagination—*New York World,* Nov. 17, 1889
211 the future—*Magazine of Business,* Chicago, Nov., 1928
211 miles of track—Sprague Report of Mar. 5, 1889; Hammond, op. cit., pp. 124–25
211 over again—McClement Audit of Sprague Company, Jan. 31, 1889
211 much mathematics—Ed. on Sprague, 1919
211 dollars a year—Sprague Report of Mar. 5, 1889
211 General Electric (fn)—Dyer and Martin, op. cit., p. 467
212 thirty years—Lab. Book, E 173-204, 1880; Mott to T. C. Martin, Dec. 20, 1910; *Ed. El. Light Co.,* Aug., 1882
212 bean pot—Jehl, *Reminiscences,* op. cit., pp. 548–50, 863

Page *Identification—Source*

213 Tesla down—Tate, op. cit., p. 149
213 utterly impractical—Ed. statement of Feb. 20, 1899
213 promised Dean—Jehl, *Reminiscences,* op. cit., pp. 676–78; Ed. to Batch., Nov. 5, 1883
214 pay it—*Upton* HRF, Oct. 6 and Nov. 1, 1883, and Oct. 30, 1884
214 new schemes—Insull to Tate, Oct. 27, 1884; Insull correspondence with W. S. Perry, May 27 and June 5, 1882
214 three-score patents—El. Light, Aug. 8, 1882
214 shut up—Electric Motors, Aug. 8, 1882; *Electric Light, Legal,* Eaton to Ed., July 31 and Nov. 3, 1882

Chapter 18. A Time of Crises

215 were Edison's—*Gramme El. Co.,* 1889; *Ed. El. Light Co.,* 1882
215 own destruction—Ed. notes on letter from S. B. Eaton, Nov. 1, 1882
215 before attained—Johnson Statement of Sept. 15, 1881
216 and contradictory—Ed. El. Light Co. trial, testimony of S. B. Eaton, pp. 3860 et seq.; *New York Herald,* July 12, 1892
216 Company participated—*Ed. El. Light Co.,* Report to Stockholders, Oct. 24, 1882
216 been lost—Ed. to Theodore Waterhouse, July 24, 1883
217 to sue—Eaton to Ed., Nov. 1, 1882; *Electric Motors,* Nov. 4, 1882
217 blocked action—Ed. to Eaton, Nov. 3, 1882
217 Sawyer and Man—Ed. El. Light Co. trial, vol. 4, p. 2414
217 than half—*Ed. El. Light Co.,* July 24, 1882; Ed. to Batch. Nov. 5, 1883
217 not paid—*Lucy F. Seyfert* v. *Thomas A. Edison,* New Jersey Supreme Court; William Seyfert to Ed., June 16 and Nov. 16, 1880
217 the verdict—Seyfert case, Feb. 20, 1884
217 Company factory—*Ed. El. Light Co.,* Dec. 4, 1883
218 St. Augustine again—Ed. to Insull, Oct. 4, 1883; *Ed. Family,* Feb. 27, 1884
218 as possible—Ed. to Insull, Apr. 7, 1884
218 very sore—Mary to Insull—Apr. 30, 1884
218 her husband—Inventory, Seyfert case; Ed. note on same
218 August 12—Tomlinson to Insull in *Legal,* July 21 and Aug. 8, 1884
218 Dr. Ward—Eugenie Stilwell letter in *Ed. Family,* 1895
219 the brain—Robert T. Lozier to John Tomlinson, Aug. 9, 1884, in *Johnny Randolph,* 1908
219 sympathetic letter—Lowrey to Ed., Aug. 10, 1884
219 the auction—Tomlinson to Insull, Nov. 6, in *Legal,* 1884; Edison Will, 1931
219 not ended—Seyfert case; Eaton to Ed., June 26, 1891
219 termed it—Eaton to Villard, Nov. 4, 1882, Baker Lib.
220 Company shared—*Ed. Lamp Co.,* 1888, Memo of Lamps Manufactured and Sold; Ed. El. Light Co., Mar. 29, 1882
220 smallest to us—Anthony J. Thomas to Upton, Jan. 27, 1884
220 than Edison—Upton to Ed., Nov. 8, 1883; Upton to Eaton, Feb. 1, 1883; Ed. to Batch., Nov. 5, 1883
220 get lamps—Upton to Ed., Jan. 4, 1883

Page *Identification—Source*

220 a settlement—Ed. to Villard, Apr. 24, 1884; *Ed. El. Light Co.*, Report of Committee of Three, June 18, 1884
220 the work—J. G. Chapman to Ed., Oct. 25, 1884
220 toward destruction—Anthony J. Thomas to Upton, Jan. 29, 1884
220 but satisfactory—Villard to Ed., June 4, 1884
221 of patents—*Ed. El. Light Co.*, Report of Committee of Three, June 18, 1884
221 patent rights—Ed. to Eaton, Jan. 19, 1883, in *Bergmann;* Welch, op. cit., p. 58
221 Isolated Company—*Ed. El. Light Co.*, Report of Committee of Three, June 18, 1884
222 to $1,080,000—Ed. to Batch., Nov. 5, 1883
222 the question—W. S. Perry to Insull, Aug. 24, 1882
222 the company—*Edison Friends*, Oct. 21, 1884; Insull to Tate, Oct. 27, 1884
222 stock each—Insull to Tate, Oct. 27, 1884; Batch. Diary, May 26, 1886
222 forbids it—Lowrey to Ed., Oct. 19, 1884
222 at last—Insull to Tate, Oct. 27, 1884
222 no account—Ed. remark to Tomlinson on Johnson note to Ed., Oct. 25, 1884

Chapter 19. An Inventive Obstetrician

225 was spotty—Recollections of Marion Edison Oeser, Mar., 1956
226 it realistic—Ed. Diary, July 12, 1885
226 a mother—Ibid.
226 at Brentano's—Recollections of Marion Edison Oeser, Mar., 1956
226 indulges in—Ed. Diary, July 12, 1885
226 vegetable kingdom—Ibid.
227 I expected—Dr. Ward to Ed., Oct. 7, 1884
227 the bank—Upton to Ed., Oct. 31, 1884
227 dirty work—Ed. to Dr. Eugene Crowell, president of Ed. Co. for Isolated Lighting, Jan. 9, 1885
227 the company—*Ed. El. Light Co.*, Nov., 1885, Annual Report
227 one time—*Ed. El. Light Co.*, Report of Committee Appointed to Recommend a Basis of Settlement, June 4, 1885
228 finance expansion—Ibid.; Report of C. H. Coster, May 22, 1885
229 in motion—*Portland Oregonian*, Feb. 15, 1886
229 between them—Henry Rogers to Ed., Nov. 1, 1878
229 as far—*Menlo Park*, July, 1885; Patent No. 465,971, applied for May 13, 1885, issued Dec. 29, 1891
229 Hotel Normandie—*Telegraph*, Dec. 11, 1885; *Telegraph, Railroad*, Feb. 3 and 6, 1886
229 Railway Telegraph—*Telegraph*, Apr. 28, 1887
230 early 1900s—*Telegraph, Phonoplex*, Oct. 30, 1885, et seq.
229 for decades (fn)—*Edison Star*, Jan. 20, 1886, et seq.
230 and Bell—Erastus Wiman to Ed., Dec. 16, 1884
230 Edison's offer—Gilliland to Ed., May 9, 1885
230 his head—T. B. A. David to Ed., Nov. 5, 1910
230 undulatory theory—Ed. to Insull, July 20, 1885

Page ***Identification—Source***

230 make-and-break current—*Telephone,* Nov. 19, 1883, in 1885 file; *Telegraph, Phonoplex,* Dec. 16, 1884; Ed.—Bergmann patent application of Nov. 13, 1883, in *Telephone,* 1886
230 on exhibit—*Exhibits,* 1884–85; *Telephone,* Feb. 3, 1885
230 of February—*Ed. Personal,* Feb. 24 and 28, 1885
231 fathered eleven—Ellwood Hendrick, *Lewis Miller,* New York, 1925
231 of Chautauqua—undated article, "Lucky Edison Draws a Prize," in *Ed. Personal,* 1886
231 fleeting encounter—*New Orleans Picayune,* Feb. 8, 1886
231 thirty feet tall—Ed. account of trip in *Fort Myers,* 1890
231 for $3000—*Ed. Personal,* Mar. 31, 1885
232 boy friends—Mary Valinda Miller to Mina, Oct. 31, 1884, C. E. Fund
232 he asserted—Nerney, op. cit., p. 272
232 later recalled—Ibid., p. 273
232 mother wrote—Mary Valinda Miller to Mina, June 7, 1885, C. E. Fund
232 of morphine—Ed. Diary, July 19, 1885
232 when a baby—Ed. Diary, July 21, 1885
233 thought of—Mary Valinda Miller to Mina, June 7, 1885, C. E. Fund
233 my intellect—Ed. Diary, July 19, 1885
233 a bore—Ibid., July 14, 1885
233 Edison enticed—Ed. to Insull, June 27, 1885
233 to himself—Ed. Diary, July 15, 1885
233 Raphaelized beauty—Ibid., July 12, 1885
233 watchful aunts—Ibid., July 16, 1885
233 his own—*Telegraph, Phonoplex,* Nov. 23, 1885
233 skate factory—*Ed. Personal,* Sept. 8, 1885
233 the obstacles—Ed. to Insull, July 20, 1885
234 $5,000 a year—*New York Tribune,* May 12, 1889
234 of stock—Agreement between Ed., Gilliland, E. H. Johnson, Bergmann, and Zalmon G. Simmons in *Telephone,* 1885
235 the benches—*Rochester Union Advertiser,* May 22, 1885
235 at a time—Meadowcroft, op. cit.
235 fine style—Ed. to Sam Edison, Jan. 29, 1879, Tannahill Lib.
235 flavoring extract—*History of St. Clair County,* op. cit., p. 631
235 years before—Sam Edison to Manager, Grand Trunk Railroad of Canada, Sept. 10, 1884, in *Ed. Personal,* 1911
235 of life—Symington to Ed., Jan. 19, 1886

Chapter 20. The Advent of Thomas Edison

236 of PLEASURE—*New York Herald,* July 5, 1885
236 proper yacht—Ed. to Insull, July 30, 1885
237 receptive listener—*Walter Mallory* HRF
237 White Mountains—Ed. to Insull, Aug. 20 and 24, 1885
237 code sigh—Milton Marmer interview of Mina Edison, Jan. 10, 1947
237 central station—*Central Stations,* Sept. 9, 1885
237 with her—Ed. Diary, July 14, 1885
237 laid siege—Gilliland to Ed., Sept. 17, 1885; Dot (Marion) Edison to Mrs. F. H. Price, Sept. 30, 1885

Page *Identification—Source*

238 good or ill—facsimile of Sept. 30 letter from Ed. to Lewis Miller in Matthew Josephson, *Edison*, New York, 1959
238 sister, Jenny—Marion Edison Oeser to Mina Edison, Aug. 9, 1924, C. E. Fund
238 to live—E. Hendrick, op. cit.
238 irreligious tartr—Ed. Diary, July 12, 1885
239 never end—Mary Valinda Miller to Mina, July 1, 1885, C. E. Fund
239 come again—Ed. to John and Theodore Miller, Dec. 24, 1885, in *Ed. Personal*
239 Mary Valinda's doubts—Madeleine Edison Sloane's history of Mina in Family Letters Scrapbook, C. E. Fund
239 began embezzling—*New York Times*, July 19, 1884
239 for $235,000—*Ed. Personal*, June 12, 1885; Edward P. Hamilton Co., New York Real Estate offering, Jan. 12, 1886, in *Glenmont*, 1887
240 thirty-five pages—Deed of Property, Jan. 20, 1886, in *Glenmont*, 1891
240 the case—Seyfert case, Jan. 20, 1886
240 the wedding—Batch. Diary, Feb. 20 and 23, 1886
240 waiting room—Unpublished manuscript, *Saga of the Taney Rainbow Trails*, Pt. 5, Ch. 6, C. E. Fund
240 had gone—*Ed. Personal*, Feb., 1886
241 to himself—Ed. Diary, July 17, 1885
241 via Tampa—Gilliland to Insull, Mar. 7, 1886
241 and late—Marmer, op. cit.
241 "A nice roommate"—Mary Emily Miller to Mina, Mar. 21, 1886
241 into Glenmont—Tom to "My dear Papa and Mama," Mar. 1, 1886
241 words imaginable—Mary Emily Miller to Mina, May 16, 1886, C. E. Fund
242 as horses—*Glenmont*, Feb. 17, 1887
242 to melancholy—Mary Emily Miller to Mina, June 12, July 16 and 22, Sept. 20, and Oct. 24, 1886; June 19, July 15, and Sept. 30, 1887, C. E. Fund

Chapter 21. A Laboratory of Grand Design

243 had expanded—Batch. Diary, May 1 and 17, 1886
243 to reconsider—Batch. Diary, May 12 and June 24, 1886
244 a reporter—*New York World*, May 20, 1886
244 $2.25 a day—Batch. Diary, May 25 and June 13, 1886
244 for $45,000—Batch. Diary, May 31 and June 6, 1886
244 new factory—Batch. Diary, Jan. 31 and Feb. 8, 1887; Dec. 4–30, 1886
244 in New York—Batch. Diary, Dec. 1 and 3, 1886
244 court judgment—*Insull*, Feb. 11, Apr. 27, May 14, June 24, and Oct. 22, 1885
244 the Works—Batch. Diary, Dec. 3, 1886
244 Drexel-Morgan (fn)—Batch. Diary, May 25 and June 6, 1886
245 and soaring—Batch. Diary, May 25 and 26 and June 11, 1886
245 ball of ours—Ed. to Upton, Jan. 21, 1886, *Upton* HRF
245 of filaments—Batch. Diary, June 11, 1886
245 commercial value—Insull Notes for Meadowcroft, Feb., 1909

Page *Identification—Source*

245 Pittsburgh hospital (fn)—*Telegraph and Telephone Age,* Nov. 16, 1910
246 the graphophone—*Am. Graphophone Co.* v. *U.S. Phono. Co. & George Tewkesbury,* testimony of Charles Tainter
246 a phonograph—Oliver Read and Walter Welch, *From Tinfoil to Stereo,* New York, 1959, p. 30
246 for them—Bruce, op. cit., p. 354; Read and Welch, op. cit., pp. 31, 36
246 about it—*Phonograph,* Dec. 29, 1885
247 dissolve it off—Gilliland notebook, N 86-06-25, Oct. 5, 1886
247 "wolf howling"—Marshall, op, cit.
247 own will—Batch. Diary, Dec. 30, 1886, Jan. 12 and 18, 1887
247 was situated—Tomlinson to Ed., Feb. 1, 1887
247 without complications—Batch. Diary, Mar. 1 and Apr. 5 and 7, 1887
248 children first—Ed. to Batch, Apr. 6, 1887; Batch. Diary, Apr. 28, 1887; Hendrick, op. cit.
248 by experimenters—Ed. to Batch., Apr. 6, 1887
248 over $150,000—*Tomlinson Legal File,* 1888
248 pull on me—*Ed. El. Light Co.,* Ed. note of July 6, 1887
248 inferior quality—Ed. to Holly, undated, in West Orange Lab, 1887
248 Joseph Taft—Holly to Ed., July 25, 1887
248 the building—Taft to J. B. Everett, Sept. 1, 1887
249 of adjustment—Batch. Record Book No. 2, Sept. 19, 1887
249 be installed—Ed. to Powers, Dec. 30, 1887
249 yet finished—*West Orange Lab Visitors,* Oct. 14, 1887; Ed. to Alfred Benjamin, Nov. 7, 1887; Ed. to Whitney, Nov. 20, 1887; correspondence with George Hopkins, Apr. 14 and 16, in *Periodicals,* 1888
249 conceivable substance—*West Orange Lab,* Oct. 24, 1887
249 and volumes—*Scientific American,* Sept. 17, 1887; Ed. to Storer How, Nov. 14, 1887; W. A. Croffert in *American,* Dec., 1887
249 without labels—Marshall, op. cit.
250 molecular telephone—*Tomlinson Legal File,* 1888; *West Orange Lab,* 1888, "Exhibit A"; Ed. Notes, N 88-01-03.2, Jan. 3, 1888
250 profits annually—Tate, op. cit., p. 121; *Mimeograph,* 1887–91
250 with factories—Tate, op. cit., p. 140
250 invest in it—Ed. to Garrison, Aug. 13, 1887
250 as he could—Garrison to Ed., Aug. 19, 1887
251 your lieutenants—Ibid.
251 active attention—Prospectus and Plan of the Ed. Industrial Co., 1887

Chapter 22. The Electric Chair

252 to 3000 volts—Lab. Book, E 173-203, No. 1
252 for an option—Upton to Ed., Jan 5, 1887, *Upton* HRF; Jehl, *Reminiscences,* op. cit., pp. 739, 831–36
253 about engines—Tate, op. cit., p. 150
253 to Edison—Hammond, op. cit., p. 106; Robert Silverberg, *Light for the World,* Princeton, 1967, p. 233
254 a curve—*Electricity, AC-DC,* Ed. memo to Johnson, Nov. 10, 1886
254 beat competitors—Ibid.
254 and cities—*New York World,* Dec. 13, 1888
254 persistent workers—Andrews to J. H. Vail, May 12, 1887

Page *Identification—Source*

254 paying business—W. J. Jenks to J. H. Vail, Nov. 12, 1887
255 this development—Johnson to Ed., Nov. 9 and 20, 1887
255 any size—*Electricity, AC-DC,* Ed. memo to Johnson, Nov. 10, 1886
255 induction coil—*New York Sun,* Dec. 1, 1878
255 an Executioner—Batch. Scrapbook, Item No. 40
255 be accomplished—Southwick to Ed., Nov. 8, 1887
255 capital punishment—Ed. to Southwick, Dec. 19, 1887
255 of barbarism—Southwick to Ed., Dec. 5, 1887
255 to Brown—*Providence Telegraph,* Jan. 12, 1889
256 a crowbar—Marshall, op. cit.
256 several days—*Brooklyn Citizen,* Nov. 4, 1888; Dickson, op. cit., p. 300
256 kill a dog—*Newark News,* July 13, 1889; *Brooklyn Citizen,* Nov. 4, 1888
256 300 volts—*Electricity, AC-DC,* Jan. 21, 1889
256 died first—*New York World,* Dec. 13, 1888; *New York Journal,* July 14, 1889
256 an elephant—*New York Sun,* Dec. 6, 1888; *Electrical Review,* Apr. 1, 1889
256 for $8,000—*Electrocution,* May 8, 1889
256 for "Westinghoused"—*Buffalo Courier,* Aug. 3, 1889
256 AC generator (fn)—Ed. to President of ASPCA, May 2, 1888
257 than hanging—*New York Times,* Aug. 7, 1890
257 Westinghouse said—Nerney, op. cit., p. 127
257 current system—Thomas A. Edison, "Dangers of Electric Lighting," in *North American Review,* Nov., 1889
257 his option—Jehl, *Reminiscences,* op. cit., pp. 834–36
257 of Isolated—Ed. to Johnson, May 10, 1886
257 a battery—*Menlo Park Lab Notes,* 1884
257 instructed them—McGowan to Ed., Nov. 16, 1887
257 and Japan—*Bamboo Search,* 1888
257 outside world (fn)—McGowan to Ed., Jan. 9, 1889
257 seen again (fn)—Tate, op. cit., p. 204 et seq.
258 displaced bamboo—Paul W. Keating, *Lamps for a Brighter America,* New York, 1954, p. 43
258 a breakthrough—Marshall, op. cit.
258 insane people—Tate to McGowan, July 2, 1888
258 the answer—Marshall, op. cit.
258 opium den—Ibid.
258 floundering about—Ibid.
258 in a calm—Maguire to Tate, Nov. 19, 1888
259 out of it—Insull to Ed., May 2 and Dec. 19, 1888
259 and leakages (fn)—*El. Light, General,* Mar. 27, 1888
259 their blood (fn)—*New York Herald* and *New York Times,* Mar. 9, 1892

Chapter 23. The Perfected Phonograph

260 New York hotel—Batch. Diary, May 8, 1887
260 acrimonious competitors—Tate memo for Ed., May 25, 1887; Bruce, op. cit., p. 353
260 indenting principle—Batch. Diary, May 16, 1887

Page *Identification—Source*

260 top of them—Batch. Diary, May 9, 17, 23–25, and 26, June 1–3 and 24–30, 1887
260 extended vacation—Gilliland to Ed., May 25, 1887; Batch. Diary, May 25, June 6 and 15, 1887
261 million dollars—Batch. Diary, Dec. 31, 1886
261 his family—Batch. Diary, July 3 to Aug. 1, 1887
261 kept building—Batch. Diary, May 16 and July 3, 1887; F. Z. Maguire to Gilliland, Mar. 21, 1888
261 *Graham Bell's*—Gouraud to Ed., July 2, 1887
261 of pirates—Ed. to Gouraud, July 21, 1887
261 the phonograph—Gouraud to Ed., Aug. 1 and 6, 1887
261 supplant letters—Phonograph caveat, Oct. 21, 1887; Patent Office to Ed., Mar. 17, 1888; *New York Press,* Oct. 21, 1887
262 been up to—Gilliland to Ed., Dec. 14, 1887
262 with business—Ed. to Gilliland, Dec. 16, 1887
262 but superior—Ibid.
262 for recording—*West Orange Lab,* Jan. 17, 1888
262 to the lab—Villard Papers, Item No. 471, Baker Lib.
262 long time—Ed. note on letter from Everett Frazar, Apr. 30, 1888
263 humorous recitals—*Cleveland Plain Dealer,* May 12, 1888
263 75 by 400 feet—*Phonograph,* June 28, 1888; Tate to McGowan, July 2, 1888
263 "Kutchy, Kutchy Coo"—*New York Sun,* May 25, 1888
263 have twins—Mary Emily Miller to Mina, Jan. 22, 1888, C. E. Fund
263 something special—Crowell, op. cit.
263 were married—Mary Emily Miller to Mina, Jan. 22, 1888, and Nov., 1887, C. E. Fund
263 at home—Ibid., Aug. 26, 1888
263 Mina remarked—*New York Sun,* May 25, 1888
264 minor defects—Ibid.
264 experimental expenses—Painter to Ed., Dec. 10, 1887; Ed.–Lippincott contract, June 28, 1888
264 of friendship—Ed. handwritten statement for Johnson, Sept. 25, 1888; *Phonograph, General,* 1888
264 business basis—Ed. note of Nov. 22, in *Phonograph,* 1887
264 his attorney—Painter to Ed., Dec. 4, 1887
264 revive them—Painter to Ed., Dec. 4 and 10, 1887
265 not be had—Johnson to Ed., Dec. 9, 1887
265 his methods—Ed. handwritten statement for Johnson, Sept. 25, 1888; Ed. to Reiff, Apr. 3, 1888
265 their machine—*North Am. Phono. Co.* Minutes, 1893
265 to sell—Gilliland to Ed., May 25, 1888
266 Tomlinson 150—Ed. handwritten statement for Johnson, Sept. 25, 1888
266 the graphophone—*Phonograph,* Ed.–Lippincott contract, June 28, 1888
266 since March—*Phonograph, General,* June 26 and 28, 1888
266 cripple us—Insull to Tate, May 23, 1888
267 Lippincott agreed—*Am. Graphophone Co.* v. *Ed. Phono. Works,* U.S. Circuit Court for New Jersey, Brief for Defendants
267 lunkhead directors—S. B. Eaton to Ed., Jan. 2, 1891; Lippincott to Ed., June 21, 1888; Lippincott statement of Nov. 28, 1888, in *Thomas A. Edison* v. *Ezra T. Gilliland & John Tomlinson*

Page *Identification—Source*

267 starting August 1—*Edison* v. *Gilliland & Tomlinson*
267 between themselves—Tate to McGowan, July 2, 1888; Lippincott statement of Sept. 21, 1888
268 for $250,000—Lippincott to Ed., Aug. 31, 1888
268 twenty-five days—Lippincott to Ed., Sept. 1 and 7, 1888
268 he responded—Ed. to Lippincott, Sept. 3, 1888
268 in Europe—Ed. to Gilliland, Sept. 11, 1888
268 wired back—Gilliland to Ed., Sept. 13, 1888
268 the stock—*Phonograph,* Sept. 22, 1888; Ed. notes of Sept. 11, in *Phonograph, Legal,* 1888; Eaton to Ed., Apr. 8, 1889
268 really level—Ed. to Lippincott, Oct. 3, 1888
269 the organization—Ed. contract with Lippincott and North American Phono. Co., Oct. 30, 1888; *Phonograph Contracts,* July 22 and 30, 1889
270 the others—Gilliland to Ed., June 6, 1888
270 first experiment—Ed. to Gouraud, June 12, 1888
270 Con Nestor—Tate, op. cit., p. 176
270 of Edison—Phonogramic Poem by the Rev. Horatio Nelson Powers, June 16, 1888
270 Lord Salisbury—*Phonograph,* Jan. 18, 1889
271 for notoriety—Ed. to Gouraud, Mar. 21, 1888
271 the mechanism—*Phonograph,* Sept. 5, 1888; Gouraud to Ed., Sept. 8, 1888
271 the country—Ed. to Gouraud, Oct. 1, 1888
271 on sapphire—George Kunz to Ed., Dec. 5, 1888; Tate, op. cit., p. 247
271 diamond cutter—Ed. handwritten notes, D 1890 N 880103, in *Phonograph,* 1890
271 held title (fn)—*Phonograph,* Dec. 8, 1887
272 weird noises—*Fred Ott* HRF
272 responded "peaking"—*Arthur E. Kennelly* HRF
272 back "pecie"—*New York Herald* article dated 1889 in *Phonograph,* 1929
272 and reproducer—*Phonograph,* Mar. 25, 1889
272 to work—*New York Times,* Apr. 21, 1889
272 music boxes—Lewis Miller to Ed., Nov. 30, 1889
272 stand it—Ed. notes for Batch., May 7, 1889

Chapter 24. Edison General Electric

273 be spent—Insull to Tate, Sept. 1, 1887
273 Edison's shares—*Insull,* June 15, 1887
273 to Edison (fn)—Insull to Ed., May 12, 1887
274 your requirements—Insull to Ed., Oct. 6, 1887
274 in America—*West Orange Lab,* Stock Holdings, 1888
274 a halt—Insull to Tate, Jan. 7, 1888
274 went into it—Upton to Tate, Jan. 17, 1888
274 $350 a week—*El. Light, General,* Jan. 30, 1888
274 the records—Upton to Laboratory, Jan. 31 and Feb. 3, 1888
274 six years—Insull to Ed., Jan. 31, 1888
274 $250 a week—Ed. Machine Works, Feb. 3, 1888
274 $100 a week—Bergmann to Ed., Feb. 11, 1888
275 of the Lab—Johnson to Ed., Feb. 21, 1888, and Ed. notes on same

Page *Identification—Source*

275 experimental account—Insull to Tate, July 5, 1888
275 engineering instruments—*West Orange Lab, Bills,* A. H. Clark & Son, Mar. 3, Goodrich, Mar. 27, Merck, Apr. 19, 1888
275 your name—Gouraud to Ed., Mar. 21, 1888
275 another party—Insull to "Dear Sirs," Apr. 6, 1888
276 Franco-Prussian war—*Ed. El. Light Co. of Europe,* 1886
276 station lighting—Upton to Ed., June 5, 7, and 10, 1886
276 ore milling—Proposed agreement with Henry Villard in *West Orange Lab,* Jan. 19, 1888
277 spring of 1889—*Ed. El. Light Co.,* 1888 Annual Report; *Central Stations,* Aug. 28, 1888
277 losing money—*Central Stations,* Aug. 28, 1888; *Philadelphia Record,* Jan. 30, 1889; *Central Stations, Lighting,* 1890
277 in Europe—Insull to Ed., Aug. 18, 1888
277 to Edison—Johnson to Ed., Aug. 22, 1888
277 manufacturing works—Insull to Ed., Aug. 18, 1888
277 their friendship—Bergmann to Ed., Sept. 26 and Nov. 20, 1888; Ed. to Bergmann, Nov. 23, 1888; Ed. handwritten twelve-page note to Johnson in *Phonograph,* 1888
277 in New Jersey—Memorandum of Consolidation Agreed upon by Mr. Edison & Mr. Villard in *Edison General Electric,* 1888; Insull to Ed., Dec. 27, 1888; *Ed. Gen. Electric,* Jan. 5, 1889
277 20 percent royalty—Villard–Ed. agreement in *Ed. Gen. Electric,* 1889
278 were nebulous—*Ed. Gen. Electric,* Feb. 9, 1889
278 appeared stalemated—*Ed. Gen. Electric,* Mar. 21 and Apr. 26, 1889
278 satisfactory to us—Villard to Eaton, Apr. 22, 1889
278 company treasury—*Ed. Machine Works* statement, Oct. 1, 1888; *Ed. El. Light Co.,* 1888 Annual Report; *Ed. Lamp Co.* Financial Statement, 1881–87, in 1888 file; Batch. Record Book No. 2, Apr. 23, 1889; Memorandum of Consolidation, *Ed. Gen. Electric,* 1888
278 the stock—*New York Times,* Apr. 23, 1878
278 to perform—Ed. note on letter from Insull, July 23, 1890

Chapter 25. The Pinnacle

279 their property—Ed. to Villard, Dec. 11, 1888
279 incandescent lighting—Ed. El. Light Co. trial, p. 266
280 to Sawyer—Consolidated El. Light Co. trial, vol. 1, Patent File wrapper
280 visit him—Westinghouse to Ed., June 7, 1888
280 business policy—Ed. to Westinghouse, June 12, 1888
280 or later—Villard Papers, Item No. 472, Feb. 2, 1889, and Ed. telegraphic reply, Baker Lib.
280 bolstered Edison—Johnson to Ed., Feb. 18, 1889
280 American patent—*Harrisburg Patriot,* Jan. 25, 1889; *Philadelphia Record,* Jan. 23, 1889
281 carbon filament—*Electrician,* May 21, 1886
281 of spiral—N 79-08-22; N 79-07-31
281 been eliminated—N 79-07-31
281 six months—Ed. El. Light Co. trial, vol. 5, pp. 3026, 3113

Page *Identification—Source*

281 deep for me—Ed. El. Light Co. trial, vol. 5, p. 2568
282 working lamp—Consolidated El. Light Co. trial, testimony of George Sawyer, vol. 1, pp. 36 et seq.
282 practical utility—Consolidated El. Light Co. trial, p. 4375
282 the courtroom—Lowrey to Ed., May 16, 1891
282 plaintiff's case—*New York Tribune,* May 23, 1889
282 her name—Glenmont, Apr. 14, 1890, and Apr. 11, 1891; Eaton to Ed., Mar. 24, 1891
282 for Edison—*Paris Exposition,* May 22, 1889
282 in place (fn)—*West Orange Lab,* Library Furnishings, 1889
283 younger sisters—Recollections of Marion Edison Oeser, Mar., 1956
283 the lab—Insull to Tate, July 30, 1889
283 August 3—Batch. Record Book No. 2, Aug. 3, 1889
283 a king—Gaston Tissandier to Ed., Sept. 21, 1889
283 snowy mounds—*New York Herald* article in *Phonograph,* 1929
283 of photography—*Ed. Personal,* Aug. 19, 1889
283 sky electrically (fn)—*New York World, Newark Advertiser, Newark Journal,* Jan. 3, 1886; *Hammer* HRF
284 highest rank—Batch. Record Book No. 2, Aug. 19 and Sept. 26, 1889
284 the piano—Tate, op. cit., pp. 234–43
284 commercial value—*New York Herald* article in *Phonograph,* 1929
284 the *something*—Ibid.
284 phonograph works—*Akron Beacon,* Aug. 10, 1889
284 Tate declared—Tate to Insull, July 23, 1889
284 an aeroplane (fn)—*New York World,* Nov. 17, 1895; *Bombay Times,* Dec. 25, 1895
285 the world—Ed. agreement with Frank Roosevelt, president of Leclanché Battery Co., Sept. 9, 1889
285 sympathetic to us—Siemens to Villard, in *Ed. Friends,* 1889
285 and projectiles—Krupp to Ed., Apr. 6, 1891
285 was forgiven—Pender to Ed., Aug. 29, 1889
285 was arranged—Tate to Ed. in *Ed. Personal,* 1889
285 the library—E 5478-4, Dec. 8, 1889
285 the Sahara (fn)—*Gouraud* HRF
286 I mean—Mary Valinda Miller to Mina, Oct. 13, 1889, C. E. Fund
286 perfect wreck—Ed. to Somerville, Oct. 9, 1889
286 had swelled—*Ed. Personal,* Oct. 6, 1889
286 in Pittsburgh—Reiff to Ed., Oct. 6, 1889
286 judge declared—Ed. El. Light Co. trial, p. 397

Chapter 26. Unprofitable Theories

288 Upper Peninsula—*Lewis Miller,* Nov. 16, 1883, and Feb. 13, 1884
288 its stock—Ed. to Tomlinson, Feb. 25, 1885
288 rebellious ores—Proposed Plan of Reorganizing the Edison Ore Milling Co., Ltd., 1887
288 the affirmative—Witherbees et al. to Ed., Oct. 7 and Nov. 14, 1887; R. B. Ayeres to Ed., Oct. 21, 1887
288 impurities separated—John Birkinbine, "The Iron Ores of the U.S.," Dec. 9, 1887
288 consulting engineer—*Ore Milling,* July 28, 1888

Page *Identification—Source*

288 be desired—Ed. note on letter from Meadowcroft, Dec. 15, 1888
288 and Minnesota—*Ore Milling,* July 25, 1888
289 were encouraging—Ed. Memo of Industrial Undertaking, in *Ore Milling,* 1889
289 material away—Ira Miller to Ed., Apr. 10, 1889; Lewis Miller to Ed., July 9, 1889
289 Southern ores—Ed. Memo of Industrial Undertaking, in *Ore Milling,* 1889
289 any parleying—Ed. note on Birkinbine letter, July 13, 1889
289 total of one hundred—Ed. Iron Concentrating Co. of New Jersey in *Ore Milling,* 1889
289 in California—Ed. Memo of Industrial Undertaking, in *Ore Milling,* 1889
290 good thing—Insull to Tate, July 30, 1889
290 to exist—Glidden to Miller, Dec. 31, 1889
290 to recover—*Ed. Family,* Jan. 3 to Feb. 19, 1889
290 lamp factory—Ed. to Force and Hickman in *Upton* HRF
291 her command—*Walter Mallory* HRF
291 common sense—*New York Sun,* May 10, 1891
291 stock-boosting scheme—Ed. to Villard, Feb. 8, 1890
292 the current—Batch. Record Book No. 2, July 31, 1889; Batch. to Ed., Aug. 19, 1889
292 nearby houses—Ed. Patent No. 493,425, "Electric Locomotive," filed Jan. 19, 1891; J. C. Henderson to Villard, June 27, 1890, Baker Lib.
292 electric traction—*Railroad, Electric,* Nov. 2 to Dec. 21, 1889, and May 3, 1890
292 for experiments—Agreement between Ed., Edison Gen. Electric, and the North American Co., Oct. 1, 1890
292 been working—Sullivan and Cromwell to Villard, Aug. 1, 1890, Baker Lib.
292 years before—*Magazine of Business,* Chicago, Nov., 1928
293 Thomson-Houston—Charles Fairchild to Villard, Feb. 23 and 25, 1890, Baker Lib.
293 streetcar field—Fairchild to Villard, Feb. 23, 1890, Baker Lib.; *Sprague,* Mar. 20, 1890
293 $100,000 a year— S. B. Eaton in *Ed. Gen. Electric, Legal,* 1890
293 in litigation—David O. Woodbury, *Elihu Thomson, Beloved Scientist,* Boston, 1960, p. 185
294 financial reasons—Ed. to Villard, Feb. 8, 1890
294 no invention—Ed. memo for Villard, Apr. 1, 1889, Baker Lib.
294 Broad Street—Insull to Tate, July 30, 1889; Eaton to Ed., Aug. 22, 1889
294 so great—Ed. to Villard, Feb. 8, 1890
295 electric lighting—Agreement between Ed. and Edison General Electric, Oct. 1, 1890
295 Hamilton McK. Twombly—Villard to Ed., Feb. 3, 1890, Baker Lib.
295 States immediately—Charles Colby et al. to Villard, Oct. 14, 1890, Baker Lib.
295 seven dollars a share—*New York World,* June 19, 1891; *Philadelphia Press,* June 20, 1891
295 juggling indefinitely—Villard Papers, Nov. 11 and 12 and Dec. 4, 1890, and Feb. 24, 1891, Baker Lib.

Page	*Identification—Source*
296	I ordered—*Ore Milling,* July 23, 1890
296	ill soon—Ed. to Mallory, July 11, 1890
296	53 percent—*New Jersey & Penna. Concentrating Works,* Sept., 1890
296	equipment installed—W. K. L. Dickson in *Ore Milling,* 1890
296	down also—*Ore Milling,* Dec. 3, 1890
296	North Carolina—*New Jersey & Penna. Concentrating Works,* Feb. 1, 1890
297	long way—Carnegie to Ed., Feb. 3, 1891
297	by Edison—*New Jersey & Penna. Concentrating Works,* Assessments
297	ore daily—*New Jersey & Penna. Concentrating Works,* Apr. 18 and May 14 and 19, 1891
297	crushing process—*Ore Milling,* May 16, 18, and 19, 1891
297	Philadelphia owners—Ibid., May 21, 1891
297	stop shipping—Ed. to Livor, June 10, 1891
298	you about—Livor to Ed., June 4, 11, 12, 15, 18, and 19, 1891
298	not quarries—Gildea to Perry, June 20, 1891
298	fired Gildea—Ed. to Perry, June 23, 1891
298	who quit—Batch. Record Book No. 2, June 29, 1891
298	to stay—Ed. to Insull, June 29, 1891
298	would submit—Henry Hart to Cutting, June 30, 1891
298	than $700—*Ore Milling,* Apr. 20, 1893
298	Ogden to Edison—Ed. to U.S. Postmaster, July 10, 1891; Appointment of U.S. Postmaster James McCarty in *Ore Milling,* Nov. 24, 1891
298	four-ton skips—*Ore Milling,* Sept. 24, 1891
298	Batchelor commented—Batch. Record Book No. 2, Nov. 5, 1891
298	melting point—Ogden Daily Report, Nov. 11 and 24, 1891
299	seriously injured—Accident Report of Travelers Detective Agency, 1892
299	pounds each—Ed. to Ogden, Dec. 30, 1891; *Ore Milling,* Mar. 17, and 23, 1892; Batch. Record Book No. 2, Apr. 15 and 22, and May 16, 1892
299	for $250,000—Petition to N.J. Board of Assessors, July 26, 1892
299	42 percent—*Ore Milling,* Oct. 22, 1892
300	the message—Insull to Ed., July 16, 1890; Ed. note on same
300	to Buffalo—Cataract Construction Co., *Niagara Falls,* Dec. 28, 1889
300	boiled crow—Report of Henry Rowland and Ed. note on same in *Niagara Falls,* Oct. 27, 1889
300	New York City—*Forty Years of Edison Service,* op. cit.
300	were innumerable—Eaton to Ed., May 7, 1890
300	central stations—Ed. to Duke of Marlborough, June 27, 1891
300	occupants complained—Letter of Aldrich, Payne, and Washburne, Dec. 22, 1891
301	additional stations—*Central Stations,* Chicago Edison Co., Jan., 1892
301	estimated $300,000—Ed. to Villard, Oct. 6, 1891, Baker Lib.
301	practical men—Ed. to Villard, Feb. 24, 1891
301	Charles Coffin—Villard to Coffin, Dec. 1, 1891, Baker Lib.
301	General Electric—Villard Papers, Feb. 3, 1892, Baker Lib.
301	in Chicago—Villard to Sprague, Feb. 8, 1892, Baker Lib.
301	a combination—*Ed. Gen. Electric,* Feb. 8, 1892
301	Eaton warned—Eaton to Ed., Feb. 8, 1892
301	the middle (fn)—*Central Stations,* Chicago Edison Co., May 6, 1890; Ed. to F. S. Gorton, Sept. 29, 1891

Page	*Identification—Source*
302	Jersey highlands—Batch. Record Book No. 2, Feb. to Mar., 1892
302	Thomson-Houston—*Boston Globe* and *Portland Oregonian,* Feb. 11, 1892
302	company treasury—Batch. Record Book No. 2, Feb. 6 and 17, 1892
302	to another—Ed. to Coffin, Aug. 16 and 20, 1892; Ed. to George Bliss, Oct. 5, 1898
302	time before—Ed. to Flood Page, Ed. and Swan United El. Co., Sept. 6, 1888
302	don't control—*Torpedo,* Feb. 11, 1891

Chapter 27. The Cyclops

303	quickly forgotten—*Ed. Personal,* 1911
303	anything electrical—Tate, op. cit., p. 278
303	any attractions—Ed. to R. D. Easterline, Aug. 2, 1893
304	and screens—Tate to Dickson, Mar. 6, 1893
304	he maintained—Ed. to Hutchinson, 1913
304	indefinite future—*Iron Age,* Jan. 11, 1894
304	without disintegration—W. S. Pilling to Ed., Jan. 23, 1893, and Mar. 9, 1894
304	he expectorated—Ed. to Joseph Harris, Oct. 5, 1893
304	leaned . . . tilted—*Owen J. Conley* v. *N.J. & Penna. Concentrating Works,* Sept. 10, 1898
305	the world—"Edison's Revolution in Iron Mining," *McClure's,* Nov., 1897
305	Edison, remarked—G. N. Morrison to Stephen Moriarty, Sept. 27, 1894
305	dollars more—*Ore Milling,* Mar. 21, 1895, and Jan. 31, 1896
305	force of 300—Owen J. Conley, op. cit.
305	no takers—Ed. to James C. Parrish, May 27 and Nov. 26, 1895; Ed. to Marks, Aug. 8 and Oct. 10, 1895
305	to work—Ed. to Fish, Oct. 11, 1895
306	every way—Fish to Ed., Feb. 21, 1893
306	own formula—Memo of Agreement between Gen. El., Ed., and Upton, 1892, in Early Edison Companies Box, Hammer Collection
306	the laboratory—*General Electric,* Oct. 14, 1892, Jan. 1, 1894, and Nov. 27, 1895; Fish to Ed., Jan. 30, 1896
306	of business—Ed. to George Hopkins, Feb. 6, 1893
306	entertains himself—Ed. to Coffin, May 16, 1901
307	at Schenectady—*Boston Advertiser,* Apr. 10, 1894; *El. Light, General,* Sept. 27, 1895; *Hammond,* op. cit., pp. 221–22, 407
307	Los Angeles—*Hammond,* op. cit., pp. 204–205, 247, 254
307	the laboratory—Morison to Moriarty in *Phono., England,* Apr. 8, 1892
307	Edison Company—Batch. Record Book No. 2, June 11, 1892

Chapter 28. An Indefensible Transaction

308	till then—Tate, op. cit., p. 62
309	of instructions—Ed. to Lippincott on note pad, Oct. 14, 1890

Page	Identification—Source
309	returned them—Insull to Ed., July 15, 1890
309	Niagara Falls—Correspondence with Josef Hofmann, Feb. 21 and Apr. 9, 1890
309	its infidelity—*Ed. Phono. Works,* May 2, 1892; Ed. to Tate, May 5, 1893
310	purposes only—Tate, op. cit., p. 253
310	the staff—Ed. to North Am. Phono. Co., Jan. 25, 1891
310	installation down—*Phonograph Recording,* Sun Spots, 1892
310	of police—*Brooklyn Times,* Sept. 21, 1890
310	Edison ruminated—*Phonograph,* Ed. note on back of letter, Nov. 15, 1890
310	Man of Promises—*Phonograph,* Sept. 5, 1890
310	in Germany—Insull to Ed., Nov. 10, 1890
310	few takers—Bush to Ed., June, 1892; Am. Graphophone Co. trial, Brief for Defendants
311	from franchisers—S. H. Latan to Ed., Dec. 13, 1890
311	near bankruptcy—Am. Graphophone Co. trial, Alfred C. Clark testimony, 1898; Eaton to Ed., Nov. 3, 1890
311	North American Company—Felix Gottschalk to Ed., Dec. 1, 1890; Eaton to Ed., Dec. 15 and 16, 1890
311	to Insull—Insull to "My dear J.C.," Jan. 3, 1885
311	by Batchelor (fn)—Batch. Record Book No. 2, Mar. 6, 1889
311	was deceased (fn)—Portland, Maine, stockholder to Tate, Oct. 29, 1890; *Toy Phonograph Co.,* 1890
312	for him—Morison to Moriarty, May 12, 1893
312	the plan—Tate, op. cit., pp. 250–51
312	all accepted—Tate to Ed., July 29, 1892; North Am. Phono. Co. Minutes, 1893
312	phonograph business—Insull to Ed., July 29, 1890
313	until lately—Ed. notes for letter to Ed. United Phono. Co., June 16, 1893
313	five cylinders—*Phono., England,* June 9, 1893; Morison to Moriarty, Nov. 9, 1894
313	only $2500—Read and Welch, op. cit., p. 108
313	into receivership—Eaton to Ed., Sept. 9, 1891; Ed. to North Am. Phono. Co., Aug. 6, 1894
313	the Works—Tate to Ed., July 13, 1893
314	been done—Ed. to Adriann Bush, Apr. 12, 1894
314	as President—Tate, op. cit., p. 292
314	the Works—Seligman to Ed., May 7, 1894
314	least strange—Seligman to Ed., May 16, 1894
314	England complained—Smith to Moriarty, May 24, 1894
314	Edison's secretary—*Phono., Legal,* 1894; Tate, op. cit., pp. 293–94
314	the bonds—Ed. to North Am. Phono. Co., Aug. 6, 1894
315	Jersey counsel—Dyer to Ed., July 22, 1895
315	criminally insane—Dyer to Ed., Apr. 4, 1909
315	only bidder—*National Phono. Co.,* Feb. 8 and 18, 1896
315	of $135,000—Ed. agreement with Charles Boston, Dec. 26, 1895
315	a battery—Ed. note on letter from Louis Glass, Aug. 23, 1894
315	monthly allowance (fn)—Adriann Bush to Ed., Sept. 5, 1894, and Ed. note on same
316	in 1896—*National Phono. Co.,* Dec. 7, 1896

Page *Identification—Source*

316 records began—*Phonograph, Bettini,* 1899; *Phonograph,* Apr. 16, 1902
316 "darky" cylinders—*Phonograph, Diamond Disk,* 1916
316 a profit—*Phonograph Recording, Europe,* 1895
317 Camden plant—McCoy's Victor Report No. 1, Jan. 31, 1909
317 in 1908—*Phonograph, Indestructible,* 1909
317 Gilmore pontificated—Gilmore to "The Trade," Sept. 29, 1903
317 received $10,000 (fn)—*Phonograph, Amberol Record,* 1909
318 with fraud—*Phono., Legal,* Sept. 25, 1900
318 I sign—*New York Herald,* Dec. 5, 1902
318 don't remember—*New York Press,* Dec. 5, 1902
318 be obtained—*New York Times,* Dec. 5, 1902
318 the documents—*New York Press, New York Herald,* and *New York Times,* Dec. 5, 1902
318 a witness—*New York Press,* Dec. 5, 1902
318 he maintained—Ed. Diary, July 13, 1885
318 since 1894—New York Phono. Co. stockholders meeting, Feb. 7, 1905, in *Phono., Legal,* 1905
318 Phonograph Company—*Phono., Legal,* Feb. 11, 1908
318 that time—Dyer to Gilmore, Feb. 7, 1908
318 50 per cent—Gilmore to Henry Auchincloss, Jan. 24, 1908
319 the country—Report of W. Maxwell in *Phonograph,* 1911
319 three years—Dyer memo of Dec., 1908
319 by Dyer—William A. Hayes statement in *Phonograph Recording, Europe,* 1895
319 for $200,000—Louis Hicks to Dyer, Feb. 11, 1909
319 be $100,000—Dyer to Ed., Mar. 9, 1909
319 on patents—Dyer to Ed., Feb. 23, 1909
319 to $500,000—*Phono., Legal,* Apr. 4 and Mar. 16, 1909
319 Tomlinson's offer—Dyer to Ed., Mar. 30 and Apr. 4, 1909

Chapter 29. The Kinetoscope

320 and apparatus—Ed. to W. P. Garrison, Dec. 21, 1887
321 $35,000 grant—*Boston Transcript,* Nov. 26, 1888
321 much further—*Scientific American,* Dec. 22, 1877
321 Anthony Drexel—Muybridge to Ed., Oct., 1888, and Nov. 28, 1888
322 been ignored—Dickson to Ed., Feb. 17, 1879
322 an offer—Dickson to Ed., Sept. 28, 1885
323 to Edison—Lewis to Ed., Oct. 10, 1888
323 the kinetograph—Chamberlain to Eaton and Lewis to Ed., Oct. 10, 1888
323 and convenient—Caveat 110, Oct. 17, 1888
323 with it—Gordon Hendricks, *The Edison Motion Picture Myth,* Berkeley, 1961, pp. 38, 42, 47, 54
323 ore building—Dyer to Randolph, Feb. 1, 1900
324 the field—G. Hendricks, op. cit., pp. 169–72
324 kinetograph caveat—Caveat of Nov. 2, 1889
324 the kinetophonograph—G. Hendricks, op. cit., p. 88
324 the ceiling—*Western Electrician,* April 12, 1890; G. Hendricks, op. cit., p. 91

Page *Identification—Source*

324 without him—Ed. to Conley, Mar. 24, 1892
326 his voice—G. Hendricks, op. cit., p. 111
326 few seconds—Randolph to George Hopkins, Feb. 4, 1893
326 the germs—*New York Sun,* May 28, 1891
326 cost sixty-five dollars—Dyer to Ed., Oct. 24 and Nov. 18, 1892
326 the patent—Tate to Seely, Nov. 29, 1892; Dyer to Ed., Dec. 1, 1892
326 instructed Dyer—*Motion Pictures,* Dec. 31, 1892, and Jan. 6, 1893
327 and disappeared—William K. L. and Antonia Dickson, "Edison's Invention of the Kineto-Phonograph," *Century,* June, 1894
327 and phonograph—Ibid.
327 two months—Ed. to A. B. Dick, Feb. 10, 1893, in reply to Dick letter of Feb. 4
327 his command—Dickson to Tate, Feb. 6, 1893
327 he remarked—Ed. to A. B. Dick, Feb. 10, 1893
327 the fair—Tate to Ed., Feb. 13, 1894; Randolph to Hopkins, May 1, 1893
327 Black Maria—Dyer to Randolph, Feb. 1, 1900
327 and sneezing—Dickson assignment of the kinetograph copyright to Ed. in *Motion Pictures,* Sept. 22, 1894
328 cost of $1000—*Motion Pictures,* Jan. 18, 1894; Tate, op. cit., pp. 284–85
328 few weeks—Tate to Ed., Feb. 13, 1894
328 invest in—Ed. to Muybridge, Feb. 21, 1894
329 taken in $120—Tate, op. cit., pp. 284–85
329 interested parties—*Motion Pictures,* Apr. 5, 1894
329 and elsewhere—Maguire to Ed., Nov. 9, 1894
329 fighting pictures—*Motion Pictures,* May 16, 1894
329 the signal—*Motion Pictures,* Sept. 29, 1930
329 in Newark—G. Wilfred Pearce letter in *Boston Advertiser,* Sept. 18, 1894
330 was revived—*Motion Pictures,* Sept. 29, 1930
330 inquiry quashed—*Motion Pictures,* Sept. 12, 1894
330 please do so—Ed. to Dickson, Oct. 15, 1894
331 the laboratory—Terry Ramsaye, *A Million and One Nights,* New York, 1926, pp. 124–26
331 better machine—*Chicago Inter-Ocean,* June 11, 1895
331 the camera—Account of Alfred C. Clark filed in *Phonograph, Am. Graphophone Co.,* 1891
331 of girls—C. E. Cowherd to his father, Dec. 8, 1894
331 go backward—J. Edmund Clark to Ed., June 30, 1893
331 a scalping—Account of Alfred C. Clark, op. cit.
332 and manufactory—Raff to Armat, Mar. 5, 1896
332 the lab—*Motion Pictures,* Aug. 3, 1901
332 projection machine—*New York Times,* June 9, 1921
332 between acts—Martin Quigley, *Magic Shadows,* New York, 1960, p. 11
332 John Ott—Ed. to Ott, July 6, 1896
332 "Edison Projectoscopes"—Maguire to Ed., July 29, 1896
332 to Armat—*Motion Pictures,* Nov. 15, 1896
333 largest theaters—Ramsaye, op. cit., pp. 210–13
333 the Vitascope—Ibid., p. 257

Chapter 30. X-Rays, Wireless, and the Conquest of Mars

Page Identification—Source

335 second wind—Ed. to Kennelly, Jan. 27, 1896
335 splendid fluorescence—Ed. to Sir John Pender, Mar. 13, 1896
335 princely gift—Pupin to Ed., Mar. 28, 1896
335 playing cards—W. R. Combs to Ed., Mar. 17, 1896
335 the junk—Henry Harrow to Ed., Feb. 23, 1896
335 locate them—Charles Wagner to Ed., May 9, 1896
335 Electrical Exposition—*St. Louis Post-Dispatch,* Mar. 25, 1896
335 human brain—Hearst to Ed., Feb. 5, 1896
335 X-ray tube—*Electromotograph,* 1902
335 is hopeless—Ed. to Henry Darling, July 7, 1910
336 late afternoon—*New York Herald,* Nov. 29, 1896
336 his screens—*X-Ray,* Apr. 7, 1896
336 *World* reported—*New York World,* Aug. 15, 1897
336 years before—*West Orange Lab,* 1896
336 miraculous shrine—*St. Louis Post-Dispatch,* Nov. 26, 1896
336 the rays—*New York Journal,* Nov. 29, 1896
336 been exposed—David Walsh in *British Medical Journal,* July 31, 1897
336 the experiments—*X-Ray,* Sept. 2, 1897
337 he died—*X-Ray,* 1902–1903; *Chicago Inter-Ocean* article, undated, in *Ed. Personal,* 1904; *Fred Ott* HRF
337 an inquiry—*X-Ray,* Sept. 6, 1902
337 were cancerous—*X-Ray,* Aug. 8, 1903
337 February, 1896—Woodbury to Ed., Feb. 17, 1896, and Ed. note on letter
337 negative pole—Upton to Ed., Oct. 19, 1882
337 little amusement—Ed. to Alex E. Outerbridge in *Ed. Effect,* 1884
337 Edward H. Johnson—Hammer Notebook, p. 31, Hammer Collection
338 of London—Edisonia, "Edison Effect," Hammer Collection
338 be long—Ed. to John S. McMillin, June 1, 1898
338 its territory—*New York Herald,* Dec. 16, 1901, and *New York Times,* Dec. 17, 1901
339 happy Christmas—Marconi to Ed., Dec. 23, 1901
339 not Edison—*Saturday Review,* Apr. 5 and May 3 and 10, 1902
339 be purchased—Ed. response of Apr. 17, 1902, to William F. Brewster letter of Apr. 9
339 was startled—*Radio,* Apr. 18, 1902
339 Rock, Massachusetts—Lewis to Ed., Apr. 28, 1902
339 wireless telegraphy—Crocker to Ed., July 12, 1902
339 the company—Agreement between Major Flood-Page and Ed., Oct. 23, 1902, filed under *Cement*
339 requires time—*Radio,* Sept. 10, 1903
339 small profit—*Radio,* 1910
340 photographic plates—Ed. notes for science-fiction book, 1890
340 an Inventor—Lathrop to Tate, Aug. 3, 1888
340 that clear—Lathrop to Ed., Aug. 10, 1891
341 highly plausible—Ed. notes for science-fiction book, 1890
341 the inventor—Lathrop to Ed., Aug. 10, 1891
341 down Tate—Ibid.

Page Identification—Source

341 his publisher—Lathrop to Ed., June 24, 1891
341 predicament again—Lathrop to Ed., Aug. 10, 1891
342 repay McClure—Ed. to Lathrop, Aug. 29, 1891
342 World's Fair—Lathrop to Ed., Oct. 5 and Nov. 9, 1891, and Ed. replies
342 autographic telegraph—Ed. to Paul Latzke, Apr. 24, 1896
342 the Universe—*Ed. Personal,* Jan. 14, 1898

Chapter 31. Invitation to Disaster

343 can make—Leonard Peckitt to Ed., Jan. 22, 1897
343 $75,000 syndicate—*Ore Milling,* Mar. 4, 1897, and Ed. notebook, Sept., 1898; Reiff to Ed., July 12, 1897; Maguire to Ed., Mar. 24, 1897; Villard Papers, Aug. 2, 1897, Baker Lib.
343 of miners (fn)—Nerney, op. cit., p. 153
344 capital investment—Villard to Bowker, Nov. 11, 1897, Baker Lib.
344 Edison serendipity—*Sand,* 1897; *Cement,* 1898; Atlas Cement Co. to Ed., Aug. 11, 1897
344 he said—Ed. handwritten note in *Ore Milling,* 1898
344 are over—Ed. to Pilling and Crane, *Upton* HRF
344 New Mexico—*Gold Ores,* Jan. 17, 1898
344 the ore—*Chicago Chronicle,* Sep. 11, 1899
344 anthracite coal—Ed. to McClure, May 4, 1898
344 Iron Mining—*McClure's,* Nov., 1897
344 come to work—Mallory to Spofford, Nov. 28, 1898; Dyer to Villard, Dec. 7, 1898, Baker Lib.
344 the floor—*Mallory* HRF
345 be surprising—Ed. to Spofford, Dec. 5, 1898, Baker Lib.
345 become obsolete—*Ore Milling,* Dec. 18, 1899, and May 28, 1902
345 earn $51,000—Statement of Expenditures, 1896–1900, in *N.J. & Penna. Concentrating Works,* 1896
345 of wood—Ore Mill, Oct. 11, 1900
345 on himself—N.J. & Penna. Concentrating Works in *Edison Portland Cement,* May 10 and 24, 1910
345 $330,000 in stock—*New York Commercial,* Sept. 27, 1899; Gordon Marks to Frank Dyer in *Dunderland Iron Ore Co.,* Dec., 1910
345 don't license—*Ed. Handwriting,* 1900
346 full charge—Ed. "Statement of Facts" in *Dunderland Iron Ore Co.,* 1909–10
346 been made—Ed. to Hermann Dick in *Ed. Handwriting,* Nov. 21, 1903
346 and money—Simpkins to Ed., Dec. 15, 1902
346 committed suicide—*Dunderland,* 1903 and 1908
346 to Dick—Ed. note on letter from Dick, May 5, 1908
346 any responsibility—Ed. "Statement of Facts," op. cit.
346 British-Krupp operation—*Dunderland,* 1909–10; *Ore Milling,* 1914
347 cement industry—*Cement,* June 8, 1899; *Philadelphia Times,* Dec. 8, 1899
347 the materials—*Cement,* July 22, 1902; *Cement and Engineering News,* Dec., 1903, p. 137
347 started up—*Cement,* Jan. 8, 1903; Villard Papers, July 22, 1902, Baker Lib.

Page *Identification—Source*

347 until fall—Villard Papers, Mar. 4, 1903, Baker Lib.
347 high-volume operation—Edison Will, Preliminary Plan for Settling Estate, Exhibit A, p. 6
347 company dissolved—Income Tax Audit, Dec. 28, 1927
347 Edison's secretary—Gilmore to Randolph, Aug. 21, 1906
347 the project—*Ed. Portland Cement,* 1906
347 and equipment—*Cement,* Dec. 15, 1908
347 concrete house—Ed. to A. K. de St. Chamas in *Cement House,* Nov. 5, 1930
347 concrete tombstone—*Cement,* Dec. 9 and 19, 1911; *West Orange Lab Experiments,* Hutchinson Report, Aug. 17, 1912
348 President Taft—Hutchinson Diary, Nov. 26–27, 1911
348 entire investment—Villard Papers, May 17, 1917, Baker Lib.; Report of Ellis Soper in *Ed. Portland Cement Co.,* Apr. 1, 1926; Edison Will, Exhibit A, p. 6, and Exhibit B, p. 2
348 $5.5 million—Edison Will, Exhibit A, p. 6
348 continued indefinitely—Edison Will, Exhibit B, p. 2

Chapter 32. The Inevitable Hour

349 set traps—Pitt to Ed., Sept. 28, 1888, and Ed. note on same
349 old died—E. R. Chadbourne to Ed., June 2, 1890; Pitt to Tate, Nov. 24, 1890; Ed. to Laboratory, Nov. 22, 1890
349 Port Huron—Pitt to Ed., Mar. 21 and Mar. 23, 1888, and Ed. response
350 wildly insane—Symington to Ed., July 26 and Sept. 14, 1891
350 Symington reported—Symington to Ed., Mar. 20, 1893
350 a year—*Ed. Personal,* Jan. 1, 1895
350 better come—Marion Page to Ed., Feb. 27, 1896
350 she stopped—Marion Page to Ed., Oct. 13, 1892
350 the grave—*Ed. Handwriting,* 1896
351 modest circumstances—Lewis Miller letter of Mar. 1, 1894, and Mina to Lewis, June 12, 1893, C. E. Fund; E. Hendrick, op. cit.
351 almost immediately—E. Hendrick, op. cit.
351 days later—Ed. telegram to Lewis Miller, July 11, 1898
351 he died—E. Hendrick, op. cit.
352 Says *almost*—W. Ormiston Ray to Ed., Oct. 16, 1930, and Ed. note on same
352 Mr. Edison—Mina to S. B. Eaton, Oct. 30, 1890
353 he shuddered—*New York American,* Oct. 6, 1904
353 articles nowadays—*Ed. Illness,* Jan. 23, 1905; Mina to Charles, undated, C. E. Fund
353 wrote him—Ed. to Arthur Reeves, Nov. 1, 1911
354 you same—*Kidnap Threat,* May 14 to June 1, 1901
354 his family—*West Orange Lab,* Nov. 16, 1908

Chapter 33. Children of the Man of the Century

356 March, 1892—*Ed. Family,* Mar. 29, 1892; Marion to Ed., Aug. 2, 1891; Mina to Johnny Randolph, Sept. 28, 1891

Page *Identification—Source*

356 her brothers—Marion to Ed., Oct. 4, 1896
357 own expenses—*Ed. Family,* Jan. 1, 1893
357 Margaret Stilwell—Margaret Stilwell to Ed., Oct. 3, 1890, and Mar. 26, 1892
357 many friends—Marion to Ed., Apr. 10, 1896
357 than myself—Marion to Ed., July 24, 1894
357 Dot answered—Marion to Ed., Aug. 24, 1894
358 time: $27.16—*Ed. Personal, Financial,* Dec. 31, 1894
358 love match—Sim Edison to Ed., Oct. 10, 1894
358 her father—Ed. to Marion, Aug. 1, 1895; Marion to Ed., Oct. 4, 1895
358 more romantic—Rose to Tom Edison, Feb. 20, 1896
358 Menlo Park, but—Marion to Ed., Nov. 16, 1895, and Jan., 1896
358 and sorrow—Eugenie Oeser to Ed., May 28, 1896
359 his father—Tom to Ed., Jan. 16, 1892
359 four records—Tom to Mina, June 9, 1892
359 the head—Tom to Mina, Apr. 25, 1892
359 stand it—Tom to Mina, Sept. 27 and Nov. 17, 1891
359 was eighteen—Tom to Mina, Aug., 1894
359 his father—Tom to Ed., Jan. 14, 1897
360 be different—Tom to Mina, Oct. 11, 1896, C. E. Fund
360 of exile—Tom to Mina, May 19, 1897, C. E. Fund
360 more life—Tom to Ed., Jan. 14, 1897
360 each one—Tom to Mina, May 19, Oct. 20, and Nov. 12, 1897, C. E. Fund
361 the attempt—Tom to Mina, Oct. 17, 1897, C. E. Fund
361 own father—Tom to Randolph, Apr. 19, 1898
361 he could—T. C. Martin to Ed., 1898
361 and otherwise—Ed. to T. C. Martin, *Ed. Family, Thomas A. Edison, Jr.,* 1898
361 Catholic priest—*Thomas A. Edison, Jr.,* HRF
361 the inventor—Will to Ed., Nov. 21, 1900
362 the city—*Ed. Family, Thomas A. Edison, Jr.,* 1904
362 Magno-Electric Vitalizer—*Ed. Personal,* Mar. 14, 1899, May 8, 1900, and 1901; *Houston Post,* Dec. 10, 1899
362 locomotor ataxia—*Ed. Personal,* Dec. 5, 1902
362 from him—McCoy Report of Dec. 2, 1902, in *Thomas A. Edison, Jr.,* HRF
363 obtaining money—Tom to Ed., Dec. 29, 1902
363 carried there—Tom to Randolph, Jan. 1, 1903
363 express approval—*Ed. Personal,* June 8, 1903
363 of mine—Tom to Ed., July 21, 1903, and Ed. note on same
363 of Orange—*Paterson Press,* Oct. 15, 1903
364 per cent mortgage—Tom to Ed., Nov. 22, 1905
364 Edison ordered—*Ed. Personal,* Sept. 27, 1907
364 wanton ways—Tom to Ed., Dec. 21, 1903, and Feb. 9, 1907; Beatrice to Ed., Dec. 14, 1911
364 as possible—Tom to Harry Miller, June 6, 1910
365 D.C. physician—*Will Edison* HRF
365 wireless experiments—*Ed. Personal,* Dec. 26, 1901, and Apr. 14, 1902
365 another $150—Will to Ed., Aug. 12 and Sept. 9 and 26, 1903

Page *Identification—Source*

365 boarding house—Blanche to Ed., Dec., 1903 in *Ed. Personal*, 1900
365 I did—Ed. to Blanche, Dec. 16, 1903
365 Edison retorted—Will to Ed., Nov. 13 and 16, 1907, and Ed. response
366 knock me—Will to Ed., Nov. 16, 1907
366 hairless dog—Will to Dyer, Oct. 2, 1908
366 the future—Will to Dyer, Feb. 17 and May 4, 1909
366 and limitations—Dyer to Will, May 4, 1909
366 his own—Will to Dyer, July 15, 1910
366 do you—Will to Dyer, Jan. 4, 1912
367 Ann Arbor, Michigan—Nellie Poyer to Ed., Dec. 26, 1907, May 12, 1910, et seq.
367 with comfort—Lizzie Wadsworth to Ed., Sept. 8, 1888, and May 6, 1905
367 needed money—John Edison to Ed., Oct. 23, 1887, Jan. 26, 1888, et seq.
367 and penniless—*Ed. Friends,* Sept. 3, 1889
367 iridium lamp—Ed. to G. H. Staite, Sept. 9, 1888
367 without support—T. C. Martin to Ed., Apr. 19, 1912
367 personal use—Ed. note on Johnson letter, Oct. 28, 1912
367 hotel room—Ed. to D. M. S. Fero, May 2, 1908
367 Edison said—Ed. to Dalgleish, June 2, 1908
368 honest living (fn)—Lillie Clifton to Mrs. Edison, Feb. 21, 1891, in *Upton* HRF
369 forgive you—Ed. to E. J. Scott, Feb. 15, 1908; Ed. to Darwin K. Pavey, Feb. 10, 1908
369 three times—*Household Accounts,* Francis Russell to Randolph, Feb. 5, 1908
369 killed himself—*Storage Battery, General,* Feb. 17, 1908
369 the family—Reiff to Ed., Feb. 17, 1908; Nellie Poyer to Gilmore, July 17, 1908
369 took sick—Bergmann to Gilmore, Mar. 11, 1908

Chapter 34. A Most Extraordinary Course

370 I signed—Meadowcroft, op. cit.
370 Welch $13,000—Eaton to Ed., Apr. 7, 1892
371 just now—Ed. to Reiff, May 17, 1905
371 such a sum—Reiff to Ed., July 18, 1906
371 its inefficiency—Ed. to Notman, June 18, 1906
371 the case—*Telegraph,* June 25, 1906
371 other scheme—Ed. to Reiff, July 3, 1906
371 plus interest—Reiff to Ed., July 30 and Dec. 11, 1906
372 any damages—Reiff to Ed., Dec. 30, 1908, and Jan. 13, 1909
372 for equity—U.S. Court of Appeals decision in *Telegraph, Reiff,* 1911
372 Edison asked—Ed. note on U.S. Court of Appeals decision, Feb. 4, 1911
372 had not—*Telegraph, Reiff,* Feb. 4, 1911
372 nearest relative—*Telegraph, Reiff,* Feb. 27 and 28, 1911
372 the funeral—Ed. to Markle in *Telegraph, Reiff,* 1911
372 than $43,000—*Telegraph, Reiff,* 1911
372 review it—*Legal,* Apr. 15 and June 16, 1911, and 1912 file

Chapter 35. Wet Electricity

Page Identification—Source

373 liver elixir—Batch. Diary, 1909
373 news hard—*Ed. Friends,* Jan. 1, 1910
373 make money—Ed. to Bergmann, Nov. 23, 1888
374 from him—Bergmann to Ed., Nov. 20, 1888
374 other's credit—*Ed. Friends,* 1897
374 lighting plants—*Ed. Friends,* Apr. 3, 1896, and Apr. 22, 1898; Jehl, *Reminiscences,* op. cit., p. 768
374 his partner—Battery, Oct. 7, 1902
374 a horse—Madeleine Edison Sloane history of Mina in Family Letters Scrapbook, C. E. Fund
374 a pet—*New York World,* Nov. 17, 1889
374 a trike—Ed. note for Kennelly in *West Orange Lab,* 1890
374 to Chicago (fn)—*Jewish Ledger,* Oct. 23, 1931
375 combustion engine—*Forty Years of Edison Service,* op. cit.; *Battery,* Feb. 21, 1899; *Automobile,* Nov. 20, 1899
375 more practical—*New York Telegraph* and *New York Journal,* June 25, 1899
375 being busted—Ed. note on W. C. Battey letter, Nov. 15, 1893
375 wet electricity—P. B. Shaw to Ed., May 28, 1910
375 storage battery—*Battery,* Dec. 4, 1896, and Apr. 11, 1900
376 lead batteries—Arthur E. Kennelly paper presented to Am. Inst. of Electrical Engineers, May 21, 1901; Twombly to Ed. in *Battery,* 1901
376 the battery—Louis Bomeisler to Ed., Feb. 18, 1901
376 the chemicals—*Battery,* June 1901
376 telephone factory—Ed. twenty-one-page handwritten proposal in *Ore Milling,* 1889
376 this company—*Ed. Industrial Works,* July 9, 1894, Statement of State Board of Assessors
376 several times—Walter Mallory to Alex Elliott, Feb. 5, 1896, et seq.
376 and France—*Battery,* Oct. 7, 1902
376 Steel Corporation—*Nickel Search,* 1902
376 ten yards—General Report of Prospecting Work, fall-winter 1901–1902
376 mineral rights—*Nickel Search,* Aug. 1901; Report of Mining, Sudbury, Oct. 19, 1903
376 *Rundschau* remarked—*Battery,* 1901
377 four months—Ed. to Bergmann, Nov. 9, 1903, and Dec. 9, 1904
377 for delay—Ed. to Bergmann, Nov. 9, 1903; Bergmann to Ed., Nov. 25, 1903, and Ed. note on same
377 round-the-clock tests—Ed. to Bergmann, Dec. 9, 1904
377 cells rebuilt—W. C. Bee to Gilmore, Aug. 9, 1907
377 secondary circuits—Ed. to Studebaker, Dec. 3, 1904; Ed. to L. C. Weir, July 7, 1905; J. A. Montpelier paper, "Alkaline Iron-Nickel Accumulator," in *Battery,* 1911
377 no fear—*Battery,* Ed. note of Feb. 21, 1905
377 and Tennessee—*Cobalt,* May 17 to June 15, 1906; *New York American* and *Buffalo Commercial,* June 14, 1906
377 into a tree—Helen Henry interview of Madeleine Edison Sloane, *Baltimore Sun,* Oct. 19, 1969

Page *Identification—Source*

377 "ginger bread" models—*New York World,* June 30, 1902
377 is unlimited—*Rochester Times,* May 30, 1902, and *St. Paul Pioneer Press,* June 7, 1902
377 *Journal* headlined—*New York Journal,* Nov. 3, 1902
377 about $200 (fn)—Ibid.
378 the man—*Automobile,* Nov. 15, 1901, Aug. 4, Sept. 3, and Dec. 18, 1903; J. R. Benton to J. M. Hill, Aug. 6, 1902
378 of batteries—Ed. to Will, Oct. 13, 1903
378 any good—Ed. to L. C. Weir, Aug. 24, 1903
378 acquisition quiet—*Lansden,* Sept. 1, 1908
378 in half—*Buffalo Commercial,* June 14, 1906
378 alkaline battery—*Electrical World,* Aug. 17, 1907
378 the batteries—Bee to Gilmore, Aug. 9, 1907; *Battery,* Nov. 24, 1908
378 run now—Ed. to Parshall, June 12, 1908
378 a machine (fn)—*American Machinist,* Aug. 10, 1911
379 potash battery—*Battery,* Report of Cells, 1908; Ed. to E. P. Earle, Jan. 25, 1908; *American Machinist,* Aug. 10, 1911
379 four years—*Battery,* Ed. note of Mar. 15, 1909
379 lead-acid batteries—E. G. Dodge to Dyer, July 18, 1908
379 when new—Montpelier, op. cit.
379 the price—Selling Talks, 1910; Insull to Charles Merz, June 14, 1910; Ed. Battery Guarantee, Oct. 7, 1908; *Automobile,* May 10, 1910
379 price differential—*American Machinist,* Aug. 10, 1910
379 with lead—*Battery,* Apr. 30, 1907
379 newspaper talk—W. C. Leslie, Columbus Buggy Co., to B. F. Arthur, Dec. 22, 1906
379 sale altogether—A. D. Hermance to Ed., June 13, 1910; Etienne Fodor to Ed., Dec. 2, 1911; *Automobile,* May 20, 1910
379 the windows (fn)—"The Edison Accumulator" in *The Autocar,* Nov. 23, 1907; Josephson, op. cit., p. 414
379 was $1,212,000 (fn)—*Battery,* Report of Cells, Nov. 30, 1908
380 poor workmanship—James A. Hearn Co. to Ed., Nov. 22, 1911; *Automobile,* Nov. 6, 1908
380 Edison exhorted—Ed. to Ira Miller, Sept. 14, 1910
380 the company—*Lansden,* June 1, 1911; *Automobile,* Sept. 12, 1911; Dyer correspondence with Wadsworth Warren, Mar. 13 to May 1, 1911
380 city traffic—Ed. handwritten statement, "Electric Automobile with the New Edison Battery," in *Automobile,* 1910
380 were mixed—*Auto Test,* 1910
380 in advertising—Ed. to Anderson, May 23, 1910
380 manufacturers combined—Dyer memos to Ed., Sept. 16 and 28, 1911
381 the tides—Ed. to Insull, June 20 and Oct. 22, 1910
381 treatment received—Ed. to Andrews, Nov. 21, 1911
381 a suitcase—*Detroit Free Press,* June 1, 1911
381 rectifier line—Anderson to Ed., June 2 and July 1, 1911
382 of yourself—Ford to Ed., Feb. 18, 1907
382 United States—*Detroit Saturday Night,* 1911 car production figures, in *Automobile,* 1911
382 West Orange—Ford to Ed., June 27, 1911, and W. G. Bee to Anderson, Apr. 6, 1911, Box 1, Ford Archives

Page *Identification—Source*

382 on the 9th—Ed. to Anderson, and Ed. note, Dec. 29, 1911
382 accommodate them—Anderson to Ed., Jan. 18, 1912; Ford to Bee, Jan. 16, 1912
382 made inquiries—Churchward to Ed., May 28, 1912; *Battery,* Apr. 27, 1912; D. M. Bliss to Ed., June 26, 1912
382 Model T—Agreement of Nov. 29, 1912, Folder 159 A, Box 4, Ford Archives
382 died entirely—Liebold to Bee, Dec. 28, 1914, Box 3, Ford Archives
383 $1.2 million—Mamber to Liebold, Jan. 23, 1918, Box 3, Ford Archives
383 the motor—Ed. to Anderson, Oct. 31, 1910
383 upon him—Dr. Thomas W. Harvey to C. Hopkins, West Orange, June 15, 1910
383 alkaline batteries—*Battery,* 1928
383 and locomotives—*Battery, Railroad,* correspondence with Baldwin Locomotive Works, May 23, June 8, and June 22, 1911
383 in Chicago—Charles Taylor to Ed., Jan. 9, 1912
384 went bankrupt—*Battery,* Mar. 16, 1915; R. H. Beach to T. J. Moncke, May 27, 1914
384 the electricity—*Miner's Lamp,* 1927; Ed. to Theodore Vail, Nov. 13, 1911; *Battery Sales,* Hutchinson memo, July 19, 1916; *Electric Canoe,* Nov. 16, 1914; Ryfylke Kraftanlaeg to Ed., Oct. 25, 1911
384 at hand—*Battery,* June 20, 1915
384 in New York—*Hutchinson* HRF
384 Edison remarked—Ed. to Brooklyn brewery owner, Dec. 31, 1904
384 a friend—Ed. to Mrs. K. G. Shepherd, Feb., 1908
385 am satisfied—Ed. to Stanley Robinson, Nov. 6, 1911, and J. G. Watson, Apr. 6, 1908
385 away inspired—Hutchinson Diary, Aug. 23, 1897
385 his genius—Du Pont to Ed., Nov. 27, 1911
385 his name—Hutchinson to Gilmore, Sept. 13, 1907
385 of ships—Hutchinson to Ed., Aug. 19, 1912
385 of batteries—Hutchinson Diary, Jan. 13 to Feb. 1, 1908, Dec. 31, 1910, July 8, 1910, Dec. 1, 3, 9, 14, and 21, 1910
385 foreign submarines—*Hutchinson* HRF, Aug. 6, 1931
385 he needed—Hutchinson Diary, Apr. 1, 1911
386 sometimes at 6—Ibid., July 5–29, 1911
386 back outside—Ibid., Nov. 27, 1911
386 library alcoves—Ibid., July 29, 1912
386 ahead of me—Ibid., Dec. 31, 1911
386 the lab—Ibid., Aug. 12, 1912, and May 5, 1913
386 community of rats—Hutchinson Report, Aug. 17, 1912; *West Orange Lab,* Feb. 12, 1920
386 the machinery—F. Devonald to Mina, 1908
386 around here—M. A. Rosanoff, "Edison in His Laboratory," *Harper's,* Sept., 1932
386 a factory—*West Orange Lab Experiments,* memo of Jan. 15, 1913
386 for himself—Hutchinson Report, Aug. 17, 1912
387 responded drily—Hutchinson to Ed., Dec. 18, 1914, and Ed. note on same
387 no experience—Hutchinson Report, Aug. 17, 1912
387 he asked—Ibid., Sept. 16, 1912, and Ed. note on same
387 kept on—Ibid., Sept. 11, 1912

Page	*Identification—Source*
387	and then—Ed. to Harry Miller, Mar. 19, 1908
387	the face—Ed. to N. F. Brady, Dec. 14, 1910
387	by noon—*Strike,* Sept. 18 and Oct. 19, 1903; *Battery,* 1905; *Newark Advertiser,* Sept. 17, 1903
387	per month—Ira Miller to John Miller, Aug. 31, 1908
387	considerably higher—Ed. to A. P. West, Oct. 6, 1905
388	incandescent lamp—Rosanoff, op. cit.
389	*don't* do it—Ibid.
389	chemical line—Ed. to Gilmore, Feb. 29, 1904
389	at 4 P.M.—Hutchinson memo, Dec., 1913, and Ed. note on same
389	his office—Hutchinson to Ed., Oct. 6, 1914
389	short while—Hutchinson to Ed., Jan. 24, 1911
389	of capacity—Hutchinson memos of Jan. 10, Sept. 23, and Sept. 30, 1911; Hutchinson to Dyer, June 22, 1911
389	submarine business—Ed. to Parshall, June 3, 1911; Ed. to Herman Harjes, Paris, 1911
390	unsuspecting customers—Hutchinson to Ed., Feb. 7, 1914
390	not purchased—Hutchinson Diary, Apr. 1, 1914; Hutchinson to Madeleine Edison Sloane, Mar. 20, 1915
390	others injured—*E-2 Disaster,* Jan. 15, 1916; *Scientific American,* Jan. 29, 1916
390	with it—*E-2 Disaster,* Apr. 2, 1917, and 1918; Memo to Secretary of the Navy, Mar. 1, 1916
390	were killed (fn)—Hutchinson Diary, Dec. 31, 1918

Chapter 36. A Trust Bust

391	acknowledged head—Dyer to Ed., Apr. 4, 1909
391	Machine Company—Hicks to Dyer, June 25, 1909
391	disk business—Dyer memo to Weber, Dec. 15, 1909
392	a year—Johnson to Dyer, July 29, 1911; Dyer to Ed., Oct. 5, 1911
392	not strong—Dyer to Ed., Oct. 5, 1911
392	$60 to $450—*Phonograph,* July 6, 1912
392	business in it—Dyer to Paul Cromelin, Aug. 29, 1911
393	musical man—Mina to Charles, May 24, 1924, C. E. Fund
393	soft records—*Phonograph,* June 9, 1911
393	every night—Henry, op. cit.; Madeleine Edison Sloane tape describing Glenmont and family life
393	different tongues—William Hayes account in *Phono. Recording, Europe,* 1895
393	West Orange—Ramsaye, op. cit., p. 408
394	yelling "Murder!"—Ibid., p. 417
394	early films—*Motion Pictures,* Apr., 1904
394	to Mina—Frank Bradley to Gilmore, June 20, 1905
394	and personnel—*Motion Pictures,* Ed. notes, 1908, July 21 and 23, 1909, and May 9, 1911
395	own order—Armat to Ed., Nov. 15, 1901
395	manufacture film—*Motion Pictures,* Feb. 1, 1908
395	New York—*Philadelphia Ledger,* Dec. 21, 1909; *New York Commercial,* Dec. 23, 1909
395	were independent—Ramsaye, op. cit., pp. 527–28

Page	*Identification—Source*
396	filed suit—*Motion Picture Patents Co.*, 1910–11
396	my throat—L. W. McChesney memo to C. H. Wilson, Jan. 2, 1915
396	moral atmosphere (fn)—*Dyer* HRF
397	is right—W. L. Wilson to Ed., Apr. 3, 1915, and C. H. Wilson comment on same
397	distinctive popularity—*Motion Pictures,* Jan. 13, 1915
397	to practice—*Motion Pictures,* Ed. notes of Aug. 15, 1908
397	he declared—*Kinetophone,* Ed. to Howenstine, Aug. 2, 1910
397	fifty kinetophones—Hutchinson Report, Aug. 17, 1912; *Kinetophone,* 1912
397	turned out—Hutchinson to Ed., Jan. 16, 1913
397	the kinetophone—*Hutchinson* HRF, Apr. 15, 1913
398	derogatory comments—*Motion Pictures,* July 17, 1913
398	their eyes—*Kinetophone,* Jan. 14, 1914
398	hysterical laughter (fn)—*Birmingham News,* Mar. 25, 1928

Chapter 37. The Faustian Soul

399	better births—Nerney, op. cit., p. 236
400	of jewelry—Madeleine Edison Sloane tape, op. cit.; Marmer, op. cit.; Henry, op. cit.
400	push buttons—Fraser, op. cit., pp. 406–407
400	Edison's enterprises—Mina Edison Statement of Expenses, 1899–1906, in *Ed. Family,* 1906; *West Orange Lab,* 1907
400	she complained—Mina to Charles, May 23, 1909, C. E. Fund
400	or Mina—Madeleine Edison Sloane tape, op. cit.; Wilson to Mina, Nov. 11, 1911
401	Mina wrote—Mina to Charles, May 17 and June 1, 1909, C. E. Fund
401	told Charles—Mina to Charles, 1931, C. E. Fund
401	storyteller—Henry, op. cit.
401	Bryn Mawr—Ibid.
401	she responded—Ibid.
401	be preordained—Madeleine to "Dear Ma," 1910, C. E. Fund
402	feel better—Madeleine to Mina, undated, C. E. Fund
402	depressed her—Madeleine to Mina, Aug., 1910, C. E. Fund
402	for grace—Mina to "My Darling, darling Mother," June 10, 1911, C. E. Fund
402	to convert—Mina to Grace Miller, July 28, 1911, C. E. Fund
402	the Sloanes—Ibid.
403	his tickets—*Newark News,* July 10, 1904; *Akron Beacon Journal,* July 19, 1909; Mina to Charles, May 17, 1909, C. E. Fund; *Theodore Edison* HRF, Sept. 9, 1912; Hutchinson Diary, Oct. 23, 1913
403	were abominable—*Ed. Family, Charles,* Apr. 28 and July 25, 1908; Hotchkiss Report Cards, Dec. 18, 1907, and June 15, 1908
403	be terrible—Mina to Charles, June 1, 1909, C. E. Fund
403	succulent morsels—Charles to Mina, Feb. 27, 1919, C. E. Fund
403	mother's heart—Mina to Charles, Sept. 24, 1917, C. E. Fund
404	the speculation—*Ed. Family, Charles,* Jan. 16, 1914
404	keep him—Mina to her mother, July, 1911, C. E. Fund
404	the church—Henry, op. cit.

Page	*Identification—Source*
404	you Enuf—"Saturday May—Dont know" to "Darling Billy" and "Thursday 12" to "Darling Billy, Report No. 2," misfiled *Ed. Personal,* 1929
404	so pinched—Mina to Charles, Feb. 29, 1924, C. E. Fund
404	put together—Charles to Mina, Aug. 4, 1919, C. E. Fund
405	else sleeping—Rosanoff, op. cit.
405	to accomplish—Meadowcroft to Dyer, June 7, 1912
406	company himself—*Dyer* HRF
406	at Glenmont—Hutchinson Diary, Jan. 1 and 3, 1913
406	of Edison—Ibid., Jan. 24, 1913
407	Silver Lake—Hutchinson to Charles, Mar. 25, 1914
407	a scenario—Hutchinson memo to Ed., Apr. 2, 1914
407	moral sense—*Paterson Press,* June 9, 1906; Ed. statement, Nov. 7, 1911
407	been substituted—Fraser, op. cit., p. 412
408	across it—Nerney, op. cit., p. 193
408	into production—The account of the fire was compiled from the following sources: *Phonograph Sales,* 1914; "Letters and Souvenirs, Edison," C. E. Fund; *Fire,* 1914; C. H. Wilson memo, Dec. 10, 1914; *Laboratory,* Feb. 25, 1915; *Laboratory Fire,* 1915; *New York Evening Post,* Dec. 10, 1914; W. G. Bee to Henry Ford, Dec. 10, 1914, and Bee to Ernest Liebold, Dec. 17, 1914, Box 2, Ford Archives

Chapter 38. Prometheus

409	the prairies—Mina to Charles, Oct. 15, 1915, C. E. Fund
409	buy a car—*New York World,* Sept. 17, 1911; Marion Edison Oeser to John Miller, Jan. 19, 1912
410	of insanity—Untitled newspaper article, Dec. 18, 1911
410	stock gambler—Hutchinson to J. F. Monnot, Sept. 1, 1911
410	all right—Ed. note of Dec. 27, 1911, on letter from *American Brewers' Review,* Dec. 21, 1911, and Ed. note of Feb. 16, 1911, on letter from International Reform Bureau, Dec. 13, 1911
410	in 1873—Schuckert to Ed., Mar. 19, 1879
410	or corporations—*Ed. Personal,* Jan. 6, 1912
411	was mailed—Letter from Wilson Drug Co., Wilson, North Carolina, Apr. 12, 1911
411	of importance—*Newark Star Eagle,* Sept. 1, 1926; *West Orange Lab,* Apr. 24, 1905; Notes of J. F. McCoy, Mar. 31, 1937
411	our friends—*West Orange Lab, Work in Progress,* McCoy memo to Insull, 1889; Notes of J. F. McCoy, Mar. 31, 1937
411	affirmative vote—*Trenton American,* Aug. 8, 1911
411	in Everything—*Burbank Visit,* Oct. 22, 1915
411	to understand—*San Francisco Examiner,* Oct. 20, 1915
412	the inventor—*San Francisco Bulletin,* Oct. 20, 1915
412	he declared—*San Francisco Chronicle,* Oct. 26, 1915
412	an attraction—Dr. Burnett Ham to Ed., June 4, 1904, and Ed. note on same
412	he said—Ed. to W. S. Nunnelly, June 19, 1912
413	forgotten, exclaimed—*San Francisco Chronicle,* Oct. 28, 1915
413	only $50,000—McChesney to Wilson, July 8, 1916

Page *Identification—Source*

413 $2.5 million—*Motion Picture Patents Co.,* 1915–16
413 West Coast—Wilson to Ed., Aug. 19, 1916; Frederick Collins to McChesney, Feb. 22, 1917
413 following Saturday—*Motion Pictures,* Jan. 25, 1918; McChesney to Wilson, Feb. 6, 1918
413 by Alysworth (fn)—*Halogen Products Co.,* 1911 and 1912
414 to xylol—*Cement,* Sept. 14, 1914; *Chemicals,* July 3 and Sept. 13, 1915
414 had said—*Manufacturers' Record,* Dec. 15, 1911
414 they economize—*Manufacturers' Record,* Dec. 7, 1911
414 the earth—World War I, Aug. 6, 1914; *Joliet Herald,* Mar. 14, 1915
414 a horror—*Indianapolis Star,* May 23, 1915
415 this board—Josephus Daniels to Ed., July 7, 1915
415 a close—*New York Electrical Review,* Feb. 27, 1886; *New York Tribune,* July 31, 1886; *Torpedo,* July 10, 1890
415 never sold—*Ed. Friends,* Mar. 12, 1914; *Phonograph,* Oct. 1, 1915
415 cities simultaneously—*Motion Pictures,* Dec. 7, 1911
415 to speak—*West Orange Lab,* Sept. 19, 1912
415 reasonable or not—*San Francisco Examiner,* Oct. 20, 1915
415 six Taft (fn)—Meadowcroft Poll, Oct. 9, 1912
416 blunders forward—Fraser, op. cit., p. 389
416 outkick you—Josephus Daniels, *The Wilson Era,* Chapel Hill, 1944, pp. 464–65
416 and submerge—*Naval Consulting Board,* 1918
416 with it—Ibid.
417 this mess—Mina to Charles, Sept. 24 and 27, 1917, C. E. Fund
417 dead stop—*Naval Consulting Board,* 1918
417 stand for—Mina to Charles, Sept. 27, 1917, C. E. Fund
417 versus submarines—*Naval Consulting Board,* 1918
417 their veins—Meadowcroft to Mina, Feb. 20, 1918, C. E. Fund
417 keep out—Mina to Charles, Nov. 23, 1917, C. E. Fund
418 manufacture helicopters—*Aerial Navigation,* July 8, 1909; Ed. to B. E. Seymour, Nov. 4, 1918
418 was accepted—*Naval Consulting Board,* 1918
418 unaware of—Ed. to Daniels, Nov. 7, 1919
418 alternating current (fn)—*New York World,* Jan. 24, 1892

Chapter 39. The Freethinker

419 as well—Ed. Will, itinerary of 1916, 1917, 1918 trips
419 umbrella mender—Burroughs to Hurley, Aug. 12, 1918
420 Hagerstown, Maryland—*Ed. Personal,* Jan. 1, 1919
420 Van Winkle—Fraser, op. cit., p. 393
420 his food—Burroughs Journal of Aug. 10, 1919, quoted in Fraser, op. cit., p. 389
420 the wilds—*Camping,* June 24, 1921
420 of Michigan—Charles Edison to Gillespie, Aug. 23, 1923
421 New Hampshire—*Camping,* Aug. 19 and 23, 1924
421 lost money—Ed. Organization conference, Mar. 6, 1916
421 May, 1917—*Phonograph,* May 18, 1917
421 one year—*Phonograph,* Aug. 22 and Dec. 27, 1917

Page	*Identification—Source*
421	lost money—*Phonograph Sales,* 1918; *Phonograph,* 1919; *Phonograph Works,* 1919–20
421	asked Charles—Wilson to Charles, Mar. 25, 1917
421	about it—Mina to Charles, Nov. 23, 1917, C. E. Fund
422	no profit—Mambert memo to Ed., Apr. 29, 1918, in *Hutchinson* HRF
422	through Hutchinson—Ibid.
422	back pay—Ed. to Charles, undated note
422	the business—Progress Report, Dec. 31, 1919
422	be accumulated—Ed. Memo, July 10, 1919
422	told Charles—Ed. to Charles, undated handwritten note
422	both ends—Jack Alexander, "The Ungovernable Governor," *Saturday Evening Post,* Jan. 23, 1943
423	the streets—Charles to Ed., Apr. 14, 1920
423	three fourths—*Ed. Financial,* Dec. 23, 1920; *Chemical Works,* Sept. 11 and Dec. 25, 1920
423	company over—*Ed. Personal,* Feb. 25, 1924
423	and Charles—Hutchinson Diary, Jan. 1, 1921
423	the devil—Charles to Ed., Sept. 12, 1921, C. E. Fund
424	are learning—Mina to Charles, Mar. 24, 1922
424	issuing invitations—Mina to Charles, May 24, 1924
424	Aylsworth noted—Aylsworth to Ed., Nov. 22, 1921
424	many instruments—R. B. Alling to Charles, Mar. 28, 1922
424	he predicted—*Radio,* Ed. note on inquiry from Charles, Feb. 17, 1922
424	November 21, 1932 (fn)—Fraser, op. cit., p. 218
425	the case—*Radio,* 1922
425	the red—Letter from Harger and Bliss, Oct. 24, 1924
425	gang of crooks—Ed. note in *Radio,* 1926
425	straight phono—Ed. to Charles, Feb. 11, 1925
425	the cities—*Phonograph,* Feb. 3, 1923; Art Walsh to Charles, Apr. 25, 1927
425	steal cash—J. W. Robbinson to Mambert, Jan. 17, 1924
425	back severely—*Storage Battery,* May 30, 1924
425	he declared—*Madison Eagle,* May 30, 1924
425	don't know—*Ed. Personal,* 1910
426	twenty-five years—*Ed. on Money,* 1921; *Ed. on Farmers,* Feb. 10, 1923
426	this subject—Baruch to Ed., May 26, 1922
426	an engineer—Ed. to Baruch, May 29, 1922
426	natural advantages—Ed. to Isaac Mertens, Nov. 15, 1911
426	rules matter—*New York Times,* June 7, 1925; *Ed. on Religion,* Dec. 10, 1910
426	mythical religion—Ed. to Joseph Lewis, June 23, 1925
427	into movement—Richard Harte to Ed., Dec. 27, 1879; H. S. Olcott to Ed., Apr. 4, 1878
427	somewhere else—*The Diary and Sundry Observations,* op. cit., pp. 212–13
427	merely transferred—*The Diary and Sundry Observations,* op. cit., pp. 214, 230
428	can appreciate—*Ed. on Memory,* 1921
428	persons' heads—*Bert Reese,* 1926; Houdini to Ed., July 29, 1926, and Ed. response
428	in Chicago—*National School of Electricity,* 1894
428	Cornell University—*Cornell University,* May 17, 1886

Page	*Identification—Source*
428	permanent usefulness—Robert H. Thurston to Ed., Sept. 7, 1894
428	the institute—*Schools & Colleges,* June 11, 1909
429	else matters—*The Diary and Sundry Observations,* op. cit., p. 115
429	decrepit mattress—Rosanoff, op. cit.
429	into blocks—untitled newspaper article, Dec. 18, 1911
429	90 percent—*The Diary and Sundry Observations,* op. cit., p. 118
430	to himself—Rosanoff, op. cit.
430	of science—*The Diary and Sundry Observations,* op. cit., pp. 112, 119
430	to think—*The Diary and Sundry Observations,* op. cit., p. 133; *New York Tribune,* June 16, 1923
430	such "radicalism"—Nerney, op. cit., pp. 285–89
430	box business—Unidentified newspaper article in *Ed. on Government,* Feb. 9, 1921

Chapter 40. Conspiracy

431	years ahead—*Ed. Personal,* 1908
431	in March—*Phono. Complaints,* Mar. 29, 1923
431	too appalling—Mina to Charles, Aug. 29, 1923, C. E. Fund
432	milling machine—Mina to Charles, Feb. 29, 1924, C. E. Fund
432	for rheumatism—Ed. to Leo Schneider, Feb. 15, 1905; Ed. to Reiff, Sept. 28, 1908
432	his stomach—Schenck to Ed., Nov. 19, 1908; Ed. to Reiff, Nov. 21, 1908
432	funny stories—Ed. to L. C. Weir, Aug. 24, 1903
432	wrong place—Rosanoff, op. cit.
432	whole system—*Diet Ideas,* Apr. 25, 1908
432	stewed prunes—Henry, op. cit.
432	no roughage—Ed. to Allan L. Benson, *Dearborn Independent,* Mar. 14, 1925
433	for months—Ed. handwritten notes on diagnosis, 1926
433	he complained—Nerney, op. cit., p. 247
433	occasional orange—Ed. note on letter, June 26, 1930
433	well enough—Ed. to Charles, May 5, 1927
433	about all—Charles to Ed., June 21, 1927, and Ed. note on same
434	Fort Myers—*Rubber,* 1924, 1926, 1927; *New York Times,* Apr. 5, 1928; *Science and Invention,* Apr., 1928
434	rubber-extracting machine—*Rubber,* 1927; Ed. on Fort Myers, in *Fort Myers,* 1890
434	dream rubber—James R. Crowell, *The American Magazine,* Feb., 1930
434	in Florida—*Rubber,* Dec. 12, 1930
435	three only—*West Orange Lab,* May 22, 1916
435	I want—Nerney, op. cit., p. 293
435	of milk—Ibid., pp. 252–53
435	he remarked—*Meadowcroft* HRF
435	communal cup—Nerney, op. cit., pp. 238–40
436	the *results—Phonograph, Cinemusic,* 1928
436	were enormous—Author's discussion with Theodore Edison
436	the trade—*Phonograph, Cinemusic,* Aug. 29, 1927
436	Edison chortled—Nerney, op. cit., p. 187
436	justify it—Charles to Ed., June 23, 1931

Page *Identification—Source*

436 in RCA—Hammond, op. cit., p. 377
437 $4 million—Charles to Ed., Nov. 1, 1929
437 the theater—Mina to Charles and Carolyn, Apr. 25, 1928

Chapter 41. The Path of Glory

438 glimpse of him—Mina to Charles, Apr. 25, 1928
439 to participate—Ed. Memorial Folder No. 339, Box 5, Ford Archives
439 of propriety—Ed. to Liebold, Oct. 25, 1927, Box 5, Ford Archives
439 the debt—Mambert to Liebold, Oct. 3, 1923, Box 4, Ford Archives; W. R. Middleton to Liebold, Folder 159 A, Box 4, Ford Archives
439 constantly stalled—E. J. Farkas to Liebold, July 26, 1921, Folder 59, Box 4, Ford Archives
439 into bankruptcy—*Ed. Family, Thomas A. Edison, Jr.*, 1929
439 above idea—Meadowcroft to Liebold, Apr. 30, 1928, Box 6, Ford Archives
439 issued it (fn)—Ed. to Ford, Dec. 31, 1917, and Liebold to Mambert, May 6, 1919, Box 3, Ford Archives
440 at Dearborn—Ford to Ed., Mar. 28, 1928, Box 6, Ford Archives
440 the countryside—Eaton to Ed., Feb. 17, 1892; Frank Dyer to Herbert Knight, July 30, 1908; *Menlo Park,* 1912 and May 8, 1918
440 he said—Ed. response to Stephen Vail, Nov. 20, 1891
440 American manufacture—Hammer to John Lieb, Aug. 13, 1928, Hammer Collection; Liebold to Meadowcroft, June 2, 1928, Box 6, Ford Archives
441 revolutionized civilization—Mellon to Ford, Oct. 8, 1928, Box 6, Ford Archives
441 national park—Ford to Mellon, Oct. 18, 1928, Box 6, Ford Archives
442 other man—*Pittsburgh Leader,* Jan. 27, 1915
442 creative work—Ford to Edward Adams, Jan. 5, 1929, Hammer Correspondence, Hammer Collection
442 he exclaimed—Nerney, op. cit., p. 240
442 to America—*Jehl* HRF
442 myth building—Edward L. Bernays, *Biography of an Idea,* New York, 1965, p. 458
443 Menlo Park (fn)—Ibid., p. 452
443 real shrine—*Mina Edison* HRF
443 backyard (fn)—*Fort Myers,* Feb. 18, 1929; Nerney, op. cit., p. 242
444 Toledo & Ironton—The Reminiscences of Edward J. Cutter, p. 24, Ford Archives
445 concluding address—*New York Times* and *Philadelphia Record,* Oct. 22, 1929
445 last time—Marion Edison Oeser to Ed., June 2, 1918
446 she exclaimed—Marion to Mina, Apr. 23, 1920, C. E. Fund
446 with money—Marion to Mina, Aug. 9, 1924, C. E. Fund
446 a country—Marion to Mina, Aug. 9 and Nov. 1, 1924, C. E. Fund
446 Norwalk, Connecticut—*Marion Edison Oeser* HRF, Apr. 21, 1926
446 any success—Marion to Mina, May 18, 1928, C. E. Fund
446 in France—Will to Ed. and Mina, Jan. 8, 1918
446 totaling $4,000—John Miller correspondence with Will, 1926–28
446 was stillborn—Beatrice to Mina, June 19, 1916, C. E. Fund

Page *Identification—Source*

446 produced electricity—Speech of Thomas A. Edison, Jr., at Knights of Columbus Technical School graduation, June 4, 1921, Box 4, Ford Archives
447 unkind to me—Beatrice to Dyer, Dec. 7, 1912
447 my Father—Tom to Wardlaw, Ed. Pioneers, Jan. 5, 1922
447 love everyone—Tom to Marion, 1929, C. E. Fund
447 at once—*Phonograph,* W. H. Miller memo of Oct. 21, 1929
447 phonograph fields—Charles Edison report to stockholders of the Splitdorf Electric Co., April 15, 1931, in *Radio,* 1930
448 business depression—Business Outlook Report of G. E. Stringfellow, Dec. 17, 1929
448 steadily worse—Charles to Mina, June 1, 1930, C. E. Fund
448 it all—Mina to Charles, Apr. 12, 1930, C. E. Fund
448 told Meadowcroft—Ed. to Meadowcroft, Apr. 12, 1930
448 an acquaintance—Ed. to Francis Nipher, Nov. 12, 1908
448 on hand—Charles to Ed., June 23, 1931
448 in January—J. V. Miller to Frank Campsall, Jan. 29, 1931, and J. V. Miller to Ed., Apr. 25, 1931, Ford Archives
449 at all—*New York Times,* Oct. 19, 1931
449 his will—Ed. Will
449 with peppermint—*New York Times,* Oct. 19, 1931
449 and peace—John Miller to Mina, Sept. 17, 1931
450 was awake—*New York Times,* Oct. 19, 1931
450 replied curtly—Hutchinson to Mina and Mina response, Oct. 16, 1931

Chapter 42. Epitaph

451 a heathen—Nerney, op. cit., p. 300
452 to Mina—Ed. Will, Preliminary Plan for Settling Estate of Thomas A. Edison
452 Madeleine's side—Ed. Will, Correspondence of Samuel P. Gilmore, counselor for Madeleine Sloane and William Edison, Dec. 15, 1931
452 with Charles—Ed. Will, Correspondence of William Edison, Oct. 31 and Dec. 5, 1932
452 did wealth—Marion to Ford, Feb. 17, 1932, Box 1, Ford Archives
452 as co-executor—Theodore to Mina, July 11, 1920, C. E. Fund; author's discussion with Theodore Edison
452 for them—Author's discussion with Theodore Edison
453 active environmentalist—Ibid.
453 to love—Mina to Charles and Carolyn, Feb. 25, 1932, C. E. Fund
453 forgetting us—Mina to Charles, May 30, 1932, C. E. Fund
453 years later (fn)—Marion to Mina, Aug. 17, 1937, C. E. Fund
454 after him—Madeleine Edison Sloane to "Dear Bill," 1932
454 and cents—Ibid.
454 of life—Mina to Charles, July 25, 1935, C. E. Fund

Chapter 43. Legend and Legacy

456 his death—*The Diary and Sundry Observations,* op. cit., p. 59
456 in life— Ibid., p. 169

Page *Identification—Source*

456 of thinking—Ibid., pp. 166, 167, 170
457 he said—Ed. to Dudley Nichols, *New York World,* May 8, 1929
458 go ahead—*London Standard,* Oct. 15, 1878
460 Edison asserted—*Brooklyn Citizen,* Nov. 4, 1888
460 is success—*New York World,* Nov. 17, 1889
461 he remarked—Ed. scratch note to E. P. Payson, 1901 or 1902
461 isn't so—Bliss to Ed., Mar. 8, 1878
462 the Mississippi—*New York Evening Telegram,* Jan. 25, 1877
462 of Edison—*The Operator,* June 1, 1878
462 Bliss said— Ibid., May 15, 1878
462 if I had—Rosanoff, op. cit.
462 to autobiography—Lathrop to Tate, Aug. 3 and Nov. 17, 1888
462 with stories—Ed. to Cyril Clemens, Jan. 10, 1927
462 back on him—Hilborne Roosevelt to Cornelius Roosevelt, Nov. 3, 1879
462 Mind Number 2—Ed. Diary, July 13, 1885
463 rather strong—Ed. to Werner School Book Co., Feb. 7, 1902
463 Edison replied—Ed. to Arthur Brisbane, Dec. 10, 1907
463 *Century* magazine—T. C. Martin to Ed., Nov. 11, 1907, and Ed. reply, Nov. 21, 1907
463 the work—*Biography,* Sept. 23, 1908
463 apparent corpse—Dyer to Ed., Feb. 23 and Mar. 8, 1909
464 he claimed—*The Diary and Sundry Observations,* op. cit., p. 45
464 dry reading—Ed. to James Swift, Dec. 2, 1908
464 of promotion—P. H. Shaughness to Ed., Dec. 21, 1909
464 his pockets—Dyer and Martin, op. cit., p. 132
464 so recklessly—Reiff to Ed., Oct. 10 and Nov. 9, 1906
464 was born—Dyer and Martin, op. cit., p. 258
464 six months—Ibid., p. 262
464 of Chicago—Ibid., p. 168
465 inventions themselves—Ibid., pp. 342–43
465 produced any—C. M. Devonshire, Bell Telephone Co., to Ed., Sept. 30, 1907
465 Wall Street sorry—Meadowcroft, op. cit.
465 to Mr. Villard—Dyer and Martin, op. cit., p. 460
466 in America—Ed. to A. K. de St. Chamas, Nov. 5, 1930; *Newark News,* Mar. 14, 1924
466 large sum—Dyer and Martin, op. cit., p. 382
466 motor unit—Henry Ford, *Edison As I Knew Him,* New York, 1930, p. 14
467 this morning—*The Diary and Sundry Observations,* op. cit., p. 324
467 I perform—Ed. note on letter from *New York Journal,* Feb. 26, 1896
467 ladies fainted—Dyer and Martin, op. cit., pp. 745–46
467 apple pie—Tate, op. cit., pp. 143–45
467 any sleep—Rosanoff, op. cit.
468 a wizard—Jehl letter of Apr. 22, 1913, Hammer Collection
468 the genius—Hughes, op. cit.
468 thing work—Rosanoff, op. cit.
469 many men—Jehl letter of Apr. 22, 1913, Hammer Collection
472 succeeds in it—Ed. to Lun A. Smith, May 12, 1885

Index

Index